JACARANDA
MATHS QUEST

MATHEMATICAL
METHODS 11
FOR QUEENSLAND
UNITS 1 & 2 | SECOND EDITION

JACARANDA

MATHS QUEST

MATHEMATICAL METHODS 11

FOR QUEENSLAND

UNITS 1 & 2 | SECOND EDITION

BEVERLY LANGSFORD WILLING

SUE MICHELL

KAHNI BURROWS

REVIEWED BY

Sally Birks | Caitlin Greenhalgh | Andrew Johnson
Joel Scott | Megan Scougall | Fiona Swan

jacaranda
A Wiley Brand

Second edition published 2024 by
John Wiley & Sons Australia, Ltd
Level 4, 600 Bourke Street, Melbourne, Vic 3000

First edition published 2018

Typeset in 10.5/13 pt TimesLTStd

ISBN: 978-1-394-26954-9

Front cover images: TWINS DESIGN STUDIO/Adobe Stock Photos, veekicl/Adobe Stock Photos, mrhighsky/Adobe Stock Photos, sapunkele/Adobe Stock Photos, NARANAT STUDIO/Adobe Stock Photos, Kullaya/Adobe Stock Photos, vectorplus/Adobe Stock Photos, WinWin/Adobe Stock Photos, valeriya_dor/Adobe Stock Photos, Ludmila/Adobe Stock Photos, Анастасия Трофимова/ Adobe Stock Photos, katarinalas/Adobe Stock Photos, izzul fikry (ijjul)/Adobe Stock Photos, Tatsiana/Adobe Stock Photos, nadiinko/Adobe Stock Photos

Illustrated by various artists, diacriTech and Wiley Composition Services

Typeset in India by diacriTech

NATIONAL LIBRARY OF AUSTRALIA

A catalogue record for this book is available from the National Library of Australia

Printed in Singapore
M130167_070824

The publisher of this series acknowledges and pays its respects to Aboriginal Peoples and Torres Strait Islander Peoples as the traditional custodians of the land on which this resource was produced.

This suite of resources may include references to (including names, images, footage or voices of) people of Aboriginal and/or Torres Strait Islander heritage who are deceased. These images and references have been included to help Australian students from all cultural backgrounds develop a better understanding of Aboriginal and Torres Strait Islander Peoples' history, culture and lived experience.

It is strongly recommended that teachers examine resources on topics related to Aboriginal and/or Torres Strait Islander Cultures and Peoples to assess their suitability for their own specific class and school context. It is also recommended that teachers know and follow the guidelines laid down by the relevant educational authorities and local Elders or community advisors regarding content about all First Nations Peoples.

All activities in this resource have been written with the safety of both teacher and student in mind. Some, however, involve physical activity or the use of equipment or tools. **All due care should be taken when performing such activities**. To the maximum extent permitted by law, the authors and publisher disclaim all responsibility and liability for any injury or loss that may be sustained when completing activities described in this resource.

The publisher acknowledges ongoing discussions related to gender-based population data. At the time of publishing, there was insufficient data available to allow for the meaningful analysis of trends and patterns to broaden our discussion of demographics beyond male and female gender identification.

Contents

> **online only**
>
> **Problem-solving and modelling task guide**

UNIT 1 SURDS, ALGEBRA, FUNCTIONS AND PROBABILITY   01

on**line**only

PRACTICE ASSESSMENT 1
Problem-solving and modelling task

PRACTICE ASSESSMENT 2
Unit 1 Examination

online only

PRACTICE ASSESSMENT 3
Unit 2 Examination

PRACTICE ASSESSMENT 4
Units 1&2 Examination

Learning with learnON

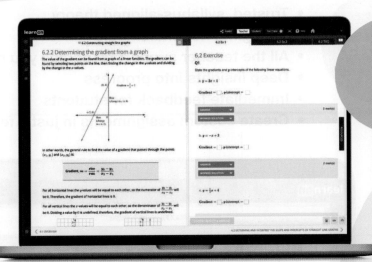

Everything you need for your students to succeed

JACARANDA MATHS QUEST
MATHEMATICAL METHODS 11

UNITS 1 AND 2 FOR QUEENSLAND | SECOND EDITION

Developed by expert teachers for students

Tried, tested and trusted. The completely revised and updated second edition of *Jacaranda Maths Quest Mathematical Methods 11 Units 1 & 2 for Queensland* continues to focus on helping teachers achieve learning success for every student — ensuring no student is left behind and no student held back.

Because both what and how students learn matter

Learning is personal

Whether students need a challenge or a helping hand, you'll find what you need to create engaging lessons.

Whether in class or at home, students can get unstuck and progress! Scaffolded lessons, with detailed worked examples, including both TI and Casio calculator support. Automatically marked, differentiated question sets are all supported by detailed worked solutions. And brand-new exam-style questions support in-depth skill acquisition in every lesson.

Learning is effortful

Learning happens when students push themselves. With learnON, Australia's most powerful online learning platform, students can challenge themselves, build confidence and ultimately achieve success.

Learning is rewarding

Through real-time results data, students can track and monitor their own progress and easily identify areas of strength and weakness.

And for teachers, Learning Analytics provide valuable insights to support student growth and drive informed intervention strategies.

Learn online with Australia's most

- Trusted, syllabus-aligned theory
- Engaging, rich multimedia
- All the teacher support resources you need
- Deep insights into progress
- Immediate feedback for students
- Create custom assignments in just a few clicks.

Practical teaching advice and ideas for each lesson provided in teachON

Each lesson linked to content points from the QCAA Mathematical Methods 2025 General senior syllabus

Reading content and rich media including embedded interactivities and calculator support.

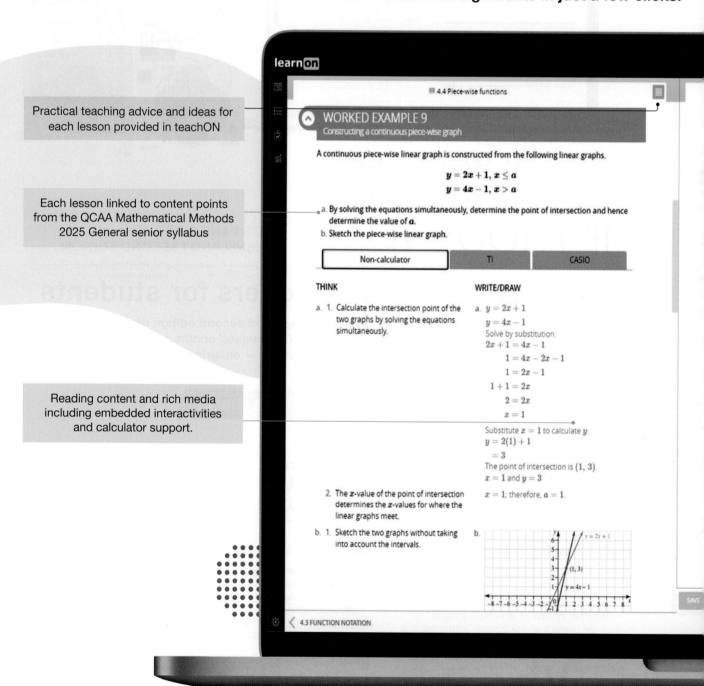

learn on

4.4 Piece-wise functions

WORKED EXAMPLE 9
Constructing a continuous piece-wise graph

A continuous piece-wise linear graph is constructed from the following linear graphs.

$$y = 2x + 1, \ x \leq a$$
$$y = 4x - 1, \ x > a$$

a. By solving the equations simultaneously, determine the point of intersection and hence determine the value of a.
b. Sketch the piece-wise linear graph.

| Non-calculator | TI | CASIO |

THINK

a. 1. Calculate the intersection point of the two graphs by solving the equations simultaneously.

2. The x-value of the point of intersection determines the x-values for where the linear graphs meet.

b. 1. Sketch the two graphs without taking into account the intervals.

WRITE/DRAW

a. $y = 2x + 1$
$y = 4x - 1$
Solve by substitution:
$2x + 1 = 4x - 1$
$1 = 4x - 2x - 1$
$1 = 2x - 1$
$1 + 1 = 2x$
$2 = 2x$
$x = 1$

Substitute $x = 1$ to calculate y:
$y = 2(1) + 1$
$= 3$
The point of intersection is $(1, 3)$.
$x = 1$ and $y = 3$

$x = 1$; therefore, $a = 1$.

b.

4.3 FUNCTION NOTATION

powerful learning tool, learnON

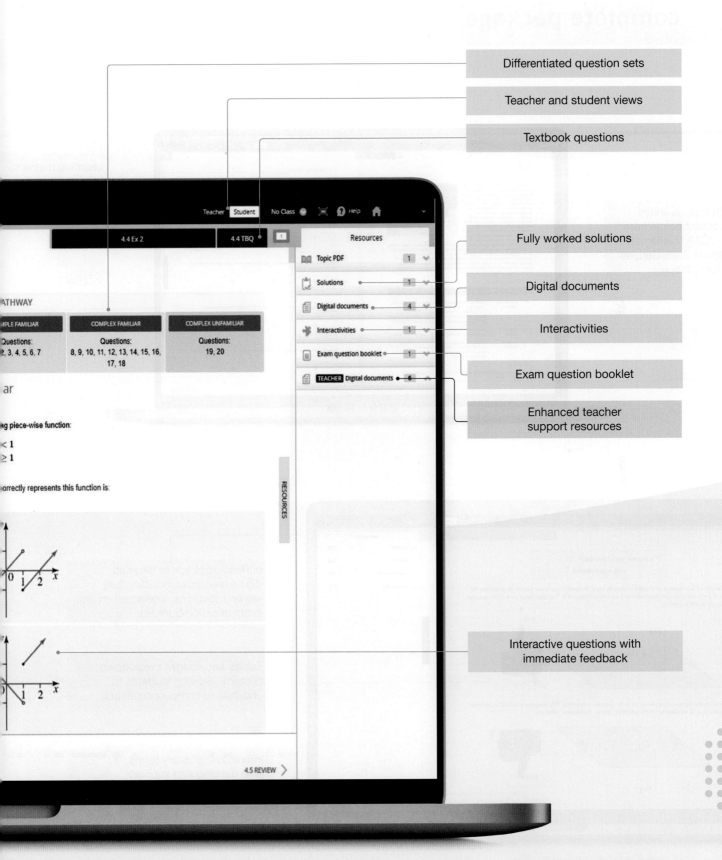

Differentiated question sets

Teacher and student views

Textbook questions

Fully worked solutions

Digital documents

Interactivities

Exam question booklet

Enhanced teacher support resources

Interactive questions with immediate feedback

Online, these new editions are the complete package

Trusted Jacaranda theory, plus tools to support teaching and make learning more engaging, personalised and visible.

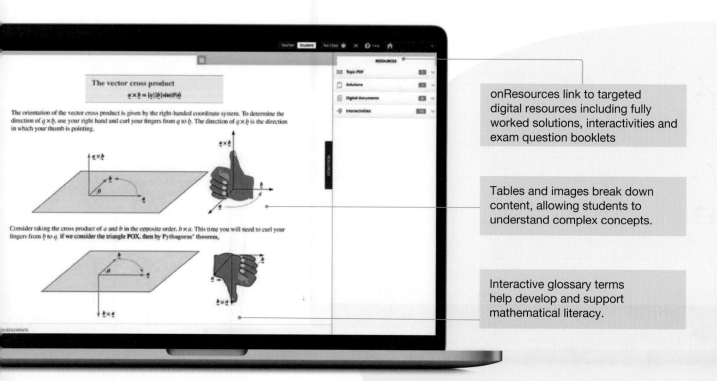

Each topic is linked to content points from the QCAA Mathematical Methods 2025 General senior syllabus

Learning matrix to monitor student's confidence level throughout topics

onResources link to targeted digital resources including fully worked solutions, interactivities and exam question booklets

Tables and images break down content, allowing students to understand complex concepts.

Interactive glossary terms help develop and support mathematical literacy.

Pink highlight boxes summarise key information.

Worked examples break down the process of answering questions using a think/write format, including calculator support.

Each unit has practice assessments (Problem solving and modelling task or examination) with sample responses to help build skills

- Online and offline question sets questions and exam-style questions with exemplary responses and marking guides.
- Every question has immediate, corrective feedback to help students to overcome misconceptions as they occur and to study independently — in class and at home.

Topic and Unit reviews

Topic and Unit reviews include online summaries and topic-level and unit review exercises that cover multiple concepts.

Get exam-ready!

Topic-level exam questions are structured just like the exams.

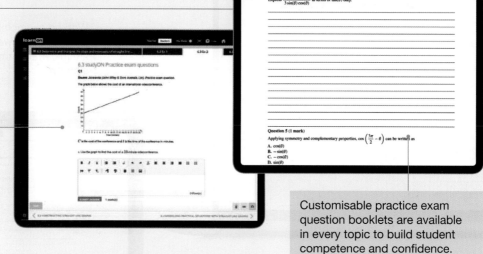

Customisable practice exam question booklets are available in every topic to build student competence and confidence.

Expert advice for problem solving and modelling tasks

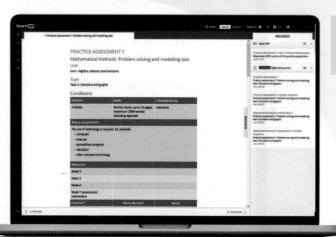

Step by step guide on how to complete problem solving and modelling tasks with tips for teachers on how to create good assessments

Teaching with learnON

Plan

Teach

Analyse

Assess

Learn

CLASS SET-UP

ONGOING SUPPORT

INTEGRATIONS

learn**on**
one resource

Enhanced teacher support resources, including:

- work programs and syllabus grids
- teaching advice and additional activities
- quarantined topic tests (with solutions)
- Units reviews
- Quarantined PSMTs and examinations
- Custom exam-builder with question differentiation (SF/CF/CU) and tech-active/Tech-free question filters

Customise and assign

A testmaker enables you to create custom tests from the complete bank of thousands of questions.

Reports and results

Data analytics and instant reports provide data-driven insights into performance across the entire course.

Show students (and their parents or carers) their own assessment data in fine detail. You can filter their esults to identify areas of strength and weakness.

Acknowledgements

The authors and publisher would like to thank the following copyright holders, organisations and individuals for their assistance and for permission to reproduce copyright material in this book.

The full list of acknowledgements can be found here: www.jacaranda.com.au/acknowledgements/#2024

Every effort has been made to trace the ownership of copyright material. Information that will enable the publisher to rectify any error or omission in subsequent reprints will be welcome. In such cases, please contact the Permissions Section of John Wiley & Sons Australia, Ltd.

UNIT 1 Surds, algebra, functions and probability

Source: Mathematical Methods Senior Syllabus 2024 © State of Queensland (QCAA) 2024; licensed under CC BY 4.0.

1 Surds

LESSON SEQUENCE

Fully worked solutions for this chapter are available online.

EXAM PREPARATION

Access exam-style questions in every lesson, available online.

on Resources

Solutions	Solutions — Chapter 1 (sol-1268)
Exam questions	Exam question booklet — Chapter 1 (eqb-0289)
Digital documents	Learning matrix — Chapter 1 (doc-4183)
	Chapter summary — Chapter 1 (doc-41796)

LESSON
1.1 Overview

1.1.1 Introduction

In 2017 the Cassini spacecraft disintegrated in Saturn's atmosphere, having successfully completed a 7-year mission to observe that planet and its moons. In all, it completed 22 orbits of Saturn, sending extraordinary images and information back to Earth. In an earlier mission, the Voyager 1 spacecraft passed Pluto in 1990, and in 2004 it left our solar system. On board Voyager 1 was a time capsule carrying information about Earth and our achievements for, should they exist, any intelligent alien life forms it may reach.

None of this would have been possible without mathematics. Mathematics and physics are essential to the launch and success of all space missions. It has been argued that should there ever be communication between Earthlings and intelligent extra-terrestrials, communication will be through mathematics. This is because mathematics is universal. It is universally true that Pythagoras' theorem, $a^2 + b^2 = c^2$, holds in every country on Earth and, by extension, in any galaxy in the universe.

The concept of numbers in counting and the introduction of symbols for numbers marked the beginning of a major intellectual development in the minds of humans. Over the course of history, different categories of numbers have been defined that collectively form the real number system that we use today.

1.1.2 Syllabus links

Lesson	Lesson title	Syllabus links
1.2	**The real number system and surds**	○ Understand the concept of a surd as an irrational number represented using a square root or a radical sign.
1.3	**Simplifying surds**	○ Simplify square roots of natural numbers which contain perfect square factors, e.g. $\sqrt{45} = \sqrt{9 \times 5} = \sqrt{9}\sqrt{5} = 3\sqrt{5}$.
1.4	**Operations with surds and rationalising the denominator**	○ Rationalise the denominator of fractional expressions involving square roots, e.g. $\dfrac{\sqrt{7}}{\sqrt{3}} = \dfrac{\sqrt{7}}{\sqrt{3}} \times \dfrac{\sqrt{3}}{\sqrt{3}} = \dfrac{\sqrt{7} \times \sqrt{3}}{\sqrt{3} \times \sqrt{3}} = \dfrac{\sqrt{21}}{3}$.
		○ Use the four operations to simplify surds, e.g. $\sqrt{5} - 2\sqrt{5} + 4\sqrt{5} = 3\sqrt{5}$ and $2\sqrt{3} \times 5\sqrt{11} = 10\sqrt{33}$.

Source: Mathematical Methods Senior Syllabus 2024 © State of Queensland (QCAA) 2024; licensed under CC BY 4.0.

LESSON
1.2 The real number system and surds

SYLLABUS LINKS

- Understand the concept of a surd as an irrational number represented using a square root or a radical sign.

Source: Mathematical Methods Senior Syllabus 2024 © State of Queensland (QCAA) 2024; licensed under CC BY 4.0.

1.2.1 The real number system

Although counting numbers are sufficient to solve equations such as $2 + x = 3$, they are not sufficient to solve, for example, $3 + x = 2$, where negative numbers are needed, or $3x = 2$, where fractions are needed.

The table below shows the classification of numbers in our real number system.

$N = \{1, 2, 3, 4, \ldots\}$	The **set of natural numbers**, that is the positive whole numbers or counting numbers
$Z = \{\ldots -2, -1, 0, 1, 2 \ldots\}$	The **set of integers**, that is all positive and negative whole numbers and the number zero
$Q = \left\{\ldots \dfrac{-9}{8}, 0.75, 0.\overline{3}, 5, \ldots\right\}$	The **set of rational numbers**: all numbers that can be expressed in the form $\dfrac{a}{b}$, $b \neq 0$ and $a, b \in Z$, that is all fractions, integers and finite and recurring decimals.
$I = \left\{\ldots -\sqrt{3}, \ldots, \pi, \ldots \sqrt{2}, \ldots\right\}$	The **set of irrational numbers**: all real numbers that are not rational. Some irrational numbers, such as π and e, are called **transcendental** numbers.
$R = Q \cup I$	The **set of real numbers**, that is the union of the set of rational numbers and the set of irrational numbers

The relationship between the sets, $N \subset Z \subset Q \subset R$, can be illustrated as shown.

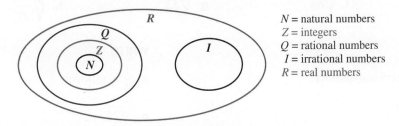

N = natural numbers
Z = integers
Q = rational numbers
I = irrational numbers
R = real numbers

The sets of the real number system can also be viewed as the following hierarchy.

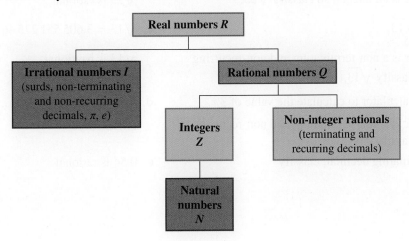

1.2.2 Irrational numbers

A **rational number** is a number that can be expressed as a ratio of two integers in the form $\dfrac{a}{b}$, where $b \neq 0$.

All **irrational numbers** have a decimal representation that is non-terminating and non-recurring. This means the decimals do not terminate and do not repeat in any particular pattern or order.

For example:

$$\sqrt{5} = 2.236\,067\,997\,5\dots$$
$$\pi = 3.141\,592\,653\,5\dots$$
$$e = 2.718\,281\,828\,4\dots$$

Rational or irrational

Rational and irrational numbers combine to form the set of **real numbers**. We can find all of these numbers somewhere on the real number line as shown below.

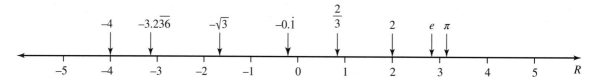

To classify a number as either rational or irrational, use the following steps:
1. Determine whether it can be expressed as a whole number, a fraction, or a terminating or recurring decimal.
2. If the answer is yes, the number is rational. If no, the number is irrational.

WORKED EXAMPLE 1 Classifying numbers as rational or irrational

Classify the following numbers as rational or irrational.

a. $\dfrac{1}{5}$

b. $\sqrt{25}$

c. $\sqrt{13}$

d. 3π

e. 0.54

f. $\sqrt[3]{64}$

g. $\sqrt[3]{32}$

h. $\sqrt[3]{\dfrac{1}{27}}$

THINK

a. $\dfrac{1}{5}$ is already a rational number.

b. 1. Evaluate $\sqrt{25}$.

   2. The answer is an integer, so classify $\sqrt{25}$.

c. 1. Evaluate $\sqrt{13}$.

   2. The answer is a non-terminating and non-recurring decimal; classify $\sqrt{13}$.

d. 1. Use your calculator to calculate the value of 3π.

   2. The answer is a non-terminating and non-recurring decimal; classify 3π.

e. 0.54 is a terminating decimal; classify it accordingly.

WRITE

a. $\dfrac{1}{5}$ is rational.

b. $\sqrt{25} = 5$

   $\sqrt{25}$ is rational.

c. $\sqrt{13} = 3.605\,551\,275\,46\dots$

   $\sqrt{13}$ is irrational.

d. $3\pi = 9.424\,777\,960\,76\dots$

   3π is irrational.

e. 0.54 is rational.

f. 1. Evaluate $\sqrt[3]{64}$.

f. $\sqrt[3]{64} = 4$

2. The answer is a whole number, so classify $\sqrt[3]{64}$.

$\sqrt[3]{64}$ is rational.

g. 1. Evaluate $\sqrt[3]{32}$.

g. $\sqrt[3]{32} = 3.174\,802\,103\,93\,...$

2. The result is a non-terminating and non-recurring decimal; classify $\sqrt[3]{32}$.

$\sqrt[3]{32}$ is irrational.

h. 1. Evaluate $\sqrt[3]{\dfrac{1}{27}}$.

h. $\sqrt[3]{\dfrac{1}{27}} = \dfrac{1}{3}$.

2. The result is a number in a rational form.

$\sqrt[3]{\dfrac{1}{27}}$ is rational.

1.2.3 Surds

A **surd** is an irrational number that is represented by a root sign or a radical sign, for example $\sqrt{}, \sqrt[3]{}, \sqrt[4]{}$.

Examples of surds include $\sqrt{7}, \sqrt{5}, \sqrt[3]{11}, \sqrt[4]{15}$.

In general, a surd can be written in the form $\sqrt[n]{x}$ where x is a rational number and n is an integer such that $n \geq 0$, and $x > 0$, when n is even.

The numbers $\sqrt{9}, \sqrt{16}, \sqrt[3]{125}$, and $\sqrt[4]{81}$ are not surds as they can be simplified to rational numbers; that is, $\sqrt{9} = 3, \sqrt{16} = 4, \sqrt[3]{125} = 5, \sqrt[4]{81} = 3, \sqrt[3]{-8} = -2$.

WORKED EXAMPLE 2 Identifying surds

Determine which of the following numbers are surds.

a. $\sqrt{16}$ **b.** $\sqrt{13}$ **c.** $\sqrt{\dfrac{1}{16}}$ **d.** $\sqrt[3]{17}$ **e.** $\sqrt[4]{63}$ **f.** $\sqrt[3]{-1728}$

THINK

a. 1. Evaluate $\sqrt{16}$.

2. The answer is rational (since it is a whole number), so state your conclusion.

b. 1. Evaluate $\sqrt{13}$.

2. The answer is irrational (since it is a non-recurring and non-terminating decimal), so state your conclusion.

c. 1. Evaluate $\sqrt{\dfrac{1}{16}}$.

2. The answer is rational (a fraction); state your conclusion.

WRITE

a. $\sqrt{16} = 4$

$\sqrt{16}$ is not a surd.

b. $\sqrt{13} = 3.605\,551\,275\,46\,...$

$\sqrt{13}$ is a surd.

c. $\sqrt{\dfrac{1}{16}} = \dfrac{1}{4}$

$\sqrt{\dfrac{1}{16}}$ is not a surd.

d. 1. Evaluate $\sqrt[3]{17}$.

d. $\sqrt[3]{17} = 2.571\,281\,590\,65\ldots$

2. The answer is irrational (a non-terminating and non-recurring decimal), so state your conclusion.

$\sqrt[3]{17}$ is a surd.

e. 1. Evaluate $\sqrt[4]{63}$.

e. $\sqrt[4]{63} = 2.817\,313\,247\,26\ldots$

2. The answer is irrational, so classify $\sqrt[4]{63}$ accordingly.

$\sqrt[4]{63}$ is a surd.

f. 1. Evaluate $\sqrt[3]{-1728}$.

f. $\sqrt[3]{-1728} = -12$

2. The answer is rational; state your conclusion. *Note:* An irrational number written in surd form gives an exact value of the number, whereas the same number written in decimal form gives an approximate value.
For example: $\sqrt{2}$ is an exact number, but $\sqrt{2} \approx 1.4142$ to 4 decimal places.

$\sqrt[3]{-1728}$ is not a surd. So **b**, **d** and **e** are surds.

Exercise 1.2 The real number system and surds

learn on

1.2 Exercise	1.2 Exam practice **on**

Simple familiar	Complex familiar	Complex unfamiliar
1, 2, 3, 4, 5, 6, 7, 8, 9, 10	11, 12, 13, 14, 15, 16	17, 18, 19, 20

Simple familiar

WE1 For questions **1** to **4**, classify the given numbers as rational (Q) or irrational (I).

1. a. $\dfrac{9}{4}$ **b.** 0.15 **c.** -2.4 **d.** $\sqrt{100}$

2. a. $\sqrt{14.4}$ **b.** $\sqrt{1.44}$ **c.** π **d.** $\sqrt{\dfrac{25}{9}}$

3. a. 7.32 **b.** $-\sqrt{21}$ **c.** $\sqrt{1000}$ **d.** $7.216\,349\,157\ldots$

4. a. $-\sqrt{81}$ **b.** 3π **c.** $\sqrt[3]{62}$ **d.** $\sqrt{\dfrac{1}{16}}$

WE2 For questions **5** to **8**, determine which of the given numbers are surds.

5. a. $\sqrt{0.16}$ **b.** $\sqrt{11}$ **c.** $\sqrt{\dfrac{3}{4}}$ **d.** $\sqrt[3]{\dfrac{3}{27}}$

6. a. $\sqrt{1000}$ **b.** $\sqrt{1.44}$ **c.** $4\sqrt{100}$ **d.** $2 + \sqrt{10}$

7. a. $\sqrt[3]{32}$ b. $\sqrt{361}$ c. $\sqrt[3]{100}$ d. $\sqrt[3]{125}$

8. a. $\sqrt{6}+\sqrt{6}$ b. 2π c. $\sqrt[3]{169}$ d. $\sqrt{\dfrac{7}{8}}$

9. **MC** Identify a rational number from the following.

   A. π **B.** $\sqrt{\dfrac{4}{9}}$ **C.** $\sqrt{\dfrac{9}{12}}$ **D.** $\sqrt[3]{3}$

10. **MC** Identify which of the following best represents an irrational number from the following numbers.

   A. $-\sqrt{81}$ **B.** $\dfrac{6}{5}$ **C.** $\sqrt[3]{343}$ **D.** $\sqrt{22}$

Complex familiar

11. **MC** Select the correct statement regarding the numbers $-0.69, \sqrt{7}, \dfrac{\pi}{3}, \sqrt{49}$.

   A. $\dfrac{\pi}{3}$ is the only rational number. **B.** $\sqrt{7}$ and $\sqrt{49}$ are both irrational numbers.

   C. -0.69 and $\sqrt{49}$ are the only rational numbers. **D.** -0.69 is the only rational number.

12. **MC** Select the correct statement regarding the numbers $2\dfrac{1}{2}, -\dfrac{11}{3}, \sqrt{624}, \sqrt[3]{99}$.

   A. $-\dfrac{11}{3}$ and $\sqrt{624}$ are both irrational numbers.

   B. $\sqrt{624}$ is an irrational number and $\sqrt[3]{99}$ is a rational number.

   C. $\sqrt{624}$ and $\sqrt[3]{99}$ are both irrational numbers.

   D. $2\dfrac{1}{2}$ is a rational number and $-\dfrac{11}{3}$ is an irrational number.

13. **MC** The correct statement regarding the set of numbers $\left\{\sqrt{\dfrac{6}{9}}, \sqrt{20}, \sqrt{54}, \sqrt[3]{27}, \sqrt{9}\right\}$ is:

   A. $\sqrt[3]{27}$ and $\sqrt{9}$ are the only rational numbers of the set.

   B. $\sqrt{\dfrac{6}{9}}$ is the only surd of the set.

   C. $\sqrt{\dfrac{6}{9}}$ and $\sqrt{20}$ are the only surds of the set.

   D. $\sqrt{20}$ and $\sqrt{54}$ are the only surds of the set.

14. **MC** Identify the numbers from the set $\left\{\sqrt{\dfrac{1}{4}}, \sqrt[3]{\dfrac{1}{27}}, \sqrt{\dfrac{1}{8}}, \sqrt{21}, \sqrt[3]{-8}\right\}$ that are surds.

   A. $\sqrt{21}$ only **B.** $\sqrt{\dfrac{1}{8}}$ only **C.** $\sqrt{\dfrac{1}{8}}$ and $\sqrt[3]{-8}$ **D.** $\sqrt{\dfrac{1}{8}}$ and $\sqrt{21}$ only

15. **MC** Select the statement regarding the set of numbers $\left\{\pi, \sqrt{\dfrac{1}{49}}, \sqrt{12}, \sqrt{16}, \sqrt{3}, +1\right\}$ that is *not* true.

   A. $\sqrt{12}$ is a surd. **B.** $\sqrt{12}$ and $\sqrt{16}$ are surds.
   C. π is irrational but not a surd. **D.** $\sqrt{12}$ and $\sqrt{3}+1$ are not rational.

16. **MC** Select the statement regarding the set of numbers $\left\{6\sqrt{7}, \sqrt{\dfrac{144}{16}}, 7\sqrt{6}, 9\sqrt{2}, \sqrt{18}, \sqrt{25}\right\}$ that is *not* true.

    A. $\sqrt{\dfrac{144}{16}}$ when simplified is an integer.
    B. $\sqrt{\dfrac{144}{16}}$ and $\sqrt{25}$ are not surds.

    C. $7\sqrt{6}$ is smaller than $9\sqrt{2}$.
    D. $9\sqrt{2}$ is smaller than $6\sqrt{7}$.

Complex unfamiliar

17. Determine the smallest value of m, where m is a positive integer, so that $\sqrt[3]{16m}$ is not a surd.

18. Determine any combination of m and n, where m and n are positive integers with $m < n$, so that $\sqrt[4]{(m+4)(16-n)}$ is not a surd.

19. π is an irrational number and so is $\sqrt{3}$. Determine whether $\left(\pi - \sqrt{3}\right)\left(\pi + \sqrt{3}\right)$ is an irrational number. Justify your decision.

20. The ancient Egyptians devised the formula $A = \dfrac{64}{81}d^2$ for calculating the area, A, of a circle of diameter d. Use this formula to obtain a rational approximation for π and evaluate it to 9 decimal places. State whether it is a better approximation than $\dfrac{22}{7}$.

Fully worked solutions for this chapter are available online.

LESSON
1.3 Simplifying surds

SYLLABUS LINKS

- Simplify square roots of natural numbers which contain perfect square factors, e.g. $\sqrt{45} = \sqrt{9 \times 5} = \sqrt{9}\sqrt{5} = 3\sqrt{5}$.

Source: Mathematical Methods Senior Syllabus 2024 © State of Queensland (QCAA) 2024; licensed under CC BY 4.0.

1.3.1 Multiplying and simplifying surds

Multiplication of surds

To multiply surds, multiply the expressions under the radical sign.

For example: $\sqrt{8} \times \sqrt{3} = \sqrt{8 \times 3} = \sqrt{24}$

If there are coefficients in front of the surds that are being multiplied, multiply the coefficients and then multiply the expressions under the radical signs.

For example: $2\sqrt{3} \times 5\sqrt{7} = (2 \times 5)\sqrt{3 \times 7} = 10\sqrt{21}$

Multiplication of surds

In order to multiply two or more surds, use the following:

$$\sqrt{a} \times \sqrt{b} = \sqrt{a \times b}$$

$$m\sqrt{a} \times n\sqrt{b} = mn\sqrt{a \times b}$$

where a and b are positive real numbers.

Simplification of surds

Surds are said to be in simplest form when the number under the square root sign contains no perfect square factors. If the radical sign is a cube root, then the simplest form has no perfect cube factors under the cube root. Simplification of a surd uses the method of multiplying surds in reverse.

For example, to express $\sqrt{128}$ in its simplest form, find perfect square factors of 128.

$$\begin{aligned}
\sqrt{128} &= \sqrt{64 \times 2} \\
&= \sqrt{64} \times \sqrt{2} \\
&= 8 \times \sqrt{2} \\
\therefore \sqrt{128} &= 8\sqrt{2}
\end{aligned}$$

If possible, try to factorise the number under the radical sign so that the largest possible perfect square is used.

Simplification of surds

$$\begin{aligned}
\sqrt{a^2 \times b} &= \sqrt{a^2} \times \sqrt{b} \\
&= a \times \sqrt{b} \\
&= a\sqrt{b}
\end{aligned}$$

WORKED EXAMPLE 3 Simplifying surds

Simplify the following surds. Assume that x and y are positive real numbers.

a. $\sqrt{45}$ b. $3\sqrt{28}$ c. $-\dfrac{1}{8}\sqrt{175}$ d. $5\sqrt{180x^3y^5}$

THINK	WRITE
a. 1. Express 45 as a product of two factors where one factor is the largest possible perfect square.	a. $\sqrt{45} = \sqrt{9 \times 5}$
2. Express $\sqrt{9 \times 5}$ as the product of two surds.	$= \sqrt{9} \times \sqrt{5}$
3. Simplify the square root from the perfect square (that is, $\sqrt{9} = 3$).	$= 3\sqrt{5}$
b. 1. Express 28 as a product of two factors, one of which is the largest possible perfect square.	b. $3\sqrt{28} = 3\sqrt{4 \times 7}$

2. Express $\sqrt{4 \times 7}$ as a product of two surds.

$$= 3\sqrt{4} \times \sqrt{7}$$

3. Simplify $\sqrt{4}$.

$$= 3 \times 2\sqrt{7}$$

4. Multiply together the whole numbers outside the square root sign (3 and 2).

$$= 6\sqrt{7}$$

c. 1. Express 175 as a product of two factors in which one factor is the largest possible perfect square.

$$\text{c. } -\frac{1}{8}\sqrt{175} = -\frac{1}{8}\sqrt{25 \times 7}$$

2. Express $\sqrt{25 \times 7}$ as a product of 2 surds.

$$= -\frac{1}{8} \times \sqrt{25} \times \sqrt{7}$$

3. Simplify $\sqrt{25}$.

$$= -\frac{1}{8} \times 5\sqrt{7}$$

4. Multiply together the numbers outside the square root sign.

$$= -\frac{5}{8}\sqrt{7}$$

d. 1. Express each of 180, x^3 and y^5 as a product of two factors where one factor is the largest possible perfect square.

$$\text{d. } 5\sqrt{180x^3y^5} = 5\sqrt{36 \times 5 \times x^2 \times x \times y^4 \times y}$$

2. Separate all the perfect squares into one surd and all other factors into the other surd.

$$= 5 \times \sqrt{36x^2y^4} \times \sqrt{5xy}$$

3. Simplify $\sqrt{36x^2y^4}$.

$$= 5 \times 6 \times x \times y^2 \times \sqrt{5xy}$$

4. Multiply together the numbers and the pronumerals outside the square root sign.

$$= 30xy^2\sqrt{5xy}$$

WORKED EXAMPLE 4 Multiplying surds

Multiply the following surds, expressing answers in the simplest form. Assume that x and y are positive real numbers.

a. $\sqrt{11} \times \sqrt{7}$ **b.** $5\sqrt{3} \times 8\sqrt{5}$ **c.** $6\sqrt{12} \times 2\sqrt{6}$ **d.** $\sqrt{15x^5y^2} \times \sqrt{12x^2y}$

THINK

a. Multiply the surds together using $\sqrt{a} \times \sqrt{b} = \sqrt{ab}$ (that is, multiply expressions under the square root sign).
Note: This expression cannot be simplified any further.

b. Multiply the coefficients together and then multiply the surds together.

WRITE

a. $\sqrt{11} \times \sqrt{7} = \sqrt{11 \times 7}$
$$= \sqrt{77}$$

b. $5\sqrt{3} \times 8\sqrt{5} = 5 \times 8 \times \sqrt{3} \times \sqrt{5}$
$$= 40 \times \sqrt{3 \times 5}$$
$$= 40\sqrt{15}$$

c. 1. Simplify $\sqrt{12}$.

c. $6\sqrt{12} \times 2\sqrt{6} = 6\sqrt{4 \times 3} \times 2\sqrt{6}$

$\qquad\qquad\qquad = 6 \times 2\sqrt{3} \times 2\sqrt{6}$

$\qquad\qquad\qquad = 12\sqrt{3} \times 2\sqrt{6}$

2. Multiply the coefficients together and multiply the surds together.

$\qquad\qquad\qquad = 24\sqrt{18}$

3. Simplify the surd.

$\qquad\qquad\qquad = 24\sqrt{9 \times 2}$

$\qquad\qquad\qquad = 24 \times 3\sqrt{2}$

$\qquad\qquad\qquad = 72\sqrt{2}$

d. 1. Simplify each of the surds.

d. $\sqrt{15x^5y^2} \times \sqrt{12x^2y}$

$\qquad = \sqrt{15 \times x^4 \times x \times y^2} \times \sqrt{4 \times 3 \times x^2 \times y}$

$\qquad = x^2 \times y \times \sqrt{15 \times x} \times 2 \times x \times \sqrt{3 \times y}$

$\qquad = x^2 y\sqrt{15x} \times 2x\sqrt{3y}$

2. Multiply the coefficients together and the surds together.

$\qquad = x^2 y \times 2x\sqrt{15x \times 3y}$

$\qquad = 2x^3 y\sqrt{45xy}$

$\qquad = 2x^3 y\sqrt{9 \times 5xy}$

3. Simplify the surd.

$\qquad = 2x^3 y \times 3\sqrt{5xy}$

$\qquad = 6x^3 y\sqrt{5xy}$

1.3.2 Ordering surds

Surds are real numbers and therefore have positions on the number line.

To estimate the position of $\sqrt{6}$, we can place it between two rational numbers by placing 6 between its closest perfect squares.

$$4 < 6 < 9$$
$$\sqrt{4} < \sqrt{6} < \sqrt{9}$$
$$\therefore 2 < \sqrt{6} < 3$$

To order the sizes of two surds such as $3\sqrt{5}$ and $5\sqrt{3}$, express each as an entire surd.

$$3\sqrt{5} = \sqrt{9} \times \sqrt{5} \qquad 5\sqrt{3} = \sqrt{25} \times \sqrt{3}$$
$$= \sqrt{9 \times 5} \quad \text{and} \qquad = \sqrt{25 \times 3}$$
$$= \sqrt{45} \qquad\qquad\qquad = \sqrt{75}$$

Since $\sqrt{45} < \sqrt{75}$, it follows that $3\sqrt{5} < 5\sqrt{3}$.

WORKED EXAMPLE 5 Ordering surds

a. Express $\left\{6\sqrt{2}, 4\sqrt{3}, 2\sqrt{5}, 7\right\}$ **in increasing order.**

THINK

WRITE

a. 1. Express each number entirely as a square root.

a. $\left\{6\sqrt{2}, 4\sqrt{3}, 2\sqrt{5}, 7\right\}$

$6\sqrt{2} = \sqrt{36} \times \sqrt{2} = \sqrt{72}$,

$4\sqrt{3} = \sqrt{16} \times \sqrt{3} = \sqrt{48}$,

$2\sqrt{5} = \sqrt{4} \times \sqrt{5} = \sqrt{20}$

and $7 = \sqrt{49}$.

2. Order the terms.

$\sqrt{20} < \sqrt{48} < \sqrt{49} < \sqrt{72}$

In increasing order, the set of numbers is

$\left\{2\sqrt{5}, 4\sqrt{3}, 7, 6\sqrt{2}\right\}$.

Exercise 1.3 Simplifying surds

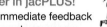

1.3 Exercise	1.3 Exam practice on

Simple familiar	Complex familiar	Complex unfamiliar
1, 2, 3, 4, 5, 6, 7, 8, 9, 10, 11, 12	13, 14, 15, 16, 17	18, 19, 20

These questions are even better in jacPLUS!
- Receive immediate feedback
- Access sample responses
- Track results and progress

Find all this and MORE in jacPLUS ▶

Simple familiar

WE3a For questions **1** to **3**, simplify the given surds.

1. a. $\sqrt{12}$
   b. $\sqrt{24}$
   c. $\sqrt{27}$
   d. $\sqrt{125}$

2. a. $\sqrt{54}$
   b. $\sqrt{112}$
   c. $\sqrt{68}$
   d. $\sqrt{180}$

3. a. $\sqrt{88}$
   b. $\sqrt{162}$
   c. $\sqrt{245}$
   d. $\sqrt{448}$

WE3b,c For questions **4** to **6**, simplify the given surds.

4. a. $2\sqrt{8}$
   b. $8\sqrt{90}$
   c. $9\sqrt{80}$
   d. $7\sqrt{54}$

5. a. $-6\sqrt{75}$
   b. $-7\sqrt{80}$
   c. $16\sqrt{48}$
   d. $\dfrac{1}{7}\sqrt{392}$

6. a. $\dfrac{1}{9}\sqrt{162}$
   b. $\dfrac{1}{4}\sqrt{192}$
   c. $\dfrac{1}{9}\sqrt{135}$
   d. $\dfrac{3}{10}\sqrt{175}$

7. Simplify each of the following.

   a. $\sqrt{32}$
   b. $2\sqrt{44}$
   c. $\sqrt{52}$

   d. $3\sqrt{80}$
   e. $\sqrt{45}$
   f. $\dfrac{\sqrt{99}}{18}$

8. Express each of the following in simplified form.

   a. $\sqrt{75}$ **b.** $5\sqrt{48}$ **c.** $\sqrt{2000}$

   d. $3\sqrt{288}$ **e.** $2\sqrt{72}$ **f.** $\sqrt[3]{54}$

9. **WE4** Express each of the following in simplest form.

   a. $\sqrt{2}\times\sqrt{7}$ **b.** $\sqrt{6}\times\sqrt{7}$ **c.** $\sqrt{8}\times\sqrt{6}$

   d. $\sqrt{10}\times\sqrt{10}$ **e.** $\sqrt{21}\times\sqrt{3}$ **f.** $\sqrt{27}\times3\sqrt{3}$

10. Express each of the following in simplest form.

   a. $5\sqrt{3}\times2\sqrt{11}$ **b.** $10\sqrt{15}\times6\sqrt{3}$ **c.** $4\sqrt{20}\times3\sqrt{5}$

   d. $10\sqrt{6}\times3\sqrt{8}$ **e.** $\dfrac{1}{4}\sqrt{48}\times2\sqrt{2}$ **f.** $\dfrac{1}{9}\sqrt{48}\times2\sqrt{3}$

11. **WE5** Express $\left\{3\sqrt{3},4\sqrt{5},5\sqrt{2},5\right\}$ with its elements in increasing order.

12. Express the set $\left\{2\sqrt{5},3\sqrt{2},4\sqrt{3},5\right\}$ with its elements in descending order.

Complex familiar

13. Express the following as entire surds.

   a. $4\sqrt{5}$ **b.** $2\sqrt[3]{6}$ **c.** $ab\sqrt{c}$ **d.** $m\sqrt[3]{n}$

WE3d For questions **14** to **16**, simplify the given surds. Assume that a, b, c, d, e, f, x and y are positive real numbers.

14. **a.** $\sqrt{16a^2}$ **b.** $\sqrt{72a^2}$ **c.** $\sqrt{90a^2b}$ **d.** $\sqrt{338a^4}$

15. **a.** $\sqrt{338a^3b^3}$ **b.** $\sqrt{68a^3b^5}$ **c.** $\sqrt{125x^6y^4}$ **d.** $5\sqrt{80x^3y^2}$

16. **a.** $6\sqrt{162c^7d^5}$ **b.** $2\sqrt{405c^7d^9}$ **c.** $\dfrac{1}{2}\sqrt{88ef}$ **d.** $\dfrac{1}{2}\sqrt{392e^{11}f^{11}}$

17. **a.** $\dfrac{1}{10}\sqrt{60}\times\dfrac{1}{5}\sqrt{40}$ **b.** $\sqrt{xy}\times\sqrt{x^3y^2}$ **c.** $\sqrt{3a^4b^2}\times\sqrt{6a^5b^3}$

   d. $\sqrt{12a^7b}\times\sqrt{6a^3b^4}$ **e.** $\sqrt{15x^3y^2}\times\sqrt{6x^2y^3}$ **f.** $\dfrac{1}{2}\sqrt{15a^3b^3}\times3\sqrt{3a^2b^6}$

Complex unfamiliar

18. Determine the exact area of an equilateral triangle with side length of 24 cm.

19. A rhombus has diagonals of length 16 cm and 24 cm. Determine the perimeter of the rhombus, expressing your answer in exact form.

20. A circle has a chord of length 18 cm. If the shortest distance from the circle's centre to the chord is 6 cm, determine the circumference of the circle. Express your answer in exact form.

Fully worked solutions for this chapter are available online.

LESSON
1.4 Operations with surds and rationalising the denominator

SYLLABUS LINKS

- Rationalise the denominator of fractional expressions involving square roots,

  e.g. $\dfrac{\sqrt{7}}{\sqrt{3}} = \dfrac{\sqrt{7}}{\sqrt{3}} \times \dfrac{\sqrt{3}}{\sqrt{3}} = \dfrac{\sqrt{7} \times \sqrt{3}}{\sqrt{3} \times \sqrt{3}} = \dfrac{\sqrt{21}}{3}$.

- Use the four operations to simplify surds, e.g. $\sqrt{5} - 2\sqrt{5} + 4\sqrt{5} = 3\sqrt{5}$ and $2\sqrt{3} \times 5\sqrt{11} = 10\sqrt{33}$.

Source: Mathematical Methods Senior Syllabus 2024 © State of Queensland (QCAA) 2024; licensed under CC BY 4.0.

1.4.1 Addition and subtraction of surds

As surds are real numbers, they obey the usual laws for addition and subtraction of like terms.

Examples of like surds include $\sqrt{7}, 3\sqrt{7}, -5\sqrt{7}$.

In some cases, surds will need to be simplified before you decide whether they are like terms and then addition and subtraction can take place.

Addition and subtraction with surds

$$a\sqrt{c} + b\sqrt{c} = (a+b)\sqrt{c}$$
$$a\sqrt{c} - b\sqrt{c} = (a-b)\sqrt{c}$$

WORKED EXAMPLE 6 Adding and subtracting surds

Simplify each of the following expressions containing surds. Assume that a and b are positive real numbers.

a. $3\sqrt{6} + 17\sqrt{6} - 2\sqrt{6}$

b. $5\sqrt{3} + 2\sqrt{12} - 5\sqrt{2} + 3\sqrt{8}$

c. $\dfrac{1}{2}\sqrt{100a^3b^2} + ab\sqrt{36a} - 5\sqrt{4a^2b}$

THINK	WRITE
a. All 3 terms are alike because they contain the same surd $\left(\sqrt{6}\right)$. Simplify.	a. $3\sqrt{6} + 17\sqrt{6} - 2\sqrt{6} = (3 + 17 - 2)\sqrt{6}$ $= 18\sqrt{6}$
b. 1. Simplify surds where possible.	b. $5\sqrt{3} + 2\sqrt{12} - 5\sqrt{2} + 3\sqrt{8}$ $= 5\sqrt{3} + 2\sqrt{4 \times 3} - 5\sqrt{2} + 3\sqrt{4 \times 2}$ $= 5\sqrt{3} + 2 \times 2\sqrt{3} - 5\sqrt{2} + 3 \times 2\sqrt{2}$
2. Add like terms to obtain the simplified answer.	$= 5\sqrt{3} + 4\sqrt{3} - 5\sqrt{2} + 6\sqrt{2}$ $= 9\sqrt{3} + \sqrt{2}$

c. 1. Simplify surds where possible.	c.	$\frac{1}{2}\sqrt{100a^3b^2} + ab\sqrt{36a} - 5\sqrt{4a^2b}$
		$= \frac{1}{2} \times 10\sqrt{a^2 \times a \times b^2} + ab \times 6\sqrt{a} - 5 \times 2 \times a\sqrt{b}$
		$= \frac{1}{2} \times 10 \times a \times b\sqrt{a} + ab \times 6\sqrt{a} - 5 \times 2 \times a\sqrt{b}$
2. Add like terms to obtain the simplified answer.		$= 5ab\sqrt{a} + 6ab\sqrt{a} - 10a\sqrt{b}$
		$= 11ab\sqrt{a} - 10a\sqrt{b}$

1.4.2 Squaring surds

When working with surds, it is sometimes necessary to multiply surds by themselves; that is, square them. Consider the following examples:

$$\left(\sqrt{2}\right)^2 = \sqrt{2} \times \sqrt{2} = \sqrt{4} = 2$$
$$\left(\sqrt{5}\right)^2 = \sqrt{5} \times \sqrt{5} = \sqrt{25} = 5$$

Observe that squaring a surd produces the number under the radical sign. This is not surprising, because squaring and taking the square root are *inverse operations* and, when applied together, leave the original unchanged.

Squaring surds

When a surd is squared, the result is the expression under the radical sign; that is:

$$\left(\sqrt{a}\right)^2 = a$$

where a is a positive real number.

WORKED EXAMPLE 7 Squaring surds

Simplify each of the following.

a. $\left(\sqrt{6}\right)^2$

b. $\left(4\sqrt{5}\right)^2$

THINK

WRITE

a. Use $\left(\sqrt{a}\right)^2 = a$, where $a = 6$.

a. $\left(\sqrt{6}\right)^2 = 6$

b. 1. Square 4 and apply $\left(\sqrt{a}\right)^2 = a$ to square $\sqrt{5}$.

b. $\left(4\sqrt{5}\right)^2 = 4^2 \times \left(\sqrt{5}\right)^2$

$= 16 \times 5$

   2. Simplify.

$= 80$

1.4.3 Multiplication and division of surds

As surds are real numbers, they obey the usual laws of multiplication and division.

Multiplication and division with surds

$$\sqrt{c} \times \sqrt{d} = \sqrt{cd}$$
$$a\sqrt{c} \times b\sqrt{d} = ab\sqrt{cd}$$
$$\frac{\sqrt{c}}{\sqrt{d}} = \sqrt{\frac{c}{d}}$$

When dividing surds, it is best to simplify them (if possible) first. Once this has been done, the coefficients are divided next and then the surds are divided.

$$\frac{m\sqrt{a}}{n\sqrt{b}} = \frac{m}{n}\sqrt{\frac{a}{b}}$$

WORKED EXAMPLE 8 Dividing surds

Divide the following surds, expressing answers in the simplest form. Assume that x and y are positive real numbers.

a $\dfrac{\sqrt{55}}{\sqrt{5}}$ b $\dfrac{\sqrt{48}}{\sqrt{3}}$ c $\dfrac{9\sqrt{88}}{6\sqrt{99}}$ d $\dfrac{\sqrt{36xy}}{\sqrt{25x^9y^{11}}}$

THINK

WRITE

a. 1. Rewrite the fraction, using $\dfrac{\sqrt{a}}{\sqrt{b}} = \sqrt{\dfrac{a}{b}}$.

a. $\dfrac{\sqrt{55}}{\sqrt{5}} = \sqrt{\dfrac{55}{5}}$

  2. Divide the numerator by the denominator (that is, divide 55 by 5). Check if the surd can be simplified any further.

$= \sqrt{11}$

b. 1. Rewrite the fraction using $\dfrac{\sqrt{a}}{\sqrt{b}} = \sqrt{\dfrac{a}{b}}$.

b. $\dfrac{\sqrt{48}}{\sqrt{3}} = \sqrt{\dfrac{48}{3}}$

  2. Divide 48 by 3.

$= \sqrt{16}$

  3. Evaluate $\sqrt{16}$.

$= 4$

c. 1. Rewrite the surds using $\dfrac{\sqrt{a}}{\sqrt{b}} = \sqrt{\dfrac{a}{b}}$.

c. $\dfrac{9\sqrt{88}}{6\sqrt{99}} = \dfrac{9}{6}\sqrt{\dfrac{88}{99}}$

  2. Simplify the fraction under the radical by dividing both numerator and denominator by 11.

$= \dfrac{9}{6}\sqrt{\dfrac{8}{9}}$

  3. Simplify surds.

$= \dfrac{9 \times 2\sqrt{2}}{6 \times 3}$

4. Multiply the whole numbers in the numerator together and those in the denominator together.

$$= \frac{18\sqrt{2}}{18}$$

5. Cancel the common factor of 18.

$$= \sqrt{2}$$

d. 1. Simplify each surd.

$$\text{d. } \frac{\sqrt{36xy}}{\sqrt{25x^9y^{11}}} = \frac{6\sqrt{xy}}{5\sqrt{x^8 \times x \times y^{10} \times y}}$$

$$= \frac{6\sqrt{xy}}{5x^4y^5\sqrt{xy}}$$

2. Cancel any common factors — in this case $\sqrt{xy}$.

$$= \frac{6}{5x^4y^5}$$

1.4.4 Rationalising denominators

If the **denominator** of a fraction is a surd, it can be changed into a rational number through multiplication. In other words, it can be rationalised.

As discussed earlier in this chapter, squaring a simple surd (that is, multiplying it by itself) results in a rational number. This fact can be used to rationalise denominators as follows.

> ### Rationalising the denominator
> $$\frac{\sqrt{a}}{\sqrt{b}} = \frac{\sqrt{a}}{\sqrt{b}} \times \frac{\sqrt{b}}{\sqrt{b}} = \frac{\sqrt{ab}}{b}$$

If both the numerator and the denominator of a fraction are multiplied by the surd contained in the denominator, the denominator becomes a rational number. The fraction takes on a different appearance but its numerical value is unchanged, because multiplying the numerator and denominator by the same number is equivalent to multiplying by 1.

WORKED EXAMPLE 9 Rationalising the denominator

Express the following in their simplest form with a rational denominator.

a. $\dfrac{\sqrt{6}}{\sqrt{5}}$

b. $\dfrac{2\sqrt{12}}{3\sqrt{54}}$

c. $\dfrac{\sqrt{17} - 3\sqrt{14}}{\sqrt{7}}$

THINK

WRITE

a. 1. Write the fraction.

a. $\dfrac{\sqrt{6}}{\sqrt{5}}$

2. Multiply both the numerator and denominator by the surd contained in the denominator (in this case $\sqrt{5}$). This has the same effect as multiplying the fraction by 1, because $\dfrac{\sqrt{5}}{\sqrt{5}} = 1$.

$$= \frac{\sqrt{6}}{\sqrt{5}} \times \frac{\sqrt{5}}{\sqrt{5}}$$

$$= \frac{\sqrt{78}}{5}$$

b. 1. Write the fraction.

b. $\dfrac{2\sqrt{12}}{3\sqrt{54}}$

2. Simplify the surds. (This avoids dealing with large numbers.)

$$\dfrac{2\sqrt{12}}{3\sqrt{54}} = \dfrac{2\sqrt{4\times 3}}{3\sqrt{9\times 6}}$$

$$= \dfrac{2\times 2\sqrt{3}}{3\times 3\sqrt{6}}$$

$$= \dfrac{4\sqrt{3}}{9\sqrt{6}}$$

3. Multiply both the numerator and denominator by $\sqrt{6}$. This has the same effect as multiplying the fraction by 1, because $\dfrac{\sqrt{6}}{\sqrt{6}} = 1$.

Note: We need to multiply only by the surd part of the denominator (that is by $\sqrt{6}$ rather than by $9\sqrt{6}$).

$$= \dfrac{4\sqrt{3}}{9\sqrt{6}} \times \dfrac{\sqrt{6}}{\sqrt{6}}$$

$$= \dfrac{4\sqrt{18}}{9\times 6}$$

4. Simplify $\sqrt{18}$.

$$= \dfrac{4\sqrt{9\times 2}}{9\times 6}$$

$$= \dfrac{4\times 3\sqrt{2}}{54}$$

$$= \dfrac{12\sqrt{2}}{54}$$

5. Divide both the numerator and denominator by 6 (cancel down).

$$= \dfrac{2\sqrt{2}}{9}$$

c. 1. Write the fraction.

c. $\dfrac{\sqrt{17}-3\sqrt{14}}{\sqrt{7}}$

2. Multiply both the numerator and denominator by $\sqrt{7}$. Use grouping symbols (brackets) to make it clear that the whole numerator must be multiplied by $\sqrt{7}$.

$$= \dfrac{\left(\sqrt{17}-3\sqrt{14}\right)}{\sqrt{7}} \times \dfrac{\sqrt{7}}{\sqrt{7}}$$

3. Apply the Distributive Law in the numerator: $a(b+c) = ab + ac$.

$$= \dfrac{\sqrt{17}\times\sqrt{7} - 3\sqrt{14}\times\sqrt{7}}{\sqrt{7}\times\sqrt{7}}$$

$$= \dfrac{\sqrt{119} - 3\sqrt{98}}{7}$$

4. Simplify $\sqrt{98}$.

$$= \dfrac{\sqrt{119} - 3\sqrt{49\times 2}}{7}$$

$$= \dfrac{\sqrt{119} - 3\times 7\sqrt{2}}{7}$$

$$= \dfrac{\sqrt{119} - 21\sqrt{2}}{7}$$

Exercise 1.4 Operations with surds and rationalising the denominator

1.4 Exercise	**1.4 Exam practice** on

Simple familiar	Complex familiar	Complex unfamiliar
1, 2, 3, 4, 5, 6, 7, 8, 9, 10, 11, 12, 13, 14	15, 16, 17	18, 19, 20

These questions are even better in jacPLUS!
- Receive immediate feedback
- Access sample responses
- Track results and progress

Find all this and MORE in jacPLUS ▶

Simple familiar

1. **WE6a** Simplify the following expressions containing surds.
   a. $3\sqrt{5} + 4\sqrt{5}$
   b. $2\sqrt{3} + 5\sqrt{3} + \sqrt{3}$
   c. $8\sqrt{5} + 3\sqrt{3} + 7\sqrt{5} + 2\sqrt{3}$
   d. $6\sqrt{11} - 2\sqrt{11}$

2. Simplify the following expressions containing surds. Assume that x and y are positive real numbers.
   a. $7\sqrt{2} + 9\sqrt{2} - 3\sqrt{2}$
   b. $9\sqrt{6} + 12\sqrt{6} - 17\sqrt{6} - 7\sqrt{6}$
   c. $12\sqrt{3} - 8\sqrt{7} + 5\sqrt{3} - 10\sqrt{7}$
   d. $2\sqrt{x} + 5\sqrt{y} + 6\sqrt{x} - 2\sqrt{y}$

3. **WE6b** Simplify the following expressions containing surds.
   a. $\sqrt{200} - \sqrt{300}$
   b. $\sqrt{125} - \sqrt{150} + \sqrt{600}$
   c. $\sqrt{27} - \sqrt{3} + \sqrt{75}$
   d. $2\sqrt{20} - 3\sqrt{5} + \sqrt{45}$

4. Simplify the following expressions containing surds.
   a. $6\sqrt{12} + 3\sqrt{27} - 7\sqrt{3} + \sqrt{18}$
   b. $\sqrt{150} + \sqrt{24} - \sqrt{96} + \sqrt{108}$
   c. $3\sqrt{90} - 5\sqrt{60} + 3\sqrt{40} + \sqrt{100}$
   d. $5\sqrt{11} + 7\sqrt{44} - 9\sqrt{99} + 2\sqrt{121}$

5. Simplify the following expressions containing surds. Assume that a and b are positive real numbers.
   a. $2\sqrt{30} + 5\sqrt{120} + \sqrt{60} - 6\sqrt{135}$
   b. $6\sqrt{ab} - \sqrt{12ab} + 2\sqrt{9ab} + 3\sqrt{27ab}$
   c. $\frac{1}{2}\sqrt{98} + \frac{1}{3}\sqrt{48} + \frac{1}{3}\sqrt{12}$
   d. $\frac{1}{8}\sqrt{32} - \frac{7}{6}\sqrt{18} + 3\sqrt{72}$

6. **WE7** Simplify each of the following.
   a. $\left(\sqrt{2}\right)^2$
   b. $\left(\sqrt{5}\right)^2$
   c. $\left(\sqrt{12}\right)^2$

7. Simplify each of the following.
   a. $\left(\sqrt{15}\right)^2$
   b. $\left(3\sqrt{2}\right)^2$
   c. $\left(4\sqrt{5}\right)^2$

8. **WE8** Divide the following surds, expressing your answers in the simplest form.
   a. $\frac{\sqrt{15}}{\sqrt{3}}$
   b. $\frac{\sqrt{8}}{\sqrt{2}}$
   c. $\frac{\sqrt{60}}{\sqrt{10}}$
   d. $\frac{\sqrt{128}}{\sqrt{8}}$

9. Divide the following surds, expressing your answers in the simplest form.
   a. $\frac{\sqrt{18}}{4\sqrt{6}}$
   b. $\frac{\sqrt{65}}{2\sqrt{13}}$
   c. $\frac{\sqrt{96}}{\sqrt{8}}$
   d. $\frac{7\sqrt{44}}{14\sqrt{11}}$

10. **WE9a,b** Express the following in their simplest form with a rational denominator.
    a. $\frac{5}{\sqrt{2}}$
    b. $\frac{7}{\sqrt{3}}$
    c. $\frac{4}{\sqrt{11}}$
    d. $\frac{8}{\sqrt{6}}$
    e. $\frac{\sqrt{12}}{\sqrt{7}}$

11. Express the following in their simplest form with a rational denominator.

a. $\dfrac{\sqrt{15}}{\sqrt{6}}$
b. $\dfrac{2\sqrt{3}}{\sqrt{5}}$
c. $\dfrac{3\sqrt{7}}{\sqrt{5}}$
d. $\dfrac{5\sqrt{2}}{2\sqrt{3}}$
e. $\dfrac{4\sqrt{3}}{3\sqrt{5}}$

12. Express the following in their simplest form with a rational denominator.

a. $\dfrac{5\sqrt{14}}{7\sqrt{8}}$
b. $\dfrac{16\sqrt{3}}{6\sqrt{5}}$
c. $\dfrac{8\sqrt{3}}{7\sqrt{7}}$
d. $\dfrac{8\sqrt{60}}{\sqrt{28}}$
e. $\dfrac{2\sqrt{35}}{3\sqrt{14}}$

13. **WE9c** Express each of the following in their simplest form with a rational denominator.

a. $\dfrac{\sqrt{6}+\sqrt{12}}{\sqrt{3}}$
b. $\dfrac{\sqrt{15}-\sqrt{22}}{\sqrt{6}}$
c. $\dfrac{6\sqrt{2}-\sqrt{15}}{\sqrt{10}}$
d. $\dfrac{2\sqrt{18}+3\sqrt{2}}{\sqrt{5}}$

14. Express each of the following in their simplest form with a rational denominator.

a. $\dfrac{7\sqrt{12}-5\sqrt{6}}{6\sqrt{3}}$
b. $\dfrac{6\sqrt{2}-\sqrt{5}}{4\sqrt{8}}$
c. $\dfrac{6\sqrt{3}-5\sqrt{5}}{7\sqrt{20}}$
d. $\dfrac{3\sqrt{5}+7\sqrt{3}}{5\sqrt{24}}$

Complex familiar

15. Simplify the following expressions containing surds. Assume that a and b are positive real numbers.

a. $\sqrt{8a^3}+\sqrt{72a^3}-\sqrt{98a^3}$
b. $\dfrac{1}{2}\sqrt{36a}+\dfrac{1}{4}\sqrt{128a}-\dfrac{1}{6}\sqrt{144a}$
c. $\sqrt{9a^3}+\sqrt{3a^5}$
d. $6\sqrt{a^5b}+\sqrt{a^3b}-5\sqrt{a^5b}$

16. Simplify the following expressions containing surds. Assume that a and b are positive real numbers.

a. $\dfrac{1}{10}\sqrt{60}\times\dfrac{1}{5}\sqrt{40}$
b. $\sqrt{xy}\times\sqrt{x^3y^2}$
c. $\sqrt{3a^4b^2}\times\sqrt{6a^5b^3}$
d. $\sqrt{12a^7b}\times\sqrt{6a^3b^4}$
e. $\sqrt{15x^3y^2}\times\sqrt{6x^2y^3}$
f. $\dfrac{1}{2}\sqrt{15a^3b^3}\times3\sqrt{3a^2b^6}$

17. Divide the following surds, expressing answers in the simplest form. Assume that a, b, x and y are positive real numbers.

a. $\dfrac{9\sqrt{63}}{15\sqrt{7}}$
b. $\dfrac{\sqrt{2040}}{\sqrt{30}}$
c. $\dfrac{\sqrt{x^4y^3}}{\sqrt{x^2y^5}}$
d. $\dfrac{\sqrt{16xy}}{\sqrt{8x^7y^9}}$
e. $\dfrac{\sqrt{xy}}{\sqrt{x^5y^7}}\times\dfrac{\sqrt{12x^8y^{12}}}{\sqrt{x^2y^3}}$
f. $\dfrac{2\sqrt{2a^2b^4}}{\sqrt{5a^3b^6}}\times\dfrac{\sqrt{10a^9b^3}}{3\sqrt{a^7b}}$

Complex unfamiliar

18. Determine the area of a triangle with base length $\dfrac{3}{\sqrt{2}}$ and perpendicular height $\dfrac{5}{\sqrt{10}}$. Express your answer with a rational denominator.

19. Determine the average of $\dfrac{1}{2\sqrt{x}}$ and $\dfrac{1}{3\sqrt{x}}$, writing your answer with a rational denominator.

20. If $x=2\sqrt{3}-\sqrt{10}$, determine the value of $x^2-4\sqrt{3}x$.

Fully worked solutions for this chapter are available online.

LESSON
1.5 Review

1.5.1 Summary

doc-
41796

1.5 Exercise

 learn on

1.5 Exercise	1.5 Exam practice on

Simple familiar	Complex familiar	Complex unfamiliar
1, 2, 3, 4, 5, 6, 7, 8, 9, 10, 11, 12, 13, 14	15, 16, 17	18, 19, 20

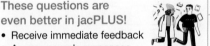
Simple familiar

1. **MC** Identify which of the given numbers are rational.

$$\sqrt{\frac{6}{12}}, \ \sqrt{0.81}, \ 5, \ -3.26, \ 0.5, \ \frac{\pi}{5}, \ \sqrt{\frac{3}{12}}$$

A. $\sqrt{0.81}, \ 5, \ -3.26, \ 0.5$ and $\sqrt{\frac{3}{12}}$

B. $\sqrt{\frac{6}{12}}$ and $\frac{\pi}{5}$

C. $\sqrt{\frac{6}{12}}, \ \sqrt{0.81}$ and $\sqrt{\frac{3}{12}}$

D. $5, \ -3.26$ and $\sqrt{\frac{6}{12}}$

2. For each of the following, state whether the number is rational or irrational and give the reason for your answer.

   a. $\sqrt{12}$ b. $\sqrt{121}$ c. $\frac{2}{9}$ d. $0.\dot{6}$ e. $\sqrt[3]{0.08}$

3. **MC** Identify which of the numbers of the given set are surds.

$$\left\{ 3\sqrt{2}, 5\sqrt{7}, 9\sqrt{4}, 6\sqrt{10}, 7\sqrt{12}, 12\sqrt{64} \right\}$$

A. $9\sqrt{4}, 12\sqrt{64}$

B. $3\sqrt{2}$ and $7\sqrt{12}$ only

C. $3\sqrt{2}, 5\sqrt{7}$ and $6\sqrt{10}$ only

D. $3\sqrt{2}, 5\sqrt{7}, 6\sqrt{10}$ and $7\sqrt{12}$

4. Simplify each of the following.

   a. $\sqrt{50}$ b. $\sqrt{180}$ c. $2\sqrt{32}$ d. $5\sqrt{80}$

5. Simplify the following, giving the answer in simplest form.

$$7\sqrt{12} + 8\sqrt{147} - 15\sqrt{27}$$

6. Simplify each of the following.
   a. $\sqrt{3} \times \sqrt{5}$
   b. $2\sqrt{6} \times 3\sqrt{7}$
   c. $3\sqrt{10} \times 5\sqrt{6}$
   d. $\left(\sqrt{5}\right)^2$

7. Simplify the following, giving answers in the simplest form.
   a. $\dfrac{1}{5}\sqrt{675} \times \sqrt{27}$
   b. $10\sqrt{24} \times 6\sqrt{12}$

8. Simplify the following.
   a. $\dfrac{\sqrt{30}}{\sqrt{10}}$
   b. $\dfrac{6\sqrt{45}}{3\sqrt{5}}$
   c. $\dfrac{3\sqrt{20}}{12\sqrt{6}}$
   d. $\dfrac{\left(\sqrt{7}\right)^2}{14}$

9. Rationalise the denominator of each of the following.
   a. $\dfrac{2}{\sqrt{6}}$
   b. $\dfrac{\sqrt{3}}{2\sqrt{6}}$
   c. $\dfrac{2\sqrt{3}}{3\sqrt{6}}$
   d. $\dfrac{\sqrt{8} + \sqrt{12}}{\sqrt{2}}$

10. Arrange the elements of the following set of numbers in decreasing order.

$$\left\{2\sqrt{6}, 4\sqrt{3}, 3\sqrt{5}, 5\sqrt{2}, 2\sqrt{10}\right\}$$

11. **MC** The expression $\sqrt{250}$ may be simplified to:
    A. $25\sqrt{10}$
    B. $5\sqrt{10}$
    C. $10\sqrt{5}$
    D. $5\sqrt{50}$

12. **MC** When expressed in its simplest form, $2\sqrt{98} - 3\sqrt{72}$ is equal to:
    A. $-4\sqrt{2}$
    B. -4
    C. $-2\sqrt{4}$
    D. $4\sqrt{2}$

13. Simplify the following.
    a. $3\sqrt{7} + 8\sqrt{3} + 12\sqrt{7} - 9\sqrt{3}$
    b. $10\sqrt{2} - 12\sqrt{6} + 4\sqrt{6} - 8\sqrt{2}$
    c. $3\sqrt{50} - \sqrt{18}$
    d. $8\sqrt{45} + 2\sqrt{125}$
    e. $\sqrt{6} + 7\sqrt{5} + 4\sqrt{24} - 8\sqrt{20}$
    f. $2\sqrt{12} - 7\sqrt{243} + \dfrac{1}{2}\sqrt{8} - \dfrac{2}{3}\sqrt{162}$

14. Evaluate the following, expressing answers in simplest form.
    a. $4\sqrt{5} \times 2\sqrt{7}$
    b. $-10\sqrt{6} \times -8\sqrt{10}$
    c. $3\sqrt{8} \times 2\sqrt{5}$
    d. $\sqrt{18} \times \sqrt{72}$
    e. $\dfrac{4\sqrt{27} \times \sqrt{147}}{2\sqrt{3}}$
    f. $5\sqrt{2} \times \sqrt{3} \times 4\sqrt{5} \times \dfrac{\sqrt{6}}{6} + 3\sqrt{2} \times 7\sqrt{10}$

15. Identify which of $\sqrt{2m}$, $\sqrt{25m}$, $\sqrt{\dfrac{m}{16}}$, $\sqrt{\dfrac{m}{20}}$, $\sqrt[3]{m}$, $\sqrt[3]{8m}$ are surds:

    a. if $m = 4$ **b.** if $m = 8$.

16. **MC** If $a = 18$, then the value of $\dfrac{6}{\sqrt{a}} + \dfrac{\sqrt{a}}{\sqrt{6}}$ is:

    A. $6\sqrt{2}$ **B.** $\sqrt{2} + \sqrt{3}$

    C. $9 + \sqrt{3}$ **D.** $\sqrt{2} + 3\sqrt{6}$

17. **MC** The expression $\sqrt{392x^8y^7}$ may be simplified to:

    A. $196x^4y^3\sqrt{2y}$ **B.** $2x^4y^3\sqrt{14y}$

    C. $14x^4y^3\sqrt{2y}$ **D.** $14x^4y^3\sqrt{2}$

18. Show that $\left(3\sqrt{2} + 2\sqrt{3}\right)^2 + \left(3\sqrt{2} - 2\sqrt{3}\right)^2$ is rational.

19. Two friends arrange to meet in a park in order to do some walking as part of their exercise regime.

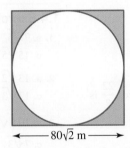

$\longleftarrow$ $80\sqrt{2}$ m $\longrightarrow$

On Monday, they walk around the square of edge $80\sqrt{2}$ metres.

On Tuesday, they walk around the circumference of the circle that the square encloses. The average speed of the friends was $\sqrt{8}$ km/h.

Determine an exact expression that shows the difference in the time, in minutes, taken on Monday compared to Tuesday.

Express your answer in the form $a(b + c\pi)$, where a, b, and c are rational numbers.

20. In the triangle PQR, the midpoints of the sides PQ, PR and RQ of the triangle are $\left(\sqrt{2}, \sqrt{3}\right)$, $\left(\sqrt{10} - \sqrt{2}, \sqrt{5}\right)$ and $\left(\sqrt{2}, \sqrt{3} - \sqrt{2}\right)$, respectively (measured in kilometres).

Determine the coordinates of P, Q and R, and determine how far it is from P to R.

Fully worked solutions for this chapter are available online.

Answers

Chapter 1 Surds

1.2 The real number system and surds

1.2 Exercise

1. a. Q b. Q c. Q d. Q
2. a. I b. Q c. I d. Q
3. a. Q b. I c. I d. I
4. a. Q b. I c. I d. Q
5. b, c and d
6. a and d
7. a and c
8. a, c and d
9. B
10. D
11. C
12. C
13. A
14. D
15. B
16. C
17. $m = 4$
18. $m = 5$, $n = 7$ and $m = 4$, $n = 14$
19. Irrational
20. $\pi \approx \dfrac{256}{81}$; $\dfrac{256}{81} = 3.160\,493\,827$ to 9 decimal places. $\dfrac{22}{7}$ is a better approximation.

1.3 Simplifying surds

1.3 Exercise

1. a. $2\sqrt{3}$ b. $2\sqrt{6}$ c. $3\sqrt{3}$ d. $5\sqrt{5}$
2. a. $3\sqrt{6}$ b. $4\sqrt{7}$ c. $2\sqrt{17}$ d. $6\sqrt{5}$
3. a. $2\sqrt{22}$ b. $9\sqrt{2}$ c. $7\sqrt{5}$ d. $8\sqrt{7}$
4. a. $4\sqrt{2}$ b. $24\sqrt{10}$ c. $36\sqrt{5}$ d. $21\sqrt{6}$
5. a. $-30\sqrt{3}$ b. $-28\sqrt{5}$ c. $64\sqrt{3}$ d. $2\sqrt{2}$
6. a. $\sqrt{2}$ b. $2\sqrt{3}$ c. $\dfrac{1}{3}\sqrt{15}$ d. $\dfrac{3}{2}\sqrt{7}$
7. a. $4\sqrt{2}$ b. $4\sqrt{11}$ c. $2\sqrt{13}$
      d. $12\sqrt{5}$ e. $3\sqrt{5}$ f. $\dfrac{\sqrt{11}}{6}$
8. a. $5\sqrt{3}$ b. $20\sqrt{3}$ c. $20\sqrt{5}$
      d. $36\sqrt{2}$ e. $12\sqrt{2}$ f. $3\sqrt[3]{2}$
9. a. $\sqrt{14}$ b. $\sqrt{42}$ c. $4\sqrt{3}$
      d. 10 e. $3\sqrt{7}$ f. 27
10. a. $10\sqrt{33}$ b. $180\sqrt{5}$ c. 120
       d. $120\sqrt{3}$ e. $2\sqrt{6}$ f. $2\dfrac{2}{3}$
11. $\left\{5, 3\sqrt{3}, 5\sqrt{2}, 4\sqrt{5}\right\}$

12. $\left\{4\sqrt{3}, 5, 2\sqrt{5}, 3\sqrt{2}\right\}$
13. a. $\sqrt{80}$ b. $\sqrt[3]{48}$
    c. $\sqrt{a^2 b^2 c}$ d. $\sqrt[3]{m^3 n}$
14. a. $4a$ b. $6a\sqrt{2}$
    c. $3a\sqrt{10b}$ d. $13a^2\sqrt{2}$
15. a. $13ab\sqrt{2ab}$ b. $2ab^2\sqrt{17ab}$
    c. $5x^3 y^2\sqrt{5}$ d. $20xy\sqrt{5x}$
16. a. $54c^3 d^2\sqrt{2cd}$ b. $18c^3 d\,^4\sqrt{5cd}$
    c. $\sqrt{22ef}$ d. $7e^5 f\,^5\sqrt{2ef}$
17. a. $\dfrac{2}{5}\sqrt{6}$ b. $x^2 y\sqrt{y}$
    c. $3a^4 b^2\sqrt{2ab}$ d. $6a^5 b^2\sqrt{2b}$
    e. $3x^2 y^2\sqrt{10xy}$ f. $\dfrac{9}{2}a^2 b^4\sqrt{5ab}$
18. $12\sqrt{3}$ cm
19. $16\sqrt{13}$ cm
20. $6\pi\sqrt{13}$ cm

1.4 Operations with surds and rationalising the denominator

1.4 Exercise

1. a. $7\sqrt{5}$ b. $8\sqrt{3}$
   c. $15\sqrt{5} + 5\sqrt{3}$ d. $4\sqrt{11}$
2. a. $13\sqrt{2}$ b. $-3\sqrt{6}$
   c. $17\sqrt{3} - 18\sqrt{7}$ d. $8\sqrt{x} + 3\sqrt{y}$
3. a. $10\left(\sqrt{2} - \sqrt{3}\right)$ b. $5\left(\sqrt{5} + \sqrt{6}\right)$
   c. $7\sqrt{3}$ d. $4\sqrt{5}$
4. a. $14\sqrt{3} + 3\sqrt{2}$ b. $3\sqrt{6} + 6\sqrt{3}$
   c. $15\sqrt{10} - 10\sqrt{15} + 10$ d. $-8\sqrt{11} + 22$
5. a. $12\sqrt{30} - 16\sqrt{15}$ b. $12\sqrt{ab} + 7\sqrt{3ab}$
   c. $\dfrac{7}{2}\sqrt{2} + 2\sqrt{3}$ d. $15\sqrt{2}$
6. a. 2 b. 5 c. 12
7. a. 15 b. 18 c. 80
8. a. $\sqrt{5}$ b. 2 c. $\sqrt{6}$ d. 4
9. a. $\dfrac{\sqrt{3}}{4}$ b. $\dfrac{\sqrt{5}}{2}$ c. $2\sqrt{3}$ d. 1
10. a. $\dfrac{5\sqrt{2}}{2}$ b. $\dfrac{7\sqrt{3}}{3}$ c. $\dfrac{4\sqrt{11}}{11}$
    d. $\dfrac{4\sqrt{6}}{3}$ e. $\dfrac{2\sqrt{21}}{7}$
11. a. $\dfrac{\sqrt{10}}{2}$ b. $\dfrac{2\sqrt{15}}{5}$ c. $\dfrac{3\sqrt{35}}{5}$
    d. $\dfrac{5\sqrt{6}}{6}$ e. $\dfrac{4\sqrt{15}}{15}$

12. a. $\dfrac{5\sqrt{7}}{14}$ b. $\dfrac{8\sqrt{15}}{15}$ c. $\dfrac{8\sqrt{21}}{49}$
    d. $\dfrac{8\sqrt{105}}{7}$ e. $\dfrac{\sqrt{10}}{3}$

13. a. $\sqrt{2}+2$ b. $\dfrac{3\sqrt{10}-2\sqrt{33}}{6}$
    c. $\dfrac{12\sqrt{5}-5\sqrt{6}}{10}$ d. $\dfrac{9\sqrt{10}}{5}$

14. a. $\dfrac{14-5\sqrt{2}}{6}$ b. $\dfrac{12-\sqrt{10}}{16}$
    c. $\dfrac{6\sqrt{15}-25}{70}$ d. $\dfrac{\sqrt{30}+7\sqrt{2}}{20}$

15. a. $a\sqrt{2a}$ b. $\sqrt{a}+2\sqrt{2a}$
    c. $3a\sqrt{a}+a^2\sqrt{3a}$ d. $(a^2+a)\sqrt{ab}$

16. a. $\dfrac{2}{5}\sqrt{6}$ b. $x^2y\sqrt{y}$
    c. $3a^4b^2\sqrt{2ab}$ d. $6a^5b^2\sqrt{2b}$
    e. $3x^2y^2\sqrt{10xy}$ f. $\dfrac{9}{2}a^2b^4\sqrt{5ab}$

17. a. $1\dfrac{4}{5}$ b. $2\sqrt{17}$ c. $\dfrac{x}{y}$
    d. $\dfrac{\sqrt{2}}{x^3y^4}$ e. $2xy\sqrt{3y}$ f. $\dfrac{4\sqrt{a}}{3}$

18. $\dfrac{3\sqrt{5}}{4}$

19. $\dfrac{5\sqrt{x}}{12x}$

20. -2

9. a. $\dfrac{\sqrt{6}}{3}$ b. $\dfrac{\sqrt{2}}{4}$ c. $\dfrac{\sqrt{2}}{3}$ d. $2+\sqrt{6}$

10. $\left\{5\sqrt{2},4\sqrt{3},3\sqrt{5},2\sqrt{10},2\sqrt{6}\right\}$

11. B

12. A

13. a. $15\sqrt{7}-\sqrt{3}$ b. $2\sqrt{2}-8\sqrt{6}$
    c. $12\sqrt{2}$ d. $34\sqrt{5}$
    e. $9\sqrt{6}-9\sqrt{5}$ f. $-59\sqrt{3}-5\sqrt{2}$

14. a. $8\sqrt{35}$ b. $160\sqrt{15}$
    c. $12\sqrt{10}$ d. 36
    e. $42\sqrt{3}$ f. $62\sqrt{5}$

15. a. $\sqrt{2m},\sqrt{\dfrac{20}{m}},\sqrt[3]{m},\sqrt[3]{8m}$
    b. $\sqrt{25m},\sqrt{\dfrac{m}{16}},\sqrt{\dfrac{20}{m}}$

16. B

17. C

18. 60, which is rational

19. $\dfrac{12}{5}(4-\pi)$ minutes

20. $\text{P}\left(\sqrt{10}-\sqrt{2},\sqrt{5}+\sqrt{2}\right)$,
    $\text{Q}\left(3\sqrt{2}-\sqrt{10},2\sqrt{3}-\sqrt{5}-\sqrt{2}\right)$,
    $\text{R}\left(\sqrt{10}-\sqrt{2},\sqrt{5}-\sqrt{2}\right)$
    $2\sqrt{2}\,\text{km}$

1.5 Review

1.5 Exercise

1. A

2. a. Irrational, since it is equal to a non-recurring and non-terminating decimal
   b. Rational, since it can be expressed as a whole number
   c. Rational, since it is given in a rational form
   d. Rational, since it is a recurring decimal
   e. Irrational, since it is equal to a non-recurring and non-terminating decimal

3. D

4. a. $5\sqrt{2}$ b. $6\sqrt{5}$ c. $8\sqrt{2}$ d. $20\sqrt{5}$

5. $25\sqrt{3}$

6. a. $\sqrt{15}$ b. $6\sqrt{42}$ c. $30\sqrt{15}$ d. 5

7. a. 27 b. $720\sqrt{2}$

8. a. $\sqrt{3}$ b. 6
   c. $\dfrac{\sqrt{10}}{4\sqrt{3}}$ or $\dfrac{\sqrt{30}}{12}$ d. $\dfrac{1}{2}$

2 Quadratic functions

LESSON SEQUENCE

Fully worked solutions for this chapter are available online.

EXAM PREPARATION

Access exam-style questions in every lesson, available online.

on Resources

Solutions	Solutions — Chapter 2 (sol-1269)
Exam questions	Exam question booklet — Chapter 2 (eqb-0290)
Digital documents	Learning matrix — Chapter 2 (doc-41784)
	Chapter summary — Chapter 2 (doc-41797)

LESSON
2.1 Overview

2.1.1 Introduction

Watching water cascade from a fountain or a ball bouncing along a pathway may be the first encounter a young child has with a parabola. The arches of water rise and fall along parabolas; the ball follows a series of parabolas, each with a maximum turning point lower than that of the previous one. It was Galileo who showed that the path of a projectile is parabolic.

In later life, that child may be lucky enough to catch a glimpse in the night sky of the return of Halley's Comet as its nearly parabolic path passes closest to Earth in 2061. Isaac Newton, having studied Halley's Comet in 1680, initially theorised that the path of the comet was consistent with that of a parabola. This was later corrected to be an ellipse, the same shape as orbits of the planets around the Sun. Nevertheless, the paths of many other comets are parabolic. The graph of a parabola is not closed; the 'arms' of the parabola go on forever, so a comet travelling on a parabolic path would never return unless other factors affected its path. Halley's Comet does return about every 75 years.

2.1.2 Syllabus links

Lesson	Lesson title	Syllabus links
2.2	**Solving quadratic equations with rational roots**	○ Solve quadratic equations algebraically using factorisation, the quadratic formula (both exact and approximate solutions), completing the square and using technology.
2.3	**Solving quadratic equations over R**	○ Solve quadratic equations algebraically using factorisation, the quadratic formula (both exact and approximate solutions), completing the square and using technology.
2.4	**Graphs of quadratic functions**	○ Recognise and determine features of the graphs of $y = x^2$, $y = ax^2 + bx + c$, $y = a(x - h)^2 + k$ and $y = a(x - x_1)(x - x_2)$, including their parabolic nature, turning points, axes of symmetry and intercepts. ○ Sketch the graphs of quadratic functions, with or without technology. ○ Determine turning points and zeros of quadratic functions, with and without technology.
2.5	**The discriminant**	○ Use the discriminant to determine the number of solutions to a quadratic equation.
2.6	**Modelling with quadratic functions**	○ Model and solve problems that involve quadratic functions, with and without technology.

Source: Mathematical Methods Senior Syllabus 2024 © State of Queensland (QCAA) 2024; licensed under CC BY 4.0.

LESSON
2.2 Solving quadratic equations with rational roots

SYLLABUS LINKS

- Solve quadratic equations algebraically using factorisation, the quadratic formula (both exact and approximate solutions), completing the square and using technology.

Source: Mathematical Methods Senior Syllabus 2024 © State of Queensland (QCAA) 2024; licensed under CC BY 4.0.

A **rational number** is any real number that can be expressed exactly as a fraction.

2.2.1 Solving quadratic equations using factorisation

The general quadratic equation can be written as $ax^2 + bx + c = 0$, where a, b, c are real constants and $a \neq 0$. If the quadratic expression on the left-hand side of this equation can be factorised, the solutions to the quadratic equation may be obtained using the **Null Factor Law**.

> ### The Null Factor Law
>
> **The Null Factor Law states that, for any a and b, if the product $ab = 0$, then $a = 0$ or $b = 0$ or both $a = 0$ and $b = 0$.**

Applying the Null Factor Law to a quadratic equation expressed in factorised form as:

$$(x - \alpha)(x - \beta) = 0$$
$$\text{then } (x - \alpha) = 0 \quad \text{or} \quad (x - \beta) = 0$$
$$\therefore x = \alpha \quad \text{or} \quad x = \beta$$

Roots, zeros and factors

The solutions of an equation are also called the **roots** of the equation or the **zeros** of the quadratic expression. This terminology applies to all algebraic and not just quadratic equations. The quadratic equation $(x - 1)(x - 2) = 0$ has roots $x = 1$, $x = 2$. These solutions are the zeros of the quadratic expression $(x - 1)(x - 2)$, since substituting either of $x = 1$, $x = 2$ in the quadratic expression $(x - 1)(x - 2)$ makes the expression equal zero.

As a converse of the Null Factor Law, it follows that if the roots of a quadratic equation, or the zeros of a quadratic, are $x = \alpha$ and $x = \beta$, then $(x - \alpha)$ and $(x - \beta)$ are linear factors of the quadratic. The quadratic would be of the form $(x - \alpha)(x - \beta)$ or any multiple of this form, $a(x - \alpha)(x - \beta)$.

WORKED EXAMPLE 1 Using factorisation to solve quadratic equations

a. Solve the equation $5x^2 - 18x = 8$.
b. Given that $x = 2$ and $x = -2$ are zeros of a quadratic, form its linear factors and expand the product of these factors.

THINK	WRITE
a. 1. Rearrange the equation to make one side of the equation equal zero.	a. $5x^2 - 18x = 8$ Rearrange: $5x^2 - 18x - 8 = 0$
2. Factorise the quadratic trinomial.	$(5x + 2)(x - 4) = 0$

▶

3. Apply the Null Factor Law.
$$5x + 2 = 0 \text{ or } x - 4 = 0$$

4. Solve these linear equations for x.
$$5x = -2 \text{ or } x = 4$$
$$x = -\frac{2}{5} \text{ or } x = 4$$

b. 1. Use the converse of the Null Factor Law.

b. Since $x = 2$ is a zero, then $(x - 2)$ is a linear factor, and since $x = -2$ is a zero, then $(x - (-2)) = (x + 2)$ is a linear factor.
Therefore, the quadratic has the linear factors $(x - 2)$ and $(x + 2)$.

2. Expand the product of the two linear factors.

The product $= (x - 2)(x + 2)$.
Expanding, $(x - 2)(x + 2) = x^2 - 4$.
The quadratic has the form $x^2 - 4$ or any multiple of this form, $a(x^2 - 4)$.

TI	THINK	WRITE

a. 1. Rearrange the given equation so that all terms are on one side.

$5x^2 - 18x = 8$
Rearrange:
$5x^2 - 18x - 8 = 0$

2. On a Calculator page, press MENU, then select:
3: Algebra
3: Polynomial Tools
1: Find Roots of Polynomial…
Complete the fields as:
Degree: 2
Roots: Real
then select OK.

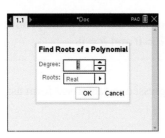

3. Complete the fields for the coefficients as:
$a_2 = 5$
$a_1 = -18$
$a_0 = -8$
then select OK.

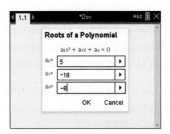

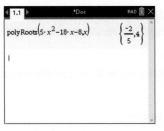

4. The answer appears on the screen.

$x = -\frac{2}{5} \text{ and } x = 4$

CASIO	THINK	WRITE

a. 1. On a Run-Matrix screen, press OPTN, then select CALC by pressing F4. Select Solve N by pressing F5, then complete the entry line as:
Solve N $(5x^2 - 18x = 8, x)$
and press EXE.

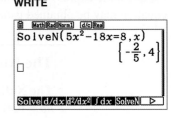

2. The answer appears on the screen.

$x = -\frac{2}{5} \text{ and } x = 4$

Resources

Interactivity Roots, zeros and factors (int-2557)

2.2.2 Solving quadratics using perfect squares

As an alternative to solving a quadratic equation by using the Null Factor Law, if the quadratic is a perfect square, solutions to the equation can be found by taking square roots of both sides of the equation. A simple illustration is:

Using the square root method

$$x^2 = 9$$
$$x = \pm\sqrt{9}$$
$$= \pm 3$$

or

Using the Null Factor Law method

$$x^2 = 9$$
$$x^2 - 9 = 0$$
$$(x-3)(x+3) = 0$$
$$x = \pm 3$$

If the square root method is used, both the positive and negative square roots must be considered.

WORKED EXAMPLE 2 Solving quadratics using perfect squares

Solve the equation $(2x+3)^2 - 25 = 0$.

THINK

1. Rearrange so that each side of the equation contains a perfect square.

2. Take the square roots of both sides.

3. Separate the linear equations and solve.

4. An alternative method uses the Null Factor Law, and factors identified by using the difference of two squares.

WRITE

$$(2x+3)^2 - 25 = 0$$
$$(2x+3)^2 = 25$$
$$2x+3 = \pm 5$$

$2x+3 = 5$ or $2x+3 = -5$
$\quad 2x = 2 \qquad\qquad 2x = -8$
$\quad x = 1$ or $\qquad\quad x = -4$

Alternatively:
$$(2x+3)^2 - 25 = 0$$
Factorise:
$$((2x+3) - 5)((2x+3) + 5) = 0$$
$$(2x-2)(2x+8) = 0$$
$2x = 2$ or $2x = -8$
$\therefore x = 1$ or $x = -4$

 Resources

Interactivity The perfect square (int-2558)

2.2.3 Equations that reduce to quadratic form

Substitution techniques can be applied to the solution of equations such as those of the form $ax^4 + bx^2 + c = 0$. Once reduced to quadratic form, progress with the solution can be made.

The equation $ax^4 + bx^2 + c = 0$ can be expressed in the form $a(x^2)^2 + bx^2 + c = 0$. Letting $u = x^2$, this becomes $au^2 + bu + c = 0$, a quadratic equation in variable u.

By solving the quadratic equation for u, then substituting back x^2 for u, any possible solutions for x can be obtained. Since x^2 cannot be negative, it would be necessary to reject negative u values, as $x^2 = u, u < 0$, would have no real solutions.

The quadratic form may be achieved from substitutions other than $u = x^2$, depending on the form of the original equation. The choice of symbol for the substitution is at the discretion of the solver.

WORKED EXAMPLE 3 Solving equations by substitution

Solve the equation $4x^4 - 35x^2 - 9 = 0$.

THINK

1. Use an appropriate substitution to reduce the equation to quadratic form.

2. Solve for μ using factorisation.

3. Substitute back, replacing μ by x^2.

4. Since x^2 cannot be negative, any negative value of a needs to be rejected.

5. Solve the remaining equation for x.

WRITE

$4x^4 - 35x^2 - 9 = 0$
Let $\mu = x^2$.
$$4\mu^2 - 35\mu - 9 = 0$$
$$(4\mu + 1)(\mu - 9) = 0$$
$$\therefore \mu = -\frac{1}{4} \text{ or } \mu = 9$$

$x^2 = -\frac{1}{4}$ or $x^2 = 9$

Reject $x^2 = -\frac{1}{4}$ since there are no real solutions.

$$x^2 = 9$$
$$x = \pm\sqrt{9}$$
$$x = \pm 3$$

TI | THINK

1. On a Calculator page, press MENU, then select:
   3: Algebra
   3: Polynomial Tools
   1: Find Roots of Polynomial…
   Complete the fields as:
   Degree: 4
   Roots: Real
   then select OK.

2. Complete the fields for the coefficients as:
   $a_4 = 4$
   $a_3 = 0$
   $a_2 = -35$
   $a_1 = -0$
   $a_0 = -9$
   then select OK.

WRITE

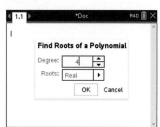

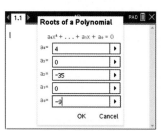

3. The answer appears on the screen. $\quad x = \pm 3$

CASIO | THINK

1. On a Run-Matrix screen, press OPTN, then select CALC by pressing F4. Select SolveN by pressing F5, then complete the entry line as:
   SolveN
   $(4x^4 - 35x^2 - 9 = 0, x)$
   and press EXE.

WRITE

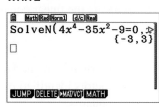

2. The answer appears on the screen. $\quad x = \pm 3$

Exercise 2.2 Solving quadratic equations with rational roots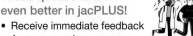

| 2.2 Exercise | 2.2 Exam practice   on |

Simple familiar	Complex familiar	Complex unfamiliar
1, 2, 3, 4, 5, 6, 7, 8	9, 10, 11, 12, 13, 14, 15	16, 17, 18, 19, 20

These questions are even better in jacPLUS!
- Receive immediate feedback
- Access sample responses
- Track results and progress

Find all this and MORE in jacPLUS ▶

Simple familiar

1. **WE1** **a.** Solve the equation $10x^2 + 23x = 21$.

   b. Given that $x = -5$ and $x = 0$ are zeros of a quadratic, determine its linear factors and expand the product of these factors.

2. **WE2** Solve the equation $(5x - 1)^2 - 16 = 0$.

3. Use the Null Factor Law to solve the following quadratic equations for x.

   a. $(3x - 4)(2x + 1) = 0$ **b.** $x^2 - 7x + 12 = 0$ **c.** $8x^2 + 26x + 21 = 0$

   d. $10x^2 = 2x$ **e.** $12x^2 + 40x - 32 = 0$ **f.** $\dfrac{1}{2}x^2 - 5x = 0$

4. Solve the following quadratic equations for x.

   a. $(x + 2)^2 = 9$ **b.** $(x - 1)^2 - 25 = 0$ **c.** $(x - 7)^2 + 4 = 0$

   d. $(2x + 11)^2 = 81$ **e.** $(7 - x)^2 = 0$ **f.** $8 - \dfrac{1}{2}(x - 4)^2 = 0$

For questions **5** and **6**, solve each of the given equations using the Null Factor Law.

5. **a.** $3x(5 - x) = 0$ **b.** $(3 - x)(7x - 1) = 0$ **c.** $(x + 8)^2 = 0$ **d.** $2(x + 4)(6 + x) = 0$

6. **a.** $6x^2 + 5x + 1 = 0$ **b.** $12x^2 - 7x = 10$ **c.** $49 = 14x - x^2$ **d.** $5x + 25 - 30x^2 = 0$

7. Obtain the solutions to the following equations.

   a. $x^2 = 121$ **b.** $9x^2 = 16$ **c.** $(x - 5)^2 = 1$

   d. $(5 - 2x)^2 - 49 = 0$ **e.** $2(3x - 1)^2 - 8 = 0$ **f.** $(x^2 + 1)^2 = 100$

8. Consider the quadratic equation $(x - \alpha)(x - \beta) = 0$.

   a. If the roots of the equation are $x = 1$ and $x = 7$, form the equation.
   b. If the roots of the equation are $x = -5$ and $x = 4$, form the equation.
   c. If the roots of the equation are $x = 0$ and $x = 10$, form the equation.
   d. If the only root of the equation is $x = 2$, form the equation.

Complex familiar

9. **WE3** Solve the equation $9x^4 + 17x^2 - 2 = 0$.

10. Determine the roots (solutions) of the following equations.

    a. $18(x - 3)^2 + 9(x - 3) - 2 = 0$ **b.** $5(x + 2)^2 + 23(x + 2) + 12 = 0$

    c. $x + 6 + \dfrac{8}{x} = 0$ **d.** $2x + \dfrac{3}{x} = 7$

11. Use a substitution technique to solve the following equations.

  a. $(3x+4)^2 + 9(3x+4) - 10 = 0$

  b. $2(1+2x)^2 + 9(1+2x) = 18$

  c. $x^4 - 29x^2 + 100 = 0$

  d. $2x^4 = 31x^2 + 16$

  e. $36x^2 = \dfrac{9}{x^2} - 77$

  f. $(x^2 + 4x)^2 + 7(x^2 + 4x) + 12 = 0$

For questions **12** and **13**, express each equation in quadratic form and hence solve the equations for x.

12. a. $x(x-7) = 8$

  b. $4x(3x-16) = 3(4x-33)$

  c. $(x+4)^2 + 2x = 0$

  d. $(2x+5)(2x-5) + 25 = 2x$

13. a. $2 - 3x = \dfrac{1}{3x}$

  b. $\dfrac{4x+5}{x+125} = \dfrac{5}{x}$

  c. $7x - \dfrac{2}{x} + \dfrac{11}{5} = 0$

  d. $\dfrac{12}{x+1} - \dfrac{14}{x-2} = 19$

14. Solve the following.

  a. $60x^2 + 113x - 63 = 0$

  b. $4x(x-7) + 8(x-3)^2 = x - 26$

15. a. Determine the roots of the equation $32x^2 - 96x + 72 = 0$.

  b. Solve $44 + 44x^2 = 250x$.

Complex unfamiliar

16. Obtain the solutions to the following equations.

  a. $x^4 = 81$

  b. $(9x^2 - 16)^2 = 20(9x^2 - 16)$

  c. $\left(x - \dfrac{2}{x}\right)^2 - 2\left(x - \dfrac{2}{x}\right) + 1 = 0$

  d. $2\left(1 + \dfrac{3}{x}\right)^2 + 5\left(1 + \dfrac{3}{x}\right) + 3 = 0$

17. Solve the equation $\left(x + \dfrac{1}{x}\right)^2 - 4\left(x + \dfrac{1}{x}\right) + 4 = 0$.

18. Solve the equation $(px + q)^2 = r^2$ for x in terms of p, q and r, $r > 0$.

19. Express the value of x in terms of the positive real numbers a and b.

  a. $(x - 2b)(x + 3a) = 0$

  b. $2x^2 - 13ax + 15a^2 = 0$

  c. $(x - b)^4 - 5(x - b)^2 + 4 = 0$

  d. $(x - a - b)^2 = 4b^2$

  e. $(x + a)^2 - 3b(x + a) + 2b^2 = 0$

  f. $ab\left(x + \dfrac{a}{b}\right)\left(x + \dfrac{b}{a}\right) = (a + b)^2 x$

20. a. If the zeros of the quadratic expression $4x^2 + bx + c$ are $x = -4$ and $x = \dfrac{3}{4}$, calculate the values of the integer constants b and c.

  b. Express the roots of $px^2 + (p+q)x + q = 0$ in terms of p and q for p, $q \in Q$, $p \neq 0$ and hence solve $p(x-1)^2 + (p+q)(x-1) + q = 0$.

Fully worked solutions for this chapter are available online.

LESSON
2.3 Solving quadratic equations over R

SYLLABUS LINKS

- Solve quadratic equations algebraically using factorisation, the quadratic formula (both exact and approximate solutions), completing the square and using technology.

Source: Mathematical Methods Senior Syllabus 2024 © State of Queensland (QCAA) 2024; licensed under CC BY 4.0.

When a quadratic expression cannot be factorised into linear factors, **surds** need to be used and the resulting factorisation is over R.

When $x^2 - 4$ is expressed as $(x - 2)(x + 2)$, it has been factorised over Q, as both of the zeros are rational numbers. However, over Q, the quadratic expression $x^2 - 3$ cannot be factorised into linear factors. Surds need to be permitted for such an expression to be factorised.

2.3.1 Factorisation over R

The quadratic $x^2 - 3$ can be expressed as the difference of two squares $x^2 - 3 = x^2 - \left(\sqrt{3}\right)^2$ using surds. This can be factorised over R because it allows the factors to contain surds.

$$x^2 - 3 = x^2 - \left(\sqrt{3}\right)^2$$
$$= \left(x - \sqrt{3}\right)\left(x + \sqrt{3}\right)$$

If a quadratic can be expressed as the difference of two squares, then it can be factorised over R. To express a quadratic trinomial as a difference of two squares, a technique called '**completing the square**' is used.

'Completing the square' technique

Expressions of the form $x^2 \pm bx + \left(\dfrac{b}{2}\right)^2 = \left(x \pm \dfrac{b}{2}\right)^2$ are perfect squares. For example, $x^2 + 4x + 4 = (x + 2)^2$.

To illustrate the 'completing the square' technique, consider the quadratic trinomial $x^2 + 4x + 1$.

If 4 is added to the first two terms $x^2 + 4x$, then this will form a perfect square $x^2 + 4x + 4$. However, 4 must also be subtracted in order not to alter the value of the expression.

$$x^2 + 4x + 1 = x^2 + 4x + 4 - 4 + 1$$

Grouping the first three terms together to form the perfect square and evaluating the last two terms,

$$= (x^2 + 4x + 4) - 4 + 1$$
$$= (x + 2)^2 - 3$$

By writing this difference of two squares form using surds, factors over R can be found.

$$= (x + 2)^2 - \left(\sqrt{3}\right)^2$$
$$= \left(x + 2 - \sqrt{3}\right)\left(x + 2 + \sqrt{3}\right)$$

Thus: $\qquad x^2 + 4x + 1 = \left(x + 2 - \sqrt{3}\right)\left(x + 2 + \sqrt{3}\right)$

'Completing the square' is the method used to factorise **monic** quadratics over R. A monic quadratic is one for which the coefficient of x^2 equals 1.

For a monic quadratic, to complete the square, add and then subtract the square of half the coefficient of x. This squaring will always produce a positive number regardless of the sign of the coefficient of x.

Completing the square

$$x^2 \pm bx = \left[x^2 \pm bx + \left(\frac{b}{2} \right)^2 \right] - \left(\frac{b}{2} \right)^2$$

$$= \left(x \pm \frac{b}{2} \right)^2 - \left(\frac{b}{2} \right)^2$$

TIP

To complete the square on $ax^2 + bx + c$, the quadratic should first be written as $a \left(x^2 + \dfrac{bx}{a} + \dfrac{c}{a} \right)$ and the technique applied to the monic quadratic in the bracket. The common factor a is carried down through all the steps in the working.

WORKED EXAMPLE 4 Completing the square

a. **Complete the square in the following.**
   i. $x^2 + 8x - 2$ ii. $3x^2 - 12x + 1$
b. **Factorise the following over R.**
   i. $x^2 - 14x - 3$ ii. $2x^2 + 7x + 4$ iii. $4x^2 - 11$

THINK	WRITE
a. i. 1. Add and subtract the square of half the coefficient of x.	**a. i.** $x^2 + 8x - 2$ $= x^2 + 8x + 4^2 - 4^2 - 2$
2. Group the first three terms together to form a perfect square and evaluate the last two terms.	$= (x^2 + 8x + 16) - 16 - 2$ $= (x + 4)^2 - 18$
ii. 1. Create a monic quadratic by taking the coefficient x^2 of out as a common factor. It may create fractions.	**ii.** $3x^2 - 12x + 1$ $= 3 \left(x^2 - 4x + \dfrac{1}{3} \right)$
2. Add and subtract the square of half the coefficient of x.	$= 3 \left(x^2 - 4x + 2^2 - 2^2 + \dfrac{1}{3} \right)$ $= 3 \left((x^2 - 4x + 4) - 4 + \dfrac{1}{3} \right)$
3. Group the first three terms together to form a perfect square and evaluate the last two terms. Write the answer.	$= 3 \left((x - 2)^2 - \dfrac{11}{3} \right)$ or $= 3 (x - 2)^2 - 11$

b. i.

1. Add and subtract the square of half the coefficient of x.

Note: The negative sign of the coefficient of x becomes positive when squared.

2. Group the first three terms together to form a perfect square and evaluate the last two terms.

3. Factorise the difference of two squares expression.

4. Express any surds in their simplest form.

5. Write the answer.

b. i. $x^2 - 14x - 3$

$= x^2 - 14x + 7^2 - 7^2 - 3$

$= \left(x^2 - 14x + 49\right) - 49 - 3$

$= (x - 7)^2 - 52$

$= (x - 7)^2 - \left(\sqrt{52}\right)^2$

$= \left(x - 7 - \sqrt{52}\right)\left(x - 7 + \sqrt{52}\right)$

$= \left(x - 7 - 2\sqrt{13}\right)\left(x - 7 + 2\sqrt{13}\right)$

Therefore, $x^2 - 14x - 3$

$= \left(x - 7 - 2\sqrt{13}\right)\left(x - 7 + 2\sqrt{13}\right)$

ii.

1. First create a monic quadratic by taking the coefficient of x^2 out as a common factor. This may create fractions.

2. Add and subtract the square of half the coefficient of x for the monic quadratic expression.

3. Within the bracket, group the first three terms together and evaluate the remaining terms.

ii. $2x^2 + 7x + 4$

$= 2\left(x^2 + \frac{7}{2}x + 2\right)$

$= 2\left(x^2 + \frac{7}{2}x + \left(\frac{7}{4}\right)^2 - \left(\frac{7}{4}\right)^2 + 2\right)$

$= 2\left[\left(x^2 + \frac{7}{2}x + \frac{49}{16}\right) - \frac{49}{16} + 2\right]$

$= 2\left[\left(x + \frac{7}{4}\right)^2 - \frac{49}{16} + 2\right]$

$= 2\left[\left(x + \frac{7}{4}\right)^2 - \frac{49}{16} + \frac{32}{16}\right]$

$= 2\left[\left(x + \frac{7}{4}\right)^2 - \frac{17}{16}\right]$

iii. The quadratic is a difference of two squares. Factorise it.

iii. $4x^2 - 11$

$= (2x)^2 - \left(\sqrt{11}\right)^2$

$= \left(2x - \sqrt{11}\right)\left(2x + \sqrt{11}\right)$

2.3.2 The quadratic formula

The **quadratic formula** is used for solving quadratic equations and is obtained by completing the square on the left-hand side of the equation $ax^2 + bx + c = 0$.

Using completing the square:

$$ax^2 + bx + c = a\left(\left(x + \frac{b}{2a}\right)^2 - \frac{b^2 - 4ac}{4a^2}\right).$$

$$ax^2 + bx + c = 0$$

$$a\left(\left(x + \frac{b}{2a}\right)^2 - \frac{b^2 - 4ac}{4a^2}\right) = 0$$

$$\left(x + \frac{b}{2a}\right)^2 - \frac{b^2 - 4ac}{4a^2} = 0$$

$$\left(x + \frac{b}{2a}\right)^2 = \frac{b^2 - 4ac}{4a^2}$$

$$x + \frac{b}{2a} = \pm\sqrt{\frac{b^2 - 4ac}{4a^2}}$$

$$x = -\frac{b}{2a} \pm \frac{\sqrt{b^2 - 4ac}}{2a}$$

$$= \frac{-b \pm \sqrt{b^2 - 4ac}}{2a}$$

The quadratic formula

The solutions of the quadratic equation $ax^2 + bx + c = 0$ are:

$$x = \frac{-b \pm \sqrt{b^2 - 4ac}}{2a}$$

Often the coefficients in the quadratic equation make the use of the formula less tedious than completing the square. Although the formula can also be used to solve a quadratic equation, factorisation is usually simpler, making it the preferred method.

 Resources

Interactivity The quadratic formula (int-2561)

WORKED EXAMPLE 5 Solving quadratics with formulas expressing roots in exact or approximate form

Apply the quadratic formula to solve each of the following equations.

a. $3x^2 + 4x + 1 = 0$ (exact answer)
b. $-3x^2 - 6x - 1 = 0$ (round to 2 decimal places)

THINK	WRITE
a. **1.** Write the equation.	a. $3x^2 + 4x + 1 = 0$
2. Write the quadratic formula.	$x = \dfrac{-b \pm \sqrt{b^2 - 4ac}}{2a}$

3. State the values for a, b and c. $a = 3$, $b = 4$, $c = 1$

4. Substitute the values into the formula.

$$x = \frac{-4 \pm \sqrt{(4)^2 - (4 \times 3 \times 1)}}{2 \times 3}$$

5. Simplify and solve for x.

$$= \frac{-4 \pm \sqrt{4}}{6}$$

$$= \frac{-4 \pm 2}{6}$$

$$x = \frac{-4 + 2}{6} \text{ or } x = \frac{-4 - 2}{6}$$

6. Write the two solutions.

$$x = -\frac{1}{3} \quad x = -1$$

b. 1. Write the equation.

b. $-3x^2 - 6x - 1 = 0$

2. Write the quadratic formula.

$$x = \frac{-b \pm \sqrt{b^2 - 4ac}}{2a}$$

3. State the values for a, b and c. $a = -3$, $b = -6$, $c = -1$

4. Substitute the values into the formula.

$$x = \frac{-(-6) \pm \sqrt{36 - 4 \times -3 \times -1}}{2 \times -3}$$

5. Simplify the fraction.

$$= \frac{6 \pm \sqrt{24}}{-6}$$

$$= \frac{6 \pm 2\sqrt{6}}{-6}$$

$$= \frac{3 \pm \sqrt{6}}{-3}$$

$$x = \frac{3 + \sqrt{6}}{-3} \text{ or } \frac{3 - \sqrt{6}}{-3}$$

6. Write the two solutions correct to 2 decimal places.
Note: When asked to give an answer in exact form, you should simplify any surds as necessary.

$x \approx -1.82$ or $x \approx -0.18$

WORKED EXAMPLE 6 Solving using the quadratic formula

Use the quadratic formula to solve the equation $x(9 - 5x) = 3$.

THINK

WRITE

1. The equation needs to be expressed in the general quadratic form $ax^2 + bx + c = 0$.

$$x(9 - 5x) = 3$$
$$9x - 5x^2 = 3$$
$$5x^2 - 9x + 3 = 0$$

2. State the values of a, b and c.

$a = 5$, $b = -9$, $c = 3$

3. State the formula for solving a quadratic equation.

$$x = \frac{-b \pm \sqrt{b^2 - 4ac}}{2a}$$

4. Substitute the a, b, c values and evaluate.

$$= \frac{-(-9) \pm \sqrt{(-9)^2 - 4 \times (5) \times (3)}}{2 \times (5)}$$

$$= \frac{9 \pm \sqrt{81 - 60}}{10}$$

$$= \frac{9 \pm \sqrt{21}}{10}$$

5. Express the roots in simplest surd form and write the answers.
   Note: If the question asked for answers correct to 2 decimal places, use technology to determine approximate answers of $x \simeq 0.44$ and $x \simeq 1.36$. Otherwise, do *not* approximate answers.

The solutions are:

$$x = \frac{9 - \sqrt{21}}{10}, \; x = \frac{9 + \sqrt{21}}{10}$$

Exercise 2.3 Solving quadratic equations over R

learn on

2.3 Exercise	2.3 Exam practice on

Simple familiar	Complex familiar	Complex unfamiliar
1, 2, 3, 4, 5, 6, 7, 8, 9, 10, 11, 12	13, 14, 15, 16, 17	18, 19, 20

These questions are even better in jacPLUS!
- Receive immediate feedback
- Access sample responses
- Track results and progress

Find all this and MORE in jacPLUS ▶

Simple familiar

1. Complete the following statements about perfect squares.
   a. $x^2 + 10x + \ldots = (x + \ldots)^2$
   b. $x^2 - 7x + \ldots = (x - \ldots)^2$
   c. $x^2 + x + \ldots = (x + \ldots)^2$
   d. $x^2 - \frac{4}{5}x + \ldots = (x - \ldots)^2$

2. **WE4** Factorise the following using completing the square.
   a. $x^2 - 10x - 7$
   b. $3x^2 + 7x + 3$
   c. $5x^2 - 9$

3. Use the 'completing the square' method to factorise $-3x^2 + 8x - 5$ and check the answer by using another method of factorisation.

4. Use the 'competing the square' method to factorise, where possible, the following over R.
   a. $x^2 - 6x + 7$
   b. $x^2 + 4x - 3$
   c. $x^2 - 2x + 6$
   d. $2x^2 + 5x - 2$
   e. $-x^2 + 8x - 8$
   f. $3x^2 + 4x - 6$

5. Write the following in the form $(x - h)^2 + k$ where h, $k \in R$ by completing the square.
   a. $x^2 + 2x$
   b. $x^2 + 7x$
   c. $x^2 - 5x$
   d. $x^2 + 4x - 2$

6. Factorise the following where possible.
   a. $3(x - 8)^2 - 6$
   b. $(xy - 7)^2 + 9$

7. Factorise the following over R where possible.
   a. $x^2 - 12$
   b. $x^2 - 12x + 4$
   c. $x^2 + 9x - 3$
   d. $2x^2 + 5x + 1$
   e. $3x^2 + 4x + 3$
   f. $1 + 40x - 5x^2$

8. Solve each of the following for x. Give exact answers.
   a. $x^2 - 10x + 23 = 0$
   b. $x^2 - 5x + 5 = 0$
   c. $x^2 + 14x + 43 = 0$
   d. $x^2 + 9x + 19 = 0$

9. Solve each of the following for x. Round answers to 2 decimal places.
   a. $x^2 - 3x - 5 = 0$
   b. $x^2 - 6x + 4 = 0$
   c. $x^2 + 7x + 12 = 0$
   d. $x^2 - 20x + 60 = 0$

10. State the values of a, b and c in each of the following quadratic expressions of the form $ax^2 + bx + c = 0$.
    a. $x^2 - 10x + 21 = 0$
    b. $10x^2 - 93x + 68 = 0$
    c. $x^2 - 9x + 20 = 0$
    d. $40x^2 + 32x + 6 = 0$

11. **WE5** Apply the quadratic formula to solve each of the following quadratic equations. Give exact answers.
    a. $3x^2 - 2x - 4 = 0$
    b. $2x^2 + 7x + 3 = 0$
    c. $-3x^2 - 6x + 4 = 0$
    d. $12x^2 - 8x - 5 = 0$

12. Use the quadratic formula to solve each of the following quadratic equations. Round answers to 2 decimal places.
    a. $-2x^2 - 5x + 4 = 0$
    b. $22x^2 - 11x - 20 = 0$
    c. $4x^2 - 29x + 19 = 0$
    d. $-12x^2 + 2x + 15 = 0$

Complex familiar

13. **MC** The solutions of the equation $15x^2 - 28x - 20 = 0$ are:

    A. 15, 20

    B. $\dfrac{-28 + \sqrt{-416}}{30}$, $\dfrac{28 - \sqrt{-416}}{30}$

    C. $\dfrac{28 + \sqrt{1984}}{30}$, $\dfrac{28 - \sqrt{1984}}{30}$

    D. $\dfrac{28 + \sqrt{1984}}{2}$, $\dfrac{28 - \sqrt{1984}}{2}$

14. **MC** The solutions of the equation $-6x^2 - 29x + 6 = 0$ are:
    A. -5.03, 0.20
    B. -0.22, -4.62
    C. -6, 6
    D. -31.62, -26.38

15. **WE6** Apply the quadratic formula $x = \dfrac{-b \pm \sqrt{b^2 - 4ac}}{2a}$ to solve the following equations, expressing solutions in simplest surd form.
    a. $3x^2 - 5x = -1$
    b. $-5x^2 + x = -5$
    c. $2x^2 + 3x = -4$
    d. $x(x + 6) = 8$

16. Use the quadratic formula to solve the equation $(2x + 1)(x + 5) - 1 = 0$.

17. Apply an appropriate substitution to reduce the following equations to quadratic form and hence obtain all solutions over R.
    a. $(x^2 - 3)^2 - 4(x^2 - 3) + 4 = 0$
    b. $5x^4 - 39x^2 - 8 = 0$

    c. $x^2(x^2 - 12) + 11 = 0$
    d. $\left(x + \dfrac{1}{x}\right)^2 + 2\left(x + \dfrac{1}{x}\right) - 3 = 0$

    e. $(x^2 - 7x - 8)^2 = 3(x^2 - 7x - 8)$

Complex unfamiliar

18. Solve the equation $3(2x + 1)^4 - 16(2x + 1)^2 - 35 = 0$ for $x \in R$.

19. Solve the equation $\sqrt{2}x^2 + 4\sqrt{3}x - 8\sqrt{2} = 0$, expressing solutions in simplest surd form.

LESSON
2.4 Graphs of quadratic functions

A quadratic **polynomial** is an algebraic expression of the form $ax^2 + bx + c$, where each power of the variable x is a positive whole number, with the highest power of x being 2. It is called a second-degree polynomial, whereas a linear polynomial of the form $ax + b$ is a first-degree polynomial, since the highest power of x is 1.

The graph of a quadratic polynomial is called a **parabola**.

2.4.1 The graph of $y = x^2$ and transformations

The simplest parabola has the equation $y = x^2$.

Key features of the graph of $y = x^2$:
- It is symmetrical about the y-axis.
- The **axis of symmetry** has the equation $x = 0$.
- The graph is **concave up** (opens upwards).
- It has a minimum turning point, or **vertex**, at the point (0, 0).

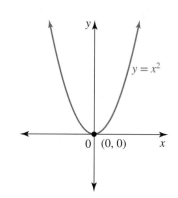

Making the graph wider or narrower

The graphs of $y = ax^2$ for $a = \dfrac{1}{3}$, 1 and 3 are drawn on the same set of axes.

Comparison of the graphs of $y = x^2$, $y = 3x^2$ and $y = \dfrac{1}{3}x^2$ shows that the graph of $y = ax^2$ will be:
- narrower than the graph of $y = x^2$ if $a > 1$
- wider than the graph of $y = x^2$ if $0 < a < 1$.

The coefficient of x^2, a, is called the **dilation factor**. It measures the amount of stretching or compression from the x-axis.

For $y = ax^2$, the graph of $y = x^2$ has been dilated by a factor of a from the x-axis or by a factor of a parallel to the y-axis.

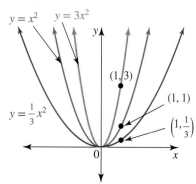

Reflecting the graph in the x-axis

The graph of $y = -x^2$ is obtained by reflecting the graph of $y = x^2$ in the x-axis.

Key features of the graph of $y = -x^2$:
- It is symmetrical about the y-axis.
- The axis of symmetry has the equation $x = 0$.
- The graph is **concave down** (opens downwards).
- It has a maximum turning point, or vertex, at the point $(0, 0)$.

A negative coefficient of x^2 indicates the graph of a parabola has been **reflected** in the x-axis.

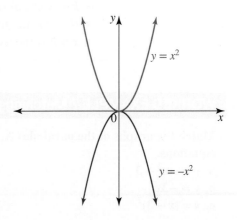

Translating the graph up or down

The graphs of $y = x^2 + k$ for $k = -2, 0$ and 2 are drawn on the same set of axes.

Comparison of the graphs of $y = x^2$, $y = x^2 + 2$ and $y = x^2 - 2$ shows that the graph of $y = x^2 + k$ will:
- have a turning point at $(0, k)$
- move the graph of $y = x^2$ vertically upwards by k units if $k > 0$
- move the graph of $y = x^2$ vertically downwards by k units if $k < 0$.

The value of k gives the **vertical translation**. For the graph of $y = x^2 + k$, the graph of $y = x^2$ has been translated by k units from the x-axis.

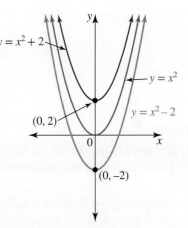

Translating the graph left or right

The graphs of $y = (x - h)^2$ for $h = -2, 0$ and 4 are drawn on the same set of axes.

Comparison of the graphs of $y = x^2$, $y = (x + 2)^2$ and $y = (x - 4)^2$ shows that the graph of $y = (x - h)^2$ will:
- have a turning point at $(h, 0)$
- move the graph of $y = x^2$ horizontally to the right by h units if $h > 0$
- move the graph of $y = x^2$ horizontally to the left by h units if $h < 0$.

The value of h gives the **horizontal translation**. For the graph of $y = (x - h)^2$, the graph of $y = x^2$ has been translated by h units from the y-axis.

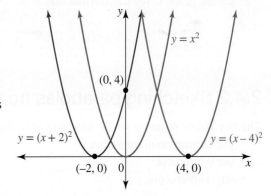

Transformations from $y = x^2$

$y = x^2$ can be transformed to $y = a(x - h)^2 + k$
where:
- $a =$ dilation from the x-axis
- $a < 0$ is a reflection in the x-axis
- $h =$ horizontal translation by h units

▶

- $k =$ vertical translation by k units
- (h, k) is the turning point
- $x = h$ is the axis of symmetry.

WORKED EXAMPLE 7 Matching parabolas to equations

Match the graphs of the parabolas A, B and C with the following equations.

a. $y = -x^2 + 3$

b. $y = -3x^2$

c. $y = (x - 3)^2$

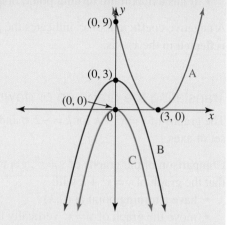

THINK

1. Compare graph A with the basic graph $y = x^2$ to identify the transformations.

2. Compare graph B with the basic graph $y = x^2$ to identify the transformations.

3. Check graph C for transformations.

WRITE

Graph A opens upwards and has been moved horizontally to the right.

Graph A matches with equation **c**, $y = (x - 3)^2$.

Graph B opens downwards and has been moved vertically upwards.

Graph B matches with equation **a**, $y = -x^2 + 3$.

Graph C opens downwards. It is narrower than both graphs A and B.

Graph C matches with equation **b**, $y = -3x^2$.

2.4.2 Sketching parabolas from their equations

The key points required when sketching a parabola are:
- the turning point, or vertex
- the y-intercept
- any x-intercepts.

The axis of symmetry is also a key feature of the graph.

The equation of a parabola allows this information to be obtained but in different ways, depending on the form of the equation.

We shall consider three forms for the equation of a parabola:
- general form
- turning point form
- x-intercept form.

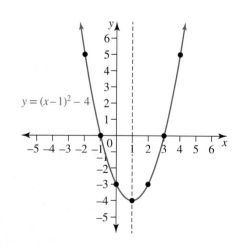

The general or polynomial form, $y = ax^2 + bx + c$

If $a > 0$, then the parabola is concave up and has a minimum turning point.

If $a < 0$, then the parabola is concave down and has a maximum turning point.

The dilation factor a, $a > 0$, determines the width of the parabola. The dilation factor is always a positive number (so it could be expressed as $|a|$).

The methods to determine the key features of the graph are as follows.
- Substitute $x = 0$ to obtain the y-intercept (the y-intercept is obvious from the equation).
- Substitute $y = 0$ and solve the quadratic equation $ax^2 + bx + c = 0$ to obtain the x-intercepts. There may be 0, 1 or 2 x-intercepts, as determined by the **discriminant**. (The discriminant is explored in lesson 2.5.)
- The equation of the axis of symmetry is $x = -\dfrac{b}{2a}$.

    This is because the formula for solving $ax^2 + bx + c = 0$ gives $x = -\dfrac{b}{2a} \pm \dfrac{\sqrt{b^2 - 4ac}}{2a}$ as the x-intercepts, and these are symmetrical about their midpoint $x = -\dfrac{b}{2a}$.

- The turning point lies on the axis of symmetry, so its x-coordinate is $x = -\dfrac{b}{2a}$. Substitute this value into the parabola's equation to calculate the y-coordinate of the turning point.

Axis of symmetry of a parabola

$$x = -\frac{b}{2a}$$

WORKED EXAMPLE 8 Sketching parabolas given in general form

Sketch the graph of $y = \dfrac{1}{2}x^2 - x - 4$ and label the key points with their coordinates.

THINK	WRITE
1. Identify the type of turning point.	Since $a > 0$, the turning point is a minimum turning point.
2. Write down the y-intercept.	$y = \dfrac{1}{2}x^2 - x - 4$ y-intercept: if $x = 0$, then $y = -4 \Rightarrow (0, -4)$
3. Obtain any x-intercepts.	x-intercepts: let $y = 0$. $\dfrac{1}{2}x^2 - x - 4 = 0$ $x^2 - 2x - 8 = 0$ $(x + 2)(x - 4) = 0$ $\therefore x = -2, 4$ $\Rightarrow (-2, 0), (4, 0)$
4. Determine the equation of the axis of symmetry.	Axis of symmetry formula: $x = -\dfrac{b}{2a}$ $a = \dfrac{1}{2}, b = -1$ $x = -\dfrac{-1}{\left(2 \times \frac{1}{2}\right)}$ $= 1$

5. Determine the coordinates of the turning point.

Turning point: when $x = 1$,

$$y = \frac{1}{2} - 1 - 4$$

$$= -4\frac{1}{2}$$

$$\Rightarrow \left(1, -4\frac{1}{2}\right) \text{ is the turning point.}$$

6. Sketch the graph using the information obtained in the previous steps. Label the key points with their coordinates.

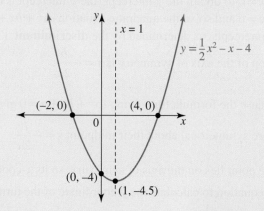

TI | THINK

1. On a Graphs page, complete the entry line as:
$$f1(x) = \frac{1}{2}x^2 - x - 4$$
Then press ENTER to view the graph.

DISPLAY/WRITE

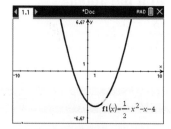

2. To view the key points, select:
MENU
6. Analyze Graph
1. Zero or
2. Minimum or
3. Maximum
Follow the prompts to show the key points.

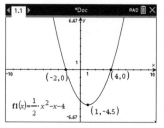

CASIO | THINK

1. On a Graph screen, complete the entry line for $y1$ as:
$$y1 = \frac{1}{2}x^2 - x - 4$$
then press EXE. Select DRAW by pressing F6.

DISPLAY/WRITE

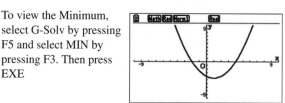

2. To view the Minimum, select G-Solv by pressing F5 and select MIN by pressing F3. Then press EXE

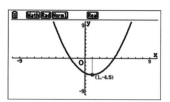

3. To view the x-intercepts, select G-Solv followed by ROOT by pressing F1. Then press EXE

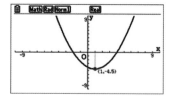

4. Use the left/right arrows to move to the other intercept and press EXE.

Turning point form, $y = a(x-h)^2 + k$

Since h represents the horizontal translation and k the vertical translation, this form of the equation readily provides the coordinates of the turning point.

- The turning point has coordinates (h, k).
  If $a > 0$, the turning point is a minimum, and if $a < 0$, it is a maximum. Depending on the nature of the turning point, the y-coordinate of the turning point gives the minimum or maximum value of the quadratic.
- Determine the y-intercept by substituting $x = 0$.
- Determine the x-intercepts by substituting $y = 0$ and solving the equation $a(x-h)^2 + k = 0$. However, before attempting to determine x-intercepts, consider the type of turning point and its y-coordinate, as this will indicate whether there are any x-intercepts.

The general form of the equation of a parabola can be converted to turning point form by the use of the 'completing the square' technique. By expanding the turning point form, the general form would be obtained.

WORKED EXAMPLE 9 Sketching parabolas given in turning point form

a. Sketch the graph of $y = -2(x+1)^2 + 8$ and label the key points with their coordinates.

b. **i.** Express $y = 3x^2 - 12x + 18$ in the form $y = a(x-h)^2 + k$ and hence state the coordinates of its vertex.

    ii. Sketch its graph.

THINK	WRITE
a. 1. Obtain the coordinates and the type of turning point from the given equation. *Note:* The x-coordinate of the turning point could also be obtained by letting $(x+1) = 0$ and solving this for x.	**a.** $y = -2(x+1)^2 + 8$ $\therefore y = -2(x-(-1))^2 + 8$ Maximum turning point at $(-1, 8)$
2. Calculate the y-intercept.	Let $x = 0$. $\therefore y = -2(1)^2 + 8$ $\qquad = 6$ $\Rightarrow (0, 6)$
3. Calculate any x-intercepts. *Note:* The graph is concave down with a maximum y-value of 8, so there will be x-intercepts.	x-intercepts: let $y = 0$. $\qquad 0 = -2(x+1)^2 + 8$ $2(x+1)^2 = 8$ $\quad (x+1)^2 = 4$ $\quad (x+1) = \pm\sqrt{4}$ $\qquad\quad x = \pm 2 - 1$ $\qquad\quad x = -3, 1$ $\Rightarrow (-3, 0), (1, 0)$
4. Sketch the graph, remembering to label the key points with their coordinates.	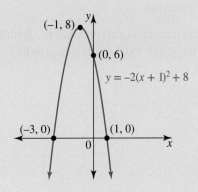

b. i. 1. Apply the 'completing the square' technique to the general form of the equation.

b.
$$y = 3x^2 - 12x + 18$$
$$= 3(x^2 - 4x + 6)$$
$$= 3[(x^2 - 4x + (2)^2) - (2)^2 + 6]$$
$$= 3[(x - 2)^2 + 2]$$

2. Expand to obtain the form $y = a(x - h)^2 + k$.

$$= 3(x - 2)^2 + 6$$
$$\therefore y = 3(x - 2)^2 + 6$$

3. State the coordinates of the vertex (turning point).

The vertex is $(2, 6)$.

ii. 1. Obtain the y-intercept from the general form.

The y-intercept is $(0, 18)$.

2. Will the graph have x-intercepts?

Since the graph is concave up with a minimum y-value of 6, there are no x-intercepts.

3. Sketch the graph.

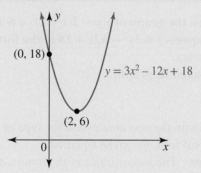

Factorised or x-intercept form, $y = a(x - x_1)(x - x_2)$

This form of the equation readily provides the x-intercepts.

- The x-intercepts occur at $x = x_1$ and $x = x_2$.
- The axis of symmetry lies halfway between the x-intercepts and its equation, $x = \dfrac{x_1 + x_2}{2}$, gives the x-coordinate of the turning point.
- The turning point is obtained by substituting $x = \dfrac{x_1 + x_2}{2}$ into the equation of the parabola and calculating the y-coordinate.
- The y-intercept is obtained by substituting $x = 0$.

If the linear factors are distinct, the graph cuts through the x-axis at each x-intercept.

If the linear factors are identical, making the quadratic a perfect square, the graph touches the x-axis at its turning point.

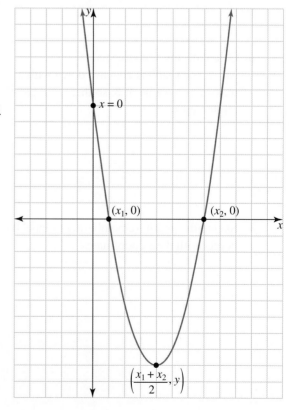

Sketch the graph of $y = -\dfrac{1}{2}(x+5)(x-1)$.

THINK	WRITE

1. Identify the x-intercepts.

$y = -\dfrac{1}{2}(x+5)(x-1)$

x-intercepts: let $y = 0$.

$\dfrac{1}{2}(x+5)(x-1) = 0$

$x+5 = 0 \quad$ or $\quad x-1 = 0$

$x = -5 \quad$ or $\qquad x = 1$

The x-intercepts are $(-5, 0), (1, 0)$.

2. Calculate the equation of the axis of symmetry.

The axis of symmetry has the equation

$x = \dfrac{-5+1}{2}$

$\therefore x = -2$

3. Obtain the coordinates of the turning point.

Turning point: substitute $x = -2$ into the equation.

$y = -\dfrac{1}{2}(x+5)(x-1)$

$\quad = -\dfrac{1}{2}(3)(-3)$

$\quad = \dfrac{9}{2}$

The turning point is $\left(-2, \dfrac{9}{2}\right)$.

4. Calculate the y-intercept.

$y = -\dfrac{1}{2}(x+5)(x-1)$

y-intercept: let $x = 0$.

$y = -\dfrac{1}{2}(5)(-1)$

$\quad = \dfrac{5}{2}$

The y-intercept is $\left(0, \dfrac{5}{2}\right)$.

5. Sketch the graph.

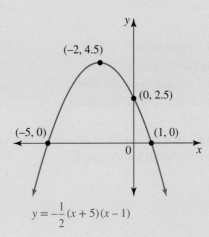

$y = -\dfrac{1}{2}(x+5)(x-1)$

2.4.3 Determining the rule of a quadratic from a graph

Whether the equation of the graph of a quadratic polynomial is expressed in $y = ax^2 + bx + c$ form, $y = a(x - h)^2 + k$ form or $y = a(x - x_1)(x - x_2)$ form, each equation contains three unknowns. Hence, three pieces of information are needed to fully determine the equation. This means that exactly one parabola can be drawn through three non-collinear points.

If the information given includes the turning point or the intercepts with the axes, one form of the equation may be preferable over another.

Identifying quadratics from a graph

As a guide:
- **if the turning point is given, use the $y = a(x - h)^2 + k$ form**
- **if the x-intercepts are given, use the $y = a(x - x_1)(x - x_2)$ form**
- **if 3 points on the graph are given, use the $y = ax^2 + bx + c$ form.**

WORKED EXAMPLE 11 Determining the equation of a parabola

Determine the rules for the following parabolas.

a.
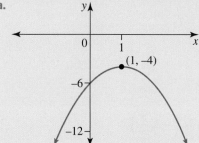

b.
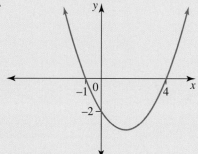

THINK

a. 1. Consider the given information to choose the form of the equation for the graph.

2. Determine the value of a.

3. Is the sign of a appropriate?

4. Write the rule for the graph.
Note: Check if the question specifies whether the rule needs to be expanded into general form.

b. 1. Consider the given information to choose the form of the equation for the graph.

WRITE

a. Let the equation be $y = a(x - h)^2 + k$.
Turning point $(1, -4)$
$\therefore y = a(x - 1)^2 - 4$

Substitute the given point $(0, -6)$.
$-6 = a(0 - 1)^2 - 4$
$-6 = a - 4$
$\therefore a = -2$

Check: the graph is concave down, so $a < 0$.

The equation of the parabola is
$y = -2(x - 1)^2 - 4$.

b. Let the equation be $y = a(x - x_1)(x - x_2)$.
Given $x_1 = -1$, $x_2 = 4$
$\therefore y = a(x + 1)(x - 4)$

---

2. Determine the value of a.

Substitute the third given point $(0, -2)$.
$$-2 = a(0+1)(0-4)$$
$$-2 = a(1)(-4)$$
$$-2 = -4a$$
$$a = \frac{-2}{-4}$$
$$= \frac{1}{2}$$

3. Is the sign of a appropriate?

Check: the graph is concave up, so $a > 0$.

4. Write the rule for the graph.

The equation of the parabola is
$$y = \frac{1}{2}(x+1)(x-4).$$

2.4.4 Using simultaneous equations

In Worked example 11 three points were available, but because two of them were key points, the x-intercepts, we chose to form the rule using the $y = a(x - x_1)(x - x_2)$ form. If the points were not key points, then simultaneous equations need to be created using the coordinates given.

WORKED EXAMPLE 12 Determining the equation given three points

Determine the equation of the parabola that passes through the points $(1, -4)$, $(-1, 10)$ and $(3, -2)$.

THINK

1. Consider the given information to choose the form of the equation for the graph.

2. Substitute the first point to form an equation in a, b and c.

3. Substitute the second point to form a second equation in a, b and c.

4. Substitute the third point to form a third equation in a, b and c.

5. Write the equations as a system of 3×3 simultaneous equations.

6. Solve the system of simultaneous equations.

WRITE

Let $y = ax^2 + bx + c$.

First point
$$(1, -4) \Rightarrow -4 = a(1)^2 + b(1) + c$$
$$\therefore -4 = a + b + c \quad [1]$$

Second point
$$(-1, 10) \Rightarrow 10 = a(-1)^2 + b(-1) + c$$
$$\therefore 10 = a - b + c \quad [2]$$

Third point
$$(3, -2) \Rightarrow -2 = a(3)^2 + b(3) + c$$
$$\therefore -2 = 9a + 3b + c \quad [3]$$

$$-4 = a + b + c \quad [1]$$
$$10 = a - b + c \quad [2]$$
$$-2 = 9a + 3b + c \quad [3]$$

Eliminate a and c from equations [1] and [2].
Equation [2] – equation [1]
$$14 = -2b$$
$$b = -7$$
Eliminate c from equations [1] and [3].
Equation [3] – equation [1]
$$2 = 8a + 2b \quad [4]$$
Substitute $b = -7$ into equation [4].
$$2 = 8a - 14$$
$$16 = 8a$$
$$a = 2$$

Substitute $a=2$, $b=-7$ into equation [1].
$$-4 = 2 - 7 + c$$
$$c = 1$$

7. Write the answer.

The equation of the parabola is $y = 2x^2 - 7x + 1$.

TI | THINK	WRITE	CASIO | THINK	WRITE
1. On a Lists & Spreadsheet page, label the first column x and the second column y. Enter the x-coordinates of the given points in the first column and the corresponding y-coordinates in the second column.		**1.** On a Statistics screen, relabel List 1 as x and List 2 as y. Enter the x-coordinates of the given points in the first column and the corresponding y-coordinates in the second column.	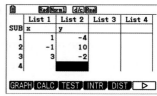
2. On a Calculator page, press MENU, then select: 4: Statistics 1: Stat Calculations 6: Quadratic Regression… Complete the fields as: X List: x Y List: y then select OK.	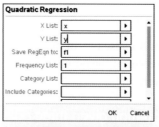	**2.** Select CALC by pressing F2, select REG by pressing F3, then select X^2 by pressing F3.	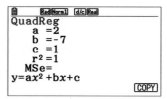
3. The answer appears on the screen.	The equation of the parabola is $y = 2x^2 - 7x + 1$.	**3.** The answer appears on the screen.	The equation of the parabola is $y = 2x^2 - 7x + 1$.

Exercise 2.4 Graphs of quadratic functions

learn on

2.4 Exercise	2.4 Exam practice on

Simple familiar	Complex familiar	Complex unfamiliar
1, 2, 3, 4, 5, 6, 7, 8, 9, 10, 11, 12	13, 14, 15, 16, 17	18, 19, 20

These questions are even better in jacPLUS!
• Receive immediate feedback
• Access sample responses
• Track results and progress

Find all this and MORE in jacPLUS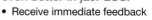

Simple familiar

1. Sketch the following parabolas on the same set of axes.

a. $y = 2x^2$

b. $y = -2x^2$

c. $y = 0.5x^2$

d. $y = -0.5x^2$

e. $y = (2x)^2$

f. $y = \left(-\dfrac{x}{2}\right)^2$

2. **WE7** Match the graphs of the parabolas A, B and C with the following equations.

  i. $y = x^2 - 2$ **ii.** $y = -2x^2$ **iii.** $y = -(x + 2)^2$

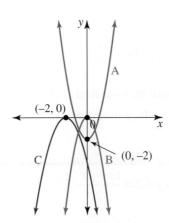

3. State the coordinates of the turning points of the parabolas with the following equations.

  a. $y = x^2 + 8$

  b. $y = x^2 - 8$

  c. $y = 1 - 5x^2$

  d. $y = -\dfrac{x^2}{4} - 7$

  e. $y = (x - 8)^2$

  f. $y = (x + 8)^2$

  g. $y = 7(x - 4)^2$

  h. $y = -\dfrac{1}{2}(x + 12)^2$

4. **WE8** Sketch the graph of $y = \dfrac{1}{3}x^2 + x - 6$ and label the key points with their coordinates.

5. Sketch the graphs of the following parabolas, labelling their key points with their coordinates.

  a. $y = 9x^2 + 18x + 8$

  b. $y = -x^2 + 7x - 10$

  c. $y = -x^2 - 2x - 3$

  d. $y = x^2 - 4x + 2$

6. **WE9** Sketch the graph of $y = -2(x + 3)^2 + 2$ and label the key points with their coordinates.

7. State the nature and the coordinates of the turning point for each of the following parabolas.

  a. $y = 4 - 3x^2$

  b. $y = (4 - 3x)^2$

8. **MC** Select which of the following is the equation of a parabola with a turning point at $(-5, 2)$.

  A. $y = -5x^2 + 2$
  B. $y = 2 - (x - 5)^2$
  C. $y = (x + 2)^2 - 5$
  D. $y = -(x + 5)^2 + 2$
  E. $y = (x + 5)^2 - 2$

9. **WE10** Sketch the graph of $y = 2x(4 - x)$.

10. Sketch the following graphs, showing all intercepts with the coordinate axes and the turning point.

  a. $y = (x + 1)(x - 3)$

  b. $y = (x - 5)(2x + 1)$

  c. $y = -\dfrac{1}{2}(2x - 7)(2x - 9)$

  d. $y = (1 - 3x)(4 + x)$

11. For each of the following parabolas, give the coordinates of:

   i. the turning point   **ii.** the y-intercept   **iii.** any x-intercepts.

   Then sketch each graph.

   a. $y = x^2 - 9$ **b.** $y = (x - 9)^2$ **c.** $y = 6 - 3x^2$

   d. $y = -3(x + 1)^2$ **e.** $y = \dfrac{1}{4}(1 - 2x)^2$ **f.** $y = -0.25(1 + 2x)^2$

12. For each of the following parabolas, give the coordinates of:

   i. the turning point   **ii.** the y-intercept   **iii.** any x-intercepts.

   Then sketch each graph.

   a. $y = (x - 5)^2 + 2$ **b.** $y = 2(x + 1)^2 - 2$ **c.** $y = -2(x - 3)^2 - 6$

   d. $y = -(x - 4)^2 + 1$ **e.** $y + 2 = \dfrac{(x + 4)^2}{2}$ **f.** $9y = 1 - \dfrac{1}{3}(2x - 1)^2$

Complex familiar

13. **a.** A parabola with equation $y = x^2 + c$ passes through the point $(1, 5)$. Determine the value of c and state the equation of the parabola.

   b. A parabola with equation $y = ax^2$ passes through the point $(6, -2)$. Calculate the value of a and state the equation of the parabola.

   c. A parabola with equation $y = a(x - 2)^2$ passes through the point $(0, -12)$. Calculate the value of a and state the equation of the parabola.

14. **a.** State the two linear factors of the equation of the parabola whose x-intercepts occur at $x = 3$ and at $x = 8$ and form a possible equation for this parabola.

   b. The x-intercepts of a parabola occur at $x = -11$ and $x = 2$. Form a possible equation for this parabola.

15. **WE11** Determine the rules for each of the following parabolas.

   a.

   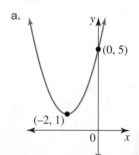

   b.

   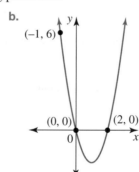

16. **WE12** Determine the equation of the parabola that passes through the points $(-1, -7)$, $(2, -10)$ and $(4, -32)$.

17. Determine the equation of each of the parabolas shown in the diagrams.

   a.

   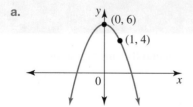

   b.

   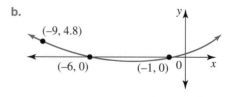

18. For each of the following graphs, two possible equations are given. Select the correct equation. Justify your decision

a.

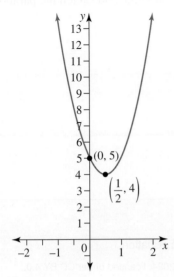

Equation A: $y = (2x - 1)^2 + 4$

Equation B: $y = \dfrac{1}{4}\left(x - \dfrac{1}{2}\right)^2 + 4$

b.

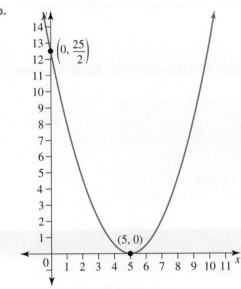

Equation A: $y = \dfrac{1}{2}(5 - x)^2$

Equation B: $y = 2(x - 5)^2$

c.

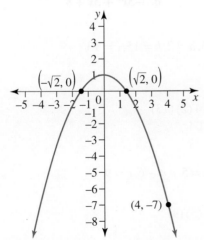

Equation A: $y = -\dfrac{1}{2}(x^2 - 2)$

Equation B: $y = -\dfrac{7}{18}(x^2 + 2)$

19. The axis of symmetry of a parabola has the equation $x = 4$. If the points $(0, 6)$ and $(6, 0)$ lie on the parabola, form its equation, expressing it in $y = ax^2 + bx + c$ form.

20. A parabola touches the x-axis at $x = p$ and passes through the points $(2, 9)$ and $(0, 36)$. Show there are two possible values for p. For each possible value, form the equation of the parabola, and sketch the parabolas on the same axes.

Fully worked solutions for this chapter are available online.

LESSON
2.5 The discriminant

SYLLABUS LINKS

- Use the discriminant to determine the number of solutions to a quadratic equation.

Source: Mathematical Methods Senior Syllabus 2024 © State of Queensland (QCAA) 2024; licensed under CC BY 4.0.

2.5.1 Defining the discriminant

The term $b^2 - 4ac$ in the quadratic formula is called the discriminant. It is denoted by the Greek letter delta, Δ.

The discriminant

$$\Delta = b^2 - 4ac$$

in the quadratic equation $ax^2 + bx + c = 0.$

WORKED EXAMPLE 13 Calculating the discriminant

For each of the following quadratics, calculate the discriminant.
a. $2x^2 + 15x + 13$ b. $5x^2 - 6x + 9$ c. $-3x^2 + 3x + 8$

THINK	WRITE
a. 1. State the values of a, b and c needed to calculate the discriminant.	a. $2x^2 + 15x + 13, a = 2, b = 15, c = 13$
2. State the formula for the discriminant.	$\Delta = b^2 - 4ac$
3. Substitute the values of a, b and c.	$\therefore \Delta = (15)^2 - 4(2)(13)$ $= 225 - 104$ $= 121$
b. 1. State the values of a, b and c, and calculate the discriminant.	b. $5x^2 - 6x + 9, a = 5, b = -6, c = 9$ $\Delta = b^2 - 4ac$ $= (-6)^2 - 4(5)(9)$ $= 36 - 180$ $= -144$

c. 1. State the values of a, b and c, and calculate the discriminant.	c. $-3x^2 + 3x + 8, a = -3, b = 3, c = 8$
	$\Delta = b^2 - 4ac$
	$\quad = (3)^2 - 4(-3)(8)$
	$\quad = 9 + 96$
	$\quad = 105$

2.5.2 The role of the discriminant in quadratic equations

The type of factors determines the type of solutions to an equation, so it is no surprise that the discriminant determines the number and type of solutions as well as the number and type of factors.

The formula for the solution to the quadratic equation $ax^2 + bx + c = 0$ can be expressed as $x = \dfrac{-b \pm \sqrt{\Delta}}{2a}$, where the discriminant $\Delta = b^2 - 4ac$.

The role of the discriminant

- If $\Delta < 0$, there are no real solutions to the equation.
- If $\Delta = 0$, there is one real solution (or two equal solutions) to the equation.
- If $\Delta > 0$, there are two distinct real solutions to the equation.

For $a, b, c \in Q$, the results are as follows:
- If $\Delta > 0$ and is a perfect square, the roots are **rational**, the factors are rational, and the quadratic factorises over Q.
- If $\Delta > 0$ and is not a perfect square, the roots are **irrational**, the factors contain surds, and the quadratic factorises over R.
- If $\Delta = 0$, the quadratic factorises to a perfect square, giving one distinct real solution.
- If $\Delta < 0$, the quadratic has no real solutions and cannot be factorised over R.

WORKED EXAMPLE 14 Determining the type and number of roots from the discriminant

a. **Use the discriminant to determine the number and type of roots for the equation:**
   i. $15x^2 + 8x - 5 = 0$ ii. $3x^2 + 8 = 4x$
b. **Determining the values of k so the equation $x^2 + kx - k + 8 = 0$ will have one real solution, and check the answer.**

THINK	WRITE
a. i. 1. Identify the values of a, b and c from the general form $ax^2 + bx + c = 0$.	a. $15x^2 + 8 - 5 = 0, a = 15, b = 8, c = -5$
2. State the formula for the discriminant.	$\Delta = b^2 - 4ac$
3. Substitute the values of a, b and c, and evaluate.	$\quad = (8)^2 - 4(15)(-5)$
	$\quad = 64 + 300$
	$\quad = 364$
4. Interpret the result.	Since the discriminant is positive but not a perfect square, the equation has two irrational roots.
ii. 1. Rewrite in the quadratic form $ax^2 + bx + c = 0$.	$3x^2 + 8 = 4x$
	$3x^2 - 4x + 8 = 0$
2. Identify the values of a, b and c.	$a = 3, b = -4, c = 8$

3. State the formula for the discriminant, substitute and evaluate.

$$\Delta = b^2 - 4ac$$
$$= (-4)^2 - 4 \times 3 \times 8$$
$$= 16 - 96$$
$$= -80$$

4. Interpret the result.

Since the discriminant is negative, the equation has no real roots.

b. 1. Express the equation in general form and identify the values of a, b and c.

b. $\quad x^2 + kx - k + 8 = 0$
$\therefore x^2 + kx + (-k + 8) = 0$
$a = 1, b = k, c = (-k + 8)$

2. Substitute the values of a, b and c, and obtain an algebraic expression for the discriminant.

$$\Delta = b^2 - 4ac$$
$$= (k)^2 - 4(1)(-k + 8)$$
$$= k^2 + 4k - 32$$

3. State the condition on the discriminant for the equation to have one solution.

For one solution, $\Delta = 0$.

4. Solve for k.

$$k^2 + 4k - 32 = 0$$
$$(k + 8)(k - 4) = 0$$
$$k = -8, k = 4$$

5. Check the solutions of the equation for each value of k.

If $k = -8$, the original equation becomes:
$x^2 - 8x + 16 = 0$
$$(x - 4)^2 = 0$$
$$\therefore x = 4$$

This equation has one solution.
If $k = 4$, the original equation becomes:
$x^2 - 4x + 4 = 0$
$$(x + 2)^2 = 0$$
$$\therefore x = -2$$
This equation has one solution.

TI	THINK	WRITE

TI | THINK

a. 1. On a Graphs page, complete the entry line for function 1 as:
$f1(x) = x^2 + k \times x - k + 8$
then press ENTER.
Select the tick box for k then select OK.
Note: Be sure to include the multiplication operator between k and x.

WRITE

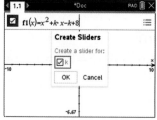

CASIO | THINK

a. 1. On a Dyna Graph screen, complete the entry line for y1 as:
$y1 = x^2 + K \times x - K + 8$
then press EXE.
Note: Be sure to include the multiplication operator between k and x.

WRITE

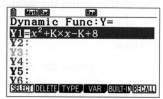

2. Press CTRL then MENU and select: 1: Settings… Complete the fields as: Variable: k Value: 1 Minimum: −10 Maximum: 10 Step Size: 1 Style: Horizontal then select OK.

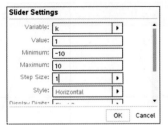

2. Select VAR by pressing by pressing F4, then select SET by pressing F2 and complete the fields as: Start: −10 End: 10 Step: 1 Press EXIT. Select SPEED by pressing F3, then select Stop&Go by pressing F1 and press EXIT.

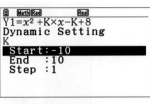

3. The graph appears on the screen.

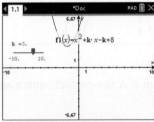

3. Select DYNA by pressing F6 to view the graph.

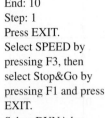

4. Use the left/right arrows to change the value of k until the turning point of the graph lies on the x-axis. Note the values of k for which this occurs.

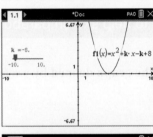

4. Use the left/right arrows to change the value of K until the turning point of the graph lies on the x-axis. Note the values of K for which this occurs.

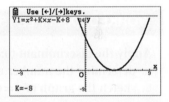

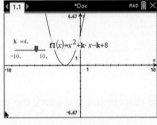

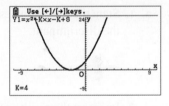

5. Write the answer.   There will be one real solution when $k = 4$ and $k = −8$.

5. Write the answer.   There will be one real solution when $k = 4$ and $k = −8$.

2.5.3 The discriminant and the x-intercepts

The zeros of the quadratic expression $ax^2 + bx + c$, the roots of the quadratic equation $ax^2 + bx + c = 0$ and the x-intercepts of the graph of a parabola with rule $y = ax^2 + bx + c$ all have the same x-values, and the discriminant determines the type and number of these values.

- If $\Delta > 0$, there are two x-intercepts. The graph *cuts* through the x-axis at two different places.
- If $\Delta = 0$, there is one x-intercept. The graph *touches* the x-axis at its turning point.
- If $\Delta < 0$, there are no x-intercepts. The graph does not intersect the x-axis and lies entirely above or entirely below the x-axis, depending on its concavity.

If $a > 0$ and $\Delta < 0$, the graph lies entirely above the x-axis and every point on it has a positive y-coordinate. The expression $ax^2 + bx + c$ is called **positive definite** in this case.

If $a < 0$ and $\Delta < 0$, the graph lies entirely below the x-axis and every point on it has a negative y-coordinate. The expression $ax^2 + bx + c$ is called **negative definite** in this case.

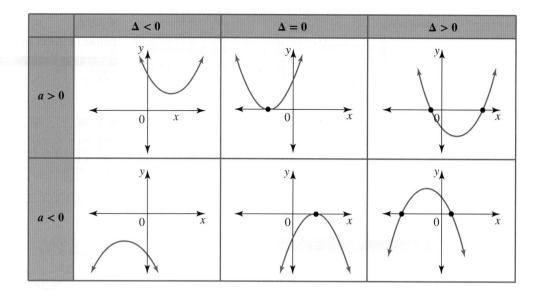

	$\Delta < 0$	$\Delta = 0$	$\Delta > 0$
$a > 0$			
$a < 0$			

When $\Delta \geq 0$ and for a, b, $c \in Q$, the x intercepts are rational if Δ is a perfect square and irrational if Δ is not a perfect square.

WORKED EXAMPLE 15 Using the discriminant to help sketch parabolas

Apply the discriminant to:
a. **determine the number and type of x-intercepts of the graph defined by $y = 64x^2 + 48x + 9$**
b. **sketch the graph of $y = 64x^2 + 48x + 9$.**

THINK

a. 1. State the a, b, c values and evaluate the discriminant.

2. Interpret the result and write the answer.

b. 1. Interpret the implication of a zero discriminant for the factors.

2. Identify the key points.

WRITE

a. $y = 64x^2 + 48x + 9$, $a = 64$, $b = 48$, $c = 9$
$\Delta = b^2 - 4ac$

$= (48)^2 - 4 \times (64) \times (9)$
$= 2304 - 2304$
$= 0$

Since the discriminant is zero, the graph has one rational x-intercept.

b. The quadratic must be a perfect square.
$y = 64x^2 + 48x + 9$

$= (8x + 3)^2$

x-intercept: let $y = 0$.
$8x + 3 = 0$

$x = -\dfrac{3}{8}$

Therefore, $\left(-\dfrac{3}{8}, 0\right)$ is both the x-intercept

and the turning point.
y-intercept: let $x = 0$ in $y = 64x^2 + 48x + 9$.
$\therefore y = 9$
Therefore, $(0, 9)$ is the y-intercept.

3. Sketch the graph.

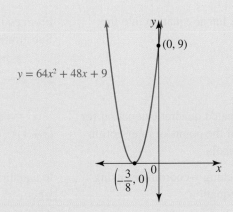

$y = 64x^2 + 48x + 9$

$(0, 9)$

$\left(-\dfrac{3}{8}, 0\right)$

2.5.4 Intersections of lines and parabolas

The possible number of points of intersection between a straight line and a parabola will be either 0, 1 or 2.

- If there is no point of intersection, the line makes no contact with the parabola.

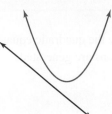

- If there is 1 point of intersection, a non-vertical line is a **tangent line** to the parabola, touching the parabola at that one point of contact.

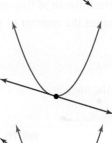

- If there are 2 points of intersection, the line cuts through the parabola at these points.

Simultaneous equations can be used to determine any points of intersection, and the discriminant can be used to predict the number of solutions. To solve a pair of linear-quadratic simultaneous equations, usually the method of substitution from the linear into the quadratic equation is used.

WORKED EXAMPLE 16 Determining points of intersection

a. Calculate the coordinates of the points of intersection of the parabola $y = x^2 - 3x - 4$ and the line $y - x = 1$.

b. Determine the number of points of intersection there will be between the graphs of $y = 2x - 5$ and $y = 2x^2 + 5x + 6$.

THINK

a. 1. Set up the simultaneous equations.

WRITE

a. $y = x^2 - 3x - 4$ [1]

$y - x = 1$ [2]

2. Substitute from the linear equation into the quadratic equation.

From equation [2], $y = x + 1$.
Substitute this into equation [1].
$$x + 1 = x^2 - 3x - 4$$
$$x^2 - 4x - 5 = 0$$

3. Solve the newly created quadratic equation for the x-coordinates of the points of intersection of the line and parabola.

$$x^2 - 4x - 5 = 0$$
$$(x + 1)(x - 5) = 0$$
$$x = -1 \text{ or } x = 5$$

4. Determine the matching y-coordinates using the simpler linear equation.

In equation [2]:
when $x = -1$, $y = 0$
when $x = 5$, $y = 6$

5. State the coordinates of the points of intersection.

The points of intersection are $(-1, 0)$ and $(5, 6)$.

b. 1. Set up the simultaneous equations.

b. $y = 2x - 5$ [1]

$y = 2x^2 + 5x + 6$ [2]

2. Create the quadratic equation from which any solutions are generated.

Substitute equation [1] in equation [2].
$$2x - 5 = 2x^2 + 5x + 6$$
$$2x^2 + 3x + 11 = 0$$

3. The discriminant of this quadratic equation determines the number of solutions.

$\Delta = b^2 - 4ac$, $a = 2$, $b = 3$, $c = 11$
$$= (3)^2 - 4 \times (2) \times (11)$$
$$= -79$$
$$\therefore \Delta < 0$$

4. Write the answer.

There are no points of intersection between the two graphs.

TI | THINK

a. 1. On a Graphs page, complete the entry line for function 1 as:
$f1(x) = x^2 - 3x - 4$
then press ENTER. Rearrange the equation $y - x = 1$ as $y = x + 1$ and complete the entry line for function 2 as:
$f2(x) = x + 1$
then press ENTER.

WRITE

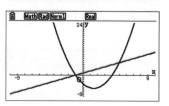

CASIO | THINK

a. 1. On a Graph screen, complete the entry line for $y1$ as:
$y1 = x^2 - 3x - 4$
then press EXE. Rearrange the equation $y - x = 1$ as $y = x + 1$ and complete the entry line for $y2$ as:
$y2 = x + 1$
then press EXE. Select DRAW by pressing F6.

WRITE

2. To find the points of intersection, press MENU, then select:
6: Analyze Graph
4: Intersection
Move the cursor to the left of the point of intersection when prompted for the lower bound, then press ENTER. Move the cursor to the right of the point of intersection when prompted for the upper bound, then press ENTER. Repeat this step to find the other point of intersection.

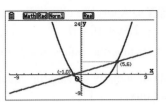

3. The answer appears on the screen. The points of intersections are $(-1, 0)$ and $(5, 6)$.

2. To find the points of intersection, select G-Solv by pressing SHIFT then F5, then select INTSECT by pressing F5. With the cursor on the first point of intersection, press EXE. Use the left/right arrows to move to the other point of intersection, then press EXE.

3. The answer appears on the screen. The points of intersections are $(-1, 0)$ and $(5, 6)$.

Exercise 2.5 The discriminant

2.5 Exercise	2.5 Exam practice **on**

Simple familiar	Complex familiar	Complex unfamiliar
1, 2, 3, 4, 5, 6, 7, 8	9, 10, 11, 12, 13, 14, 15, 16	17, 18, 19, 20

These questions are even better in jacPLUS!
• Receive immediate feedback
• Access sample responses
• Track results and progress

Find all this and MORE in jacPLUS ▶

Simple familiar

1. **WE13** For each of the following quadratics, calculate the discriminant.
   a. $4x^2 + 5x + 10$
   b. $169x^2 - 78x + 9$
   c. $-3x^2 + 11x - 10$
   d. $\frac{1}{3}x^2 - \frac{8}{3}x + 2$

2. a. Calculate the discriminant for the equation $3x^2 - 4x + 1 = 0$.
   b. Use the result of **a** to determine the number and nature of the roots of the equations $3x^2 - 4x + 1 = 0$.

3. Apply the discriminant to determine the number and type of solutions to the given equation.
   a. $-x^2 - 4x + 3 = 0$
   b. $2x^2 - 20x + 50 = 0$
   c. $x^2 + 4x + 7 = 0$
   d. $1 = x^2 + 5x$

4. For each of the following, calculate the discriminant and hence state the number and type of linear factors.
   a. $5x^2 + 9x - 2$
   b. $12x^2 - 3x + 1$
   c. $121x^2 + 110x + 25$
   d. $x^2 + 10x + 23$

5. **WE14** a. Apply the discriminant to determine the number and type of roots to the equation $0.2x^2 - 2.5x + 10 = 0$.

   b. Determine the values of k so the equation $kx^2 - (k+3)x + k = 0$ will have one real solution.

6. Without actually solving the equations, determine the number and the nature of the roots of the following equations.

   a. $-5x^2 - 8x + 9 = 0$

   b. $4x^2 + 3x - 7 = 0$

   c. $4x^2 + x + 2 = 0$

   d. $28x - 4 - 49x^2 = 0$

   e. $4x^2 + 25 = 0$

   f. $3\sqrt{2}x^2 + 5x + \sqrt{2} = 0$

7. **WE15** Apply the discriminant to:

   a. determine the number and type of x-intercepts of the graph defined by $y = 42x - 18x^2$

   b. sketch the graph of $y = 42x - 18x^2$.

8. Use the discriminant to determine the number and type of intercepts each of the following graphs makes with the x-axis.

   a. $y = 9x^2 + 17x - 12$

   b. $y = -5x^2 + 20x - 21$

   c. $y = -3x^2 - 30x - 75$

   d. $y = 0.02x^2 + 0.5x + 2$

Complex familiar

9. Determine the values of k for which the graph of $y = 5x^2 + 10x - k$ has:

   a. one x-intercept

   b. two x-intercepts

   c. no x-intercepts.

10. **WE16** a. Calculate the coordinates of the points of intersection of the parabola $y = x^2 + 3x - 10$ and the line $y + x = 2$.

    b. Determine how many points of intersection there will be between the graphs of $y = 6x + 1$ and $y = -x^2 + 9x - 5$.

11. Show that the line $y = 4x$ is a tangent to the parabola $y = x^2 + 4$ and sketch the line and parabola on the same diagram, labelling the coordinates of the point of contact.

12. Solve each of the following pairs of simultaneous equations.

    a. $y = 5x + 2$
       $y = x^2 - 4$

    b. $4x + y = 3$
       $y = x^2 + 3x - 5$

    c. $2y + x - 4 = 0$
       $y = (x - 3)^2 + 4$

    d. $\dfrac{x}{3} + \dfrac{y}{5} = 1$
       $x^2 - y + 5 = 0$

13. Obtain the coordinates of the point(s) of intersection of:

    a. the line $y = 2x + 5$ and the parabola $y = -5x^2 + 10x + 2$
    b. the line $y = -5x - 13$ and the parabola $y = 2x^2 + 3x - 5$
    c. the line $y = 10$ and the parabola $y = (5 - x)(6 + x)$
    d. the line $19x - y = 46$ and the parabola $y = 3x^2 - 5x + 2$.

14. Use the discriminant to determine the number of intersections of:

    a. the line $y = 4 - 2x$ and the parabola $y = 3x^2 + 8$
    b. the line $y = 2x + 1$ and the parabola $y = -x^2 - x + 2$
    c. the line $y = 0$ and the parabola $y = -2x^2 + 3x - 2$.

15. Consider the line $2y - 3x = 6$ and the parabola $y = x^2$.

    a. Calculate the coordinates of their points of intersection, correct to 2 decimal places.
    b. Sketch the line and the parabola on the same diagram.

16. a. Determine the values of m for which $mx^2 - 2x + 4$ is positive definite.
   b. i. Show that there is no real value of p for which $px^2 + 3x - 9$ is positive definite.
      ii. If $p = 3$, obtain the equation of the axis of symmetry of the graph of $y = px^2 + 3x - 9$.

Complex unfamiliar

17. a. Determine the values of m so the equation $x^2 + (m + 2)x - m + 5 = 0$ has one root.
   b. Determine the values of m so the equation $(m + 2)x^2 - 2mx + 4 = 0$ has one root.
   c. Determine the values of p so the equation $3x^2 + 4x - 2(p - 1) = 0$ has no real roots.

18. Show that the equation $kx^2 - 4x - k = 0$ always has two solutions for $k \in R \backslash \{0\}$.

19. Show that for $p, q \in Q$ the equation $px^2 + (p + q)x + q = 0$ always has rational roots.

20. Show that the equation $mx^2 + (m - 4)x = 4$ will always have real roots for any real value of m.

Fully worked solutions for this chapter are available online.

LESSON
2.6 Modelling with quadratic functions

SYLLABUS LINKS

- Model and solve problems that involve quadratic functions, with and without technology.

Source: Mathematical Methods Senior Syllabus 2024 © State of Queensland (QCAA) 2024; licensed under CC BY 4.0.

Quadratic polynomials can be used to model a number of situations, for example the motion of a falling object or the time of flight of a projectile. They can be used to model the shape of physical objects such as bridges, and they can also occur in economic models of cost and revenue.

2.6.1 Solving problems using quadratic equations

Quadratic equations may occur in problem solving and when setting up mathematical models of real-life situations.

In setting up equations:
- define the symbols used for the variables, specifying units where appropriate
- ensure any units used are consistent
- check whether mathematical solutions are feasible in the context of the problem
- express answers in the context of the problem.

WORKED EXAMPLE 17 Using quadratic equations in problem solving

A rectangle with an area of $400\,\text{m}^2$ is shaped such that the width is 10 metres shorter than the length. Model this situation with a quadratic equation and determine the exact dimensions of the rectangle.

THINK	WRITE
1. Define the key variables.	Let x metres be the length of the rectangle. The width is $(x - 10)$ metres.
2. Determine an expression for the area of the rectangle.	Area of rectangle $=$ length $\times$ width

▶

3. Create an equation using the given area.

$A = x(x - 10)$

4. Solve the quadratic equation, using completing the square (or the quadratic formula).

$$A = x(x - 10)$$
$$400 = x(x - 10)$$
$$x^2 - 10x - 400 = 0$$
$$x^2 - 10x + 5^2 - 5^2 - 400 = 0$$
$$(x - 5)^2 - 425 = 0$$
$$(x - 5)^2 = 425$$
$$x - 5 = \pm\sqrt{425}$$
$$x = 5 \pm \sqrt{25 \times 17}$$
$$x = 5 \pm 5\sqrt{17}$$

5. Justify the reasonableness of the solution.

Reject $x = 5 - 5\sqrt{17}$, since x must be positive since it is a length.

6. Determine the width of the rectangle.

$$\text{Width} = \left(5 + 5\sqrt{17}\right) - 10$$
$$= \left(-5 + 5\sqrt{17}\right)$$

7. Write the answer in context.

Therefore, the dimensions of the rectangle are $\left(5 + 5\sqrt{17}\right)$ m and $\left(-5 + 5\sqrt{17}\right)$ m.

2.6.2 Quadratically related variables

The formula for the area, A, of a circle in terms of its radius, r, is $A = \pi r^2$. This is of the form $A = kr^2$ as π is a constant.

The area varies directly as the square of its radius with the **constant of proportionality** $k = \pi$. This is a quadratic relationship between A and r.

Plotting the graph of A against r gives a curve that is part of a parabola.

r	0	1	2	3
A	0	π	4π	9π

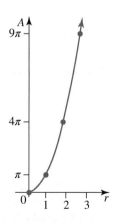

WORKED EXAMPLE 18 Determining constants of proportionality to solve problems

The volume of a cone of fixed height is directly proportional to the square of the radius of its base. When the radius is 3 cm, the volume is 30π cm³. Calculate the radius when the volume is 480π cm³.

THINK

1. Write the variation equation, defining the symbols used.

WRITE

$V = kr^2$ where V is the volume of a cone of fixed height and radius r.

k is the constant of proportionality.

2. Use the given information to determine k.		$r = 3, V = 30\pi \Rightarrow 30\pi = 9k$
		$\therefore k = \dfrac{30\pi}{9}$
		$= \dfrac{10\pi}{3}$
3. Write the rule connecting V and r.		$V = \dfrac{10\pi}{3} r^2$
4. Substitute $V = 480\pi$ and determine r.		$480\pi = \dfrac{10\pi}{3} r^2$
		$10\pi r^2 = 480\pi \times 3$
		$r^2 = \dfrac{480\pi \times 3}{10\pi}$
		$r^2 = 144$
		$r = \pm\sqrt{144}$
		$r = \pm 12$
5. Check the feasibility of the mathematical solutions.		Reject $r = -12$, since r must be positive.
		$\therefore r = 12$
6. Write the answer in context.		The radius of the cone is 12 cm.

2.6.3 Maximum and minimum values

The greatest or least value of the quadratic model is often of interest.

A quadratic reaches its local maximum or local minimum value at its turning point. The y-coordinate of the turning point represents the maximum or minimum value, depending on the nature of the turning point (and whether the graph is restricted).

a < 0	The quadratic is concave down, giving $a(x-h)^2 + k \leq k$, so the maximum value of the quadratic is k.	(h, k)
a > 0	The quadratic is concave up, giving $a(x-h)^2 + k \geq k$, so the minimum value of the quadratic is k.	(h, k)

A stone is thrown vertically into the air so that its height h metres above the ground after t seconds is given by $h = 1.5 + 5t - 0.5t^2$.

a. Determine the greatest height the stone reaches.

b. Determine how many seconds it takes for the stone to reach its greatest height.

c. Determine when the stone is 6 metres above the ground. Explain why there are two times.

d. Sketch the graph and give the time to return to the ground to 1 decimal place.

THINK	WRITE
a. 1. The turning point is required. Calculate the coordinates of the turning point and state its type.	a. $h = 1.5 + 5t - 0.5t^2$ $a = -0.5, b = 5, c = 1.5$ Turning point: The axis of symmetry has the equation $t = -\dfrac{b}{2a}$. $t = -\dfrac{5}{2 \times (-0.5)}$ $ = 5$ When $t = 5$, $h = 1.5 + 5(5) - 0.5(5)^2$ $ = 14$ The turning point is $(5, 14)$. This is a maximum turning point as $a < 0$.
2. Write the answer. *Note:* The turning point is in the form (t, h) as t is the independent variable and h the dependent variable. The greatest height is the h-coordinate.	Therefore, the greatest height the stone reaches is 14 metres above the ground.
b. The required time is the t-coordinate of the turning point. Write the answer.	b. The stone reaches its greatest height after 5 seconds.
c. 1. Substitute the given height and solve for t.	c. $h = 1.5 + 5t - 0.5t^2$ When $h = 6$, $6 = 1.5 + 5t - 0.5t^2$ $0.5t^2 - 5t + 4.5 = 0$ $t^2 - 10t + 9 = 0$ $(t - 1)(t - 9) = 0$ $\therefore t = 1 \text{ or } t = 9$
2. Write the answer.	Therefore, the first time the stone is 6 metres above the ground is 1 second after it has been thrown into the air and is rising upwards. It is again 6 metres above the ground after 9 seconds when it is falling down.
d. 1. Calculate the time the stone returns to the ground.	d. Returns to ground when $h = 0$ $0 = 1.5 + 5t - 0.5t^2$ $t^2 - 10t - 3 = 0$ $t^2 - 10t = 3$ $t^2 - 10t + 25 = 3 + 25$

$$(t-5)^2 = 28$$
$$t = 5 \pm \sqrt{28}$$
$$t \simeq 10.3 \text{ (reject negative value)}$$
The stone reaches the ground after 10.3 seconds.

2. Sketch the graph, from its initial height to when the stone hits the ground. Label the axes appropriately.

When $t = 0, h = 1.5$, so the stone is thrown from a height of 1.5 metres.
Initial point: $(0, 1.5)$
Maximum turning point: $(5, 14)$
End point: $(10.3, 0)$

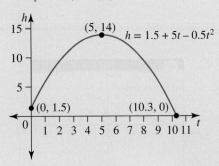

TI | THINK

WRITE

a. 1. On a Graphs page, complete the entry line for function 1 as:
$f1(x) = 1.5 + 5x - 0.5x^2$
then press ENTER.

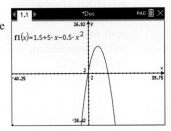

2. To find the maximum, press MENU, then select:
6: Analyze Graph
3: Maximum
Move the cursor to the left of the maximum when prompted for the lower bound, then press ENTER. Move the cursor to the right of the maximum when prompted for the upper bound, then press ENTER.

3. The answer appears on the screen.

The greatest height the stone reaches is 14 metres above the ground.

CASIO | THINK

WRITE

a. 1. On a Graph screen, complete the entry line for y1 as:
$y1 = 1.5 + 5x - 0.5x^2$
then press EXE. Select DRAW by pressing F6.

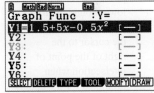

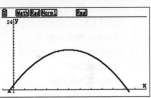

2. To find the maximum, select G-Solv by pressing SHIFT then F5, then select MAX by pressing F2. Press EXE.

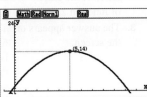

3. The answer appears on the screen.

The greatest height the stone reaches is 14 metres above the ground.

b. **1.** See **a.2.**

The stone reaches its greatest height after 5 seconds.

c. **1.** Press TAB to bring up the function entry line and complete the entry line for function 2 as:
$f2(x) = 6$
then press ENTER.

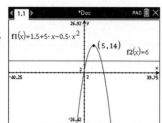

2. To find the points of intersection, press MENU, then select:
6: Analyze Graph
4: Intersection
Move the cursor to the left of the point of intersection when prompted for the lower bound, then press ENTER. Move the cursor to the right of the point of intersection when prompted for the upper bound, then press ENTER.
Repeat this step to find the other point of intersection.
Press TAB and unselect the graph of function 2.

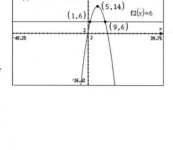

3. The answer appears on the screen.

The first time the stone is 6 metres above the ground is 1 second after it has been thrown into the air and is rising upwards. It is again 6 metres above the ground after 9 seconds when it is falling down.

d. **1.** To find the y-intercept, press MENU, then select:
5: Trace
1: Graph Trace
Type ' 0 ' then press ENTER twice.

b. **1.** See **a.2.**

The stone reaches its greatest height after 5 seconds.

c. **1.** Return to the function entry screen by pressing SHIFT then F6, then complete the entry line for y2 as:
$y2 = 6$
Press EXE.
Select DRAW by pressing F6.

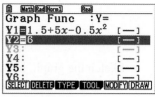

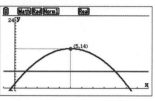

2. To find the points of intersection, select G-Solv by pressing SHIFT then F5, then select INTSECT by pressing F5.
With the cursor on the first point of intersection, press EXE. Use the left/right arrows to move to the other point of intersection, then press EXE.

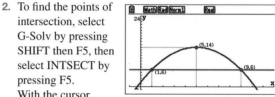

3. The answer appears on the screen.

The first time the stone is 6 metres above the ground is 1 second after it has been thrown into the air and is rising upwards. It is again 6 metres above the ground after 9 seconds when it is falling down.

d. **1.** To find the y-intercept, select G-Solv by pressing SHIFT then F5, then select Y-ICEPT by pressing F4. Press EXE.

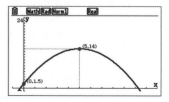

2. To find the
   x-intercept, press
   MENU, then select:
   6: Analyze Graph
   1: Zero
   Move the cursor to the
   left of the *x*-intercept
   when prompted for
   the lower bound, then
   press ENTER.
   Move the cursor to the
   right of the *x*-intercept
   when prompted for
   the upper bound, then
   press ENTER.
   Note: Only the
   positive *x*-intercept
   needs to be located
   due to the implied
   domain $0 \leq x \leq 10.3$.

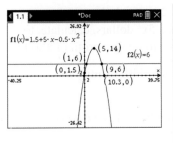

2. To find the
   x-intercept, select
   G-Solv by pressing
   SHIFT then F5, then
   select ROOT by
   pressing F1.
   Use the left/right
   arrows to move to the
   positive *x*-intercept,
   then press EXE.
   Note: Only the
   positive
   x-intercept needs
   to be located due to
   the implied domain
   $0 \leq x \leq 10.3$.

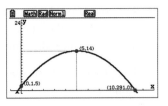

3. The answer appears on   The stone reaches the ground
   the screen.   after 10.3 seconds.

3. The answer appears   The stone reaches the ground
   on the screen.   after 10.3 seconds.

Exercise 2.6 Modelling with quadratic functions

learn on

2.6 Exercise	2.6 Exam practice on

Simple familiar	Complex familiar	Complex unfamiliar
1, 2, 3, 4, 5, 6, 7, 8, 9, 10, 11	12, 13, 14, 15, 16	17, 18, 19, 20

These questions are
even better in jacPLUS!
• Receive immediate feedback
• Access sample responses
• Track results and progress
Find all this and MORE in jacPLUS ▶

Simple familiar

1. **WE17** A rectangle with an area of $200 \, \text{m}^2$ is shaped such that the length is 8 metres longer than the width. Model this situation with a quadratic equation and determine the exact dimensions of the rectangle.

2. **WE18** The surface area of a sphere is directly proportional to the square of its radius. When the radius is 5 cm, the area is $100\pi \, \text{cm}^2$. Calculate the radius when the area is $360\pi \, \text{cm}^2$.

3. a. The area of an equilateral triangle varies directly as the square of its side length. A triangle of side length $2\sqrt{3} \, \text{cm}$ has an area of $3\sqrt{3} \, \text{cm}^2$. Calculate the side length if the area is $12\sqrt{3} \, \text{cm}^2$.
   b. The distance a particle falls from rest is in direct proportion to the square of the time of fall. Determine the effect on the distance fallen if the time of fall is doubled.
   c. The number of joules of heat produced in a wire in a given time varies as the square of the voltage. If the voltage is reduced by 20%, determine the effect on the number of joules of heat produced.

4. The cost of producing x hundred litres of olive oil is $20 + 5x$ dollars. If the revenue from the sale of x hundred litres of the oil is $1.5x^2$ dollars, calculate, to the nearest litre, the number of litres that must be sold to make a profit of $800.

5. The product of two consecutive even natural numbers is 440. Identify the numbers.

6. The sum of the squares of two consecutive natural numbers plus the square of their sums is 662. Identify the numbers.

7. The hypotenuse of a right-angled triangle is $(3x + 3)$ cm, and the other two sides are $3x$ cm and $(x - 3)$ cm. Determine the value of x and calculate the perimeter of the triangle.

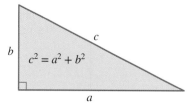

8. A gardener has 30 metres of edging to enclose a rectangular area using the back fence as one edge.
   a. Show the area function is $A = 30x - 2x^2$ where A square metres is the area of the garden bed of width x metres.
   b. Calculate the dimensions of the garden bed for maximum area.
   c. Determine the maximum area that can be enclosed.

9. **WE19** A missile is fired vertically into the air from the top of a cliff so that its height h metres above the ground after t seconds is given by $h = 100 + 38t - \dfrac{19}{12}t^2$.

   a. Determine the greatest height the missile reaches.
   b. Determine how many seconds it takes for the missile to reach its greatest height.
   c. Sketch the graph and give the time to return to the ground to 1 decimal place.

10. A child throws a ball vertically upwards so that after t seconds its height h metres above the ground is given by $10h = 16t + 4 - 9t^2$.

    a. Calculate how long the ball takes to reach the ground.
    b. Determine whether the ball will strike the foliage overhanging from a tree if the foliage is 1.6 metres vertically above the ball's point of projection.
    c. Calculate the greatest height the ball could reach.

11. In a game of volleyball, a player serves from the back of a playing court of length 18 metres. The path of the ball can be considered to be part of the parabola $y = 1.2 + 2.2x - 0.2x^2$ where x (metres) is the horizontal distance travelled by the ball from where it was hit and y (metres) is the vertical height the ball reaches.

    a. Determine how high the volleyball reaches.
    b. The net is 2.43 metres high and is placed in the centre of the playing court. Show that the ball clears the net and calculate by how much.

12. The cost of hiring a chainsaw is $10 plus an amount that is proportional to the square of the number of hours for which the chainsaw is hired. If it costs $32.50 to hire the chainsaw for 3 hours, determine, to the nearest half hour, the length of time for which the chainsaw was hired if the cost of hire was $60.

13. A photograph, 17 cm by 13 cm, is placed in a rectangular frame. If the border around the photograph is of uniform width and has an area of 260 cm², measured to the nearest cm², determine the dimensions of the frame measured to the nearest cm.

14. The daily cost, C dollars, of producing x kg of plant fertiliser for use in market gardens is $C = 15 + 10x$. The manufacturer decides that the fertiliser will be sold for v dollars per kg where $v = 50 - x$. Determine an expression for the daily profit in terms of x and hence determine the price per kilogram that should be charged for maximum daily profit.

15. Georgie has a large rectangular garden area with dimensions l metres by w meters, which she wishes to divide into three sections so she can grow different vegetables. She plans to put a watering system along the perimeter of each section. This will require a total of 120 metres of hosing.

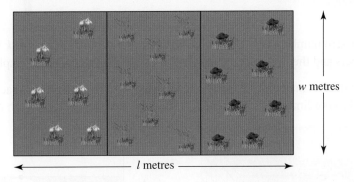

w metres

l metres

a. Show the total area of the three sections, A m², is given by $A = 60w - 2w^2$ and hence calculate the dimensions when the total area is a maximum.

b. Using the maximum total area, Georgie decides she wants the areas of the three sections to be in the ratio $1 : 2 : 3$. Determine the lengths of hosing for the watering system that is required for each section.

16. The number of bacteria in a slowly growing culture at time t hours after 8:00 am is given by $N = 100 + 46t + 2t^2$.

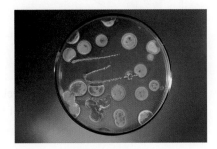

a. Determine how long it takes for the initial number of bacteria to double.

b. Calculate how many bacteria are present at 1:00 pm.

c. At 1:00 pm a virus is introduced that initially starts to destroy the bacteria so that t hours after 1:00 pm the number of bacteria is given by $N = 380 - 180t + 30t^2$. Determine the minimum number the population of bacteria reaches and the time at which this occurs.

17. A gardener has 16 metres of edging to place around three sides of a rectangular garden bed, the fourth side of which is bounded by the backyard fence.

Backyard fence

The gardener decides the area of the garden bed is to be 15 square metres. Given that the gardener would also prefer to use as much of the backyard fence as possible as a boundary to the garden bed, calculate the dimensions of the rectangle in this case, correct to 1 decimal place.

18. The cost, C dollars, of manufacturing n dining tables is the sum of three parts. One part represents the fixed overhead costs, c; another represents the cost of raw materials and is directly proportional to n; and the third part represents the labour costs, which are directly proportional to the square of n.

If 5 tables cost $195 to manufacture, 8 tables cost $420 to manufacture and 10 tables cost $620 to manufacture, determine the maximum number of dining tables that can be manufactured if costs are not to exceed $1000.

19. The sum of two non-zero numbers is k. Determine whether there are any values of k for which the sum of the squares of the numbers and their product are equal. If so, state the values; if not, explain why.

20. The arch of a bridge over a small creek is parabolic in shape with its feet evenly spaced from the ends of the bridge. Relative to the coordinate axes, the points A, B and C lie on a parabola with AC = 8 metres.

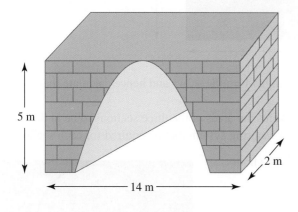

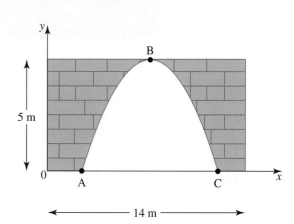

Following heavy rainfall, the creek floods and overflows its bank, causing the water level to reach 1.5 metres above AC. Calculate the width of the water level, correct to 1 decimal place.

Fully worked solutions for this chapter are available online.

LESSON
2.7 Review

doc-41797

2.7.1 Summary

2.7 Exercise

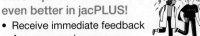 learn on

2.7 Exercise	2.7 Exam practice on

Simple familiar	Complex familiar	Complex unfamiliar
1, 2, 3, 4, 5, 6, 7, 8, 9, 10, 11	12, 13, 14, 15, 16	17, 18, 19, 20

Simple familiar

1. **MC** The solutions of the equation $(x-2)(x+1) = 4$ are:

   A. $x = 2, x = -1$ **B.** $x = 6, x = -1$
   C. $x = -6, x = 1$ **D.** $x = 3, x = -2$

2. **MC** The parabola with equation $y = x^2$ is translated so that its image has its vertex at $(-4, 3)$. The equation of the image is:

   A. $y = (x-4)^2 + 3$ **B.** $y = (x-3)^2 + 4$
   C. $y = (x+4)^2 + 3$ **D.** $y = (x+3)^2 - 4$

3. **MC** The maximum value of $4 - 2x - x^2$ is:

   A. 5   **B.** 4
   C. 3   **D.** 1

4. **MC** The equation of the parabola shown is:

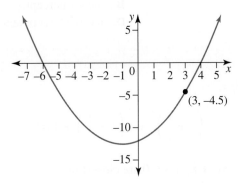

(3, −4.5)

   A. $y = x^2 + 2x - 24$ **B.** $y = 0.5x^2 + x - 12$ **C.** $y = x^2 - 2x - 24$ **D.** $y = 0.5x^2 - x - 12$

5. **MC** A quadratic graph touches the x-axis at $x = -6$ and cuts the y-axis at $y = -10$. Its equation is:

A. $y = (x + 6)(x + 10)$

B. $y = \dfrac{5}{18}x^2 - 10$

C. $y = \dfrac{5}{18}(x + 6)^2$

D. $y = -\dfrac{5}{18}(x + 6)^2$

6. Solve for x.

   a. $\left(x^2 + 4\right)^2 - 7\left(x^2 + 4\right) - 8 = 0$

   b. $2x^2 = 3x(x - 2) + 1$

   c. $x = \dfrac{12}{x - 2} - 2$

   d. $3 + \sqrt{x} = 2x$

7. Sketch the graphs of the following, showing all key points.

   a. $y = 2(x - 3)(x + 1)$

   b. $y = 1 - (x + 2)^2$

   c. $y = x^2 + x + 9$

8. **MC** The x-coordinates of the points of intersection of the parabola $y = 3x^2 - 10x + 2$ with the line $2x - y = 1$ can be determined from the equation:

A. $3x^2 - 10x + 2 = 0$

B. $3x^2 - 12x + 3 = 0$

C. $x^2 - 6x + 1 = 0$

D. $3x^2 - 8x + 1 = 0$

9. **MC** If $x^2 + 4x - 6$ is expressed in the form $(x - h)^2 + k$, then the values of h and k would be:

A. $h = -2, k = -10$

B. $h = 2, k = -10$

C. $h = -4, k = -2$

D. $h = 4, k = -2$

10. **MC** For the graph of the parabola $y = ax^2 + bx + c$ shown, with $\Delta = b^2 - 4ac$, identify the correct statement.

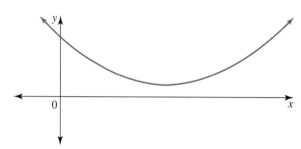

A. $a > 0$ and $\Delta > 0$
B. $a > 0$ and $\Delta < 0$
C. $a < 0$ and $\Delta < 0$
D. $a < 0$ and $\Delta > 0$

11. **MC** The graph of $-4x^2 - x - 3$ has:

A. no x-intercepts.

B. one x-intercept.

C. two x-intercepts.

D. one x-intercept and one y-intercept.

Complex familiar

12. Solve the following equation for x.

$$\left(x + \frac{1}{x}\right)^2 + 4\left(x + \frac{1}{x}\right) + 4 = 0$$

13. **MC** The values of x for which $-5x^2 + 8x + 3 = 0$ are closest to:

A. $-0.6, -1$

B. $0.6, -1$

C. $-0.3, 1.9$

D. $0.3, -1.9$

14. Factorise by completing the square.

   a. $-x^2 + 20x + 24$

   b. $4x^2 - 2x - 9$

15. Determine the values of k for which the equation $kx^2 - 4x(k+2) + 36 = 0$ has no real roots.

16. a. Use an algebraic method to determine the coordinates of the points of intersection of the parabola $y = x^2 + 2x$ and the line $y = x + 2$.

   b. Sketch $y = x^2 + 2x$ and $y = x + 2$ on the same set of axes.

Complex unfamiliar

17. At a winter skiing championship, two competitors, one from Japan and the other from Canada, compete for the gold medal in one of the jump events.

Each competitor leaves the ski run at point S and travels through the air, landing back on the ground at some point G. The winner will be the competitor who covers the greater horizontal distance OG.

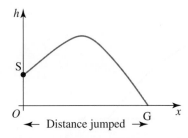

The Canadian skier jumps first. Her height, h metres, above the ground is described by $h = -\dfrac{1}{35}\left(x^2 - 60x - 700\right)$, where x metres is the horizontal distance travelled.

The Japanese skier jumps next. She reaches a maximum height of 35 metres above the ground after a horizontal distance of 30 metres has been covered.

Assuming that both skiers started at point S and that the path of the Japanese skier is a parabola, decide which competitor receives the gold medal. Your decision should be supported with appropriate mathematical reasoning.

18. The diagram shows the arch of a bridge where the shape of the curve, OAB, is a parabola. OB is the horizontal road level. Taking O as the origin, the equation of the curve OAB is $y = 2.5x - 0.3125x^2$. All measurements are in metres.

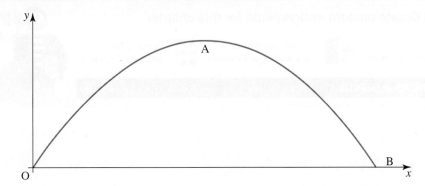

A car towing a caravan needs to drive under the bridge. The caravan is 5 metres wide and has a height of 2 metres. Only one single lane of traffic can pass under the bridge. Explain clearly, using mathematical analysis, whether the caravan can be towed under this bridge.

19. ABCD is a rectangle of length one more unit than its width. Point F lies on AB and divides AB in the ratio $x : 1$ so that AF is x units in length and FB is 1 unit in length. Point G lies on DC and divides DC in the same ratio, $x : 1$.

If the area of the square AFGD is one more square unit than the area of the rectangle FBCG, determine the value of x in simplest surd form.

20. Ignoring air resistance, the path of a cricket ball hit by a batsman can be considered to travel on a parabolic path that starts at the point $(0, 0)$ where the ball is struck by the batsman.

Let x metres measure the horizontal distance of the ball from the batsman in the direction the ball travels, and let y metres measure the vertical height above the ground that the ball reaches.

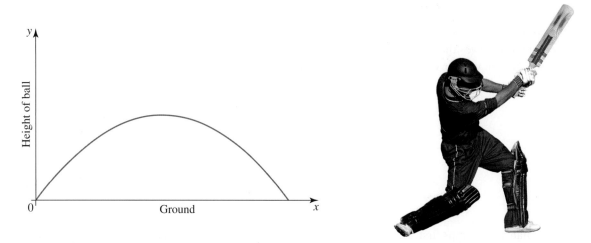

A batsman hits a cricket ball towards a fielder who is 65 metres away. The ball is struck with a horizontal speed of 28 m/s, which is assumed to remain constant throughout the flight of the ball. On its way, the ball reaches a maximum height of 4.9 metres after 1 second.

The fielder starts running forward at the instant the ball is hit and catches it at a height of 1.3 metres above the ground.

Determine the time it takes the fielder to reach the ball and obtain the uniform speed in m/s with which the fielder runs in order to catch the ball.

Fully worked solutions for this chapter are available online.

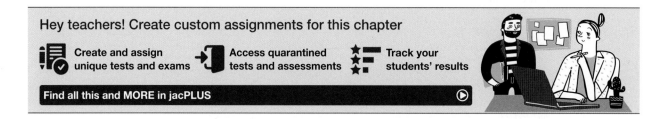

Hey teachers! Create custom assignments for this chapter

Create and assign unique tests and exams

Access quarantined tests and assessments

Track your students' results

Find all this and MORE in jacPLUS

Answers

Chapter 2 Quadratic functions

2.2 Solving quadratic equations with rational roots

2.2 Exercise

1. a. $x = -3, \dfrac{7}{10}$ b. $(x+5)x = x^2 + 5x$

2. $x = -\dfrac{6}{10}, 1$

3. a. $x = \dfrac{4}{3}, -\dfrac{1}{2}$ b. $x = 4, 3$

 c. $x = -\dfrac{3}{2}, x = -\dfrac{7}{4}$ d. $x = 0, \dfrac{1}{5}$

 e. $x = \dfrac{2}{3}, -4$ f. $x = 0, 10$

4. a. $x = -5, 1$ b. $x = -4, 6$
 c. No real solutions d. $x = -10, -1$
 e. $x = 7$ f. $x = 0, 8$

5. a. $0, 5$ b. $\dfrac{1}{7}, 3$

 c. -8 d. $-6, -4$

6. a. $-\dfrac{1}{2}, -\dfrac{1}{3}$ b. $-\dfrac{2}{3}, \dfrac{5}{4}$

 c. 7 d. $-\dfrac{5}{6}, 1$

7. a. ± 11 b. $\pm \dfrac{4}{3}$ c. $4, 6$

 d. $-1, 6$ e. $-\dfrac{1}{3}, 1$ f. ± 3

8. a. $(x-1)(x-7) = 0$ b. $(x+5)(x-4) = 0$
 c. $x(x-10) = 0$ d. $(x-2)^2 = 0$

9. $x = \pm \dfrac{1}{3}$

10. a. $x = \dfrac{19}{6}, \dfrac{7}{3}$ b. $x = -6, -\dfrac{13}{5}$

 c. $x = -2, -4$ d. $x = \dfrac{1}{2}, 3$

11. a. $-\dfrac{14}{3}, -1$ b. $-\dfrac{7}{2}, \dfrac{1}{4}$ c. $\pm 2, \pm 5$

 d. ± 4 e. $\pm \dfrac{1}{3}$ f. $-3, -2, -1$

12. a. $-1, 8$ b. $\dfrac{11}{6}, \dfrac{9}{2}$

 c. $-8, -2$ d. $0, \dfrac{1}{2}$

13. a. $\dfrac{1}{3}$ b. $\pm \dfrac{25}{2}$

 c. $-\dfrac{5}{7}, \dfrac{2}{5}$ d. $0, \dfrac{17}{19}$

14. a. $x = -\dfrac{7}{3}$ or $x = \dfrac{9}{20}$

 b. $x = -\dfrac{7}{4}$ or $x = \dfrac{14}{3}$

15. a. $x = \dfrac{3}{2}$ b. $x = -\dfrac{2}{11}$ or $x = \dfrac{11}{2}$

16. a. ± 3 b. $\pm \dfrac{4}{3}, \pm 2$

 c. $-1, 2$ d. $-\dfrac{3}{2}, -\dfrac{6}{5}$

17. $x = 1$

18. $x = \dfrac{r-q}{p}, \quad x = -\dfrac{(r+q)}{p}$

19. a. $x = -3a, x = 2b$

 b. $x = \dfrac{3a}{2}, x = 5a$

 c. $x = b-1, x = b+1, x = b-2, x = b+2$

 d. $x = a-b, x = a+3b$

 e. $x = b-a, x = 2b-a$

 f. $x = 1$

20. a. $b = 13, c = -12$

 b. Roots $-\dfrac{q}{p}, -1$; solutions $x = \dfrac{p-q}{p}, 0$

2.3 Solving quadratic equations over R

2.3 Exercise

1. a. $x^2 + 10x + 25 = (x+5)^2$

 b. $x^2 - 7x + \dfrac{49}{4} = \left(x - \dfrac{7}{2}\right)^2$

 c. $x^2 + x + \dfrac{1}{4} = \left(x + \dfrac{1}{2}\right)^2$

 d. $x^2 - \dfrac{4}{5}x + \dfrac{4}{25} = \left(x - \dfrac{2}{5}\right)^2$

2. a. $\left(x - 5 - 4\sqrt{2}\right)\left(x - 5 + 4\sqrt{2}\right)$

 b. $3\left(x + \dfrac{7 - \sqrt{13}}{6}\right)\left(x + \dfrac{7 + \sqrt{13}}{6}\right)$

 c. $\left(\sqrt{5}x - 3\right)\left(\sqrt{5}x + 3\right)$

3. $(-3x + 5)(x - 1)$

4. a. $\left(x - 3 - \sqrt{2}\right)\left(x - 3 + \sqrt{2}\right)$

 b. $\left(x + 2 - \sqrt{7}\right)\left(x + 2 + \sqrt{7}\right)$

 c. No linear factors over R

 d. $2\left(x + \dfrac{5}{4} - \dfrac{\sqrt{41}}{4}\right)\left(x + \dfrac{5}{4} + \dfrac{\sqrt{41}}{4}\right)$

 e. $-\left(x - 4 - 2\sqrt{2}\right)\left(x - 4 + 2\sqrt{2}\right)$

 f. $3\left(x + \dfrac{2}{3} - \dfrac{\sqrt{22}}{3}\right)\left(x + \dfrac{2}{3} + \dfrac{\sqrt{22}}{3}\right)$

5. a. $x^2 + 2x$
 $= x^2 + 2x + 1 - 1$
 $= (x + 1)^2 - 1$

b. $x^2 + 7x$

$\quad = x^2 + 7x + \dfrac{49}{4} - \dfrac{49}{4}$

$\quad = \left(x + \dfrac{7}{2}\right)^2 - \dfrac{49}{4}$

c. $x^2 - 5x$

$\quad = x^2 - 5x + \dfrac{25}{4} - \dfrac{25}{4}$

$\quad = \left(x - \dfrac{5}{2}\right)^2 - \dfrac{25}{4}$

d. $x^2 + 4x - 2$

$\quad = x^2 + 4x + 4 - 4 - 2$

$\quad = (x + 2)^2 - 6$

6. a. $3\left(x - 8 - \sqrt{2}\right)\left(x - 8 + \sqrt{2}\right)$

b. No real factors

7. a. $\left(x - 2\sqrt{3}\right)\left(x + 2\sqrt{3}\right)$

b. $\left(x - 6 - 4\sqrt{2}\right)\left(x - 6 + 4\sqrt{2}\right)$

c. $\left(x + \dfrac{9 - \sqrt{93}}{2}\right)\left(x + \dfrac{9 + \sqrt{93}}{2}\right)$

d. $2\left(x + \dfrac{5 - \sqrt{17}}{4}\right)\left(x + \dfrac{5 + \sqrt{17}}{4}\right)$

e. $3\left(\left(x + \dfrac{2}{3}\right)^2 + \dfrac{5}{9}\right)$, no linear factors over R

f. $-5\left(x - 4 - \dfrac{9\sqrt{5}}{5}\right)\left(x - 4 + \dfrac{9\sqrt{5}}{5}\right)$

8. a. $x = 5 - \sqrt{2}$ or $5 + \sqrt{2}$

b. $x = \dfrac{5}{2} + \dfrac{\sqrt{5}}{2}$ or $\dfrac{5}{2} - \dfrac{\sqrt{5}}{2}$

c. $x = -7 + \sqrt{6}$ or $-7 - \sqrt{6}$

d. $x = \dfrac{-9}{2} + \dfrac{\sqrt{5}}{2}$ or $\dfrac{-9}{2} - \dfrac{\sqrt{5}}{2}$

9. a. -1.19 or 4.19 b. 0.76 or 5.24
   c. -4 or -3 d. 3.68 or 16.32

10. a. $a = 1$, $b = -10$, $c = 21$
    b. $a = 10$, $b = -93$, $c = 68$
    c. $a = 1$, $b = -9$, $c = 20$
    d. $a = 40$, $b = 32$, $c = 6$

11. a. $x = \dfrac{1}{3} + \dfrac{\sqrt{13}}{3}$ or $\dfrac{1}{3} - \dfrac{\sqrt{13}}{3}$

b. $x = -3$ or $-\dfrac{1}{2}$

c. $x = -1 + \dfrac{\sqrt{21}}{3}$ or $-1 - \dfrac{\sqrt{21}}{3}$

d. $x = \dfrac{1}{3} + \dfrac{\sqrt{19}}{6}$ or $\dfrac{1}{3} - \dfrac{\sqrt{19}}{6}$

12. a. $x = -3.14$ or 0.64 b. $x = -0.74$ or 1.24
    c. $x = 0.73$ or 6.52 d. $x = -1.04$ or 1.20

13. C

14. A

15. a. $x = \dfrac{5 \pm \sqrt{13}}{6}$

b. $x = \dfrac{1 \pm \sqrt{101}}{10}$

c. No real solutions

d. $-3 \pm \sqrt{17}$

16. $x = \dfrac{-11 \pm \sqrt{89}}{4}$

17. a. $\pm \sqrt{5}$ b. $\pm 2\sqrt{2}$ c. $\pm \sqrt{11}, \pm 1$

d. $\dfrac{-3 \pm \sqrt{5}}{2}$ e. $-1, 8, \dfrac{7 \pm \sqrt{93}}{2}$

18. $x = \dfrac{-1 \pm \sqrt{7}}{2}$

19. $x = -\sqrt{6} \pm \sqrt{14}$

20. a. $x = -3\sqrt{2}$

b. $x = \dfrac{3\sqrt{2} \pm \sqrt{10}}{4}$

2.4 Graphs of quadratic functions

2.4 Exercise

1.

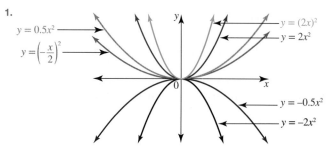

$y = 0.5x^2$

$y = \left(-\dfrac{x}{2}\right)^2$

$y = (2x)^2$

$y = 2x^2$

$y = -0.5x^2$

$y = -2x^2$

2. A i. $y = x^2 - 2$
   B i. $y = -2x^2$
   C i. $y = -(x + 2)^2$

3. a. $(0, 8)$ b. $(0, -8)$
   c. $(0, 1)$ d. $(0, -7)$
   e. $(8, 0)$ f. $(-8, 0)$
   g. $(4, 0)$ h. $(-12, 0)$

4. Axis intercepts $(-6, 0)$, $(3, 0)$, $(0, -6)$; minimum turning point $(-1.5, -6.75)$

$(-6, 0)$ $(3, 0)$ $y = x^2$ $(0, -6)$ $(-1.5, -6.75)$

5. a.

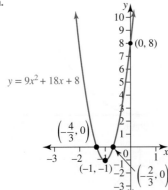

$y = 9x^2 + 18x + 8$

$(0, 8)$

$\left(-\dfrac{4}{3}, 0\right)$

$\left(-\dfrac{2}{3}, 0\right)$

$(-1, -1)$

b.

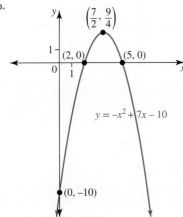

$\left(\dfrac{7}{2}, \dfrac{9}{4}\right)$

$(2, 0)$

$(5, 0)$

$y = -x^2 + 7x - 10$

$(0, -10)$

c.

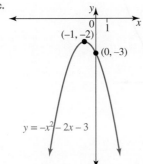

$(-1, -2)$

$(0, -3)$

$y = -x^2 - 2x - 3$

d.

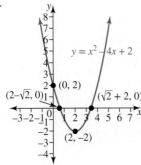

$y = x^2 - 4x + 2$

$(0, 2)$

$(2 - \sqrt{2}, 0)$

$(\sqrt{2} + 2, 0)$

$(2, -2)$

6. Maximum turning point $(-3, 2)$; axis intercepts $(0, -16)$, $(-4, 0)$, $(-2, 0)$

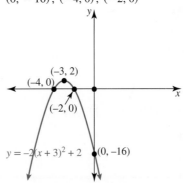

$(-3, 2)$

$(-4, 0)$

$(-2, 0)$

$y = -2(x + 3)^2 + 2$

$(0, -16)$

7. a. Maximum turning point at $(0, 4)$

b. Minimum turning point at $\left(\dfrac{4}{3}, 0\right)$

8. D

9. Axis intercepts $(0, 0)$, $(4, 0)$; maximum turning point $(2, 8)$

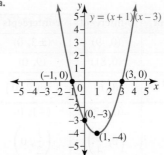

$(2, 8)$

$y = 2x(4 - x)$

$(0, 0)$

$(4, 0)$

10. a.

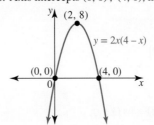

$y = (x + 1)(x - 3)$

$(-1, 0)$

$(3, 0)$

$(0, -3)$

$(1, -4)$

b.

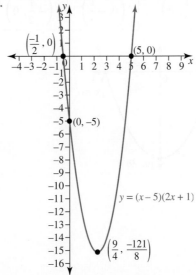

$\left(\dfrac{-1}{2}, 0\right)$

$(5, 0)$

$(0, -5)$

$y = (x - 5)(2x + 1)$

$\left(\dfrac{9}{4}, \dfrac{-121}{8}\right)$

c.

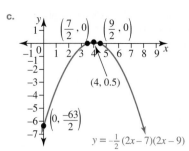

$\left(\dfrac{7}{2}, 0\right)$ $\left(\dfrac{9}{2}, 0\right)$

$(4, 0.5)$

$\left(0, \dfrac{-63}{2}\right)$

$y = -\dfrac{1}{2}(2x - 7)(2x - 9)$

d.

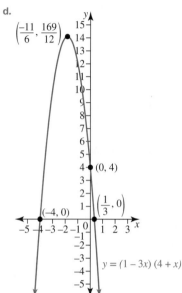

$\left(\dfrac{-11}{6}, \dfrac{169}{12}\right)$

$(0, 4)$

$\left(\dfrac{1}{3}, 0\right)$

$(-4, 0)$

$y = (1 - 3x)(4 + x)$

b.

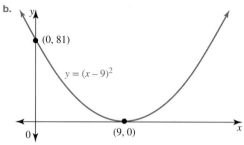

$(0, 81)$

$y = (x - 9)^2$

$(9, 0)$

c.

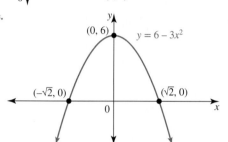

$(0, 6)$

$y = 6 - 3x^2$

$(-\sqrt{2}, 0)$ $(\sqrt{2}, 0)$

d.

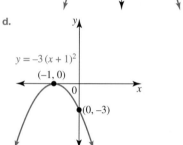

$y = -3(x + 1)^2$

$(-1, 0)$

$(0, -3)$

e.

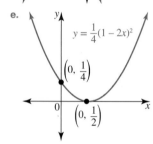

$y = \dfrac{1}{4}(1 - 2x)^2$

$\left(0, \dfrac{1}{4}\right)$

$\left(0, \dfrac{1}{2}\right)$

f.

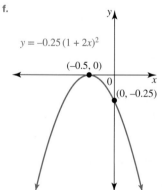

$y = -0.25(1 + 2x)^2$

$(-0.5, 0)$

$(0, -0.25)$

11.

	Turning point	y-intercept	x-intercepts
a	$(0, -9)$	$(0, -9)$	$(\pm 3, 0)$
b	$(9, 0)$	$(0, 81)$	$(9, 0)$
c	$(0, 6)$	$(0, 6)$	$\left(\pm\sqrt{2}, 0\right)$
d	$(-1, 0)$	$(0, -3)$	$(-1, 0)$
e	$\left(\dfrac{1}{2}, 0\right)$	$\left(0, \dfrac{1}{4}\right)$	$\left(\dfrac{1}{2}, 0\right)$
f	$\left(-\dfrac{1}{2}, 0\right)$	$\left(0, -\dfrac{1}{4}\right)$	$\left(-\dfrac{1}{2}, 0\right)$

a.

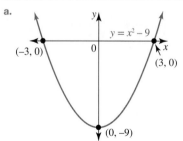

$y = x^2 - 9$

$(-3, 0)$

$(3, 0)$

$(0, -9)$

12.

	Turning point	y-intercept	x-intercepts
a	$(5, 2)$	$(0, 27)$	none
b	$(-1, -2)$	$(0, 0)$	$(0, 0), (-2, 0)$
c	$(3, -6)$	$(0, -24)$	none
d	$(4, 1)$	$(0, -15)$	$(3, 0), (5, 0)$
e	$(-4, -2)$	$(0, 6)$	$(-6, 0), (-2, 0)$
f	$\left(\dfrac{1}{2}, \dfrac{1}{9}\right)$	$\left(0, \dfrac{2}{27}\right)$	$\left(\dfrac{1 \pm \sqrt{3}}{2}, 0\right)$

a.

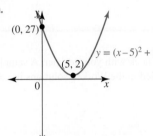

$(0, 27)$
$(5, 2)$
$y = (x-5)^2 + 2$

b.

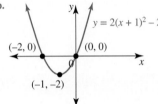

$y = 2(x+1)^2 - 2$
$(-2, 0)$ $(0, 0)$
$(-1, -2)$

c.

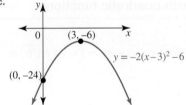

$(3, -6)$
$y = -2(x-3)^2 - 6$
$(0, -24)$

d.

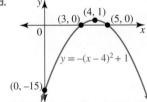

$(3, 0)$ $(4, 1)$ $(5, 0)$
$y = -(x-4)^2 + 1$
$(0, -15)$

e.

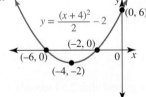

$y = \dfrac{(x+4)^2}{2} - 2$
$(0, 6)$
$(-2, 0)$
$(-6, 0)$
$(-4, -2)$

f.

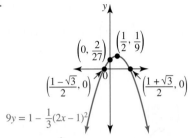

$\left(0, \dfrac{2}{27}\right)$ $\left(\dfrac{1}{2}, \dfrac{1}{9}\right)$
$\left(\dfrac{1-\sqrt{3}}{2}, 0\right)$ $\left(\dfrac{1+\sqrt{3}}{2}, 0\right)$
$9y = 1 - \dfrac{1}{3}(2x-1)^2$

13. a. $c = 4, y = x^2 + 4$

b. $a = -\dfrac{1}{18}, y = -\dfrac{1}{18}x^2$

c. $a = -3, y = -3(x-2)^2$

14. a. $x - 3, x - 8, y = (x-3)(x-8)$

b. $y = (x+11)(x-2)$

15. a. $y = (x+2)^2 + 1$ **b.** $y = 2x(x-2)$

16. $y = -2x^2 + x - 4$

17. a. $y = -2x^2 + 6$

b. $y = 0.2(x+6)(x+1)$

18. a. Equation A   **b.** Equation A   **c.** Equation A

19. $y = \dfrac{1}{2}x^2 - 4x + 6$

20. $p = \dfrac{4}{3} \Rightarrow y = \dfrac{9}{4}(3x-4)^2$

$p = 4 \Rightarrow y = \dfrac{9}{4}(x-4)^2$

2.5 The discriminant

2.5 Exercise

1. a. $\Delta = -135$

b. $\Delta = 0$

c. $\Delta = 1$

d. $\Delta = \dfrac{40}{9}$

2. a. 4

b. Two rational roots

3. a. Two irrational solutions

b. One rational solution

c. No real solutions

d. Two irrational solutions

4. a. $\Delta = 121$, 2 rational factors

b. $\Delta = -39$, no real factors

c. $\Delta = 0$, 1 repeated rational factor

d. $\Delta = 8$, 2 irrational factors

5. a. There are no real roots.

b. $k = -1, \ k = 3$

6. a. 2 irrational roots   **b.** 2 rational roots

   c. No real roots   **d.** 1 rational root

   e. No real roots   **f.** 2 irrational roots

7. a. 2 rational x-intercepts

b. Axis intercepts $(0, 0)$, $\left(\dfrac{7}{3}, 0\right)$; maximum turning point $\left(\dfrac{7}{6}, \dfrac{49}{2}\right)$

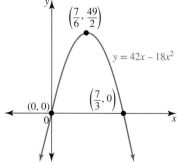

8. a. 2 irrational x-intercepts

b. No x-intercepts

c. 1 rational x-intercept

d. 2 rational x-intercepts

9. a. $k = -5$　　**b.** $k > -5$　　**c.** $k < -5$

10. a. $(-6, 8)$ and $(2, 0)$

b. No intersections

11. $(2, 8)$;

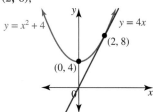

12. a. $x = 6$, $y = 32$ or $x = -1$, $y = -3$

b. $x = -8$, $y = 35$ or $x = 1$, $y = -1$

c. No solution

d. $x = 0$, $y = 5$ or $x = -\dfrac{5}{3}$, $y = \dfrac{70}{9}$

13. a. $\left(\dfrac{3}{5}, \dfrac{31}{5}\right)$, $(1, 7)$

b. $(-2, -3)$

c. $(-5, 10)$, $(4, 10)$

d. $(4, 30)$

14. a. No intersections

b. 2 intersections

c. No intersections

15. a. $(-1.14, 1.29)$, $(2.64, 6.96)$

b.

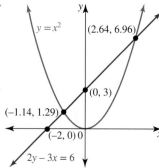

16. a. $m > \dfrac{1}{4}$

b. i. Proof required – check with your teacher. A sample response is provided in the Worked solutions in your online resources

ii. $x = -\dfrac{1}{2}$

17. a. $m = -4 \pm 4\sqrt{2}$

b. $m = 2 \pm 2\sqrt{3}$

c. $p < \dfrac{1}{3}$

18. $\Delta > 0$

19. Δ is a perfect square.

20. $\Delta = (m + 4)^2 \Rightarrow \Delta \geq 0$

2.6 Modelling with quadratic functions

2.6 Exercise

1. Length $= \left(4 + 6\sqrt{6}\right)$ m and width $= \left(-4 + 6\sqrt{6}\right)$ m

2. $3\sqrt{10}$ cm

3. a. $4\sqrt{3}$ cm

b. Distance is quadrupled.

c. Heat is reduced by 36%.

4. 2511 litres

5. 20 and 22

6. 10, 11

7. $x = 24$; perimeter $= 168$ cm

8. a. Sample responses can be found in the worked solutions in the online resources.

b. Width 7.5 metres; length 15 metres

c. 112.5 square metres

9. a. 328 metres

b. 12 seconds

c. The missile reaches the ground after 2.64 seconds.

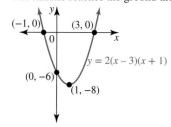

10. a. 2 seconds

b. The ball does not strike the foliage.

c. $\dfrac{10}{9}$ metres

11. a. 7.25 metres

b. The net is cleared by 2.37 metres.

12. $4\dfrac{1}{2}$ hours

13. Length 24 cm; width 20 cm

14. Profit $= -x^2 + 40x - 15$, $30 per kg

15. a. Sample responses can be found in the worked solutions in the online resources; length 30 metres, width 15 metres

b. 40 metres, 35 metres, 45 metres

16. a. 2 hours

b. 380

c. 110 bacteria at 4 pm

17. Width 1.1 metres and length 13.8 metres

18. 13

19. No values possible

20. 6.7 metres

2.7 Review

2.7 Exercise

1. D

2. C

3. A

4. B

5. D

6. a. $x = \pm 2$ **b.** $x = 3 \pm 2\sqrt{2}$

     c. $x = \pm 4$ **d.** $x = \dfrac{9}{4}, 1$

7. a. x-intercepts $(3, 0), (-1, 0)$; y-intercept $(0, -6)$; turning point $(1, -8)$

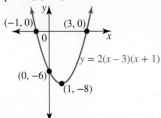

b. x-intercepts $(-3, 0), (-1, 0)$; y-intercept $(0, -3)$; turning point $(-2, 1)$

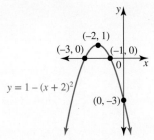

c. No x-intercepts; y-intercept $(0, 9)$; turning point $(-0.5, 8.75)$

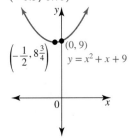

8. B

9. A

10. B

11. A

12. $x = -1$

13. C

14. a. $-\left(x - 10 - 2\sqrt{31}\right)\left(x - 10 + 2\sqrt{31}\right)$

b. $4\left(x - \dfrac{1 + \sqrt{37}}{4}\right)\left(x - \dfrac{1 - \sqrt{37}}{4}\right)$

15. $1 < k < 4$

16. a. $(-2, 0), (1, 3)$

b. Parabola: x-intercepts $(-2, 0), (0, 0)$; turning point $(-1, -1)$; line through $(-2, 0), (0, 2)$. Both graphs meet at $(-2, 0), (1, 3)$.

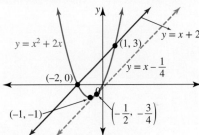

17. The Japanese competitor wins.

18. The caravan can be towed under the bridge.

19. $\dfrac{1 + \sqrt{5}}{2}$

20. $\dfrac{13}{7}$ seconds and 7 metres/second

3 Binomial expansions and cubic functions

LESSON SEQUENCE

Fully worked solutions for this chapter are available online.

EXAM PREPARATION

Access exam-style questions in every lesson, available online.

on Resources

Solutions	Solutions — Chapter 3 (sol-41807)
Exam questions	Exam questions booklet — Chapter 3 (eqb-41808)
Digital documents	Learning matrix — Chapter 3 (doc-41785)
	Chapter summary — Chapter 3 (doc-41798)

LESSON
3.1 Overview

3.1.1 Introduction

Mathematicians are driven by curiosity and a desire to generalise. One historical example of this is: 'Since quadratic equations can be solved by a formula, is the same true for cubic equations?' Mathematicians were able to solve second degree polynomial equations (quadratic equations) in 400 BCE. However, it wasn't until the early 16th century that progress was made on solving equations involving higher powers of x. In 1732 Leonhard Euler devised a general method for solving equations of the type $ax^3 + bx^2 + cx + d = 0$, and soon after Lodovico Ferrari found a general solution to fourth degree equations of the type $ax^4 + bx^3 + cx^2 + dx + e = 0$.

Many important developments in mathematics, engineering, science and medicine have occurred through the use of polynomial functions. Polynomial modelling functions play an important part in solving real-life problems. For example, they are used to design roller coasters, roads, buildings and other structures, and to describe and predict traffic patterns. Economists use polynomial functions to predict growth patterns, medical researchers use them to describe the growth of bacterial colonies, and meteorologists used them to understand weather patterns.

3.1.2 Syllabus links

Lesson	Lesson title	Syllabus links
3.2	**Pascal's triangle and binomial expansions**	○ Understand the notion of a combination as an unordered set of r objects taken from a set of n distinct objects. ○ Recognise and use the link between Pascal's triangle and the notation $\binom{n}{r}$.
3.3	**The binomial theorem**	○ Use the binomial theorem $(x + y)^n = x^n + \binom{n}{1} x^{n-1} y + \ldots + \binom{n}{r} x^{n-r} y^r + \ldots + y^n$ to expand expressions, e.g. $(2x - 1)^3$.
3.4	**Polynomials and cubic functions**	○ Identify the coefficients and the degree of a polynomial. ○ Expand quadratic and cubic polynomials from factors.
3.5	**Graphing cubic functions**	○ Recognise and determine features of the graphs of $y = x^3$, $y = a(x - h)^3 + k$ and $y = a(x - x_1)(x - x_2)(x - x_3)$, including shape, intercepts, and behaviour as $x \to \infty$ and $x \to -\infty$. ○ Sketch the graphs of cubic functions, with or without technology.
3.6	**Solving cubic equations**	○ Solve cubic equations using technology, and algebraically in cases where the equation is factorised.
3.7	**Modelling cubic functions**	○ Model and solve problems that involve cubic functions, with and without technology.

Source: Mathematical Methods Senior Syllabus 2024 © State of Queensland (QCAA) 2024; licensed under CC BY 4.0.

LESSON
3.2 Pascal's triangle and binomial expansions

SYLLABUS LINKS

- Understand the notion of a combination as an unordered set of r objects taken from a set of n distinct objects.
- Recognise and use the link between Pascal's triangle and the notation $\begin{pmatrix} n \\ r \end{pmatrix}$.

Source: Mathematical Methods Senior Syllabus 2024 © State of Queensland (QCAA) 2024; licensed under CC BY 4.0.

3.2.1 Factorial notation

In this and other topics, calculations such as $7 \times 6 \times 5 \times 4 \times 3 \times 2 \times 1$ will be encountered. This expression can be written in shorthand as $7!$, which is read as '7 factorial'. There is a factorial key on most calculators.

Definition

$n! = n \times (n-1) \times (n-2) \times \ldots \times 3 \times 2 \times 1$ for any natural number n.

It is also necessary to define $0! = 1$.

$7!$ is equal to 5040. It can also be expressed in terms of other factorials such as:

$$7! = 7 \times (6 \times 5 \times 4 \times 3 \times 2 \times 1) \quad \text{or} \quad 7! = 7 \times 6 \times (5 \times 4 \times 3 \times 2 \times 1)$$
$$= 7 \times 6! \qquad\qquad\qquad\qquad\qquad = 7 \times 6 \times 5!$$

This is useful when working with fractions containing factorials. For example:

$$\frac{7!}{6!} = \frac{7 \times \cancel{6!}}{\cancel{6!}} \quad \text{or} \quad \frac{5!}{7!} = \frac{\cancel{5!}}{7 \times 6 \times \cancel{5!}}$$
$$= 7 \qquad\qquad\qquad = \frac{1}{42}$$

By writing the larger factorial in terms of the smaller factorial, the fractions were simplified.

Factorial notation is just an abbreviation, so factorials cannot be combined arithmetically.

For example, $3! - 2! \neq 1!$. This is verified by evaluating $3! - 2!$.

$$3! - 2! = 3 \times 2 \times 1 - 2 \times 1$$
$$= 6 - 2$$
$$= 4$$
$$\neq 1$$

WORKED EXAMPLE 1 Evaluating an expression with factorial notation

Evaluate $5! - 3! + \dfrac{50!}{49!}$.

THINK	WRITE
1. Expand the two smaller factorials.	$5! - 3! + \dfrac{50!}{49!}$
	$= 5 \times 4 \times 3 \times 2 \times 1 - 3 \times 2 \times 1 + \dfrac{50!}{49!}$

▶

2. To simplify the fraction, write the larger factorial in terms of the smaller factorial.

$$= 5 \times 4 \times 3 \times 2 \times 1 - 3 \times 2 \times 1 + \frac{50 \times 49!}{49!}$$

3. Calculate the answer.

$$= 120 - 6 + \frac{50 \times \cancel{49!}}{\cancel{49!}}$$
$$= 120 - 6 + 50$$
$$= 164$$

TI	THINK	DISPLAY/WRITE	CASIO	THINK	DISPLAY/WRITE

TI | THINK

1. On a Calculator page, complete the entry line as:
$$5! - 3! + \frac{50!}{49!}$$
Then press ENTER.
Note: The factorial symbol is located in CTRL Menu, Symbols OR
   • Menu
   • Probability
   • factorial

DISPLAY/WRITE

2. The answer appears on the screen.   164

CASIO | THINK

1. On a Run-Matrix screen, complete the entry line as:
$$5! - 3! + \frac{50!}{49!}$$
Then press EXE.
Note: The factorial symbol is located by pressing OPTN then F6 to find more options. Then press F3 for PROB and then F1 for x!

DISPLAY/WRITE

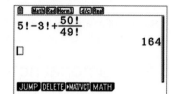

2. The answer appears on the screen.   164

3.2.2 Combinations or selections

Consider choosing r objects from a set of n distinct objects and placing them in a row.

The number of arrangements is shown in the box table.

n	$n-1$	$n-2$	… … … … … … … … … … … … … …	$n-r+1$

↑   ↑   ↑   ↑

1st   2nd   3rd rth object

The number of arrangements equals $n(n-1)(n-2)\ldots(n-r+1)$.

This can be expressed using factorial notation as:

$$n(n-1)(n-2)\ldots(n-r+1) = n(n-1)(n-2)\ldots(n-r+1) \times \frac{(n-r)(n-r-1) \times \ldots 2 \times 1}{(n-r)(n-r-1) \times \ldots 2 \times 1}$$
$$= \frac{n!}{(n-r)!}$$

Now consider the situation where order is unimportant. This is the situation where the selection AB is the same as the selection BA. For example, in a mixed doubles tennis match, the entry Alan and Bev is no different to the entry Bev and Alan; they are the same entry.

The number of combinations of r objects from a set of n distinct objects is calculated by counting the number of arrangements of the objects r at a time and then dividing that by the number of ways each group of these r objects can rearrange between themselves. This is done in order to cancel out counting these as different selections.

The arrangements of r objects between themselves is shown in the box table.

r	$r-1$	$r-2$	....	2	1

This can be expressed using factorial notation as:

$$r! = r(r-1)(r-2)\ldots\ldots(2)(1)$$

The number of combinations of r objects from a set of n distinct objects is therefore:

$$\left(\frac{n!}{(n-r)!}\right) \div r! = \frac{n!}{(n-r)!\,r!}.$$

This symbol for the number of ways of choosing r objects from a total of n objects is ^nC_r or $\binom{n}{r}$.

Combinations

The number of combinations of r objects from a total of n objects is

$$^nC_r = \frac{n!}{r!\,(n-r)!}, 0 \leq r \leq n$$

where r and n are non-negative integers.

Drawing on that knowledge, we have the following:

- $^nC_0 = 1 = {}^nC_n$, there being only one way to choose none or all of the n objects.
- $^nC_1 = n$, there being n ways to choose one object from a group of n objects.
- $^nC_r = {}^nC_{n-r}$, since choosing r objects must leave behind a group of $(n-r)$ objects and vice versa.
- $0! = 1$, by definition.

The formula is always used for calculations to determine the number of combinations or selections. Most calculators have a ^nC_r key to assist with the evaluation when the figures become large.

WORKED EXAMPLE 2 Evaluating combinatorial notation

Evaluate $\binom{8}{3}$.

THINK

1. Apply the formula.

2. Write the largest factorial in terms of the next largest factorial and simplify.

WRITE

$$\binom{n}{r} = \frac{n!}{r!\,(n-r)!}$$

Let $n = 8$ and $r = 3$.

$$\binom{8}{3} = \frac{8!}{3!\,(8-3)!}$$

$$= \frac{8!}{3!\,5!}$$

$$= \frac{8 \times 7 \times 6 \times \cancel{5!}}{3!\,\cancel{5!}}$$

$$= \frac{8 \times 7 \times 6}{3!}$$

3. Calculate the answer.

$$= \frac{8 \times 7 \times \cancel{6}}{\cancel{3} \times \cancel{2} \times 1}$$
$$= 8 \times 7$$
$$= 56$$

TI	THINK	WRITE
1.	On a Calculator page, press MENU, then select: 5: Probability 3: Combinations. Complete the entry line as: nCr (8, 3) then press ENTER.	
2.	The answer appears on the screen.	56

CASIO	THINK	WRITE
a.1.	On a Run-Matrix screen, type '8', press OPTN, then press F6 to scroll across to more menu options. Select PROB by pressing F3 then select nCr by pressing F3. Type '3' then press EXE.	
2.	The answer appears on the screen.	56

Combinations can be applied to practical situations such as selecting a group of students to form a committee. The following worked example demonstrates this concept.

WORKED EXAMPLE 3 Applying combinations

A committee of 5 students is to be chosen from 7 boys and 4 girls.
a. Calculate how many committees can be formed.
b. Determine how many of the committees contain exactly 2 boys and 3 girls.

THINK	WRITE
a. 1. As there is no restriction, choose the committee from the total number of students.	**a.** There are 11 students in total from whom 5 students are to be chosen. This can be done in $^{11}C_5$ ways.
2. Use the formula $^nC_r = \dfrac{n!}{r! \times (n-r)!}$ to calculate the answer.	$^{11}C_5 = \dfrac{11!}{5! \times (11-5)!}$ $= \dfrac{11!}{5! \times 6!}$ $= \dfrac{11 \times 10 \times 9 \times 8 \times 7 \times 6!}{5! \times 6!}$ $= \dfrac{11 \times \cancel{10}^2 \times 9 \times \cancel{8}^3 \times 7}{\cancel{5} \times 4 \times 3 \times 2 \times 1}$ $= 462$ There are 462 possible committees.
b. 1. Select the committee to satisfy the given restriction.	**b.** The 2 boys can be chosen from the 7 boys available in $^7C_2$ ways. The 3 girls can be chosen from the 4 girls available in $^4C_3$ ways.
2. Use the multiplication principle to form the total number of committees. *Note*: The upper numbers sum to the total available, $7 + 4 = 11$, and the lower numbers sum to the number that must be on the committee, $2 + 3 = 5$.	The total number of committees that contain two boys and three girls is $^7C_2 \times {}^4C_3$.

3.2.3 Pascal's triangle

Pascal's triangle contains many fascinating patterns. Each row from row 1 onwards begins and ends with '1'. Each other number along a row is formed by adding the two terms to its left and right from the preceding row.

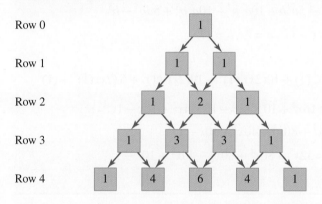

The numbers in each row are called **binomial coefficients**.

The number 1 in row 0 is the coefficient of the expansion of $(a + b)^0$.

$$(a + b)^0 = 1$$

The numbers 1, 1 in row 1 are the coefficients of the expansion of $(a + b)^1$.

$$(a + b)^1 = 1a + 1b$$

The numbers 1, 2, 1 in row 2 are the coefficients of the terms in the expansion of $(a + b)^2$.

$$(a + b)^2 = 1a^2 + 2ab + 1b^2$$

The numbers 1, 3, 3, 1 in row 3 are the coefficients of the terms in the expansion of $(a + b)^3$.

$$(a + b)^3 = 1a^3 + 3a^2b + 3ab^2 + 1b^3$$

To expand $(a + b)^4$ we would use the binomial coefficients 1, 4, 6, 4, 1 from row 4 to obtain:

$$(a + b)^4 = 1a^4 + 4a^3b + 6a^2b^2 + 4ab^3 + 1b^4$$
$$= a^4 + 4a^3b + 6a^2b^2 + 4ab^3 + b^4$$

Notice that the powers of a decrease by 1 as the powers of b increase by 1, with the sum of the powers of a and b always totalling 4 for each term in the expansion of $(a+b)^4$.

For the expansion of $(a-b)^4$ the signs would alternate:

$$(a-b)^4 = a^4 - 4a^3b + 6a^2b^2 - 4ab^3 + b^4$$

By extending Pascal's triangle, higher powers of such binomial expressions can be expanded.

WORKED EXAMPLE 4 Using Pascal's triangle in binomial expansions

Form the rule for the expansion of $(a-b)^5$ and hence expand $(2x-1)^5$.

THINK	WRITE
1. Choose the row in Pascal's triangle that contains the required binomial coefficients.	For $(a-b)^5$, the power of the binomial is 5. Therefore, the binomial coefficients are in row 5. The binomial coefficients are: $1, 5, 10, 10, 5, 1$.
2. Write down the required binomial expansion.	Alternate the signs: $(a-b)^5 = a^5 - 5a^4b + 10a^3b^2 - 10a^2b^3 + 5ab^4 - b^5$
3. State the values to substitute in place of a and b.	To expand $(2x-1)^5$, $a=2x$, $b=1$.
4. Write down the expansion.	$(2x)^5 - 5(2x)^4(1) + 10(2x)^3(1)^2 - 10(2x)^2(1)^3 + 5(2x)(1)^4 - (1)^5$
5. Evaluate the coefficients and write the answer.	$= 32x^5 - 5 \times 16x^4 + 10 \times 8x^3 - 10 \times 4x^2 + 10x - 1$ $= 32x^5 - 80x^4 + 80x^3 - 40x^2 + 10x - 1$ $\therefore (2x-1)^5 = 32x^5 - 80x^4 + 80x^3 - 40x^2 + 10x - 1$

3.2.4 Formula for binomial coefficients

Each of the terms in the rows of Pascal's triangle can be expressed using factorial notation.

For example, row 3 contains the coefficients 1, 3, 3, 1.

These can be written as $\dfrac{3!}{0! \times 3!}, \dfrac{3!}{1! \times 2!}, \dfrac{3!}{2! \times 1!}, \dfrac{3!}{3! \times 0!}$.

(Remember that 0! was defined to equal 1.)

The coefficients in row 5 (1, 5, 10, 10, 5, 1) can be written as:

$$\frac{5!}{0! \times 5!}, \frac{5!}{1! \times 4!}, \frac{5!}{2! \times 3!}, \frac{5!}{3! \times 2!}, \frac{5!}{4! \times 1!}, \frac{5!}{5! \times 0!}$$

The third term of row 4 would equal $\dfrac{4!}{2! \times 2!}$ and so on.

The $(r+1)$th term of row n would equal $\dfrac{n!}{r! \times (n-r)!}$.

This is normally written using the notations ^nC_r or $\dbinom{n}{r}$.

These expressions for the binomial coefficients are referred to as **combinatoric coefficients**. They occur frequently in other branches of mathematics, including probability theory.

Combinatoric coefficients

$$\binom{n}{r} = \frac{n!}{r!\,(n-r)!} = {}^nC_r$$

where $r \leq n$ and r and n are non-negative whole numbers.

WORKED EXAMPLE 5 Identifying terms in Pascal's triangle

Identify and evaluate the 10th term in row 13 of Pascal's triangle.

THINK	WRITE
1. Write down the values of n and r. Remember, numbering starts at 0.	$n = 13$ and $r = 9$
2. Write in ${}^nC_r$ notation.	The 10th term in row 13 is ${}^{13}C_9$ or $\binom{13}{9}$.
3. Evaluate using formula.	$\binom{13}{9} = \dfrac{13!}{9! \times 4!}$ $= \dfrac{13 \times 12 \times 11 \times 10 \times \cancel{9!}}{\cancel{9!} \times 4 \times 3 \times 2 \times 1}$ $= \dfrac{13 \times 11 \times 10}{2}$ $= 715$
4. Write the answer.	The 10th term in row 13 is 715.

3.2.5 Pascal's triangle with combinatoric coefficients

Pascal's triangle can now be expressed using this notation.

Row 0 $\qquad\qquad\qquad\qquad\qquad \binom{0}{0}$

Row 1 $\qquad\qquad\qquad\qquad \binom{1}{0} \qquad\quad \binom{1}{1}$

Row 2 $\qquad\qquad\qquad \binom{2}{0} \qquad \binom{2}{1} \qquad \binom{2}{2}$

Row 3 $\qquad\qquad \binom{3}{0} \qquad \binom{3}{1} \qquad \binom{3}{2} \qquad \binom{3}{3}$

Row 4 $\qquad \binom{4}{0} \qquad \binom{4}{1} \qquad \binom{4}{2} \qquad \binom{4}{3} \qquad \binom{4}{4}$

Binomial expansions can be expressed using this notation for each of the binomial coefficients.

The expansion $(a+b)^3 = \binom{3}{0} a^3 + \binom{3}{1} a^2 b + \binom{3}{2} ab^2 + \binom{3}{3} b^3$.

Note the following patterns:

- $\binom{n}{0} = 1 = \binom{n}{n}$ (the start and end of each row of Pascal's triangle)
- $\binom{n}{1} = n = \binom{n}{n-1}$ (the second from the start and the second from the end of each row)
- $\binom{n}{r} = \binom{n}{n-r}$.

WORKED EXAMPLE 6 Evaluating combinatoric coefficients

a. Calculate the value of $\binom{7}{5}$.

b. Verify that $\binom{16}{5} = \binom{16}{11}$.

THINK	WRITE
a. 1. Use the formula.	a. $\binom{7}{5} = \dfrac{7!}{(7-5)!\,5!}$
	$= \dfrac{7 \times 6}{2!}$
	$= 21$
2. Verify using technology.	$^{7}C_5 = 21$
b. 1. Use the formula to express one side.	b. $\binom{16}{11} = \dfrac{16!}{(16-11)!\,11!}$
	$= \dfrac{16!}{5!\,11!}$
	$= \dfrac{16!}{11!\,5!}$
	$= \binom{16}{5}$
2. Verify your answer by evaluating each side.	$\binom{16}{5} = 4368$
	$\binom{16}{11} = 4368$

Exercise 3.2 Pascal's triangle and the binomial expansions

learn on

3.2 Exercise	3.2 Exam practice on

Simple familiar	Complex familiar	Complex unfamiliar
1, 2, 3, 4, 5, 6, 7, 8, 9, 10, 11, 12	13, 14, 15, 16	17, 18, 19, 20

These questions are even better in jacPLUS!
- Receive immediate feedback
- Access sample responses
- Track results and progress

Find all this and MORE in jacPLUS ▶

Simple familiar

1. Evaluate the following.

   a. $3!$ b. $4!$ c. $5!$ d. $2!$ e. $0! \times 1!$ f. $\dfrac{7!}{6!}$

2. Rewrite the following using factorial notation.

   a. $7 \times 6 \times 5 \times 4 \times 3 \times 2 \times 1$ b. $8 \times 7 \times 6 \times 5 \times 4 \times 3 \times 2 \times 1$

   c. $8 \times 7!$ d. $9 \times 8!$

3. **WE1** Evaluate $6! + 4! - \dfrac{10!}{9!}$.

4. Evaluate the following.

   a. $\dfrac{26!}{24!}$ b. $\dfrac{42!}{43!}$ c. $\dfrac{49!}{50!} \div \dfrac{69!}{70!}$ d. $\dfrac{11! + 10!}{11! - 10!}$

5. **WE2** Evaluate $\begin{pmatrix} 7 \\ 4 \end{pmatrix}$.

6. Evaluate the following.

   a. $\begin{pmatrix} 5 \\ 2 \end{pmatrix}$ b. $\begin{pmatrix} 5 \\ 3 \end{pmatrix}$ c. $\begin{pmatrix} 12 \\ 12 \end{pmatrix}$ d. $^{20}C_3$ e. $\begin{pmatrix} 7 \\ 0 \end{pmatrix}$ f. $\begin{pmatrix} 13 \\ 10 \end{pmatrix}$

7. **WE3** A committee of 5 students is to be chosen from 6 boys and 8 girls.

   a. How many committees can be formed?
   b. How many of the committees contain exactly 2 boys and 3 girls?

8. A panel of 8 is to be selected from a group of 8 men and 10 women. Determine how many panels can be formed if:

   a. there are no restrictions
   b. there are 5 men and 3 women on the panel

9. **WE4** Copy and complete the following table by making use of Pascal's triangle.

Binomial power	Expansion	Number of terms in the expansion	Sum of indices in each term
$(x+a)^2$			
$(x+a)^3$			
$(x+a)^4$			
$(x+a)^5$			

10. Expand the following using the binomial coefficients from Pascal's triangle.
    a. $(x+4)^5$
    b. $(x-4)^5$
    c. $(xy+2)^5$

11. Expand the following using the binomial coefficients from Pascal's triangle.
    a. $(3x-5y)^4$
    b. $\left(3-x^2\right)^4$

12. **WE3&6** a. Write down the notation for the 4th coefficient in the 7th row of Pascal's triangle.

    b. Evaluate $\dbinom{9}{6}$.

    c. Show that $\dbinom{18}{12} = \dbinom{18}{6}$.

    d. Show that $\dbinom{15}{7} + \dbinom{15}{6} = \dbinom{16}{7}$.

Complex familiar

13. Expand and simplify $(1+x)^6 - (1-x)^6$.

14. Determine an algebraic expression for $\dbinom{n}{2}$ and use this to evaluate $\dbinom{21}{2}$.

15. Simplify the following.
    a. $\dbinom{n}{3}$
    b. $\dbinom{n}{n-3}$
    c. $\dbinom{n+3}{3}$
    d. $\dbinom{2n+1}{2n-1}$
    e. $\dbinom{n}{2} + \dbinom{n}{3}$
    f. $\dbinom{n+1}{3}$

16. For each of the following, determine the term specified in the expansion.
    a. The 3rd term in $(2w-3)^5$
    b. The 5th term in $\left(3-\dfrac{1}{b}\right)^7$
    c. The constant term in $\left(y-\dfrac{3}{y}\right)^4$

Complex unfamiliar

17. If the coefficient of x^2y^2 in the expansion of $(x+ay)^4$ is 3 times the coefficient of x^2y^3 in the expansion of $\left(ax^2-y\right)^4$, determine the value of a.

18. Determine the coefficient of x in the expansion of $(1+2x)(1-x)^5$.

19. Expand $(x+1)^5 - (x+1)^4$ and hence show that $(x+1)^5 - (x+1)^4 = x(x+1)^4$.

20. A section of Pascal's triangle is shown. Determine the values of a, b and c.

    $$\begin{array}{ccccccc} & & 45 & & & a & \\ & b & & 165 & & & 330 \\ & & 220 & & & c & \end{array}$$

Fully worked solutions for this chapter are available online.

LESSON
3.3 The binomial theorem

SYLLABUS LINKS

- Use the binomial theorem $(x+y)^n = x^n + \binom{n}{1}x^{n-1}y + \ldots + \binom{n}{r}x^{n-r}y^r + \ldots + y^n$ to expand expressions, e.g. $(2x-1)^3$.

Source: Mathematical Methods Senior Syllabus 2024 © State of Queensland (QCAA) 2024; licensed under CC BY 4.0.

3.3.1 The binomial theorem

The binomial coefficients in row n of Pascal's triangle can be expressed as $\binom{n}{0}, \binom{n}{1}, \binom{n}{2}, \ldots \binom{n}{n}$, and hence the expansion of $(x+y)^n$ can be formed.

> **The binomial theorem**
>
> **The binomial theorem gives the rule for the expansion of $(x+y)^n$ as:**
>
> $$(x+y)^n = x^n + \binom{n}{1}x^{n-1}y + \ldots + \binom{n}{r}x^{n-r}y^r + \ldots + y^n$$
>
> **since**
>
> $$\binom{n}{0} = 1 = \binom{n}{n}.$$

Features of the binomial theorem formula for the expansion of $(x+y)^n$

- In each successive term, the powers of x decrease by 1 as the powers of y increase by 1.
- For each term, the powers of x and y add up to n.
- For the expansion of $(x-y)^n$, the signs of each term alternate $+ - + - + \ldots$ with every even term assigned the $-$ sign and every odd term assigned the $+$ sign.

WORKED EXAMPLE 7 Expanding using the binomial theorem

Use the binomial theorem to expand $(3x+2)^4$.

THINK	WRITE
1. Write out the expansion using the binomial theorem. *Note:* There should be 5 terms in the expansion.	$(3x+2)^4 = (3x)^4 + \binom{4}{1}(3x)^3(2) + \binom{4}{2}(3x)^2(2)^2 + \binom{4}{3}(3x)(2)^3 + (2)^4$
2. Evaluate the binomial coefficients.	$= (3x)^4 + 4 \times (3x)^3(2) + 6 \times (3x)^2(2)^2 + 4 \times (3x)(2)^3 + (2)^4$
3. Complete the calculations and write the answer.	$= 81x^4 + 4 \times 27x^3 \times 2 + 6 \times 9x^2 \times 4 + 4 \times 3x \times 8 + 16$
	$\therefore (3x+2)^4 = 81x^4 + 216x^3 + 216x^2 + 96x + 16$

3.3.2 Using the binomial theorem

The binomial theorem is very useful for expanding $(x+y)^n$. However, for powers $n \geq 7$ the calculations can become quite tedious. If a particular term is of interest, forming an expression for the general term of the expansion is an easier option.

The general term of the binomial theorem

Consider the terms of the expansion:

$$(x+y)^n = x^n + \binom{n}{1} x^{n-1}y + \binom{n}{2} x^{n-2}y^2 + \ldots + \binom{n}{r} x^{n-r}y^r + \ldots + y^n$$

$$\text{Term 1: } t_1 = \binom{n}{0} x^n y^0$$

$$\text{Term 2: } t_2 = \binom{n}{1} x^{n-1}y^1$$

$$\text{Term 3: } t_3 = \binom{n}{2} x^{n-2}y^2$$

Following the pattern gives:

$$\text{Term } (r+1): t_{r+1} = \binom{n}{r} x^{n-r}y^r$$

The general term of the binomial theorem

For the expansion of $(x+y)^n$, the general term is $t_{r+1} = \binom{n}{r} x^{n-r}y^r$.

For the expansion of $(x-y)^n$, the general term can be expressed as

$t_{r+1} = \binom{n}{r} x^{n-r}(-y)^r$.

The general term formula enables a particular term to be evaluated without the need to carry out the full expansion. As there are $(n+1)$ terms in the expansion, if the middle term is sought, there will be two middle terms if n is odd and one middle term if n is even.

WORKED EXAMPLE 8 Determining a term in the binomial expansion

Determine the fifth term in the expansion of $\left(\dfrac{x}{2} - \dfrac{y}{3} \right)^9$.

THINK

1. State the general term formula of the expansion.

WRITE

$\left(\dfrac{x}{2} - \dfrac{y}{3} \right)^9$

The $(r+1)$th term is $t_{r+1} = \binom{n}{r} \left(\dfrac{x}{2} \right)^{n-r} \left(-\dfrac{y}{3} \right)^r$.

Since the power of the binomial is 9, $n = 9$.

$\therefore t_{r+1} = \binom{9}{r} \left(\dfrac{x}{2} \right)^{9-r} \left(-\dfrac{y}{3} \right)^r$

2. Choose the value of r for the required term.

For the fifth term, t_5:
$$r + 1 = 5$$
$$r = 4$$

$$t_5 = \binom{9}{4} \left(\frac{x}{2}\right)^{9-4} \left(-\frac{y}{3}\right)^4$$

$$= \binom{9}{4} \left(\frac{x}{2}\right)^5 \left(-\frac{y}{3}\right)^4$$

3. Evaluate to obtain the required term.

$$= 126 \times \frac{x^5}{32} \times \frac{y^4}{81}$$

$$= \frac{7x^5 y^4}{144}$$

3.3.3 Identifying a term in the binomial expansion

The general term can also be used to determine which term has a specified property, such as the term independent of x or the term containing a particular power of x.

WORKED EXAMPLE 9 Determining a particular term

Identify which term in the expansion of $\left(3x^2 - 4\right)^7$ would contain x^8 and express the coefficient of x^8 as an integer.

THINK	WRITE
1. Write down the general term for this expansion.	$\left(3x^2 - 4\right)^7$ General term, $t_{r+1} = \binom{7}{r} \left(3x^2\right)^{7-r} (-4)^r$
2. Rearrange the general term by grouping numerical parts and algebraic parts together.	$t_{r+1} = \binom{7}{r} (3)^{7-r} (-4)^r (x^2)^{7-r}$ $= \binom{7}{r} (3)^{7-r} (-4)^r x^{(14-2r)}$
3. Determine the value of r to form the given power of x.	For x^8: $14 - 2r = 8$ $\therefore r = 3$
4. Identify which term is required.	Hence, it is the fourth term that contains x^8.
5. State the required coefficient.	The coefficient is $\binom{7}{3} (3)^4 (-4)^3$.
6. Evaluate using technology and write the answer.	The coefficient of $x^8 = 35 \times 81 \times (-64)$. Hence, the coefficient of x^8 is $-127\,545$.

Exercise 3.3 The binomial theorem

learn on

3.3 Exercise	**3.3 Exam practice** on

Simple familiar	Complex familiar	Complex unfamiliar
1, 2, 3, 4, 5, 6, 7, 8, 9, 10, 11, 12, 13	14, 15, 16, 17	18, 19, 20

These questions are even better in jacPLUS!
- Receive immediate feedback
- Access sample responses
- Track results and progress

Find all this and MORE in jacPLUS ▶

Simple familiar

1. **WE7** Use the binomial theorem to expand $(2x + 3)^5$.

2. Use the binomial theorem to expand $(x - 2)^7$.

3. Expand and simplify each of the following.
   a. $(x + y)^3$
   b. $(a + 2)^4$
   c. $(m - 3)^4$
   d. $(2 - x)^5$

4. Expand and simplify each of the following.
   a. $\left(1 - \dfrac{2}{x}\right)^3$
   b. $\left(1 + \dfrac{p}{q}\right)^4$
   c. $\left(3 - \dfrac{m}{2}\right)^4$
   d. $\left(2x - \dfrac{1}{x}\right)^3$

5. Expand the following.
   a. $(x + 1)^5$
   b. $(2 - x)^5$
   c. $(2x + 3y)^6$

6. Expand the following.
   a. $\left(\dfrac{x}{2} + 2\right)^7$
   b. $\left(x - \dfrac{1}{x}\right)^8$
   c. $\left(x^2 + 1\right)^{10}$

7. **WE8** Determine the fourth term in the expansion of $\left(\dfrac{x}{3} - \dfrac{y}{2}\right)^7$.

8. For each of the following, determine the term specified in the expansion.
   a. The 3rd term in $(2w - 3)^5$
   b. The 5th term in $\left(3 - \dfrac{1}{b}\right)^7$

9. Determine the constant term in $\left(y - \dfrac{3}{y}\right)^4$.

10. Determine the term independent of x in the expansion of $\left(x + \dfrac{2}{x}\right)^6$.

11. Obtain each of the following terms.
    a. The fourth term in the expansion of $(5x + 2)^6$
    b. The third term in the expansion of $\left(3x^2 - 1\right)^6$
    c. The middle term(s) in the expansion of $(x + 2y)^7$

12. **WE9** Identify which term in the expansion of $\left(8 - 3x^2\right)^{12}$ would contain x^8 and express the coefficient of x^8 as a product of its prime factors.

13. Identify which term in the expansion of $\left(4 + 3x^3\right)^8$ would contain x^{15} and express the coefficient of x^{15} as a product of its prime factors.

14. Determine the middle term in the expansion of $\left(x^2 + \dfrac{y}{2} \right)^{10}$.

15. Determine the term independent of x in the expansion of $\left(x^2 + \dfrac{1}{x^3} \right)^{10}$.

16. Obtain the coefficient of x^6 in the expansion of $\left(1 - 2x^2 \right)^9$.

17. Express the coefficient of x^5 in the expansion of $(3 + 4x)^{11}$ as a product of its prime factors.

Complex unfamiliar

18. The ratio of the coefficients of the 4th and 5th terms in the expansion of $(ax + 2y)^5$ is $3 : 1$. Determine the value of a.

19. The first 3 terms in the expansion of $(1 + kx)^n$ are 1, $2x$ and $\dfrac{3}{2}x^2$. Determine the values of k and n.

20. Determine the values of a and b given $\left(3 + \sqrt{2} \right)^9 = a + b\sqrt{2}$.

Fully worked solutions for this chapter are available online.

LESSON
3.4 Polynomials and cubic functions

SYLLABUS LINKS

- Identify the coefficients and the degree of a polynomial.
- Expand quadratic and cubic polynomials from factors.

Source: Mathematical Methods Senior Syllabus 2024 © State of Queensland (QCAA) 2024; licensed under CC BY 4.0.

A **polynomial** is an algebraic expression in which the power of the variable is a positive whole number. For example, $3x^2 + 5x - 1$ is a quadratic polynomial in the variable x, but $\dfrac{3}{x^2} + 5x - 1$, that is $3x^{-2} + 5x - 1$, is not a polynomial because of the $3x^{-2}$ term. Similarly, $\sqrt{3}x + 5$ is a linear polynomial, but $3\sqrt{x} + 5$, that is $3x^{\frac{1}{2}} + 5$, is not a polynomial because the power of the variable x is not a whole number. Note that the coefficients of x, x^2 etc. can be positive or negative integers, or rational or irrational real numbers.

3.4.1 Classification of polynomials

- The **degree of a polynomial** is the highest power of the variable.
  For example, linear polynomials have degree 1, quadratic polynomials have degree 2 and cubic polynomials have degree 3.
- The **leading term** is the term containing the highest power of the variable.
- If the coefficient of the leading term is 1, then the polynomial is said to be monic.
- The **constant term** is the term that does not contain the variable.

General form of a polynomial

A polynomial of degree n has the form $a_n x^n + a_{n-1} x^{n-1} + \ldots + a_1 x + a_0$ where $n \in N$ and the coefficients $a_n, a_{n-1}, \ldots a_1, a_0 \in R$. The leading term is $a_n x^n$ and the constant term is a_0.

WORKED EXAMPLE 10 Classifying polynomials

Select the polynomials from the following list of algebraic expressions and for these polynomials, state the degree, the coefficient of the leading term, the constant term and the type of coefficients.

A. $5x^3 + 2x^2 - 3x + 4$

B. $5x - x^3 + \dfrac{x^4}{2}$

C. $4x^5 + 2x^2 + 7x^{-3} + 8$

THINK	WRITE
1. Check the powers of the variable x in each algebraic expression.	A and B are polynomials since all the powers of x are positive integers. C is not a polynomial due to the $7x^{-3}$ term.
2. For polynomial A, state the degree, the coefficient of the leading term and the constant term.	Polynomial A: the leading term of $5x^3 + 2x^2 - 3x + 4$ is $5x^3$. Therefore, the degree is 3 and the coefficient of the leading term is 5. The constant term is 4.
3. Classify the coefficients of polynomial A as elements of a subset of R.	The coefficients in polynomial A are integers. Therefore, A is a polynomial over Z.
4. For polynomial B, state the degree, the coefficient of the leading term and the constant term.	Polynomial B: the leading term of $5x - x^3 + \dfrac{x^4}{2}$ is $\dfrac{x^4}{2}$. Therefore, the degree is 4 and the coefficient of the leading term is $\dfrac{1}{2}$. The constant term is 0.
5. Classify the coefficients of polynomial B as elements of a subset of R.	The coefficients in polynomial B are rational numbers. Therefore, B is a polynomial over Q.

3.4.2 Polynomial notation

- The polynomial in variable x is often referred to as $P(x)$.
- The value of the polynomial $P(x)$ when $x = a$ is written as $P(a)$.
- $P(a)$ is evaluated by substituting a in place of x in the $P(x)$ expression.

WORKED EXAMPLE 11 Evaluating polynomials

a. If $P(x) = 5x^3 + 2x^2 - 3x + 4$, calculate $P(-1)$.
b. If $P(x) = ax^2 - 2x + 7$ and $P(4) = 31$, obtain the value of a.

THINK	WRITE
a. Substitute -1 in place of x and evaluate.	a. $P(x) = 5x^3 + 2x^2 - 3x + 4$ $P(-1) = 5(-1)^3 + 2(-1)^2 - 3(-1) + 4$ $= -5 + 2 + 3 + 4$ $= 4$

b. 1. Determine an expression for $P(4)$ by substituting 4 in place of x, and then simplify.

2. Equate the expression for $P(4)$ with 31.

3. Solve for a.

b.
$$P(x) = ax^2 - 2x + 7$$
$$P(4) = a(4)^2 - 2(4) + 7$$
$$= 16a - 1$$

$$P(4) = 31$$
$$\Rightarrow 16a - 1 = 31$$

$$16a = 32$$
$$a = 2$$

TI	THINK	WRITE

a. 1. On a Calculator page, press MENU, then select:
1: Actions
1: Define
Complete the entry line as:
Define $p(x) =$ $5x^3 + 2x^2 - 3x + 4$ then press ENTER.

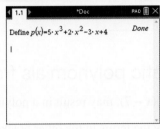

2. Complete the next entry line as:
$p(-1)$
then press ENTER.

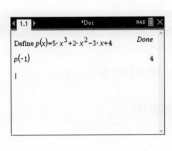

3. The answer appears on the screen.

$P(-1) = 4$.

CASIO	THINK	WRITE

a. 1. On a Graph screen, complete the entry line for $y1$ as:
$y1 = 5x^3 + 2x^2 - 3x + 4$
then press EXE.

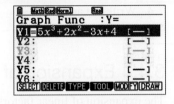

2. On a Run-Matrix screen, press VARS, then select GRAPH by pressing F4. Select Y by pressing F1 and complete the entry line as:
$Y1(-1)$
then press EXE.

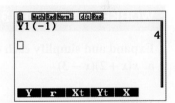

3. The answer appears on the screen.

$P(-1) = 4$.

3.4.3 Identity of polynomials

If two polynomials are **identically equal**, then the coefficients of like terms are equal. **Equating coefficients** means that if $ax^2 + bx + c \equiv 2x^2 + 5x + 7$, then $a = 2$, $b = 5$ and $c = 7$. The identically equal symbol '$\equiv$' means the equality holds for all values of x. For convenience, however, we shall replace this symbol with the equality symbol '$=$' in working steps.

WORKED EXAMPLE 12 Equating coefficients

Calculate the values of a, b and c so that $x(x - 7) \equiv a(x - 1)^2 + b(x - 1) + c$.

THINK

1. Expand each bracket and express both sides of the equality in expanded polynomial form.

WRITE

$$x(x - 7) \equiv a(x - 1)^2 + b(x - 1) + c$$
$$\therefore x^2 - 7x \equiv a(x^2 - 2x + 1) + bx - b + c$$
$$\therefore x^2 - 7x \equiv ax^2 + (-2a + b)x + (a - b + c)$$

2. Equate the coefficients of like terms.

Equate the coefficients.
$$x^2: 1 = a \qquad [1]$$
$$x: -7 = -2a + b \qquad [2]$$
$$\text{Constant: } 0 = a - b + c \qquad [3]$$

3. Solve the system of simultaneous equations.

Since $a = 1$, substitute $a = 1$ into equation [2].
$$-7 = -2(1) + b$$
$$b = -5$$
Substitute $a = 1$ and $b = -5$ into equation [3].
$$0 = 1 - (-5) + c$$
$$c = -6$$

4. Write the answer.

$$\therefore a = 1, b = -5, c = -6$$

3.4.4 Expansion of cubic and quadratic polynomials from factors

The expansion of linear factors, for example $(x + 1)(x + 2)(x - 7)$, may result in a polynomial.

To expand, each term in one bracket must be multiplied by the terms in the other brackets, then like terms collected to simplify the expression.

WORKED EXAMPLE 13 Expanding and simplifying polynomials

Expand and simplify each of the following.
a. $x(x + 2)(x - 3)$

b. $(x - 1)(x + 5)(x + 2)$

THINK

WRITE

a. 1. Write the expression.

a. $x(x + 2)(x - 3)$

2. Expand the last two linear factors.

$$= x(x^2 - 3x + 2x - 6)$$
$$= x(x^2 - x - 6)$$

3. Multiply the expression in the grouping symbols by x.

$$= x^3 - x^2 - 6x$$

b. 1. Write the expression.

b. $(x - 1)(x + 5)(x + 2)$

2. Expand the last two linear factors.

$$= (x - 1)(x^2 + 2x + 5x + 10)$$

3. Multiply the expression in the second pair of grouping symbols by x and then by -1.

$$= (x - 1)(x^2 + 7x + 10)$$
$$= x^3 + 7x^2 + 10x - x^2 - 7x - 10$$

4. Collect like terms.

$$= x^3 + 6x^2 + 3x - 10$$

3.4.5 Operations on polynomials

The addition, subtraction and multiplication of two or more polynomials results in another polynomial.

For example, if $P(x) = x^2$ and $Q(x) = x^3 + x^2 - 1$, then $P(x) + Q(x) = x^3 + 2x^2 - 1$, a polynomial of degree 3; $P(x) - Q(x) = -x^3 + 1$, a polynomial of degree 3; and $P(x)Q(x) = x^5 + x^4 - x^2$, a polynomial of degree 5.

WORKED EXAMPLE 14 Using operations with polynomials

Given $P(x) = 3x^3 + 4x^2 + 2x + m$ and $Q(x) = 2x^2 + kx - 5$, determine the values of m and k for which $2P(x) - 3Q(x) = 6x^3 + 2x^2 + 25x - 25$.

THINK	WRITE
1. Form a polynomial expression for $2P(x) - 3Q(x)$ by collecting like terms together.	$2P(x) - 3Q(x)$ $= 2(3x^3 + 4x^2 + 2x + m) - 3(2x^2 + kx - 5)$ $= 6x^3 + 2x^2 + (4 - 3k)x + (2m + 15)$
2. Equate the two expressions for $2P(x) - 3Q(x)$.	Hence, $6x^3 + 2x^2 + (4 - 3k)x + (2m + 15)$ $\qquad = 6x^3 + 2x^2 + 25x - 25$
3. Calculate the values of m and k.	Equate the coefficients of x. $4 - 3k = 25$ $\quad k = -7$ Equate the constant terms. $2m + 15 = -25$ $\qquad m = -20$
4. Write the answer.	Therefore, $m = -20, k = -7$.

Exercise 3.4 Polynomials and cubic functions

learn on

3.4 Exercise	3.4 Exam practice on

These questions are even better in jacPLUS!
- Receive immediate feedback
- Access sample responses
- Track results and progress

Find all this and MORE in jacPLUS ▶

Simple familiar	Complex familiar	Complex unfamiliar
1, 2, 3, 4, 5, 6, 7, 8, 9, 10, 11, 12, 13, 14, 15, 16	17, 18, 19, 20	N/A

Simple familiar

1. **WE10** Select the polynomials from the following list of algebraic expressions and state their degree, the coefficient of the leading term, the constant term and the type of coefficients.

   A. $30x + 4x^5 - 2x^3 + 12$

   B. $\dfrac{3x^2}{5} - \dfrac{2}{x} + 1$

   C. $5.6 + 4x - 0.2x^2$

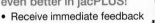
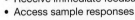
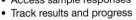

2. Identify which of the following expressions are polynomials. State their degree. For those that are not polynomials, state a reason why not.

   a. $7x^4 + 3x^2 + 5$

   b. $9 - \dfrac{5}{2}x - 4x^2 + x^3$

   c. $-9x^3 + 7x^2 + 11\sqrt{x} - \sqrt{5}$

   d. $\dfrac{6}{x^2} + 6x^2 + \dfrac{x}{2} - \dfrac{2}{x}$

3. Consider the following list of algebraic expressions.

   A. $3x^5 + 7x^4 - \dfrac{x^3}{6} + x^2 - 8x + 12$

   B. $9 - 5x^4 + 7x^2 - \sqrt{5}x + x^3$

   C. $\sqrt{4x^5} - \sqrt{5x^3} + \sqrt{3}x - 1$

   D. $2x^2(4x - 9x^2)$

   E. $\dfrac{x^6}{10} - \dfrac{2x^8}{7} + \dfrac{5}{3x^2} - \dfrac{7x}{5} + \dfrac{4}{9}$

   F. $(4x^2 + 3 + 7x^3)^2$

   a. Select the polynomials from the list and for each of the polynomials state:
      i. its degree
      iii. the leading term
      ii. the type of coefficients
      iv. the constant term.
   b. Give a reason why each of the remaining expressions is not a polynomial.

4. a. If $P(x) = -x^3 + 2x^2 + 5x - 1$, calculate $P(1)$.
   b. If $P(x) = 2x^3 - 4x^2 + 3x - 7$, calculate $P(-2)$.
   c. If $P(x) = 3x^3 - x^2 + 5$, calculate $P(3)$ and $P(-x)$.
   d. If $P(x) = x^3 + 4x^2 - 2x + 5$, calculate $P(-1)$ and $P(2a)$.

5. Given $P(x) = 2x^3 + 3x^2 + x - 6$, evaluate the following.
   a. $P(3)$
   b. $P(-2)$
   c. $P(1)$
   d. $P(0)$
   e. $P\left(-\dfrac{1}{2}\right)$
   f. $P(0.1)$

6. If $P(x) = x^2 - 7x + 2$, obtain expressions for the following.
   a. $P(a) - P(-a)$
   b. $P(1 + h)$
   c. $P(x + h) - P(x)$

7. a. **WE11** If $P(x) = 7x^3 - 8x^2 - 4x - 1$ calculate $P(2)$.
   b. If $P(x) = 2x^2 + kx + 12$ and $P(-3) = 0$, determine the value of k.

8. a. If $P(x) = ax^2 + 9x + 2$ and $P(1) = 3$, determine the value of a.
   b. Given $P(x) = -5x^2 + bx - 18$, calculate the value of b if $P(3) = 0$.
   c. Given $P(x) = -2x^3 + 3x^2 + kx - 10$, calculate the value of k if $P(-1) = -7$.
   d. If $P(x) = x^3 - 6x^2 + 9x + m$ and $P(0) = 2P(1)$, determine the value of m.

9. If $P(x) = -2x^3 + 9x + m$ and $P(1) = 2P(-1)$, determine the value of m.

10. a. If $P(x) = 4x^3 + kx^2 - 10x - 4$ and $P(1) = 15$, obtain the value of k.
    b. If $Q(x) = ax^2 - 12x + 7$ and $Q(-2) = -5$, obtain the value of a.
    c. If $P(x) = x^3 - 6x^2 + nx + 2$ and $P(2) = 3P(-1)$, obtain the value of n.
    d. If $Q(x) = -x^2 + bx + c$ and $Q(0) = 5$ and $Q(5) = 0$, obtain the values of b and c.

11. **WE12** Calculate the values of a, b and c so that $(2x + 1)(x - 5) \equiv a(x + 1)^2 + b(x + 1) + c$.

12. a. Determine the values of a and b so that $x^2 + 10x + 6 \equiv x(x + a) + b$.
    b. Express $8x - 6$ in the form $ax + b(x + 3)$.
    c. Express the polynomial $6x^2 + 19x - 20$ in the form $(ax + b)(x + 4)$.
    d. Obtain the values of a, b and c so that $x^2 - 8x = a + b(x + 1) + c(x + 1)^2$ for all values of x.

13. Express $(x+2)^3$ in the form $px^2(x+1) + qx(x+2) + r(x+3) + t$.

14. **WE13a** Expand and simplify each of the following.

    a. $x(x-9)(x+2)$

    b. $-3x(x-4)(x+4)$

15. **WE13b** Expand and simplify each of the following.

    a. $(x-2)(x+4)(x-5)$

    b. $(x+6)(x-1)(x+1)$

    c. $(x+2)(x-7)^2$

    d. $(x+1)(x-1)(x+1)$

16. If $P(x) = 2x^2 - 7x - 11$ and $Q(x) = 3x^4 + 2x^2 + 1$, determine each of the following, expressing the terms in descending power of x.

    a. $Q(x) - P(x)$

    b. $3P(x) + 2Q(x)$

    c. $P(x)Q(x)$

Complex familiar

17. **WE14** Given $P(x) = 4x^3 - px^2 + 8$ and $Q(x) = 3x^2 + qx - 7$, determine the values of p and q for which $P(x) + 2Q(x) = 4x^3 + x^2 - 8x - 6$.

18. If $x^3 - 2x^2 - 3x + 10 \equiv (x+2)\left(ax^2 + bx + c\right)$, determine the values of a, b and c.

19. Determine the values of a, b and c if

$$3x^3 + ax^2 + bx + c \equiv (x+1)\left(3x^2 - 12x + 10\right) - (x-2)(x+3)$$

20. If $P(x)$ is a polynomial of degree m and $Q(x)$ is a polynomial of degree n where $m > n$, state the degree of the following and justify your decisions.

    a. $P(x) + Q(x)$

    b. $P(x) - Q(x)$

    c. $P(x)Q(x)$

Fully worked solutions for this chapter are available online.

LESSON
3.5 Graphing cubic functions

Cubic functions can be represented in various forms. These include:

- $y = x^3$, the simplest cubic function
- $y = a(x-h)^3 + k$, the stationary point of inflection form
- $y = a(x - x_1)(x - x_2)(x - x_3)$, the intercept form
- $y = ax^3 + bx^2 + cx + d, a \neq 0$, the general form.

Since a cubic function may have up to three linear factors, its graph may have up to three x-intercepts. The shape of the graph is affected by the number of x-intercepts.

3.5.1 The graph of $y = x^3$ and transformations

The graph of the simplest cubic function has the equation $y = x^3$.

The 'maxi–min' point at the origin is sometimes referred to as a 'saddle point'. Formally, it is called a **stationary point of inflection** (or inflexion as a variation of spelling).

The key features of the graph of $y = x^3$ are as follows:
- $(0, 0)$ is a stationary point of inflection.
- The shape of the graph changes from concave down to concave up at the stationary point of inflection.
- There is only one x-intercept.
- As the values of x increase positively, the behaviour of the graph shows that its y-values also increase positively. This means that as $x \to \infty$, $y \to \infty$. This is read as 'as x approaches infinity, y approaches infinity'.
- As the values of x increase negatively, the behaviour of the graph shows that its y-values also increase negatively. This means that as $x \to -\infty$, $y \to -\infty$.
- The graph starts from below the x-axis and increases as x increases.

Once the basic shape is known, the graph can be dilated, reflected and translated in much the same way as the parabola $y = x^2$.

Dilation

The graph of $y = 4x^3$ will be narrower than the graph of $y = x^3$ due to the dilation factor of 4 from the x-axis. Similarly, the graph of $y = \frac{1}{4}x^3$ will be wider than the graph of $y = x^3$ due to the dilation factor of $\frac{1}{4}$ from the x-axis or parallel to the y-axis.

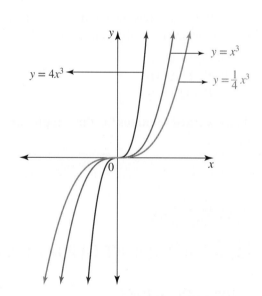

Reflection

The graph of $y = -x^3$ is the reflection of the graph of $y = x^3$ in the x-axis.

For the graph of $y = -x^3$, note that:
- as $x \to \infty$, $y \to -\infty$ and as $x \to -\infty$, $y \to \infty$
- the graph starts from above the x-axis and decreases as x increases
- at $(0, 0)$, the stationary point of inflection, the graph changes from concave up to concave down.

Note: Reflection in the y-axis has the same result, $y = (-x)^3 = -x^3$.

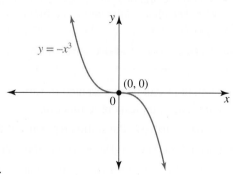

Translation

The graph of $y = x^3 + 4$ is obtained when the graph of $y = x^3$ is translated vertically upwards by 4 units. The stationary point of inflection is at the point $(0, 4)$.

The graph of $y = (x + 4)^3$ is obtained when the graph of $y = x^3$ is translated horizontally 4 units to the left. The stationary point of inflection is at the point $(-4, 0)$.

The transformations from the basic parabola $y = x^2$ are recognisable from the equation $y = a(x - h)^2 + k$, and the equation of the graph of $y = x^3$ can be transformed to a similar form.

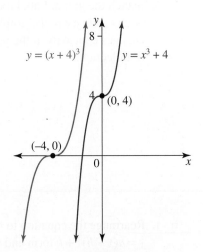

The graph of $y = a(x - h)^3 + k$

The key features of the graph of $y = a(x - h)^3 + k$ are as follows:
- There is a stationary point of inflection at (h, k).
- There is a change of concavity at the stationary point of inflection.
- if $a > 0$, the graph starts below the x-axis and increases, like $y = x^3$.
- if $a < 0$, the graph starts above the x-axis and decreases, like $y = -x^3$.
- The one x-intercept is found by solving $a(x - h)^3 + k = 0$.
- The y-intercept is found by substituting $x = 0$.

The graph of $y = 2(x + 1)^3 + 3$ illustrates these features.

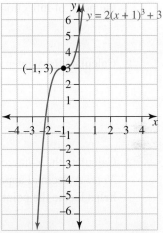

WORKED EXAMPLE 15 Sketching cubic functions

Sketch the graphs of the following.

a. $y = (x + 1)^3 + 8$

b. $y = 6 - \dfrac{1}{2}(x - 2)^3$

THINK	WRITE
a. 1. State the point of inflection.	a. $y = (x + 1)^3 + 8$ The point of inflection is $(-1, 8)$.
2. Calculate the y-intercept.	y-intercept: let $x = 0$. $y = (1)^3 + 8$ $\quad = 9$ $\Rightarrow (0, 9)$
3. Calculate the x-intercept.	x-intercept: let $y = 0$. $(x + 1)^3 + 8 = 0$ $\quad (x + 1)^3 = -8$ Take the cube root of both sides: $x + 1 = \sqrt[3]{-8}$ $x + 1 = -2$ $\quad\quad x = -3$ $\Rightarrow (-3, 0)$

4. Sketch the graph. Label the key points and ensure the graph changes concavity at the point of inflection.

The coefficient of x^3 is positive, so the graph starts below the x-axis and increases.

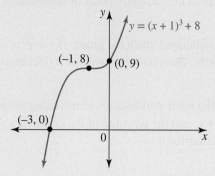

b. 1. Rearrange the equation to the $y = a(x-h)^3 + k$ form and state the point of inflection.

b. $y = 6 - \dfrac{1}{2}(x-2)^3$

$\quad = -\dfrac{1}{2}(x-2)^3 + 6$

Point of inflection: $(2, 6)$

2. Calculate the y-intercept.

y-intercept: let $x = 0$.

$y = -\dfrac{1}{2}(-2)^3 + 6$

$\quad = 10$

$\Rightarrow (0, 10)$

3. Calculate the x-intercept.
Note: A decimal approximation helps locate the point.

x-intercept: let $y = 0$.

$-\dfrac{1}{2}(x-2)^3 + 6 = 0$

$\dfrac{1}{2}(x-2)^3 = 6$

$(x-2)^3 = 12$

$x-2 = \sqrt[3]{12}$

$x = 2 + \sqrt[3]{12}$

$\Rightarrow \left(2 + \sqrt[3]{12}, 0\right) \approx (4.3, 0)$

4. Sketch the graph, showing all key features.

$a < 0$, so the graph starts above the x-axis and decreases.

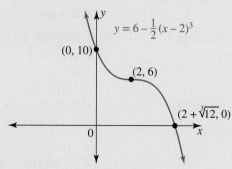

3.5.2 Cubic graphs with three x-intercepts

For the graph of a cubic polynomial to have three x-intercepts, the polynomial must have three distinct linear factors. This is the case when the cubic polynomial expressed as the product of a linear factor and a quadratic factor is such that the quadratic factor has two distinct linear factors.

This means that the graph of a cubic with an equation of the form $y = a(x - x_1)(x - x_2)(x - x_3)$ where $x_1 < x_2 < x_3$ will have the shape of the graph shown.

If the graph is reflected in the x-axis, its equation is of the form $y = -a(x - x_1)(x - x_2)(x - x_3)$ and the shape of its graph satisfies the long-term behaviour that as $x \to \pm\infty$, $y \to \mp\infty$.

It is important to note the graph is not a quadratic, so the maximum and minimum turning points do not lie halfway between the x-intercepts.

To sketch the graph, it is usually sufficient to identify the x- and y-intercepts and to ensure the shape of the graph satisfies the long-term behaviour requirement determined by the sign of the leading term.

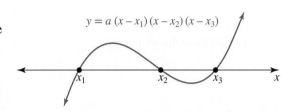

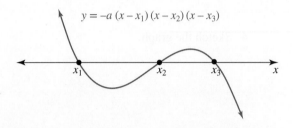

WORKED EXAMPLE 16 Sketching cubics in factorised form

Sketch the following without attempting to locate turning points.

a. $y = (x - 1)(x - 3)(x + 5)$

b. $y = (x + 1)(2x - 5)(6 - x)$

THINK	WRITE
a. 1. Calculate the x-intercepts.	a. $y = (x - 1)(x - 3)(x + 5)$ x-intercepts: let $y = 0$. $(x - 1)(x - 3)(x + 5) = 0$ $x = 1$, $x = 3$, $x = -5$ $\Rightarrow (-5, 0), (1, 0), (3, 0)$ are the x-intercepts.
2. Calculate the y-intercept.	y-intercept: let $x = 0$. $y = (-1)(-3)(5)$ $\quad = 15$ $\Rightarrow (0, 15)$ is the y-intercept.
3. Determine the shape of the graph.	Multiplying together the terms in x from each bracket gives x^3, so its coefficient is positive. The shape is of a positive cubic.
4. Sketch the graph.	

b. 1. Calculate the x-intercepts.

b. $y = (x + 1)(2x - 5)(6 - x)$

x-intercepts: let $y = 0$.

$(x + 1)(2x - 5)(6 - x) = 0$

$x + 1 = 0, 2x - 5 = 0, 6 - x = 0$

$x = -1, \ x = 2.5, \ x = 6$

$\Rightarrow (-1, 0), (2.5, 0), (6, 0)$ are the x-intercepts.

2. Calculate the y-intercept.

y-intercept: let $x = 0$.

$y = (1)(-5)(6)$

$\quad = -30$

$\Rightarrow (0, -30)$ is the y-intercept.

3. Determine the shape of the graph.

Multiplying the terms in x from each bracket gives $(x) \times (2x) \times (-x) = -2x^3$, so the shape is of a negative cubic.

4. Sketch the graph.

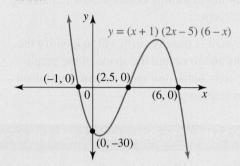

3.5.3 Cubic graphs with two x-intercepts

If a cubic has two x-intercepts, one at $x = x_1$ and one at $x = x_2$, then in order to satisfy the long-term behaviour required of any cubic, the graph either touches the x-axis at $x = x_1$ and turns, or it touches the x-axis at $x = x_2$ and turns. One of the x-intercepts must be a turning point.

Thinking of the cubic polynomial as the product of a linear and a quadratic factor, for its graph to have two instead of three x-intercepts, the quadratic factor must have two identical factors. Either the factors of the cubic are $(x - x_1)(x - x_1)(x - x_2) = (x - x_1)^2(x - x_2)$ or the factors are $(x - x_1)(x - x_2)(x - x_2) = (x - x_1)(x - x_2)^2$. The repeated factor identifies the x-intercept that is the turning point. The repeated factor is said to be of multiplicity 2 and the single factor of multiplicity 1.

The graph of a cubic polynomial with equation of the form $y = (x - x_1)^2(x - x_2)$ has a turning point on the x-axis at $(x_1, 0)$ and a second x-intercept at $(x_2, 0)$. The graph is said to *touch* the x-axis at $x = x_1$ and *cut* it at $x = x_2$.

Although the turning point on the x-axis must be identified when sketching the graph, there will be a second turning point that cannot yet be located without technology.

Sketch the graphs of the following.

a. $y = \dfrac{1}{4}(x-2)^2(x+5)$

b. $y = -2(x+1)(x+4)^2$

THINK

a. 1. Calculate the x-intercepts and interpret the multiplicity of each factor.

2. Calculate the y-intercept.

3. Sketch the graph.

b. 1. Calculate the x-intercepts and interpret the multiplicity of each factor.

2. Calculate the y-intercept.

3. Sketch the graph.

WRITE

a. $y = \dfrac{1}{4}(x-2)^2(x+5)$

x-intercepts: let $y = 0$.

$\dfrac{1}{4}(x-2)^2(x+5) = 0$

$\therefore x = 2$ (touch), $x = -5$ (cut)

x-intercept at $(-5, 0)$ and turning-point

x-intercept at $(2, 0)$

y-intercept: let $x = 0$.

$y = \dfrac{1}{4}(-2)^2(5)$

$= 5$

$\Rightarrow (0, 5)$

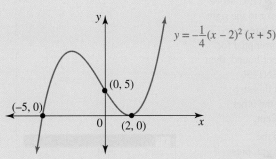

b. $y = -2(x+1)(x+4)^2$

x-intercepts: let $y = 0$.

$-2(x+1)(x+4)^2 = 0$

$(x+1)(x+4)^2 = 0$

$\therefore x = -1$ (cut), $x = -4$ (touch)

x-intercept at $(-1, 0)$ and turning point

x-intercept at $(-4, 0)$

y-intercept: let $x = 0$.

$y = -2(1)(4)^2$

$= -32$

y-intercept at $(0, -32)$

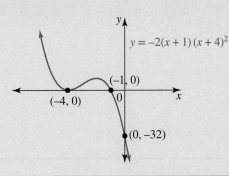

TI	THINK	WRITE	CASIO	THINK	WRITE

b. 1. On a Graphs page, complete the entry line for function 1 as: $f1(x) = -2(x+1)(x+4)^2$ then press ENTER.

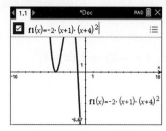

b. 1. On a Graph screen, complete the entry line for y1 as: $y1 = -2(x+1)(x+4)^2$ then press EXE. Select DRAW by pressing F6.

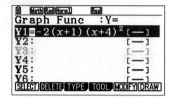

2. To find the x-intercepts, press MENU, then select:
6: Analyze Graph
1: Zero
Move the cursor to the left of the x-intercept when prompted for the lower bound, then press ENTER. Move the cursor to the right of the x-intercept when prompted for the upper bound, then press ENTER. Repeat this step to find the other x-intercept.

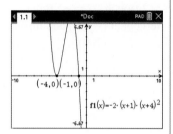

2. To find the x-intercepts, select G-Solv by pressing SHIFT then F5, then select ROOT by pressing F1. Press EXE. Use the left/right arrows to move across to the next x-intercept, then press EXE.

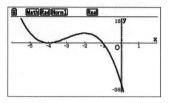

3. To find the y-intercept, press MENU, then select:
5: Trace
1: Graph
Trace Type '0', then press ENTER twice.

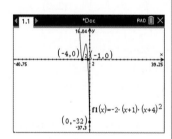

3. To find the y-intercept, select G-Solv by pressing SHIFT then F5, then select Y-ICEPT by pressing F4. Press EXE.

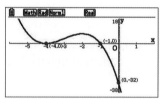

3.5.4 Determining the equation of a cubic graph

Depending on the information given, one form of the cubic equation may be preferable over another to determine the equation from either a graph or key points on the graph.

Equations of cubic functions

As a guide:
- if there is a stationary point of inflection given, use the $y = a(x-h)^3 + k$ form
- if the three x-intercepts are given, use the $y = a(x - x_1)(x - x_2)(x - x_3)$ form
- if there is a turning point at one of the x-intercepts, use the $y = a(x - x_1)^2(x - x_2)$ repeated factor form, where the turning point is at $x = x_1$.

WORKED EXAMPLE 18 Determining the equation

Determine the equation for each of the following graphs.
a. The graph of a cubic polynomial that has a stationary point of inflection at the point $(-7, 4)$ and an x-intercept at $(1, 0)$

b.

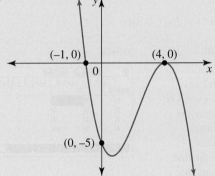

c.

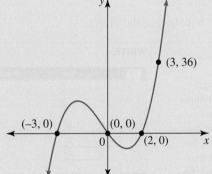

THINK	WRITE
a. 1. Consider the given information and choose the form of the equation to be used.	**a.** A stationary point of inflection is given. Let $y = a(x - h)^3 + k$. The point of inflection is $(-7, 4)$. $\therefore y = a(x + 7)^3 + 4$.
2. Calculate the value of a. *Note:* The coordinates of the given points show the y-values decrease as the x-values increase, so a negative value for a is expected.	Substitute the given x-intercept point $(1, 0)$. $0 = a(8)^3 + 4$ $(8)^3 a = -4$ $a = \dfrac{-4}{8 \times 64}$ $a = -\dfrac{1}{128}$
3. Write the equation of the graph.	The equation is $y = -\dfrac{1}{128}(x + 7)^3 + 4$.

b. 1. Consider the given information and choose the form of the equation to be used.

b. Two x-intercepts are given.
One shows a turning point at $x = 4$ and the other a cut at $x = -1$.
Let the equation be $y = a(x+1)(x-4)^2$.

2. Calculate the value of a.

Substitute the given y-intercept point $(0, -5)$.
$$-5 = a(1)(-4)^2$$
$$-5 = a(16)$$
$$a = -\frac{5}{16}$$

3. Write the equation of the graph.

The equation is $y = -\frac{5}{16}(x+1)(x-4)^2$.

c. 1. Consider the given information and choose the form of the equation to be used.

c. Three x-intercepts are given.
Let the equation be
$$y = a(x+3)(x-0)(x-2)$$
$$\therefore y = ax(x+3)(x-2)$$

2. Calculate the value of a.

Substitute the given point $(3, 36)$.
$$36 = a(3)(6)(1)$$
$$36 = 18a$$
$$a = 2$$

3. Write the equation of the graph.

The equation is $y = 2x(x+3)(x-2)$.

TI | THINK   **WRITE**   **CASIO | THINK**   **WRITE**

c. 1. On a Lists & Spreadsheet page, label the first column x and the second column y. Enter the x-coordinates of the given points in the first column and the corresponding y-coordinates in the second column.

c. 1. On a Statistics screen, relabel List 1 as x and List 2 as y. Enter the x-coordinates of the given points in the first column and the corresponding y-coordinates in the second column.

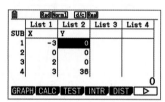

2. On a Calculator page, press MENU, then select:
6: Statistics
1: Stat Calculations
7: Cubic Regression…
Complete the fields as
X List: x
Y List: y
then select OK.

2. Select CALC by pressing F2, select REG by pressing F3, then select X^3 by pressing F4.

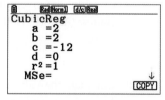

3. The answer appears on the screen.

$y = 2x^3 + 2x^2 - 12x$

3. The answer appears on the screen.

$y = 2x^3 + 2x^2 - 12x$

Exercise 3.5 Graphing cubic functions

learn on

3.5 Exercise	**3.5 Exam practice** on

Simple familiar	Complex familiar	Complex unfamiliar
1, 2, 3, 4, 5, 6, 7, 8, 9, 10, 11, 12	13, 14, 15, 16, 17, 18	19, 20

These questions are even better in jacPLUS!
- Receive immediate feedback
- Access sample responses
- Track results and progress

Find all this and MORE in jacPLUS ⊙

Simple familiar

1. State the coordinates of the point of inflection for each of the following.
   a. $y = (x - 7)^3$
   b. $y = x^3 - 7$
   c. $y = -7x^3$
   d. $y = 2 - (x - 2)^3$
   e. $y = \frac{1}{6}(x + 5)^3 - 8$
   f. $y = -\frac{1}{2}(2x - 1)^3 + 5$

2. **WE15** Sketch the graphs of the following polynomials.
   a. $y = (x - 1)^3 - 8$
   b. $y = 1 - \frac{1}{36}(x + 6)^3$

3. a. Sketch the graph of $y = -x^3 + 1$. Include all important features.
   b. Sketch the graph of $y = 2(3x - 2)^3$. Include all important features.
   c. Sketch the graph of $y = 2(x + 3)^3 - 16$. Include all important features.
   d. Sketch the graph of $y = (3 - x)^3 + 1$. Include all important features.

4. **WE16** Sketch the following without attempting to locate turning points.
   a. $y = (x + 1)(x + 6)(x - 4)$
   b. $y = (x - 4)(2x + 1)(6 - x)$

5. Sketch the graphs of the following without attempting to locate any turning points that do not lie on the coordinate axes.
   a. $y = (x - 2)(x + 1)(x + 4)$
   b. $y = -0.5x(x + 8)(x - 5)$
   c. $y = (x + 3)(x - 1)(4 - x)$
   d. $y = \frac{1}{4}(2 - x)(6 - x)(4 + x)$
   e. $y = 0.1(2x - 7)(x - 10)(4x + 1)$
   f. $y = 2\left(\frac{x}{2} - 1\right)\left(\frac{3x}{4} + 2\right)\left(x - \frac{5}{8}\right)$

6. **WE17** Sketch the graphs of the following polynomials.
   a. $y = \frac{1}{9}(x - 3)^2(x + 6)$
   b. $y = -2(x - 1)(x + 2)^2$

7. Sketch the graphs of the following without attempting to locate any turning points that do not lie on the coordinate axes.

   a. $y = -(x+4)^2(x-2)$

   b. $y = 2(x+3)(x-3)^2$

   c. $y = (x+3)^2(4-x)$

   d. $y = \dfrac{1}{4}(2-x)^2(x-12)$

   e. $y = 3x(2x+3)^2$

   f. $y = -0.25x^2(2-5x)$

8. Sketch the graphs of the following, showing any intercepts with the coordinate axes and any stationary point of inflection.

   a. $y = (x+3)^3$

   b. $y = (x+3)^2(2x-1)$

   c. $y = (x+3)(2x-1)(5-x)$

   d. $2(y-1) = (1-2x)^3$

   e. $4y = x(4x-1)^2$

   f. $y = -\dfrac{1}{2}(2-3x)(3x+2)(3x-2)$

9. **WE18** Determine the equation of each of the following graphs.

   a. The graph of a cubic polynomial that has a stationary point of inflection at the point $(3, -7)$ and an x-intercept at $(10, 0)$

   b.
   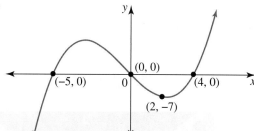

   c.
   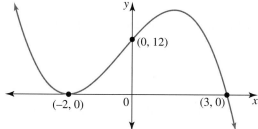

10. Determine the equation for each of the following graphs of cubic polynomials.

    a.
    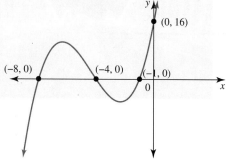

    b.
    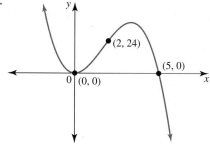

    c.
    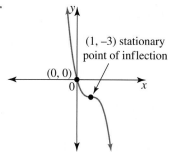

    d.
    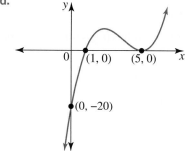

11. a. The graph of a cubic function of the form $y = a(x - h)^3 + k$ has a stationary point of inflection at $(3, 9)$ and passes through the origin. Form the equation of the graph.

b. The graph of the form $y = a(x - h)^3 + k$ has a stationary point of inflection at $(-2, 2)$ and a y-intercept at $(0, 10)$. Determine the equation.

c. The graph of the form $y = a(x - h)^3 + k$ has a stationary point of inflection at $(0, 4)$ and passes through the x-axis at $\left(\sqrt[3]{2}, 0 \right)$. Determine the equation.

12. a. The graph of $y = x^3$ is translated 5 units to the left and 4 units upwards. State its equation after these translations take place.

b. After the graph $y = x^3$ has been reflected about the x-axis, translated 2 units to the right and translated downwards 1 unit, determine what its equation becomes.

c. The graph shown has a stationary point of inflection at $(3, -1)$. Determine its equation.

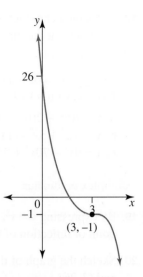

Complex familiar

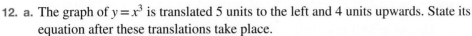

13. Factorise and sketch the graphs of the cubic polynomials with equations given by the following.

a. $y = 9x^2 - 2x^3$

b. $y = 9x^3 - 4x$

c. $y = 9x^2 - 3x^3 + x - 3$

d. $y = 9x(x^2 + 4x + 3)$

e. $y = 9x^3 + 27x^2 + 27x + 9$

f. $y = -9x^3 - 9x^2 + 9x + 9$

14. a. Determine the x- and y-intercepts of the cubic graph $y = -x^3 - 3x^2 + 16x + 48$.
Factorise the cubic equation to express it as a product of linear factors by grouping and hence sketch the graph.

b. The partially factorised form of $2x^3 + x^2 - 13x + 6$ is $(x - 2)(2x^2 + 5x - 3)$. Complete the factorisation and sketch the graph of $y = 2x^3 + x^2 - 13x + 6$, showing all intercepts with the coordinate axes.

15. a. Determine the x- and y-intercepts of the cubic graph $y = x^3 + 5x^2 - x - 5$. (To determine the linear factors, use the grouping technique.)
Hence, sketch the graph.

b. Partially factorised, $-x^3 - 5x^2 - 3x + 9$ is expressed as $(x - 1)(-x^2 - 6x - 9)$. Complete the factorisation and sketch the graph of $y = -x^3 - 5x^2 - 3x + 9$ showing all intercepts with the coordinate axes.

16. Using the factorised forms of each of the following, sketch the graph without attempting to locate any turning points that do not lie on the coordinate axes.

a. $y = 2x^3 - 3x^2 - 17x - 12$
Factorised $y = (x + 1)(2x + 3)(x - 4)$

b. $y = 6 - 55x + 57x^2 - 8x^3$
Factorised partially $y = (x - 1)(-8x^2 + 49x - 6)$

c. $y = 6x^3 - 13x^2 - 59x - 18$
Factorised partially $(x + 2)(6x^2 - 25x - 9)$

d. $y = -\dfrac{1}{2}x^3 + 14x - 24$
Factorised partially $-2y = (x - 2)(x^2 + 2x - 24)$

17. a. Using technology sketch, locating turning points, the graph of $y = x^3 + 4x^2 - 44x - 96$.
 b. Show that the turning points are not placed symmetrically in the interval between the adjoining x-intercepts.

18. Using technology sketch the following, locating intercepts with the coordinate axes and any turning points. Express values to 1 decimal place where appropriate.
 a. $y = 10x^3 - 20x^2 - 10x - 19$
 b. $y = -x^3 + 5x^2 - 11x + 7$
 c. $y = 9x^3 - 70x^2 + 25x + 500$

Complex unfamiliar

19. By expressing $y = x^3 + 3x^2 + 3x + 2$ in the form $a(x - h)^3 + k$, determine the coordinates of the stationary point of inflection of the graph and sketch the graph.

20. Sketch the graph of the curve with equation $y = (x + a)^3 + b$ and passing through the points $(0, 0)$, $(1, 7)$ and $(2, 26)$.

Fully worked solutions for this chapter are available online.

LESSON
3.6 Solving cubic equations

SYLLABUS LINKS

- Solve cubic equations using technology, and algebraically in cases where the equation is factorised.

Source: Mathematical Methods Senior Syllabus 2024 © State of Queensland (QCAA) 2024; licensed under CC BY 4.0.

3.6.1 Polynomial equations

If a cubic or any polynomial is expressed in factorised form, then the polynomial equation can be solved using the Null Factor Law.

$$(x - a)(x - b)(x - c) = 0$$
$$\therefore (x - a) = 0, \ (x - b) = 0, \quad (x - c) = 0$$
$$\therefore x = a, \qquad x = b \ \text{ or } \qquad x = c$$

$x = a$, $x = b$ and $x = c$ are called the roots or the solutions to the equation $P(x) = 0$.

If the equation cannot be expressed in factorised form, technology can be used to solve the equation.

3.6.2 Solving cubic equations using the Null Factor Law

To solve cubic equations using the Null Factor Law, the equations need to be expressed in factorised form, $(x - a)(x - b)(x - c) = 0$. It may be necessary to rearrange an equation before applying the Null Factor Law and use normal algebraic techniques to factorise an expression.

Solving cubic equations

To solve a cubic equation, the following steps may be useful.
- **Rearrange so that one side is 0.**
- **Factorise the cubic expression by applying algebraic techniques:**
  - **common factors**
  - **difference of two squares**
  - **quadratic factorisation**
  - **grouping in pairs**
- **Use technology to factorise the cubic expression.**
- **Calculate the zeros by applying the Null Factor Law.**

WORKED EXAMPLE 19 Solving a cubic equation given in factorised form

a. **Solve the equation $(x - 2)(3x + 4)(5 - x) = 0$.**
b. **Determine the zeros of the equation $3x(x + 6)(x - 7) = 0$.**

THINK	WRITE
a. 1. Write the equation.	$(x - 2)(3x + 4)(5 - x) = 0$
2. Apply the Null Factor Law	$x - 2 = 0$ or $3x + 4 = 0$ or $5 - x = 0$
3. Determine the solutions.	$x = 2$ or $x = -\dfrac{4}{3}$ or $x = 5$
4. State the solutions.	$\therefore x = -\dfrac{4}{3}$ or $x = 2$ or $x = 5$
b. 1. Write the equation.	$3x(x + 6)(x - 7) = 0$
2. Apply the Null Factor Law	$3x = 0$ or $x + 6 = 0$ or $x - 7 = 0$
3. Determine the zeros or solutions.	$x = 0$ or $x = -6$ or $x = 7$
4. State the zeros.	$\therefore x = -6$ or $x = 0$ or $x = 7$

Worked example 20 illustrates using basic algebraic techniques to express the cubic equation in factorised form first before applying the Null Factor Law.

WORKED EXAMPLE 20 Factorising and solving cubic equations

Solve the following.
a. $x^3 = 9x$
b. $-2x^3 + 4x^2 + 70x = 0$
c. $x^3 - 4x^2 - 9x + 36 = 0$

THINK	WRITE
a. 1. Write the equation.	a. $x^3 = 9x$
2. Rearrange so all terms are on the left.	$x^3 - 9x = 0$
3. Take out a common factor of x.	$x(x^2 - 9) = 0$
4. Factorise the expression in the grouping symbols using the difference of two squares rule.	$x(x + 3)(x - 3) = 0$
5. Use the Null Factor Law to solve.	$x = 0, \; x + 3 = 0$ or $x - 3 = 0$ $x = 0, \quad\; x = -3$ or $x = 3$

b. **1.** Write the equation.

2. Take out a common factor of $-2x$.

3. Factorise the expression in the grouping symbols.

4. Use the Null Factor Law to solve.

b. $-2x^3 + 4x^2 + 70x = 0$

$-2x(x^2 - 2x - 35) = 0$

$2x(x - 7)(x + 5) = 0$

$-2x = 0, \ x - 7 = 0 \quad \text{or} \ x + 5 = 0$

$\qquad x = 0, \qquad x = 7 \quad \text{or} \qquad x = -5$

c. **1.** Write the equation.

2. Group the expression in pairs.

3. Take out the common factor.

4. Use the difference of two squares.

5. Apply the Null Factor Law and solve.

c. $x^3 - 4x^2 - 9x + 36 = 0$

$x^2(x - 4) - 9(x - 4) = 0$

$(x - 4)\left(x^2 - 9\right) = 0$

$(x - 4)(x - 3)(x + 3) = 0$

$x - 4 = 0 \quad \text{or} \ x - 3 = 0 \quad \text{or} \ x + 3 = 0$

$\qquad x = 4 \qquad\qquad x = 3 \qquad\qquad x = -3$

$\therefore x = -3 \ \text{or} \ x = 3 \ \text{or} \ x = 4$

Technology may be used to solve cubic equations that are not given in factorised form or are not easily factorised with simple algebraic tools. The following example illustrates using technology.

WORKED EXAMPLE 21 Solving cubic equations using technology

Solve $2x^3 + 18x = 11x^2 + 9$ using technology.

THINK

1. Write the equation.

2. Rearrange to make one side equal to zero.

3. Use chosen technology to solve.

WRITE

$$2x^3 + 18x = 11x^2 + 9$$

$$2x^3 - 11x^2 + 18x - 9 = 0$$

$$x = 1, \ x = \frac{3}{2} \ \text{or} \ x = 3$$

TI | THINK

c. **1.** On a Calculator page, press MENU, then select:
3. Algebra
3: Polynomial Tools
1: Find Roots of Polynomial…
Complete the fields as:
Degree: 3
Roots: Real
then select OK.

WRITE

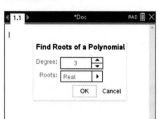

CASIO | THINK

c. **1.** On a Run-Matrix screen, press OPTN, then select CALC by pressing F4.
Select Solve N by pressing F5, then complete the entry line as:
Solve N($2x^3 -$ $11x^2 + 18x - 9 =$ $0, x$) then press EXE.

WRITE

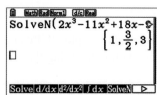

2. Complete the fields for the coefficients as:

$a_3 = 2$
$a_2 = -11$
$a_1 = 18$
$a_0 = -9$

Select OK then press Enter.

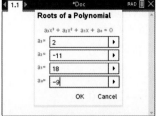

2. The answer appears on the screen.

$$x = 1, x = \frac{3}{2} \text{ or } x = 3$$

3. The answer appears on the screen.

$$x = 1, x = \frac{3}{2} \text{ or } x = 3$$

3.6.3 Intersections of cubic graphs with linear and quadratic graphs

If $P(x)$ is a cubic polynomial and $Q(x)$ is either a linear or a quadratic polynomial, then the intersection of the graphs of $y = P(x)$ and $y = Q(x)$ occurs when $P(x) = Q(x)$. Hence, the x-coordinates of the points of intersection are the roots of the equation $P(x) - Q(x) = 0$. This is a cubic equation since $P(x) - Q(x)$ is a polynomial of degree 3.

WORKED EXAMPLE 22 Determining points of intersection

Sketch the graphs of $y = x(x - 1)(x + 1)$ and $y = x$, and calculate the coordinates of the points of intersection.

THINK

1. Sketch the graphs.

WRITE

$y = x(x - 1)(x + 1)$

This is a positive cubic.

x-intercepts: let $y = 0$.
$x(x - 1)(x + 1) = 0$
$x = 0, \; x = \pm 1$
$(-1, 0), (0, 0), (1, 0)$ are the three x-intercepts.
The y-intercept is $(0, 0)$.

Line: $y = x$ passes through $(0, 0)$ and $(1, 1)$.

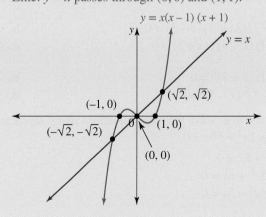

2. Calculate the coordinates of the points of intersections.

At intersection:
$$x(x-1)(x+1) = x$$
$$x(x^2 - 1) - x = 0$$
$$x^3 - 2x = 0$$
$$x(x^2 - 2) = 0$$
$$x = 0, \; x^2 = 2$$
$$x = 0, \; x = \pm\sqrt{2}$$

3. Write the answer.

Substituting these x-values in the equation of the line $y = x$, the points of intersection are
$$(0, 0), \left(\sqrt{2}, \sqrt{2}\right), \left(-\sqrt{2}, -\sqrt{2}\right).$$

Exercise 3.6 Solving cubic equations

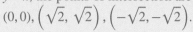 **learn on**

Simple familiar

For questions **1** to **3**, solve the cubic equations shown.

1. **a.** $(x + 7)(x + 2)(x + 3) = 0$
   b. $(x - 2)(x + 4)(x - 5) = 0$
   c. $(x - 1)(x - 4)(x + 8) = 0$
   d. $(x - 1)(x + 1)(x - 6) = 0$

2. **a.** $(2x - 7)(3x + 1)(x - 3) = 0$ **b.** $(x - 2)(5x + 4)(3x - 5) = 0$
   c. $(x - 1)(x - 4)(3x + 8) = 0$ **d.** $(2x - 1)(2x + 1)(x - 6) = 0$

3. **a.** $(x - 1)(x + 2)^2 = 0$ **b.** $2(x - 3)^2(x + 5) = 0$
   c. $(2x - 3)^2(x - 5) = 0$ **d.** $3x^2(x + 8) = 0$

For questions **4** to **6**, solve the cubic equations using algebraic factorising techniques.

4. **WE20** **a.** $2x^3 - 50x = 0$
   b. $-4x^3 + 8x = 0$
   c. $x^3 - 5x^2 + 6x = 0$
   d. $x^3 - 4x^2 - 5x = 0$

5. a. $x^3 - x^2 - 16x + 16 = 0$
   b. $x^3 + 4x^2 - 4x - 16 = 0$
   c. $x^3 - x^2 - 25x + 25 = 0$
   d. $x^3 - 3x^2 - 6x + 18 = 0$

6. a. $x^3 + 2x^2 = 25x + 50$
   b. $x^3 = 9x$
   c. $2x^3 + 2x = 3x^2 + 3$
   d. $3x^3 = x^2 + 12x - 4$

For questions **7** to **9**, use technology to solve the cubic equations.

7. **WE21** a. $x^3 - 4x^2 + x + 6 = 0$

   b. $x^3 - 4x^2 - 7x + 10 = 0$

   c. $x^3 + 6x^2 + 11x + 6 = 0$

   d. $x^3 - 6x^2 - 15x + 100 = 0$

8. a. $x^3 - 3x^2 = 6x - 8$
   b. $x^3 + 2x^2 = 29x - 42$
   c. $2x^3 + 15x^2 + 19x + 6 = 0$
   d. $16x^2 = 9 + 9x + 4x^3$

9. a. $x^3 = 7x - 6$
   b. $-18x^3 + 9x^2 = 4 - 23x$
   c. $2x^3 + 7x^2 = 9$
   d. $x^3 = 13x^2 - 34x - 48$

Complex familiar

10. **WE22** Sketch the graphs of $y = (x + 2)(x - 1)^2$ and $y = -3x$, and calculate the coordinates of the points of intersection.

11. Sketch the graphs of $y = 2x^3$ and $y = x^2$, and calculate the coordinates of the points of intersection.

12. Sketch the graphs of $y = 2x^3$ and $y = x - 1$, and calculate the coordinates of the points of intersection.

13. The number of solutions to the equation $x^3 + 2x - 5 = 0$ can be found by determining the number of intersections of the graphs of $y = x^3$ and a straight line. Determine the equation of this line and how many solutions $x^3 + 2x - 5 = 0$ has.

14. Use a graph of a cubic and a linear polynomial to determine the number of solutions to the equation $x^3 + 3x^2 - 4x = 0$.

15. Use a graph of a cubic and a quadratic polynomial to determine the number of solutions to the equation $x^3 + 3x^2 - 4x = 0$.

16. Determine the values of x that make the following equation true. If necessary, express your answer correct to 2 decimal places.

$$x^3 + 8 = x(5x - 2)$$

17. Determine the values of x that make the following equation true.

$$2(x^3 + 5) = 13x(x - 1)$$

Complex unfamiliar

18. Show that the line $y = 3x + 2$ is a tangent to the curve $y = x^3$ at the point $(-1, -1)$ and determine the coordinates of any other point of intersection.

19. Determine the value of p if $x = \dfrac{p}{2}$ is a solution of $x^3 - 5x^2 + 2x + 8 = 0$.

20. Sketch on the same axes the graphs of $y = 4 - x^2$ and $y = 4x - x^3$, clearly indicating all important features, and identify all points where the graphs intersect.

Fully worked solutions for this chapter are available online.

LESSON
3.7 Modelling cubic functions

SYLLABUS LINKS

- Model and solve problems that involve cubic functions, with and without technology.

Source: Mathematical Methods Senior Syllabus 2024 © State of Queensland (QCAA) 2024; licensed under CC BY 4.0.

Practical situations that use cubic polynomials as models are likely to require a restriction of the possible values the variable may take. This is called a **domain** restriction. The domain is the set of possible values of the variable that the polynomial may take. We shall look more closely at domains in other chapters.

3.7.1 Maximum and minimum values and cubic models

Similar methods to modelling with linear and quadratic relationships are used when using cubic polynomials, with the polynomial model expressed in terms of one variable.

In setting up equations:
- define the symbols used for the variables, specifying units where appropriate
- ensure any units used are consistent and lengths of objects are positive
- determine the links between the variables
- check whether mathematical solutions are feasible in the context of the problem
- express answers in the context of the problem.

Applications of cubic models where a maximum or minimum value of the model is sought will require identification of turning point coordinates. In a later chapter we will see how this is done. For now, obtaining turning points may require the use of graphing or other technology.

WORKED EXAMPLE 23 Determining cubic models

A rectangular storage container is designed to have an open top and a square base.
The base has side length x cm and the height of the container is h cm. The sum of its dimensions (the sum of the length, width and height) is 48 cm.
a. Express h in terms of x.
b. Show that the volume, V cm³, of the container is given by
$V = 48x^2 - 2x^3$.
c. State any restrictions on the values x can take.
d. Sketch the graph of V against x for appropriate values of x, given its maximum turning point has coordinates (16, 4096).
e. Calculate the dimensions of the container with the greatest possible volume.

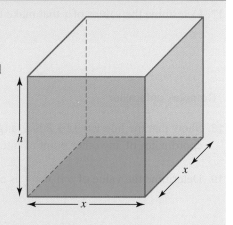

THINK	**WRITE**
a. Write the given information as an equation connecting the two variables.	**a.** The sum of the dimensions is 48 cm. $$x + x + h = 48$$ $$h = 48 - 2x$$
b. Use the result from part **a** to express the volume in terms of one variable and prove the required statement.	**b.** The formula for the volume of a cuboid is $$V = lwh$$ $$\therefore V = x^2 h$$ Substitute $h = 48 - 2x$. $$V = x^2(48 - 2x)$$ $\therefore V = 48x^2 - 2x^3$, as required
c. State the restrictions. *Note*: It could be argued that the restriction is $0 < x < 24$ because when $x = 0$ or $x = 24$ there is no storage container, but we are adopting the closed convention.	**c.** Length cannot be negative, so $x \geq 0$. Height cannot be negative, so $h \geq 0$. $$48 - 2x \geq 0$$ $$-2x \geq -48$$ $$\therefore x \leq 24$$ Hence, the restriction is $0 \leq x \leq 24$.
d. Draw the cubic graph but only show the section of the graph for which the restriction applies. Label the axes with the appropriate symbols and label the given turning point.	**d.** $V = 48x^2 - 2x$ $\quad = 2x^2(24 - x)$ x-intercepts: let $V = 0$. $\quad x^2 = 0$ or $24 - x = 0$ $\therefore x = 0$ (touch), $x = 24$ (cut) $(0, 0), (24, 0)$ are the x-intercepts. This is a negative cubic. Maximum turning point $(16, 4096)$ Draw the section for which $0 \leq x \leq 24$. 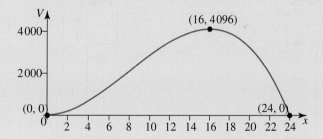
e. 1. Calculate the required dimensions. *Note:* The maximum turning point (x, V) gives the maximum value of V and the value of x when this maximum occurs.	**e.** The maximum turning point is $(16, 4096)$. This means the greatest volume is 4096 cm^3. It occurs when $x = 16$. $\therefore h = 48 - 2(16) \Rightarrow h = 16$ Dimensions: length $= 16$ cm, width $= 16$ cm, height $= 16$ cm
2. Write the answer.	The container has the greatest volume when it is a cube of edge 16 cm.

Exercise 3.7 Modelling cubic functions

| **3.7 Exercise** | **3.7 Exam practice** on |

Simple familiar	**Complex familiar**	**Complex unfamiliar**
1, 2, 3, 4, 5	6, 7, 8, 9, 10, 11, 12	13, 14, 15

Simple familiar

1. **WE23** A rectangular storage container is designed to have an open top and a square base.
   The base has side length x metres and the height of the container is h metres. The total length of its 12 edges is 6 metres.

   a. Express h in terms of x.
   b. Show that the volume, $V\,\text{m}^3$, of the container is given by $V = 1.5x^2 - 2x^3$.
   c. State any restrictions on the values x can take.
   d. Sketch the graph of V against x for appropriate values of x, given its maximum turning point has coordinates $(0.5, 0.125)$.
   e. Calculate the dimensions of the container with the greatest possible volume.

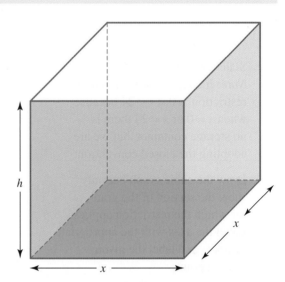

2. A rectangular box with an open top is to be constructed from a rectangular sheet of cardboard measuring 20 cm by 12 cm by cutting equal squares of side length x cm out of the four corners and folding the flaps up.

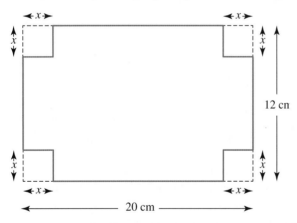

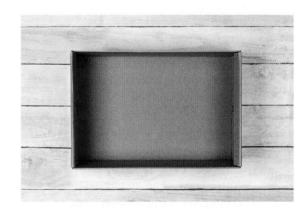

   The box has length l cm, width w cm and volume $V\,\text{cm}^3$.

   a. Express l, w and h in terms of x and hence express V in terms of x.
   b. State any restrictions on the values of x.
   c. Sketch the graph of V against x for appropriate values of x, given the unrestricted graph would have turning points at $x = 2.43$ and $x = 8.24$.
   d. Calculate the length and width of the box with maximum volume and give this maximum volume to the nearest whole number.

3. The cost C dollars for an artist to produce x sculptures by contract is given by $C = x^3 + 100x + 2000$. Each sculpture is sold for $500 and as the artist only makes the sculptures by order, every sculpture produced will be paid for. However, too few sales will result in a loss to the artist.

a. Show the artist makes a loss if only 5 sculptures are produced and a profit if 6 sculptures are produced.

b. Show that the profit, P dollars, from the sale of x sculptures is given by $P = -x^3 + 400x - 2000$.

c. Explain what will happen to the profit if a large number of sculptures are produced and why this effect occurs.

d. Calculate the profit (or loss) from the sale of:

   i. 16 sculptures
   ii. 17 sculptures.

e. Use the above information to sketch the graph of the profit P for $0 \le x \le 20$. Place its intersection with the x-axis between two consecutive integers but don't attempt to obtain its actual x-intercepts.

f. In order to guarantee a profit is made, determine how many sculptures the artist should produce.

4. The number of bacteria in a slow-growing culture at time t hours after 9 am is given by $N = 54 + 23t + t^3$.

a. Calculate the initial number of bacteria at 9 am.

b. Determine how long it takes for the initial number of bacteria to double.

c. Calculate how many bacteria there are by 1 pm.

d. Once the number of bacteria reaches 750, the experiment is stopped. Determine the time of the day at which this happens.

5. A new playground slide for children is to be constructed at a local park. At the foot of the slide the children climb a vertical ladder to reach the start of the slide. The slide must start at a height of 2.1 metres above the ground and end at a point 0.1 metres above the ground and 4 metres horizontally from its foot. A model for the slide is $h = ax^3 + bx^2 + cx + d$, where h metres is the height of the slide above ground level at a horizontal distance of x metres from its foot. The foot is at the origin.

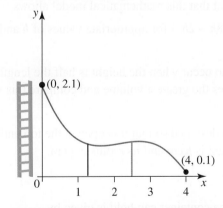

The ladder supports the slide at one end, and the slide also requires two vertical struts as support. One strut of length 1 metre is placed at a point 1.25 metres horizontally from the foot of the slide and the other is placed at a point 1.5 metres horizontally from the end of the slide and is of length 1.1 metres.

a. Give the coordinates of 4 points that lie on the cubic graph of the slide.

b. State the value of d in the equation of the slide.

c. Form a system of 3 simultaneous equations, the solutions to which give the coefficients a, b, c in the equation of the slide.

d. The equation of the slide can be shown to be $y = -0.164x^3 + x^2 - 1.872x + 2.1$. Use this equation to calculate the length of a third strut thought necessary at $x = 3.5$. Give your answer to 2 decimal places.

6. Identify the smallest positive integer and the largest negative integer for which the difference between the square of 5 more than this number and the cube of 1 more than the number exceeds 22.

7. A tent used by a group of bushwalkers is in the shape of a square-based right pyramid with a slant height of 8 metres.
   For the figure shown, let OV, the height of the tent, be h metres and the edge of the square base be $2x$ metres.

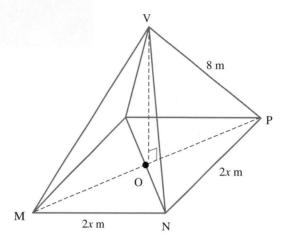

a. Use Pythagoras' theorem to express the length of the diagonal of the square base of the tent in terms of x.
b. Use Pythagoras' theorem to show $2x^2 = 64 - h^2$.

c. The volume, V, of a pyramid is found using the formula $V = \frac{1}{3}Ah$, where A is the area of the base of the pyramid. Use this formula to show that the volume of space contained within the bushwalkers' tent is given by $V = \frac{1}{3}\left(128h - 2h^3\right)$.

d. i. If the height of the tent is 3 metres, calculate the volume.
   ii. Identify the values for height that this mathematical model allows.

e. Sketch the graph of $V = \frac{1}{3}\left(128h - 2h^3\right)$ for appropriate values of h and estimate the height for which the volume is greatest.

f. The greatest volume is found to occur when the height is half the length of the base. Use this information to calculate the height that gives the greatest volume and compare this value with your estimate from your graph in part e.

8. A cylindrical storage container is designed so that it is open at the top and has a surface area of 400π cm^2. Its height is h cm and its radius is r cm.

a. Show that $h = \dfrac{400 - r^2}{2r}$.

b. Show that the volume V cm^3 the container can hold is given by
   $V = 200\pi r - \frac{1}{2}\pi r^3$.

c. State any restrictions on the values r can take.
d. Sketch the graph of V against r for appropriate values of r.
e. Determine the radius and height of the container if the volume is 396π cm^3.

f. State the maximum possible volume to the nearest cm^3 if the maximum turning point on the graph of $y = 200\pi x - \frac{1}{2}\pi x^3$ has an x-coordinate of $\dfrac{20}{\sqrt{3}}$.

9. Engineers are planning to build an underground tunnel through a city to ease traffic congestion. The cross-section of their plan is bounded by the curve shown.

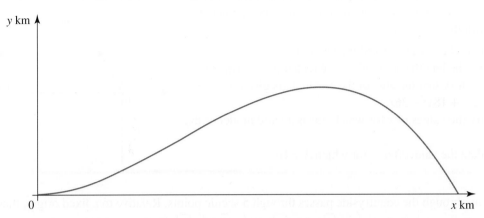

The equation of the bounding curve is $y = ax^2(x - b)$, and all measurements are in kilometres.

It is planned that the greatest breadth of the bounding curve will be 6 km and the greatest height will be 1 km above this level at a point 4 km from the origin.

a. Determine the equation of the bounding curve.

b. If the greatest breadth of the curve was extended to 7 km, determine the greatest height of the curve above this new lowest level.

10. Since 1988, the world record times for the men's 100-m sprint can be roughly approximated by the cubic model $T(t) = -0.000\,05(t - 6)^3 + 9.85$, where T is the time in seconds and t is the number of years since 1988.

a. In 1991 the world record was 9.86 seconds and in 2008 the record was 9.72 seconds. Compare these times with those predicted by the cubic model.

b. Sketch the graph of T versus t from 1988 to 2008.

c. Determine what the model predicts for 2016. Explain whether the model is likely to be a good predictor beyond 2016.

11. A rectangle is inscribed under the parabola $y = 9 - (x-3)^2$ so that two of its corners lie on the parabola and the other two lie on the x-axis at equal distances from the intercepts the parabola makes with the x-axis.

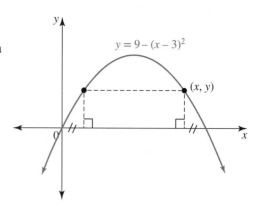

$y = 9 - (x-3)^2$

(x, y)

a. Calculate the x-intercepts of the parabola.
b. Express the length and width of the rectangle in terms of x.
c. Hence, show that the area of the rectangle is given by
$A = -2x^3 + 18x^2 - 36x$.
d. Identify the values of x for which this is a valid model of the area.
e. Calculate the value(s) of x for which $A = 16$.

12. A pathway through the countryside passes through 5 scenic points. Relative to a fixed origin, these points have coordinates $A(-3, 0)$, $B\left(-\sqrt{3}, -12\sqrt{3}\right)$, $C\left(\sqrt{3}, 12\sqrt{3}\right)$, $D(3, 0)$; the fifth scenic point is the origin, $O(0, 0)$. The two-dimensional shape of the path is a cubic polynomial.

a. Determine the equation of the pathway through the 5 scenic points.
b. Sketch the path, given that points B and C are turning points of the cubic polynomial graph.
c. It is proposed that another pathway be created to link B and C by a direct route. Show that if a straight-line path connecting B and C is created, it will pass through O and give the equation of this line.
d. An alternative plan is to link B and C by a cubic path that has a stationary point of inflection at O. Determine the equation of this path.

Complex unfamiliar

13. Determine the maximum volume of a cuboid container with a base length twice its width and made with $48 \, \text{m}^2$ of metal. Give your answer correct to 1 decimal place.

14. The total surface area of a closed cylinder is $200 \, \text{cm}^2$. Determine, to the nearest integer, the minimum volume if the radius is restricted to between 2 cm and 4 cm.

15. A piece of wire 240 cm long is used to reinforce the edges of a rectangular box where the length of the box is three times the width. If the length of the box needs to be restricted to 12 cm or less, determine the maximum volume of the box and state its dimensions.

Fully worked solutions for this chapter are available online.

LESSON
3.8 Review

doc-41798

3.8.1 Summary

3.8 Exercise

learn on

3.8 Exercise	**3.8 Exam practice** on

Simple familiar	Complex familiar	Complex unfamiliar
1, 2, 3, 4, 5, 6, 7, 8, 9, 10, 11, 12	13, 14, 15, 16	17, 18, 19, 20

Simple familiar

1. a. Evaluate $\begin{pmatrix} 10 \\ 6 \end{pmatrix}$.

   b. i. Write down row 5 of Pascal's triangle.

     ii. Hence, or otherwise, expand $(1 + 2xy)^5$.

2. **MC** State which of the following expressions is a polynomial.

   A. $x^3 + 4x - 7x^{-1} + x^{-3}$ **B.** $\left(2\sqrt{x} + 7\right)^3$ **C.** $\dfrac{4x - 7x^5}{x^3}$ **D.** $\left(\sqrt{3} - 5x\right)^3$

3. Consider the polynomial $P(x) = x^3 - ax^2 + bx - 3$.
   Determine the values of a and b if $P(1) = 2$ and $P(-1) = -4$.

4. Solve the following.

   a. $(x + 2)(x - 4)(5 - x) = 0$ **b.** $(3x + 7)(x - 8)^2 = 0$

5. **MC** A possible equation for the curve shown is:

   A. $y = (2 - x)(x - 3)^2$
   B. $y = -(x + 2)(x - 3)^2$
   C. $y = -(x + 2)^2(x - 3)$
   D. $y = (x + 2)(x - 3)^2$

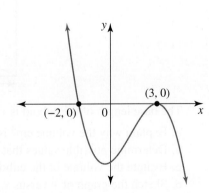

6. **MC** Select which of the following shows the graph of $y = -2(x+5)^3 - 12$.

A.

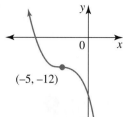

$(-5, -12)$

B.

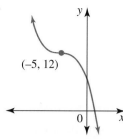

$(-5, 12)$

C.

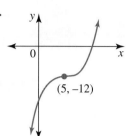

$(5, -12)$

D.

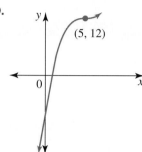

$(5, 12)$

7. Calculate the coordinates of the points of intersection of $y = 4x - x^3$ and $y = -2x$.

8. **MC** The solutions to the equation $2x^3 = 14x^2 + 16x$ are:
   A. $x = -1, x = 2, x = 8$
   B. $x = -8, x = 2, x = 7$
   C. $x = -8, x = 0, x = 7$
   D. $x = -1, x = 0, x = 8$

9. **MC** If $x^3 - 2x^2 - 3x + 10 \equiv (x+2)(ax^2 + bx + c)$, then the values of a, b and c are, respectively:
   A. $1, -2, 5$
   B. $1, 0, 5$
   C. $-2, -3, 10$
   D. $1, -4, 5$

10. **MC** If $P(x) = 3 + kx - 5x^2 + 2x^3$ and $P(-1) = 8$, then k is equal to:
    A. 0
    B. 4
    C. -12
    D. 12

11. a. Expand $\left(x^2 - \dfrac{3}{x}\right)^4$ in decreasing powers of x.

    b. Identify the term independent of y in the expansion of $\left(\dfrac{3}{y^2} - y\right)^{12}$.

12. A girl uses 140 cm of wire to make a frame of a cuboid with a square base as shown.

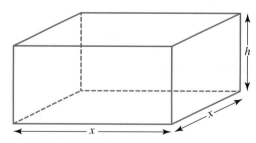

The base length of the cuboid is x cm and the height is h cm.

a. Explain why the volume cm³ is given by $V = 35x^2 - 2x^3$.
b. Determine possible values that x can assume.
c. Evalute the volume of the cuboid when the base area is 81 cm².
d. Sketch the graph of V versus x.
e. Use technology to determine the coordinates of the maximum turning point.
   Explain what these coordinates mean.

13. The cubic polynomial $P(x) = 8x^3 - 34x^2 + 33x - 9$ can also be expressed as $P(x) = (x-3)(2x-1)(4x-3)$.

    a. The graph of the polynomial $y = P(x) = 8x^3 - 34x^2 + 33x - 9$ has turning points at $(0.62, 0.3)$ and $(22, -15.8)$. Sketch the graph, labelling all key points with their coordinates.

    b. Calculate $\{x : P(x) = -9\}$.

    c. Determine the values of k for which the line $y = k$ will intersect the graph of $y = P(x)$ in:

        i. 3 places   **ii.** 2 places   **iii.** 1 place.

14. The revenue ($) from the sale of x thousand items is given by $R(x) = 6\left(2x^2 + 10x + 3\right)$ and the manufacturing cost ($) of x thousand items is $C(x) = x\left(6x^2 - x + 1\right)$.

    a. State the degree of $R(x)$ and of $C(x)$.

    b. Calculate the revenue and the cost if 1000 items are sold, and explain whether a profit is made.

    c. Show that the profit ($) from the scale of x thousand items is given by $P(x) = -6x^3 + 13x^2 + 59x + 18$.

    d. Sketch the graph of $y = P(x)$ for appropriate values of x.

    e. If a loss occurs when the number of items manufactured is d, state the smallest value of d.

15. Relative to a reference point O, two towns, A and B, are located at the points $(1, 20)$ and $(5, 12)$, respectively.

A freeway passing through A and B can be considered to be a straight line.

    a. Determine the equation of the line modelling the freeway.

    b. Before the freeway was built, the road between A and B followed a scenic route modelled by the equation

$y = a(2x - 1)(x - 6)(x + b)$ for $0 \le x \le 8$.

Using the fact this road goes through towns A and B, show

that $a = \dfrac{2}{3}$ and $b = -7$.

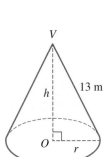

    c. Determine the coordinates of the end points where the scenic route starts and finishes.

    d. On the same diagram, sketch the scenic route and the freeway. Any end points and intercepts with the axes should be given, and the positions of the points A and B should be marked on your graph.

    e. The freeway meets the scenic route at three places. Calculate the coordinates of these three points.

    f. Determine which of the three points found in part **e** is closest to the reference point O.

16. The slant height of a right conical tent has a length of 13 metres.

For the figure shown, O is the centre of the circular base of radius r metres. OV, the height of the tent, is h metres.

    a. Calculate the height of the cone if the radius of the base is $\dfrac{13\sqrt{6}}{3}$ metres.

    b. Express the volume V in terms of h, given that the formula for the volume of a cone is

$V = \dfrac{1}{3}\pi r^2 h$.

    c. State any restrictions on the values h can take and sketch the graph of V against h for these restrictions.

    d. Express the volume as multiples of π for $h = 7$, $h = 8$, $h = 9$ and hence obtain the integer a so that the greatest volume occurs when $a < h < a + 1$.

    e. Specify the greatest volume to the nearest whole number.

17. A rectangular prism is to be made from 64 m of wire. If the lengths of the sides of the rectangular base are in the ratio of 5 : 1, determine the maximum volume of the cuboid, correct to two decimal places.

18. A sheet of cardboard measures 12 cm by 16 cm. Four equal squares are cut out of the corners and the sides are turned up to form an open rectangular box. If the length of the squares is restricted to be from 1 cm to 3 cm, determine the minimum volume of the box and its dimensions.

19. The diagonal of the base of a box in the shape of a rectangular prism has a length of 15 cm. One side of the base has a length of a cm. If the other side of the base and the height of the box are equal in length, determine the maximum volume of the box, to the nearest cm^3.

20. Calculate the maximum volume, to the nearest integer, of a cylinder that can be inscribed in a sphere with a diameter of 18 cm.

Fully worked solutions for this chapter are available online.

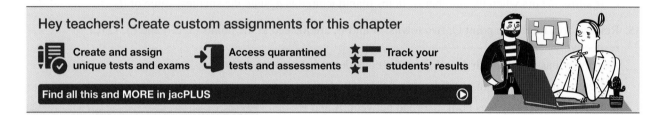

Answers

Chapter 3 Binomial expansions and cubic functions

3.2 Pascal's triangle and binomial expansions

3.2 Exercise

1. a. 6   b. 24   c. 120   d. 2   e. 1   f. 7

2. a. 7!   b. 8!   c. 8!   d. 9!

3. 734

4. a. 650   b. $\dfrac{1}{43}$ c. $\dfrac{7}{5}$ d. $\dfrac{6}{5}$

5. 35

6. a. 10   b. 10   c. 1   d. 1140   e. 1   f. 286

7. a. 2002   b. 840

8. a. 43 758   b. 6720

9. See the table at the bottom of the page.*

10. a. $x^5 + 20x^4 + 160x^3 + 640x^2 + 1280x + 1024$

    b. $x^5 - 20x^4 + 160x^3 - 640x^2 + 1280x - 1024$

    c. $x^5y^5 + 10x^4y^4 + 40x^3y^3 + 80x^2y^2 + 80xy + 32$

11. a. $81x^4 - 540x^3y + 1350x^2y^2 - 1500xy^3 + 625y^4$

    b. $81 - 108x^2 + 54x^4 - 12x^6 + x^8$

12. a. $^7C_3$

    b. 84

    c. Sample responses can be found in the worked solutions in the online resources.

    d. Sample responses can be found in the worked solutions in the online resources.

13. $12x + 40x^3 + 12x^5$

14. $\dbinom{n}{2} = \dfrac{n(n-1)}{2}$ and $\dbinom{21}{2} = 210$

15. a. $\dfrac{n(n-1)(n-2)}{6}$ b. $\dfrac{n(n-1)(n-2)}{6}$

    c. $\dfrac{(n+3)(n+2)(n+1)}{6}$ d. $n(2n+1)$

    e. $\dfrac{n(n-1)(n+1)}{6}$ f. $\dfrac{n(n-1)(n+1)}{6}$

16. a. $720w^3$ b. $\dfrac{945}{b^4}$ c. 54

17. $a = -2$

18. -3

19. $(x+1)^5 - (x+1)^4 = x^5 + 4x^4 + 6x^3 + 4x^2 + x$; sample responses can be found in the worked solutions in the online resources.

20. $a = 120, b = 55, c = 495$

3.3 The binomial theorem

3.3 Exercise

1. $32x^5 + 240x^4 + 720x^3 + 1080x^2 + 810x + 243$

2. $x^7 - 14x^6 + 84x^5 - 280x^4 + 560x^3 - 672x^2 + 448x - 128$

3. a. $x^3 + 3x^2y + 3xy^2 + y^3$

    b. $a^4 + 8a^3 + 24a^2 + 32a + 16$

    c. $m^4 - 12m^3 + 54m^2 - 108m + 81$

    d. $32 - 80x + 80x^2 - 40x^3 + 10x^4 - x^5$

4. a. $1 - \dfrac{6}{x} + \dfrac{12}{x^2} - \dfrac{8}{x^3}$

    b. $1 + \dfrac{4p}{q} + \dfrac{6p^2}{q^2} + \dfrac{4p^3}{q^3} + \dfrac{p^4}{q^4}$

    c. $81 - 54m + \dfrac{27}{2}m^2 - \dfrac{3}{2}m^3 + \dfrac{1}{16}m^4$

    d. $8x^3 - 12x + \dfrac{6}{x} - \dfrac{1}{x^3}$

5. a. $x^5 + 5x^4 + 10x^3 + 10x^2 + 5x + 1$

    b. $32 - 80x + 80x^2 - 40x^3 + 10x^4 - x^5$

    c. $64x^6 + 576x^5y + 2160x^4y^2 + 4320x^3y^3 + 4860x^2y^4 + 2916xy^5 + 729y^6$

6. a. $\dfrac{x^7}{128} + \dfrac{7x^6}{32} + \dfrac{21x^5}{8} + \dfrac{35x^4}{2} + 70x^3 + 168x^2 + 224x + 128$

    b. $x^8 - 8x^6 + 28x^4 - 56x^2 + 70 - \dfrac{56}{x^2} + \dfrac{28}{x^4} - \dfrac{8}{x^6} + \dfrac{1}{x^8}$

    c. $x^{20} + 10x^{18} + 45x^{16} + 120x^{14} + 210x^{12} + 252x^{10} + 210x^8 + 120x^6 + 45x^4 + 10x^2 + 1$

7. $-\dfrac{35}{648}x^4y^3$

8. a. $720w^3$ b. $\dfrac{945}{b^4}$

9. 54

10. $t_4 = 160$

11. a. $t_4 = 20\,000x^3$

    b. $t_3 = 1215x^8$

    c. $t_4 = 280x^4y^3, t_5 = 560x^3y^4$

*9.

Binomial power	Expansion	Number of terms in the expansion	Sum of indices in each term
$(x+a)^2$	$x^2 + 2xa + a^2$	3	2
$(x+a)^3$	$x^3 + 3x^2a + 3xa^2 + a^3$	4	3
$(x+a)^4$	$x^4 + 4x^3a + 6x^2a^2 + 4xa^3 + a^4$	5	4
$(x+a)^5$	$x^5 + 5x^4a + 10x^3a^2 + 10x^2a^3 + 5xa^4 + a^5$	6	5

12. The coefficient of x^8 is $11 \times 5 \times 3^6 \times 2^{24}$.

13. Sixth term; the coefficient of x^{15} is $2^9 \times 3^5 \times 7$.

14. $\dfrac{63x^{10}y^5}{8}$

15. 210

16. -672

17. $11 \times 7 \times 3^7 \times 2^{11}$

18. $a = 3$ (assuming a is non-zero)

19. $n = 4$ and $k = \dfrac{1}{2}$

20. $a = 318\,195; b = 224\,953$

3.4 Polynomials and cubic functions

3.4 Exercise

1. A: Degree 5; leading coefficient 4; constant term 12; coefficients $\in Z$
C: Degree 2; leading coefficient -0.2; constant term 5.6; coefficients $\in Q$

2. a. Polynomial of degree 4

   b. Polynomial of degree 3

   c. Not a polynomial due to the $\sqrt{x}$ term

   d. Not a polynomial due to the $\dfrac{6}{x^2}$ term and the $\dfrac{2}{x}$ term

3. a. A, B, D, F are polynomials.

	Degree	Type of coefficient	Leading term	Constant term
A	5	Q	$3x^5$	12
B	4	R	$-5x^4$	9
D	4	Z	$-18x^4$	0
F	6	N	$49x^6$	9

   b. C is not a polynomial due to $\sqrt{4x^5} = 2x^{\frac{5}{2}}$ term.
   E is not a polynomial due to $\dfrac{5}{3x^2} = \dfrac{5}{3}x^{-2}$ term.

4. a. $P(1) = 5$

   b. $P(-2) = -45$

   c. $P(3) = 77$, $P(-x) = -3x^3 - x^2 + 5$

   d. $P(-1) = 10$, $P(2a) = 8a^3 + 16a^2 - 4a + 5$.

5. a. 78   **b.** -12
   c. 0   **d.** -6
   e. -6 **f.** -5.868

6. a. $-14a$ **b.** $h^2 - 5h - 4$
   c. $2xh + h^2 - 7h$

7. a. 15   **b.** 10

8. a. $a = -8$ **b.** $b = 21$
   c. $k = 2$ **d.** $m = -8$

9. 21

10. a. $k = 25$ **b.** $a = -9$

   c. $n = -\dfrac{1}{5}$ **d.** $b = 4; c = 5$

11. $a = 2; b = -13; c = 6$

12. a. $a = 10, b = 6$

   b. $8x - 6 = 10x - 2(x + 3)$

   c. $6x^2 + 19x - 20 = (6x - 5)(x + 4)$

   d. $a = 9, b = -10, c = 1$

13. $(x + 2)^3 = x^2(x + 1) + 5x(x + 2) + 2(x + 3) + 2$

14. a. $x^3 - 7x^2 - 18x$

   b. $48x - 3x^3$

15. a. $x^3 - 3x^2 - 18x + 40$

   b. $x^3 + 6x^2 - x - 6$

   c. $x^3 - 12x^2 + 21x + 98$

   d. $x^3 + x^2 - x - 1$

16. a. $3x^4 + 7x + 12$

   b. $6x^4 + 10x^2 - 21x - 31$

   c. $6x^6 - 21x^5 - 29x^4 - 14x^3 - 20x^2 - 7x - 11$

17. $p = 5; q = -4$

18. $\therefore a = 1, b = -4, c = 5$

19. $a = -10, b = -3, c = 16$

20. a. m **b.** m **c.** $m + n$

3.5 Graphing cubic functions

3.5 Exercise

1. a. $(7, 0)$ **b.** $(0, -7)$
   c. $(0, 0)$ **d.** $(2, 2)$
   e. $(-5, -8)$ **f.** $\left(\dfrac{1}{2}, 5\right)$

2.

	Inflection point	y-intercept	x-intercept
a.	$(1, -8)$	$(0, -9)$	$(3, 0)$
b.	$(-6, 1)$	$(0, -5)$	$(-2.7, 0)$ approx.

a.

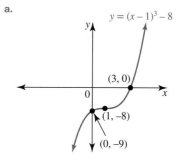

$y = (x - 1)^3 - 8$

b.

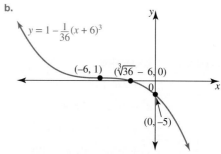

$y = 1 - \dfrac{1}{36}(x + 6)^3$

3. a.

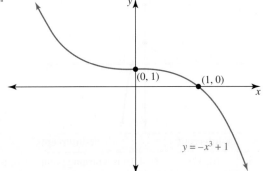

$y = -x^3 + 1$

(0, 1)
(1, 0)

b.

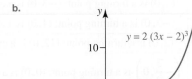

$y = 2(3x - 2)^3$

10

$\left(\frac{2}{3}, 0\right)$

(0, −16)

c.

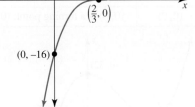

$y = 2(x + 3)^3 - 16$

38

(−1, 0)

−3   −1   0

−16

(−3, −16)

d.

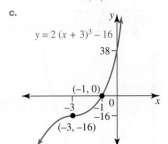

(0, 28)

10

(3, 1)

0   2   (4, 0)

$y = (3 - x)^3 + 1$

4.

	y-intercept	x-intercepts
a.	(0, −24)	(−6, 0), (−1, 0), (4, 0)
b.	(0, −24)	$\left(-\frac{1}{2}, 0\right)$, (4, 0), (6, 0)

a.

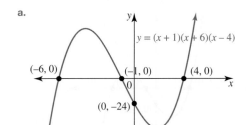

$y = (x + 1)(x + 6)(x - 4)$

(−6, 0)   (−1, 0)   (4, 0)

0

(0, −24)

b.

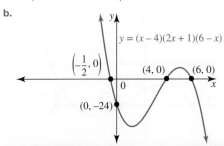

$y = (x - 4)(2x + 1)(6 - x)$

$\left(-\frac{1}{2}, 0\right)$

(4, 0)   (6, 0)

0

(0, −24)

5.

	y-intercept	x-intercepts
a.	(0, −8)	(−4, 0), (−1, 0), (2, 0)
b.	(0, 0)	(−8, 0), (0, 0), (5, 0)
c.	(0, −12)	(−3, 0), (1, 0), (4, 0)
d.	(0, 12)	(−4, 0), (2, 0), (6, 0)
e.	(0, 7)	$\left(-\frac{1}{4}, 0\right)$, $\left(\frac{7}{2}, 0\right)$, (10, 0)
f.	$\left(0, \frac{5}{2}\right)$	$\left(-\frac{8}{3}, 0\right)$, $\left(\frac{5}{8}, 0\right)$, (2, 0)

a.

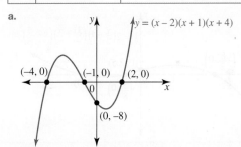

$y = (x - 2)(x + 1)(x + 4)$

(−4, 0)   (−1, 0)   (2, 0)

0

(0, −8)

b.

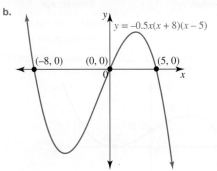

$y = -0.5x(x + 8)(x - 5)$

(−8, 0)   (0, 0)   (5, 0)

0

c.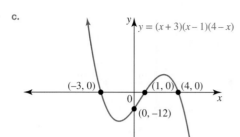

$y = (x+3)(x-1)(4-x)$

$(-3, 0)$ $(1, 0)$ $(4, 0)$ $(0, -12)$

d.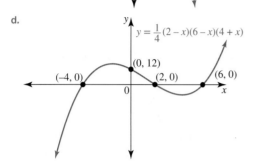

$y = \frac{1}{4}(2-x)(6-x)(4+x)$

$(-4, 0)$ $(0, 12)$ $(2, 0)$ $(6, 0)$

e.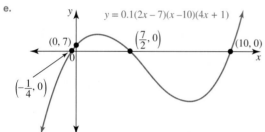

$y = 0.1(2x-7)(x-10)(4x+1)$

$(0, 7)$ $\left(\frac{7}{2}, 0\right)$ $(10, 0)$ $\left(-\frac{1}{4}, 0\right)$

f.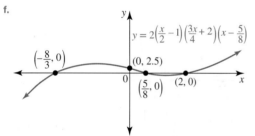

$y = 2\left(\frac{x}{2}-1\right)\left(\frac{3x}{4}+2\right)\left(x-\frac{5}{8}\right)$

$\left(-\frac{8}{3}, 0\right)$ $(0, 2.5)$ $\left(\frac{5}{8}, 0\right)$ $(2, 0)$

b.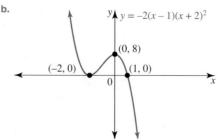

$y = -2(x-1)(x+2)^2$

$(0, 8)$ $(-2, 0)$ $(1, 0)$

7.

	y-intercept	**x-intercepts**
a.	$(0, 32)$	$(-4, 0)$ is a turning point; $(2, 0)$ is a cut
b.	$(0, 54)$	$(3, 0)$ is a turning point; $(-3, 0)$ is a cut
c.	$(0, 36)$	$(-3, 0)$ is a turning point; $(4, 0)$ is a cut
d.	$(0, -12)$	$(2, 0)$ is a turning point; $(12, 0)$ is a cut
e.	$(0, 0)$	$\left(-\frac{3}{2}, 0\right)$ is a turning point; $(0, 0)$ is a cut
f.	$(0, 0)$	$(0, 0)$ is a turning point; $(0.4, 0)$ is a cut

a.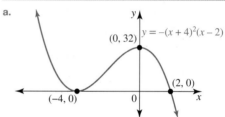

$y = -(x+4)^2(x-2)$

$(0, 32)$ $(-4, 0)$ $(2, 0)$

b.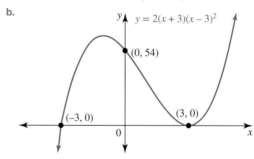

$y = 2(x+3)(x-3)^2$

$(0, 54)$ $(-3, 0)$ $(3, 0)$

c.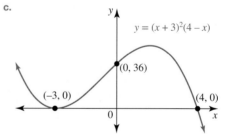

$y = (x+3)^2(4-x)$

$(0, 36)$ $(-3, 0)$ $(4, 0)$

d.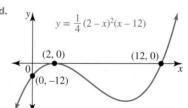

$y = \frac{1}{4}(2-x)^2(x-12)$

$(2, 0)$ $(12, 0)$ $(0, -12)$

6.

	y-intercept	**x-intercept**
a.	$(0, 6)$	$(-6, 0)$ and $(3, 0)$, which is a turning point
b.	$(0, 8)$	$(-2, 0)$ is a turning point and $(1, 0)$

a.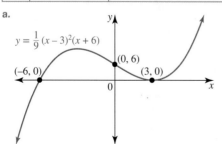

$y = \frac{1}{9}(x-3)^2(x+6)$

$(-6, 0)$ $(0, 6)$ $(3, 0)$

e.

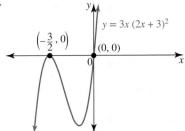

$y = 3x(2x+3)^2$

$\left(-\frac{3}{2}, 0\right)$ $(0, 0)$

f.

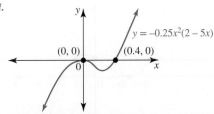

$y = -0.25x^2(2 - 5x)$

$(0, 0)$ $(0.4, 0)$

8.

	Stationary point of inflection	y-intercept	x-intercepts
a.	$(-3, 0)$	$(0, 27)$	$(-3, 0)$
b.	None	$(0, -9)$	$(-3, 0)$ is a turning point; $\left(\frac{1}{2}, 0\right)$ is a cut
c.	None	$(0, -15)$	$(-3, 0), \left(\frac{1}{2}, 0\right), (5, 0)$
d.	$\left(\frac{1}{2}, 1\right)$	$\left(0, \frac{3}{2}\right)$	$(1.1, 0)$ approx.
e.	None	$(0, 0)$	$\left(\frac{1}{4}, 0\right)$ is a turning point; $(0, 0)$ is a cut
f.	None	$(0, 4)$	$\left(\frac{2}{3}, 0\right)$ is a turning point; $\left(-\frac{2}{3}, 0\right)$ is a cut

a.

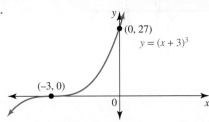

$(0, 27)$
$y = (x+3)^3$
$(-3, 0)$

b.

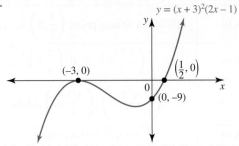

$y = (x+3)^2(2x-1)$
$(-3, 0)$ $\left(\frac{1}{2}, 0\right)$
$(0, -9)$

c.

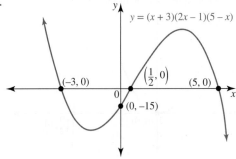

$y = (x+3)(2x-1)(5-x)$

$(-3, 0)$ $\left(\frac{1}{2}, 0\right)$ $(5, 0)$
$(0, -15)$

d.

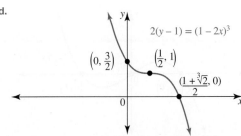

$2(y-1) = (1-2x)^3$

$\left(0, \frac{3}{2}\right)$ $\left(\frac{1}{2}, 1\right)$
$\dfrac{(1 + \sqrt[3]{2}, 0)}{2}$

e.

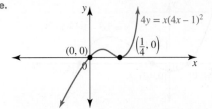

$4y = x(4x-1)^2$

$(0, 0)$ $\left(\frac{1}{4}, 0\right)$

f.

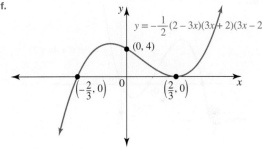

$y = -\frac{1}{2}(2 - 3x)(3x + 2)(3x - 2)$
$(0, 4)$
$\left(-\frac{2}{3}, 0\right)$ $\left(\frac{2}{3}, 0\right)$

9. a. $y = \dfrac{1}{49}(x-3)^3 - 7$

   b. $y = 0.25x(x+5)(x-4)$

   c. $y = -(x+2)^2(x-3)$

10. a. $y = \dfrac{1}{2}(x+8)(x+4)(x+1)$

   b. $y = -2x^2(x-5)$

   c. $y = -3(x-1)^3 - 3$

   d. $y = \dfrac{4}{5}(x-1)(x-5)^2$

11. a. $y = \dfrac{1}{3}(x-3)^3 + 9$

   b. $y = (x+2)^3 + 2$

   c. $y = -2x^3 + 4$

12. a. $y = (x+5)^3 + 4$

   b. $y = -(x-2)^3 - 1$

   c. $y = -(x-3)^3 - 1$

13. See the table at the bottom of the page.*

a.

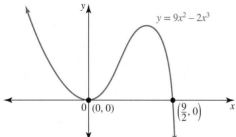

$y = 9x^2 - 2x^3$

$(0, 0)$, $\left(\frac{9}{2}, 0\right)$

b.

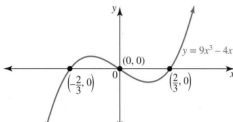

$y = 9x^3 - 4x$

$(0, 0)$, $\left(-\frac{2}{3}, 0\right)$, $\left(\frac{2}{3}, 0\right)$

c.

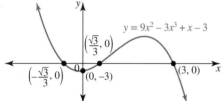

$y = 9x^2 - 3x^3 + x - 3$

$\left(\frac{\sqrt{3}}{3}, 0\right)$, $\left(-\frac{\sqrt{3}}{3}, 0\right)$, $(0, -3)$, $(3, 0)$

d.

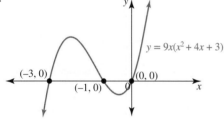

$y = 9x(x^2 + 4x + 3)$

$(-3, 0)$, $(-1, 0)$, $(0, 0)$

e.

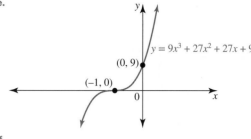

$y = 9x^3 + 27x^2 + 27x + 9$

$(0, 9)$, $(-1, 0)$

f.

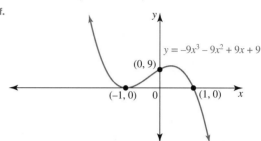

$y = -9x^3 - 9x^2 + 9x + 9$

$(0, 9)$, $(-1, 0)$, $(1, 0)$

14. a.

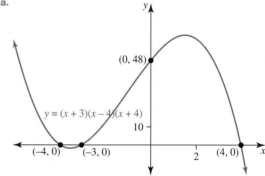

$(0, 48)$

$y = (x + 3)(x - 4)(x + 4)$

10

$(-4, 0)$, $(-3, 0)$, 2, $(4, 0)$

b. $y = (x - 2)(2x - 1)(x + 3)$

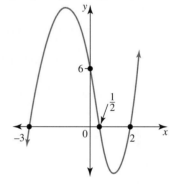

6, $\frac{1}{2}$, -3, 2

***13.**

	Factorised form	**Stationary point of inflection**	**y-intercept**	**x-intercepts**
a.	$y = x^2(9 - 2x)$	None	$(0, 0)$	$(0, 0)$ is a turning point; $\left(\frac{9}{2}, 0\right)$ is a cut
b.	$y = x(3x - 2)(3x + 2)$	None	$(0, 0)$	$\left(-\frac{2}{3}, 0\right), (0, 0) \left(\frac{2}{3}, 0\right)$
c.	$(x - 3)\left(1 - \sqrt{3}x\right)\left(1 + \sqrt{3}x\right)$	None	$(0, -3)$	$\left(-\frac{\sqrt{3}}{3}, 0\right), \left(\frac{\sqrt{3}}{3}, 0\right), (3, 0)$
d.	$y = 9x(x + 1)(x + 3)$	None	$(0, 0)$	$(-3, 0), (-1, 0), (0, 0)$
e.	$y = 9(x + 1)^3$	$(-1, 0)$	$(0, 9)$	$(-1, 0)$
f.	$y = -9(x + 1)^2(x - 1)$	None	$(0, 9)$	$(-1, 0)$ is a turning point; $(1, 0)$ is a cut

15. a.

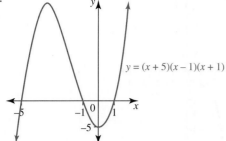

$y = (x + 5)(x - 1)(x + 1)$

b.

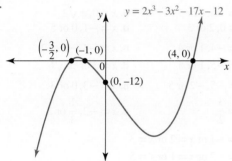

$(0, 9)$
$y = -(x - 1)(x - 3)^2$
$(-3, 0)$
$(1, 0)$

16. See the table at the bottom of the page.*

a.

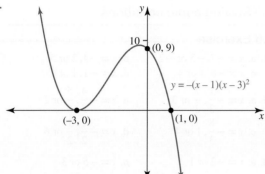

$y = 2x^3 - 3x^2 - 17x - 12$

$\left(-\frac{3}{2}, 0\right)$ $(-1, 0)$ $(4, 0)$
$(0, -12)$

b.

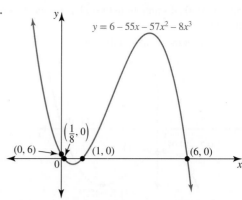

$y = 6 - 55x - 57x^2 - 8x^3$

$\left(\frac{1}{8}, 0\right)$
$(0, 6)$ $(1, 0)$ $(6, 0)$

c.

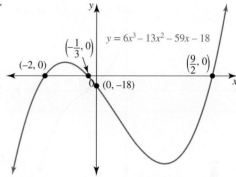

$y = 6x^3 - 13x^2 - 59x - 18$

$\left(-\frac{1}{3}, 0\right)$
$(-2, 0)$
$\left(\frac{9}{2}, 0\right)$
$(0, -18)$

d.

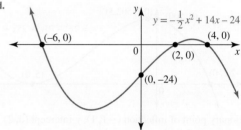

$y = -\frac{1}{2}x^2 + 14x - 24$

$(-6, 0)$ $(4, 0)$
$(2, 0)$
$(0, -24)$

17. a. Maximum turning point $(-5.4, 100.8)$; minimum turning point $(2.7, -166.0)$; y-intercept $(0, -96)$; x-intercepts $(-8, 0), (-2, 0), (6, 0)$

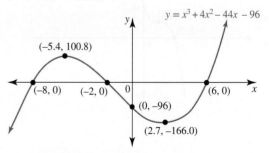

$y = x^3 + 4x^2 - 44x - 96$

$(-5.4, 100.8)$
$(-8, 0)$ $(-2, 0)$ $(6, 0)$
$(0, -96)$
$(2.7, -166.0)$

***16.**

	Factorised form	y-intercept	x-intercepts
a.	$y = (x + 1)(2x + 3)(x - 4)$	$(0, -12)$	$\left(-\frac{3}{2}, 0\right), (-1, 0), (4, 0)$
b.	$y = -(x - 1)(8x - 1)(x - 6)$	$(0, 6)$	$\left(\frac{1}{8}, 0\right), (1, 0), (6, 0)$
c.	$y = (x + 2)(3x + 1)(2x - 9)$	$(0, -18)$	$(-2, 0), \left(-\frac{1}{3}, 0\right), \left(\frac{9}{2}, 0\right)$
d.	$y = -\frac{1}{2}(x - 2)(x + 6)(x - 4)$	$(0, -24)$	$(-6, 0), (2, 0), (4, 0)$

b. See the worked solutions in the online resources for the proof.

18. See the table at the bottom of the page.*

a.

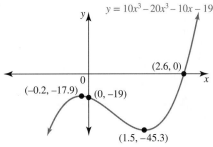

$y = 10x^3 - 20x^3 - 10x - 19$

(2.6, 0)

(−0.2, −17.9) (0, −19)

(1.5, −45.3)

b.

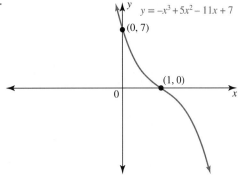

$y = -x^3 + 5x^2 - 11x + 7$

(0, 7)

(1, 0)

c.

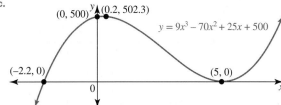

(0, 500) (0.2, 502.3)

$y = 9x^3 - 70x^2 + 25x + 500$

(−2.2, 0)

(5, 0)

19. Stationary point of inflection $(-1, 1)$; y-intercept $(0, 2)$; x-intercept $(-2, 0)$

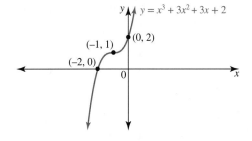

$y = x^3 + 3x^2 + 3x + 2$

(−1, 1) (0, 2)

(−2, 0)

20.

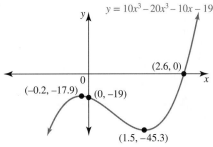

$y = (x + 1)^3 - 1$

(0, 0)

(−1, −1)

3.6 Solving cubic equations

3.6 Exercise

1. a. $x = -7, -3$ or -2 **b.** $x = -4, 2$ or 5
  c. $x = -8, 1$ or 4 **d.** $x = -1, 1$ or 6

2. a. $x = -\dfrac{1}{3}, \dfrac{7}{2}$ or 3 **b.** $x = -\dfrac{4}{5}, \dfrac{5}{3}$ or 2
  c. $x = -\dfrac{8}{3}, 1$ or 4 **d.** $x = -\dfrac{1}{2}, \dfrac{1}{2}$ or 6

3. a. $x = -2$ or 1 **b.** $x = -5$ or 3
  c. $x = \dfrac{3}{2}$ or 5 **d.** $x = -8$ or 0

4. a. $x = -5, 0$ or 5 **b.** $x = 0, \pm\sqrt{2}$
  c. $x = 0, 2$ or 3 **d.** $x = -1, 0$ or 5

5. a. $x = -4, 1$ or 4 **b.** $x = -4, -2$ or 2
  c. $x = -5, 1$ or 5 **d.** $x = 3$ or $\pm\sqrt{6}$

6. a. $x = -5, -2$ or 5 **b.** $x = -3, 0$ or 3
  c. $x = \dfrac{3}{2}$ **d.** $x = -2, \dfrac{1}{3}$ or 2

7. a. $x = -1$ or $x = 2$ or $x = 3$
  b. $x = -2$ or $x = 1$ or $x = 5$
  c. $x = -3$ or $x = -2$ or $x = -1$
  d. $x = -4$ or $x = 5$

8. a. $x = -2$ or $x = 1$ or $x = 4$
  b. $x = -7$ or $x = 2$ or $x = 3$
  c. $x = -6$ or $x = -1$ or $x = -\dfrac{1}{2}$
  d. $x = -\dfrac{1}{2}$ or $x = \dfrac{3}{2}$ or $x = 3$

9. a. $x = -3$ or $x = 1$ or $x = 2$
  b. $x = -1$ or $x = \dfrac{1}{6}$ or $x = \dfrac{4}{3}$
  c. $x = -3$ or $x = -\dfrac{3}{2}$ or $x = 1$
  d. $x = -1$ or $x = 6$ or $x = 8$

*18.

	Maximum turning point	Minimum turning point	y-intercept	x-intercepts
a.	$(-0.2, -17.9)$	$(1.5, -45.3)$	$(0, -19)$	$(2.6, 0)$
b.	None	None	$(0, 7)$	$(1, 0)$
c.	$(0.2, 502.3)$	$(5, 0)$	$(0, 500)$	$(-2.2, 0), (5, 0)$

10. The point of intersection is $\left(-\sqrt[3]{2}, 3\sqrt[3]{2}\right)$. When $x > -\sqrt[3]{2}$, $-3x < (x+2)(x-1)^2$.

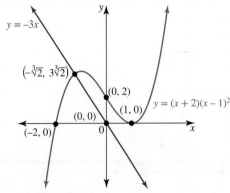

11. $(0,0), \left(\dfrac{1}{2}, \dfrac{1}{4}\right)$

12. $(-1, -2)$

13. $y = -2x + 5$; 1 solution

14. $y = x^3 + 3x^2, y = 4x$; 3 solutions

15. There are 3 solutions. One method is to use $y = x^3$, $y = -3x^2 + 4x$.

16. $x = -1$ or $x = 2$ or $x = 4$

17. $x = -\dfrac{1}{2}$ or $x = 2$ or $x = 5$

18. See the worked solutions in the online resources for a sample proof.

19. $p = -2$ or $p = 4$ or $p = 8$

20.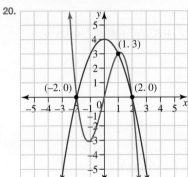

3.7 Modelling cubic functions

3.7 Exercise

1. a. $h = \dfrac{3 - 4x}{2}$

   b. See the worked solutions in the online resources for a sample proof.

   c. $0 \le x \le \dfrac{3}{4}$

d.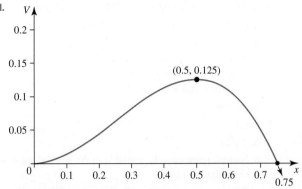

x-intercepts at $x = 0$ (touch), $x = 0.75$ (cut); shape of a negative cubic

e. Cube of edge 0.5 m

2. a. $l = 20 - 2x; w = 12 - 2x; V = (20 - 2x)(12 - 2x)x$

   b. $0 \le x \le 6$

   c.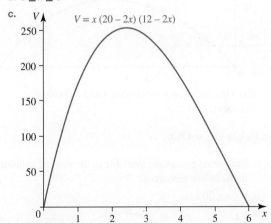

   x-intercepts at $x = 10, x = 6, x = 0$, but since $0 \le x \le 6$, the graph won't reach $x = 10$; the shape is of a positive cubic.

   d. Length 15.14 cm; width 7.14 cm; height 2.43 cm; greatest volume 263 cm^3

3. a. Making 5 sculptures gives a loss of $125.
   Making 6 sculptures gives a profit of $184.

   b. Sample responses are available in the worked solutions in the online resources.

   c. Too many and the costs outweigh the revenue from the sales. A negative cubic tends to $-\infty$ as x becomes very large.

   d. i. Profit $304

   ii. Loss $113

   e.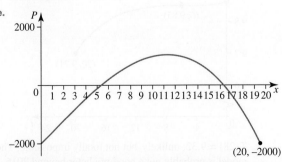

   x-intercepts lie between 5 and 6 and between 16 and 17.

   f. Between 6 and 16 inclusive

4. a. 54 **b.** 2 hours **c.** 210 **d.** 5 pm

5. a. $(0, 2.1), (1.25, 1), (2.5, 1.1), (4, 0.1)$

  b. $d = 2.1$

  c. $125a + 100b + 80c = -70.4$
  $125a + 50b + 20c = -8$
  $64a + 16b + 4c = -2$

  d. $0.77\,\text{m}$

6. $-4, 1$

7. a. $2\sqrt{2x}$

  b. Sample responses are available in the worked solutions in the online resources.

  c. Sample responses are available in the worked solutions in the online resources.

  d. i. $110\,\text{m}^3$

  ii. Mathematically $0 \le h \le 8$

  e.

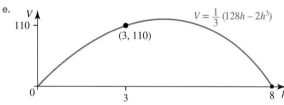

  Max volume when h is between 4 and 5 (estimates will vary).

  f. Height $\dfrac{8}{\sqrt{3}} \approx 4.6\,\text{m}$

8. a, b. Sample responses are available in the worked solutions in the online resources.

  c. $0 \le r \le 20$

  d.

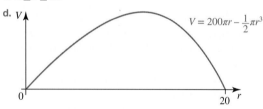

  e. Radius 2 cm, height 99 cm or radius 18.9 cm, height 1.1 cm

  f. $4837\,\text{cm}^3$

9. a. $y = -\dfrac{1}{32}x^2(x-6)$ **b.** $\dfrac{81}{32}\,\text{km}$ or $2\dfrac{17}{32}\,\text{km}$

10. a. $T(3) = 9.85$, $T(20) = 9.71$

  b.

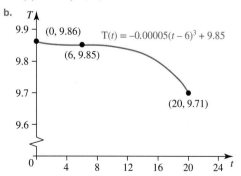

  c. $T(28) = 9.32$; unlikely, but not totally impossible. The model is probably not a good predictor beyond 2016.

11. a. $x = 0, x = 6$

  b. Length $2x - 6$; width $6x - x^2$

  c. Sample responses are available in the worked solutions in the online resources.

  d. $3 \le x \le 6$

  e. $x = 4, x = \dfrac{5 + \sqrt{33}}{2}$

12. a. $y = -2x(x^2 - 9)$

  b.

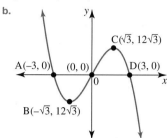

  c. $y = 12x$

  d. $y = 4x^3$

13. $21.3\,\text{m}^3$

14. $175\,\text{m}^3$

15. Maximum volume $2112\,\text{cm}^3$
Dimensions 12 cm by 4 cm and height of 44 cm

3.8 Review

3.8 Exercise

1. a. 210

  b. i. 1 5 10 10 5 1

  ii. $1 + 10xy + 40x^2y^2 + 80x^3y^3 + 80x^4y^4 + 32x^5y^5$

2. D

3. $a = -2$, $b = 2$

4. a. $x = -2$ or $x = 4$ or $x = 5$

  b. $x = -\dfrac{7}{3}$ or $x = 8$

5. B

6. A

7. $(0, 0), \left(\sqrt{6}, -2\sqrt{6}\right), \left(-\sqrt{6}, 2\sqrt{6}\right)$

8. D

9. D

10. C

11. a. $x^8 - 12x^5 + 54x^2 - 108x^{-1} + 81x^{-4}$

  b. 40 095

12. a. Sample responses are available in the worked solutions in the online resources.

  b. $0 < x < 17.5$

  c. $1377\,\text{cm}^3$

d. See the figure at the bottom of the page.*

e. Sample responses are available in the worked solutions in the online resources.

13. a.

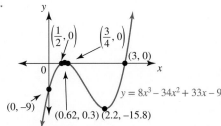

$y = 8x^3 - 34x^2 + 33x - 9$

(0, −9)
(0.62, 0.3) (2.2, −15.8)

b. $\left\{0, \dfrac{3}{2}, \dfrac{11}{4}\right\}$

c. **i.** $-15.8 < k < 0.3$

　　ii. $k = 0.3, k = -15.8$

　　iii. $k < -15.8$ or $k > 0.3$

14. a. R has degree 2; C has degree 3

b. Revenue \$90; cost \$6; profit \$84

c. The profit is revenue $R - $ cost C.

$$\therefore P(x) = R(x) - C(x)$$
$$= 6(2x^2 + 10x + 3) - x(6x^2 - x + 1)$$
$$= 12x^2 + 60x + 18 - 6x^3 + x^2 - x$$
$$\therefore P(x) = -6x^3 + 13x^2 + 59x + 18$$

d. Restriction $x \ge 0$; x-intercept $(4.5, 0)$

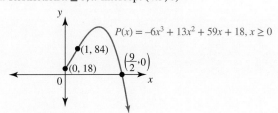

$P(x) = -6x^3 + 13x^2 + 59x + 18, x \ge 0$

(1, 84)
(0, 18)
$\left(\dfrac{9}{2}, 0\right)$

e. $d = 4501$

15. a. $y = -2x + 22$

b. $a = \dfrac{2}{3}, b = -7$

c. End points: $(0, -28), (8, 20)$

d.

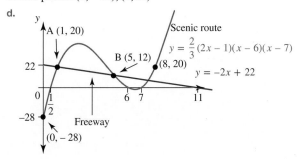

A (1, 20)
22
B (5, 12)
(8, 20)
Scenic route
$y = \dfrac{2}{3}(2x - 1)(x - 6)(x - 7)$
$y = -2x + 22$
−28
Freeway
(0, −28)

e. $(1, 20), (5, 12)$ and $\left(\dfrac{15}{2}, 7\right)$

f. $\left(\dfrac{15}{2}, 7\right)$

16. a. $\dfrac{13\sqrt{3}}{3}$ metres

b. $V = \dfrac{1}{3}\pi h(169 - h^2)$

c. $0 \le h \le 13$

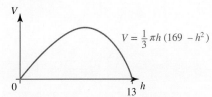

$V = \dfrac{1}{3}\pi h (169 - h^2)$

d. $V(7) = 280\pi = V(8), V(9) = 264\pi, a = 7$

e. $886\,\text{m}^3$

17. $84.28\,\text{m}^3$

18. $140\,\text{cm}^3$, dimensions: $10\,\text{cm}$ by $14\,\text{cm}$ with height of $1\,\text{cm}$

19. $1299\,\text{cm}^3$

20. $1763\,\text{cm}^3$

---

*12. d.

Turning point
(11.6667, 1587.963)

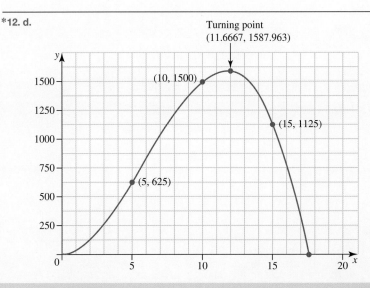

(10, 1500)
(15, 1125)
(5, 625)

4 Introduction to functions and relations

LESSON SEQUENCE

Fully worked solutions for this chapter are available online.

EXAM PREPARATION

Access exam-style questions in every lesson, available online.

on Resources

Solutions	Solutions — Chapter 4 (sol-1271)
Exam questions	Exam question booklet — Chapter 4 (eqb-0292)
Digital documents	Learning matrix — Chapter 4 (doc-41786)
	Chapter summary — Chapter 4 (doc-41799)

LESSON
4.1 Overview

4.1.1 Introduction

Functions are one of the most fundamental ideas in modern mathematics. Concepts related to functions have been developed over centuries by many famous mathematicians, including Leibniz, Euler and Fourier. Defining a function on a basic level allows for analysis of situations that appear to be complex but can often be modelled by an equation or set of equations. More thorough investigation can occur by looking at derivatives and integrals of functions using methods of calculus, allowing for a deeper understanding of the model or optimisation of processes.

These concepts and skills are particularly important in numerous careers including many engineering disciplines, medical research and computer science. In these careers, functions are used to develop safer structures, evaluate drug efficacy and design, and optimise programs, among many other uses. Functions are also used extensively in astrophysics to calculate trajectories for space travel! Calculating timing and direction is vital for successful launches and re-entries of space shuttles. When exploring further away from Earth, functions can be used to model gravitational slingshot manoeuvres around stars, planets and moons, allowing us to reach further into the cosmos.

4.1.2 Syllabus links

Lesson	Lesson title	Syllabus links
4.2	**Functions and relations**	○ Understand the concept of a relation as a mapping between sets, a graph and as a rule or a formula that defines one variable quantity in terms of another. ○ Recognise the distinction between functions and relations and use the vertical line test to determine whether a relation is a function. ○ Recognise and use function notation, domain and range, and independent and dependent variables.
4.3	**Function notation**	○ Recognise and use function notation, domain and range, and independent and dependent variables.
4.4	**Piece-wise functions**	○ Recognise and use piece-wise functions as a combination of multiple sub-functions with restricted domains. ○ Model and solve problems that involve piece-wise functions with and without technology.

Source: Mathematical Methods Senior Syllabus 2024 © State of Queensland (QCAA) 2024; licensed under CC BY 4.0.

LESSON
4.2 Functions and relations

SYLLABUS LINKS

- Understand the concept of a relation as a mapping between sets, a graph and as a rule or a formula that defines one variable quantity in terms of another.
- Recognise the distinction between functions and relations and use the vertical line test to determine whether a relation is a function.
- Recognise and use function notation, domain and range, and independent and dependent variables.

Source: Mathematical Methods Senior Syllabus 2024 © State of Queensland (QCAA) 2024; licensed under CC BY 4.0.

4.2.1 Set and interval notation

Set notation

A **set** is a collection of things, and in mathematics sets are usually used to represent a group of numbers. Each number within a set is called an **element**, and these elements can be listed individually or described by a rule. Elements within a set are separated by commas. Some important symbols and pre-defined common sets are listed below.

{...} refers to a set of something.

$\in$ means 'is an element of'.

$\notin$ means 'is not an element of'.

$\subset$ means 'is a subset of'.

$\not\subset$ means 'is *not* a subset (or is not contained in)'.

$\cap$ means 'intersection with'.

$\cup$ means 'union with'.

\ means 'excluding'.

$\varnothing$ refers to 'the null set or empty set'.

$\{a, b, c\}$ is a set of three letters.

$\{(a, b), (c, d), ...\}$ is an infinite set of ordered pairs.

N refers to the set of natural numbers.

Z refers to the set of integers.

Q refers to the set of rational numbers.

R refers to the set of real numbers.

R^+ refers to the set of positive real numbers.

R^- refers to the set of negative real numbers.

Interval notation

Interval notation is a convenient way to represent an **interval** using only the end values and indicating whether those end values are included or excluded. When using interval notation, a rounded bracket is used to indicate a value that is excluded and a square bracket is used to indicate a value that is included in the interval. Recall that on a number line and on a Cartesian plane, excluded values are represented by an open circle and included values by a closed circle.

If a and b are real numbers and $a < b$, then the following intervals are defined with accompanying number lines.

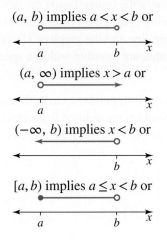

(a, b) implies $a < x < b$ or

(a, ∞) implies $x > a$ or

$(-\infty, b)$ implies $x < b$ or

$[a, b)$ implies $a \le x < b$ or

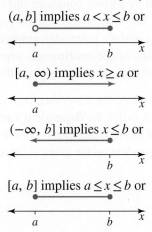

$(a, b]$ implies $a < x \le b$ or

$[a, \infty)$ implies $x \ge a$ or

$(-\infty, b]$ implies $x \le b$ or

$[a, b]$ implies $a \le x \le b$ or

WORKED EXAMPLE 1 Identifying subsets of the real numbers

Identify each of the following subsets of the real numbers using interval notation.

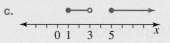

a.
b.
c.

THINK

a. The interval is $x < 2$ (2 is not included).

b. The interval is $-3 \leq x < 5$ (-3 is included).

c. The interval is both $1 \leq x < 3$ (1 is included, 3 is not included) and $x \geq 5$ (5 is included). The symbol $\cup$ indicates the combination of the two intervals.

WRITE

a. $(-\infty, 2)$

b. $[-3, 5)$

c. $[1, 3) \cup [5, \infty)$

WORKED EXAMPLE 2 Constructing intervals

Construct a number line to represent each of the following intervals.
a. $(-2, 10]$ b. $[1, \infty)$

THINK

a. The interval is $-2 < x \leq 10$ (-2 is not included, 10 is).

b. The interval is $x \geq 1$ (1 is included).

WRITE

a.

b.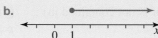

4.2.2 Relations

A mathematical **relation** may be described by:
- a set of ordered pairs
- a graph
- a rule or formula that defines one variable quantity in terms of another
- a mapping between sets.

The ordered pairs may be listed or described by a rule or presented as a graph. Examples of relations include $A = \{(-2, 4), (1, 5), (3, 4)\}$, where the ordered pairs have been listed; $B = \{(x, y) : y = 2x\}$, where the ordered pairs are described by a linear equation; and $C = \{(x, y) : y \leq 2x\}$, where the ordered pairs are described by a linear inequation. These relations could be presented visually by being graphed on coordinate axes. The graph of A would consist of three points, the graph of B would be a straight line and the graph of C would be a closed half-plane. Relations can be **continuous**, where all values of a variable are possible within a specified interval, or **discrete**, where only fixed values are permitted.

In a set of ordered pairs, the first value, or x-value, is referred to as the **independent variable** and the second value, or y-value, is called the **dependent variable**. The possible x-values are defined first, then the resulting y-values are found through substitution of these x-values into the rule that describes the relation. As such, the values of y are dependent on the given x-values.

Construct the graph representing each of the following relations and identify whether each is discrete or continuous.

a. $y = x^2$, where $x \in \{1, 2, 3, 4\}$

b. $y = 2x + 1$, where $x \in R$

THINK	WRITE
a. 1. Use the rule to calculate y and state the ordered pairs by letting $x = 1, 2, 3$ and 4.	a. When $x = 1$, $y = 1^2$ $= 1$ (1, 1) $x = 2$, $y = 2^2$ $= 4$ (2, 4) $x = 3$, $y = 3^2$ $= 9$ (3, 9) $x = 4$, $y = 4^2$ $= 16$ (4, 16)
2. Plot the points $(1, 1)$, $(2, 4)$, $(3, 9)$ and $(4, 16)$ on a set of axes.	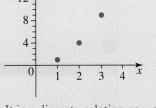
3. Do not join the points, as x is a discrete variable (x could only have the values 1, 2, 3 and 4).	It is a discrete relation as x can only have the values 1, 2, 3 and 4.
b. 1. Use the rule to calculate y. Select values of x, say $x = 0$, 1 and 2 (or determine the intercepts). State the ordered pairs.	b. When $x = 0$, $y = 2(0) + 1$ $= 1$ (0, 1) $x = 1$, $y = 2(1) + 1$ $= 3$ (1, 3) $x = 2$, $y = 2(2) + 1$ $= 5$ (2, 5)
2. Plot the points $(0, 1)$, $(1, 3)$ and $(2, 5)$ on a set of axes.	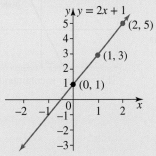
3. Join the points with a straight line, continuing in both directions as x is a continuous variable (any real number).	It is a continuous relation as x can be any real number.

TI | THINK	WRITE	CASIO | THINK	WRITE

a. 1. In a Lists & Spreadsheet page, label the first column as x and the second column as y.
Enter the values 1–4 in the first column.

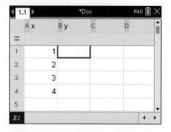

a. 1. On a Graph & Table screen, complete the entry line for Y1 as:
$Y1 = x^2$
then press EXE.

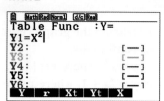

2. In the function cell below the label y, complete the entry line as:
$= x^2$
then press ENTER.
Select the Variable Reference for x when prompted.

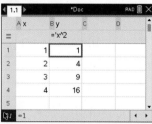

2. Select SET by pressing F5, then complete the fields as:
Start: 1
End: 4
Step: 1
then press EXE.
Select TABLE by pressing F6.

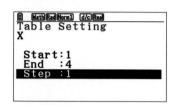

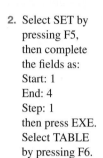

3. In a Data & Statistics page, click on the label of the horizontal axis and select x.
Click on the label of the vertical axis and select y.
The graph appears on the screen.

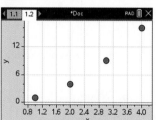

This relation is discrete.

3. Select GPH-PLT by pressing F6.

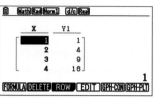

This relation is discrete.

b. 1. On a Graphs page, complete the entry line for function 1 as:
$f1(x) = 2x + 1$
then press ENTER.
The graph appears on the screen.

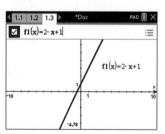

This relation is continuous.

b. 1. On a Graph screen, complete the entry line for Y1 as:
$Y1 = 2x + 1$
then press EXE.

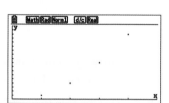

2. Select DRAW by pressing F6.

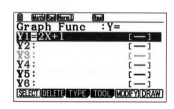

This relation is continuous.

Types of relations

Relations are classified according to the correspondence between the coordinates of their ordered pairs. Note that the word *many* in this context means more than one, and the precise number is not considered.

One-to-one relations

A one-to-one relation exists if for any x-value there is only one corresponding y-value and vice versa.

Examples are:

a. $\{(1, 1), (2, 2), (3, 3), (4, 4)\}$

b.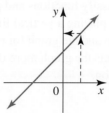

One-to-many relations

A one-to-many relation exists if for any x-value there is more than one y-value, but for any y-value there is only one x-value.

Examples are:

a. $\{(1, 1), (1, 2), (2, 3), (3, 4)\}$

b.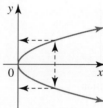

Many-to-one relations

A many-to-one relation exists if there is more than one x-value for any y-value but for any x-value there is only one y-value.

Examples are:

a. $\{(-1, 1), (0, 1), (1, 2)\}$

b.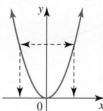

Many-to-many relations

A many-to-many relation exists if there is more than one x-value for any y-value and vice versa.

Examples are:

a. $\{(0, -1), (0, 1), (1, 0), (-1, 0)\}$

b.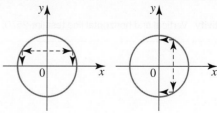

4.2.3 Functions

Relations that are one-to-one or many-to-one are called **functions**. That is, a function is a relation where for any x-value there is only one y-value.

This means that all functions are relations, but not all relations are functions.

The vertical line test

Line tests can be used to help classify functions and relations from a graph. A vertical line test is used to identify a function and can be applied by placing a vertical line (parallel to the y-axis) through the graph. If there is only one intersection between this line and the graph for each possible x-value, then the graph is a function. If the line can be placed such that it intersects the graph more than once while remaining parallel to the y-axis, then the graph does not represent a function.

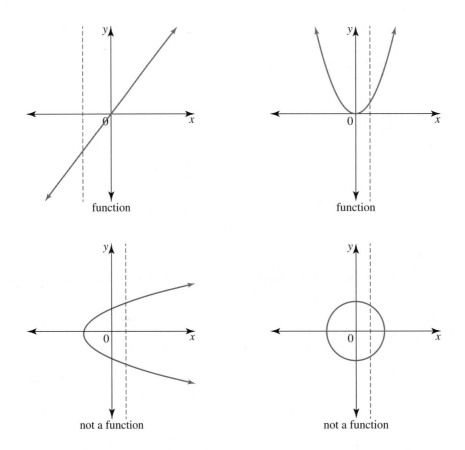

Notice that the first two graphs above pass the vertical line test (shown in pink), but the bottom two graphs do not. All four graphs are relations, but only the top two are functions.

WORKED EXAMPLE 4 Determining functions

Determine whether each relation is a function or not.

a. $y = (x + 3)(x - 1)(x - 6)$

b. $\{(1, 3), (2, 4), (1, 5)\}$

THINK

a. 1. Draw the graph.

WRITE

a. $y = (x + 3)(x - 1)(x - 6)$
x-intercepts: $(-3, 0), (1, 0), (6, 0)$
y-intercept: $(0, 18)$
The graph is a positive cubic.

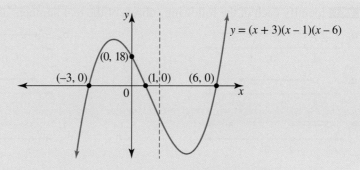

2. Apply the vertical line test.

A vertical line cuts the graph exactly once.

3. State whether the relation is a function.

$y = (x + 3)(x - 1)(x - 6)$ is a function.

b. 1. Look to see if there are points with the same x- or y-coordinates.

b. $\{(1, 3), (2, 4), (1, 5)\}$
$x = 1$ is paired to both $y = 3$ and $y = 5$. This is a relation. It is not a function.

2. Alternatively, or as a check, plot the points and apply the vertical line test.

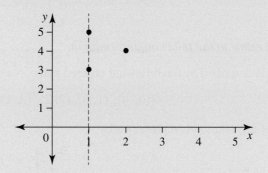

A vertical line cuts the graph in more than one place. This is a relation but not a function.

4.2 Exercise	4.2 Exam practice on

Simple familiar	Complex familiar	Complex unfamiliar
1, 2, 3, 4, 5, 6, 7, 8, 9, 10, 11, 12, 13, 14, 15, 16	17, 18, 19, 20	N/A

Simple familiar

1. 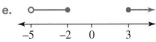 Identify each of the following subsets of the real numbers using interval notation.

   a.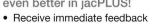

   b.

   c.

   d.

   e.

   f.

2. **WE2** Construct a number line to represent each of the following intervals.
   a. $[-6, 2)$
   b. $(-9, -3)$
   c. $(-\infty, 2]$
   d. $(1, 10]$
   e. $(-\infty, -2) \cup [1, 3)$
   f. $[-8, 0) \cup (2, 6]$

3. Identify each of the following sets using interval notation.
   a. $\{x : -4 \le x < 2\}$
   b. $\{y : -1 < y < \sqrt{3}\}$
   c. $\{x : x > 3\}$
   d. $\{x : x \le -3\}$
   e. R
   f. $R \setminus \{0\}$

Questions 4, 5 and 6 relate to the following information.

A particular relation is described by the following ordered pairs:

$$\{(0, 4), (1, 3), (2, 2), (3, 1)\}.$$

4. **MC** The graph of this relation is represented by:

   A.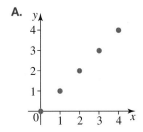

   B.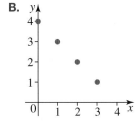

   C.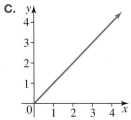

   D.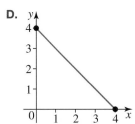

5. **MC** The elements of the dependent variable are:

A. $\{1, 2, 3, 4\}$ B. $\{1, 2, 3\}$

C. $\{0, 1, 2, 3, 4\}$ D. $\{0, 1, 2, 3\}$

6. **MC** The rule for the relation is correctly described by:

A. $y = 4 - x, \ x \in R$ B. $y = x - 4, \ x \in N$

C. $y = 4 - x, \ x \in N$ D. $y = 4 - x, \ x \in \{0, 1, 2, 3\}$

7. **WE3** Sketch the graph representing each of the following relations, and state whether each is discrete or continuous.

a.

Day	Mon	Tues	Wed	Thur	Fri	Sat	Sun
Cost of petrol (c/L)	168	167.1	166.5	164.9	167	168.5	170

b. $\{(0, 0), (1, 1), (2, 4), (3, 9)\}$

c. $y = -x^2$, where $x \in \{-2, -1, 0, 1, 2\}$

d. $y = x - 2$, where $x \in R$

e. $y = 2x + 3$, where $x \in J$

f. $y = x^2 + 2$, where $-2 \leq x \leq 2$ and $x \in R$

8. **WE4** Identify whether each relation is a function or not.

a. 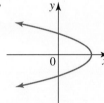   b. 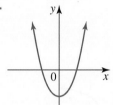   c.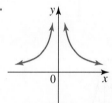

d. 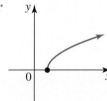   e.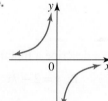

9. Identify whether each relation is a function or not.

a. 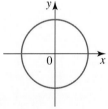   b. 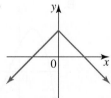   c.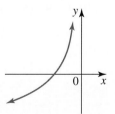

d. 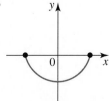   e. 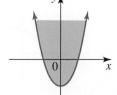   f.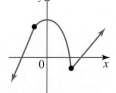

10. Consider the relations below and identify which are functions.

a.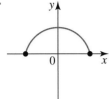

b.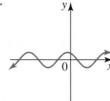

c.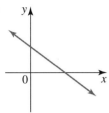

d.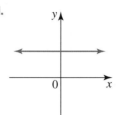

11. Consider the relations below and identify which are functions.

a.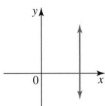

b.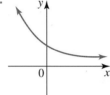

c.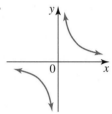

d.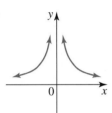

e.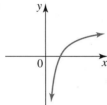

12. **MC** Identify which of the following rules does not describe a function.

A. $y = \dfrac{x}{5}$

B. $y = 2 - 7x$

C. $x = 5$

D. $y = 10x^2 + 3$

13. **MC** Identify which of the following relations is not a function.

A. $\{(5, 8), (6, 9), (7, 9), (8, 10), (9, 12)\}$

B.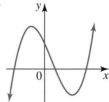

C. $y^2 = x$

D.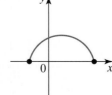

14. **MC** During one week, the number of people travelling on a particular train at a certain time progressively increases from Monday through to Friday. Identify which graph below best represents this information.

A.
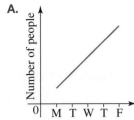

B.
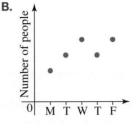

C.
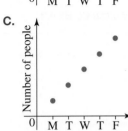

D.
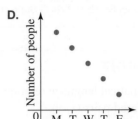

15. The table below shows the temperature of a cup of coffee, $T°C$, t minutes after it is poured.

t (min)	0	2	4	6	8
T (°C)	80	64	54	48	44

a. Plot the points on a graph.
b. Join the points with a smooth curve.
c. Explain why this can be done.
d. Use the graph to determine how long it takes the coffee to reach half of its initial temperature.

16. A salesperson in a computer store is paid a base salary of $300 per week plus $40 commission for each computer she sells. If n is the number of computers she sells per week and P dollars is the total amount she earns per week:

a. copy and complete the following table

n	0	1	2	3	4	5	6
P							

b. plot the information on a graph
c. explain why the points cannot be joined together
d. write an equation in terms of P and n to represent this situation.

Complex familiar

17. Consider the relation $y = 2$.
Identify whether the relation is a function. Justify your decision.

18. Consider the relation $x + 5 = 0$.
Identify whether the relation is a function. Justify your decision.

19. Consider the relation $y = (x + 5)^2 + 2$.
Identify whether the relation is a function. Justify your decision.

20. Consider the relation $y = (x - 2)(x + 5)^2$.
Identify whether the relation is a function. Justify your decision.

Fully worked solutions for this chapter are available online.

LESSON
4.3 Function notation

SYLLABUS LINKS

- Recognise the distinction between functions and relations and use the vertical line test to determine whether a relation is a function.
- Recognise and use function notation, domain and range, and independent and dependent variables.

Source: Mathematical Methods Senior Syllabus 2024 © State of Queensland (QCAA) 2024; licensed under CC BY 4.0.

4.3.1 Domain and range

Special terminology and mathematical language relating to this topic is listed below.

- A relation is any set of ordered pairs (x, y).
- The **domain** is the set of all x-values of the ordered pairs.
- The **range** is the set of all y-values of the ordered pairs.
- A function is a set of ordered pairs in which every x-value is paired to a unique y-value.
- An **implied domain** (or maximal domain) is the set of x-values for which the rule has meaning.
- A **restricted domain** occurs when restrictions are placed on values of variables in some practical situations and models. A restricted domain usually affects the range.

Consider the following examples of sets of ordered pairs.

$A = \{(-2, 4), (1, 5), (3, 4)\}$	$B = \{(x, y) : y = 2x\}$	$C = \{(x, y) : y \leq 2x\}$
This is a set of ordered pairs.	This set describes a linear equation.	This set describes a linear inequation.
The domain is $x \in \{-2, 1, 3\}$.	The domain is $x \in R$.	The domain is $x \in R$.
The range is $y \in \{4, 5\}$.	The range is $y \in R$.	The range is $y \in R$.
The relation A is a function as every x-value has a unique y-value.	The relation B is a function as every x-value has a unique y-value.	The relation C is not a function as every x-value does not have a unique y-value. $(1, 0)$, $(1, -1)$ and $(1, -2)$ are ordered pairs in the region with the same x-value.

These examples illustrate that functions can be recognised from their graphs by applying the vertical line test. Set notation or interval notation should be used for domains and ranges.

For each of the following, identify the domain and range, and whether the relation is a function or not.

a. $\{(1, 4), (2, 0), (2, 3), (5, -1)\}$

b.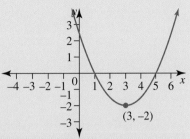

c.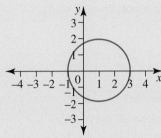

d. $\{(x, y) : y = 4 - x^3\}$

THINK

a. 1. State the domain.

2. State the range.
3. Are there any ordered pairs that have the same x-coordinate?

b. 1. Reading from left to right horizontally in the direction of the x-axis, the graph uses every possible x-value.
State the domain.

2. Reading from bottom to top vertically in the direction of the y-axis, the graph's y-values start at -2 and increase from there.
State the range.

3. Use the vertical line test.

c. 1. State the domain and range.
2. Use the vertical line test.

WRITE

a. $\{(1, 4), (2, 0), (2, 3), (5, -1)\}$
The domain is the set of x-values: $\{1, 2, 5\}$.

The range is the set of y-values: $\{-1, 0, 3, 4\}$.

The relation is not a function, since there are two different points with the same x-coordinate: $(2, 0)$ and $(2, 3)$.

b.

The domain is $(-\infty, \infty)$ or R.

The range is $[-2, \infty)$ or $\{y : y \geq -2\}$.
This is a function since any vertical line cuts the graph exactly once.

c. The domain is $[-1, 3]$; the range is $[-2, 2]$.

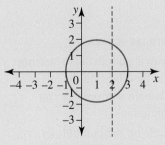

This is not a function, as a vertical line can cut the graph more than once.

d. 1. State the domain.

d. $y = 4 - x^3$

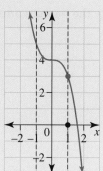

This is the equation of a polynomial, so its domain is R.

2. It is the equation of a cubic polynomial with a negative coefficient of its leading term, so as $x \to \pm\infty, y \to \mp\infty$.

State the range.

The range is R or $(-\infty, \infty)$.

3. Is the relation a function?

This is a function because all polynomial relations are functions, and it passes the vertical line test.

WORKED EXAMPLE 6 Identifying domain and range using interval notation

For each relation given, sketch its graph and identify the domain and range using interval notation.

a. $\left\{ (x, y) : y = \sqrt{x - 1} \right\}$

b. $\left\{ (x, y) : y = x^2 - 4, x \in [0, 4] \right\}$

THINK

WRITE

a. 1. The rule has meaning for $x \geq 1$ because if $x < 1, y = \sqrt{\text{negative number}}$.

a.

2. Therefore, calculate the value of y when $x = 1, 2, 3, 4$ and 5, and state the coordinate points.

When $x = 1, \quad y = \sqrt{0}$
$\qquad\qquad = 0 \qquad (1, 0)$
$x = 2, \quad y = \sqrt{1}$
$\qquad\qquad = 1 \qquad (2, 1)$
$x = 3, \quad y = \sqrt{2} \qquad (3, \sqrt{2})$
$x = 4, \quad y = \sqrt{3} \qquad (4, \sqrt{3})$
$x = 5, \quad y = \sqrt{4}$
$\qquad\qquad = 2 \qquad (5, 2)$

3. Plot the points on a set of axes.

4. Join the points with a smooth curve starting from $x = 1$, extending it beyond the last point. Since no domain is given, we can assume $x \in R$ (continuous).

5. Place a closed circle on the point $(1, 0)$ and put an arrow on the other end of the curve.

6. The domain is the set of values covered horizontally by the graph or implied by the rule.

Domain $= [1, \infty)$

7. The range is the set of values covered vertically by the graph.

Range $= [0, \infty)$

b. 1. Calculate the value of y when $x = 0, 1, 2, 3$ and 4, as the domain is [0, 4]. State the coordinate points.

b. When $x = 0$, $\quad y = 0^2 - 4$
$\qquad\qquad\qquad = -4 \qquad (0, -4)$

$\qquad x = 1$, $\quad y = 1^2 - 4$
$\qquad\qquad\qquad = -3 \qquad (1, -3)$

$\qquad x = 2$, $\quad y = 2^2 - 4$
$\qquad\qquad\qquad = 0 \qquad (2, 0)$

$\qquad x = 3$, $\quad y = 3^2 - 4$
$\qquad\qquad\qquad = 5 \qquad (3, 5)$

$\qquad x = 4$, $\quad y = 4^2 - 4$
$\qquad\qquad\qquad = 12 \qquad (4, 12)$

2. Plot these points on a set of axes.

3. Join the dots with a smooth curve from $x = 0$ to $x = 4$.

4. Place a closed circle on the points $(0, -4)$ and $(4, 12)$.

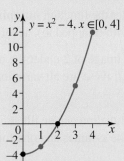

5. The domain is the set of values covered by the graph horizontally.

Domain $= [0, 4]$

6. The range is the set of values covered by the graph vertically.

Range $= [-4, 12]$

Technology can be used to check the graphs.

TI | THINK	WRITE	CASIO | THINK	WRITE
b. 1. On a Graphs page, complete the entry line for function 1 as: $f1(x) = x^2 - 4 \mid 0 \le x \le 4$ then press ENTER. The graph appears on the screen.	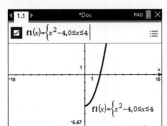	b. 1. On a Graph & Table screen, complete the entry line for Y1 as: $Y1 = x^2 - 4 \mid 0 \le x \le 4$ then press EXE. Select DRAW by pressing F6.	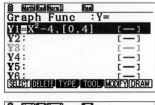 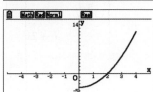

2. To calculate the end points of the graph, press MENU, then select: 5: Trace 1: Graph Trace. Type '0' then press ENTER twice. Type '4' then press ENTER twice.	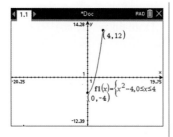	2. To calculate the end points of the graph, select Trace by pressing F1. Type '0' then press EXE twice. Type '4' then press EXE twice.	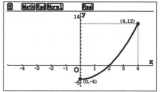
3. The domain and range can be read from the graph.	The domain is $[0, 4]$ and the range is $[-4, 12]$.	3. The domain and range can be read from the graph.	The domain is $[0, 4]$ and the range is $[-4, 12]$.

4.3.2 Function notation

The rule for a function such as $y = x^2$ will often be written as $f(x) = x^2$. This is read as 'f of x equals x^2'. We shall also refer to a function as $y = f(x)$, particularly when graphing a function as the set of ordered pairs (x, y) with x as the independent variable and y as the dependent variable.

$f(x)$ is called the **image** of x under the function **mapping**, which means that if, for example, $x = 2$, then $f(2)$ is the y-value that $x = 2$ is paired with (mapped to), according to the function rule.

For $f(x) = x^2$, $f(2) = 2^2 = 4$. The image of 2 under the mapping f is 4; the ordered pair (2, 4) lies on the graph of $y = f(x)$; 2 is mapped to 4 under f: these are all variations of the mathematical language that could be used for this function.

The ordered pairs that form the function can be illustrated on a mapping diagram. The mapping diagram shown uses two number lines, one for the x-values and one for the y-values, but there are varied ways to show mapping diagrams.

Below are two mapping diagrams that correspond to the functions $y = 2x$ and $y = x^2$.

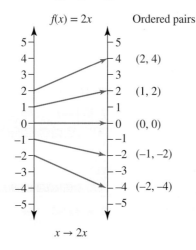

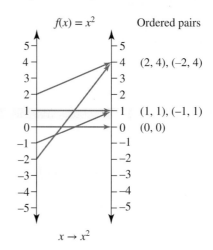

WORKED EXAMPLE 7 Using function notation

If $f(x) = x^2 - 3$, determine:

a. $f(-2)$ b. $f(a)$ c. $f(2a)$ d. $f(a+1)$.

THINK	WRITE
a. 1. Write the rule.	a. $f(x) = x^2 - 3$
2. Substitute $x = -2$ into the rule.	$f(-2) = (-2)^2 - 3$
3. Simplify the expression if possible.	$= 4 - 3$
	$= 1$
b. 1. Write the rule.	b. $f(x) = x^2 - 3$
2. Substitute $x = a$ into the rule.	$f(a) = a^2 - 3$
c. 1. Write the rule.	c. $f(x) = x^2 - 3$
2. Substitute $x = 2a$ into the rule.	$f(2a) = (2a)^2 - 3$
3. Simplify the expression if possible.	$= 2^2 a^2 - 3$
	$= 4a^2 - 3$
d. 1. Write the rule.	d. $f(x) = x^2 - 3$
2. Substitute $x = a + 1$ into the rule.	$f(a+1) = (a+1)^2 - 3$
3. Simplify the expression if possible.	$f(a+1) = a^2 + 2a + 1 - 3$
	$= a^2 + 2a - 2$

4.3.3 Formal mapping notation

The set of all the available y-values, whether used in the mapping or not, is called the **codomain**. Only the set of those y-values that are used for the mapping form the range.

The mapping $x \rightarrow x^2$ is written formally as:

$$f: \quad R \quad \rightarrow \quad R, \quad f(x) = x^2$$
$$\downarrow \qquad \downarrow \qquad \downarrow \qquad \downarrow$$

name of function   domain of f codomain   rule for, or equation of, f

The domain of the function must always be specified when writing functions formally.

We will always use R as the codomain. Mappings will be written as $f : D \rightarrow R$, where D is the domain. Usually a graph of the function is required in order to determine its range.

Note that f is a symbol for the name of the function or mapping, whereas $f(x)$ is an element of the range of the function: $f(x)$ gives the image of x under the mapping f. Although f is the commonly used symbol for a function, other symbols may be used.

WORKED EXAMPLE 8 Evaluating and using function notation

Consider $f : R \rightarrow R, f(x) = a + bx$, where $f(1) = 4$ and $f(-1) = 6$.
a. Calculate the values of a and b and determine the function rule.
b. Evaluate $f(0)$.
c. Calculate the value of x for which $f(x) = 0$.
d. Determine the image of -5.
e. Determine the mapping for a function g that has the same rule as f but a domain restricted to R^+. ▶

THINK	WRITE
a. 1. Use the given information to set up a system of simultaneous equations.	**a.** $f(x) = a + bx$ $f(1) = 4 \Rightarrow 4 = a + b \times 1$ $\therefore a + b = 4 \ldots\ldots\ldots\ldots(1)$ $f(-1) = 6 \Rightarrow 6 = a + b \times -1$ $\therefore a - b = 6 \ldots\ldots\ldots\ldots(2)$
2. Solve the system of simultaneous equations to obtain the values of a and b.	Equation (1) + equation (2): $2a = 10$ $a = 5$ Substitute $a = 5$ into equation (1) $\therefore b = -1$
3. State the answer.	$a = 5, \ b = -1$ $f(x) = 5 - x$
b. Substitute the given value of x.	**b.** $f(x) = 5 - x$ $f(0) = 5 - 0$ $= 5$
c. Substitute the rule for $f(x)$ and solve the equation for x.	**c.** $f(x) = 0$ $5 - x = 0$ $\therefore x = 5$
d. Write the expression for the image and then evaluate it.	**d.** The image of -5 is $f(-5)$. $f(x) = 5 - x$ $f(-5) = 5 - (-5)$ $= 10$ The image is 10.
e. Change the name of the function and change the domain.	**e.** $g : R^+ \to R, \ g(x) = 5 - x$

Exercise 4.3 Function notation

4.3 Exercise	**4.3 Exam practice** on

Simple familiar	Complex familiar	Complex unfamiliar
1, 2, 3, 4, 5, 6, 7, 8, 9, 10, 11, 12, 13, 14, 15, 16, 17, 18	19, 20	N/A

Simple familiar

1. **MC** The domain of the relation shown in the diagram is:
   A. $[-4, 4]$
   B. $(-4, 7)$
   C. $[-1, 7]$
   D. $(-4, 4)$

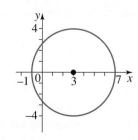

2. **MC** The range of the relation $\{(x, y) : y = 2x + 5, x \in [-1, 4]\}$ is:

A. $[7, 13]$ **B.** $[3, 13]$ **C.** $[3, \infty)$ **D.** R

3. **MC** A relation has the $y = x + 3$, where $x \in R^+$. The range of this relation is:

A. R^+ **B.** $[3, \infty)$ **C.** R **D.** $(3, \infty)$

4. **MC** The function $f : \{x : x = 0, 1, 2\} \to R$, where $f(x) = x - 4$, may be expressed as:

A. $\{(0, -4), (1, -3), (2, -2)\}$ **B.** $\{0, 1, 2\}$

C. $\{(0, 4), (1, 3), (2, 2)\}$ **D.** $\{(-1, -5), (1, -3), (2, -2)\}$

5. **WE5** For each of the following, identify the domain and range, and whether the relation is a function or not.

a. $\{(4, 4), (3, 0), (2, 3), (0, -1)\}$ **b.** $\{(x \cdot y) : y = 4 - x^2\}$

c.

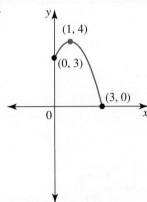

d.

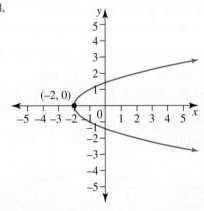

6. Identify the domain and range for each of the following relations.

a.

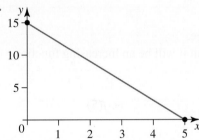

b.

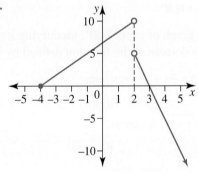

c.

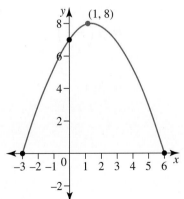

d.

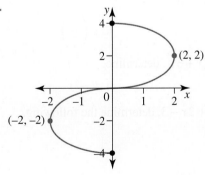

e.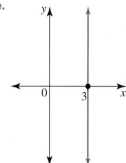

f.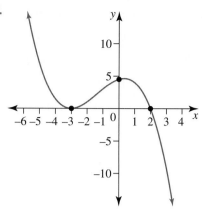

7. Consider each of the graphs in question 6.
   Identify any of the relations that are not functions.

8. Determine the domain and range of each of the following relations.
   a. $\{(3, 8), (4, 10), (5, 12), (6, 14), (7, 16)\}$
   b. $\{(1.1, 2), (1.3, 1.8), (1.5, 1.6), (1.7, 1.4)\}$

   c.
Time (min)	3	4	5	6
Distance (m)	110	130	150	170

   d.
Day	Monday	Tuesday	Wednesday	Thursday	Friday
Cost ($)	25	35	30	35	30

   e. $y = 5x - 2$, where x is an integer greater than 2 and less than 6
   f. $y = x^2 - 1$, $x \in R$

9. a. Sketch the graph of $y = (x - 2)^2$, identifying its domain and range.
   b. Restrict the domain of the function defined by $y = (x - 2)^2$ so that it will be an increasing function.

10. **WE7** a. If $f(x) = 3x + 1$, determine:
    i. $f(0)$ ii. $f(2)$ iii. $f(-2)$ iv. $f(5)$

    b. If $g(x) = \sqrt{x + 4}$, determine:
    i. $g(0)$ ii. $g(-3)$ iii. $g(5)$ iv. $g(-4)$

    c. If $g(x) = 4 - \dfrac{1}{x}$, determine:
    i. $g(1)$ ii. $g\left(\dfrac{1}{2}\right)$ iii. $g\left(-\dfrac{1}{2}\right)$ iv. $g\left(-\dfrac{1}{5}\right)$

    d. If $f(x) = (x + 3)^2$, determine:
    i. $f(0)$ ii. $f(-2)$ iii. $f(1)$ iv. $f(a)$

11. If $f(x) = x^2 + 2x - 3$, determine the following.
    a. i. $f(-2)$ ii. $f(9)$
    b. i. $f(2a)$ ii. $f(1 - a)$
    c. $f(x + h) - f(x)$
    d. $\{x : f(x) > 0\}$
    e. The values of x for which $f(x) = 12$
    f. The values of x for which $f(x) = 1 - x$

12. Determine the value (or values) of x for which each function has the value given.

   a. $f(x) = 3x - 4$, $f(x) = 5$
   b. $g(x) = x^2 - 2$, $g(x) = 7$
   c. $f(x) = \dfrac{1}{x}$, $f(x) = 3$

   d. $h(x) = x^2 - 5x + 6$, $h(x) = 0$
   e. $g(x) = x^2 + 3x$, $g(x) = 4$
   f. $f(x) = \sqrt{8 - x}$, $f(x) = 3$

13. **WE8** Consider $f : R \to R$, $f(x) = ax + b$, where $f(2) = 7$ and $f(3) = 9$.

   a. Calculate the values of a and b and determine the function rule.
   b. Calculate the value of x for which $f(x) = 0$.
   c. Determine the image of -3.
   d. Determine the mapping for a function g that has the same rule as f but a domain restricted to $(-\infty, 0]$.

14. **MC** The range of the function $f(x) = 2\sqrt{4 - x}$ is:

   A. R^+
   B. R^-
   C. $[0, \infty)$
   D. $(2, \infty)$

15. Express $y = x^2 - 6x + 10$, $0 \le x < 7$ in mapping notation and identify its domain and range.

16. The maximum side length of the rectangle shown is 10 metres.

   $(x + 4)$ m
   $(x - 1)$ m

   a. Construct a function which gives the perimeter, P metres, of the rectangle.
   b. Identify the domain and range of this function.

17. **WE6** For each relation given, sketch its graph, and identify the domain and range using interval notation.

   a. $\{(x, y) : y = 2 - x^2\}$
   b. $\{(x, y) : y = x^3 + 1, x \in [-2, 2]\}$
   c. $\{(x, y) : y = x^2 + 3x + 2\}$
   d. $\{(x, y) : y = x^2 - 4, x \in [-2, 1]\}$
   e. $\{(x, y) : y = 2x - 5, x \in [-1, 4)\}$
   f. $\{(x, y) : y = 2x^2 - x - 6\}$

18. Identify the implied domain for each relation defined by the following rules.

   a. $y = 10 - x$
   b. $y = 3\sqrt{x}$
   c. $y = -\sqrt{16 - x^2}$

   d. $y = x^2 + 3$
   e. $y = \dfrac{1}{x}$
   f. $y = 10 - 7x^2$

Complex familiar

19. The number of koalas remaining in a parkland t weeks after a virus strikes is given by the function $N(t) = 15 + \dfrac{96}{t + 3}$ koalas per hectare.

   a. Calculate how many koalas per hectare there were before the virus struck.
   b. Calculate how many koalas per hectare there were 13 weeks after the virus strikes.
   c. Calculate how soon after the virus strikes there are 23 koalas per hectare.
   d. Determine whether the virus will kill off all the koalas. Explain your answer.

20. Consider the functions f and g where $f(x) = a + bx + cx^2$ and $g(x) = f(x - 1)$.

   a. Given $f(-2) = 0$, $f(5) = 0$ and $f(2) = 3$, determine the rule for the function f.
   b. Express the rule for g as a polynomial in x.
   c. Calculate any values of x for which $f(x) = g(x)$.
   d. On the same axes, sketch the graphs of $y = f(x)$ and $y = g(x)$, and describe the relationship between the two graphs.

Fully worked solutions for this chapter are available online.

LESSON
4.4 Piece-wise functions

SYLLABUS LINKS

- Recognise and use piece-wise functions as a combination of multiple sub-functions with restricted domains.
- Model and solve problems that involve piece-wise functions with and without technology.

Source: Mathematical Methods Senior Syllabus 2024 © State of Queensland (QCAA) 2024; licensed under CC BY 4.0.

4.4.1 Piece-wise functions

A piece-wise function is one in which the rule may take a different form over different sections of the domain.

An example of a simple piece-wise function is given below.

Consider the function defined by the rule:

$$y = \begin{cases} x, & x \geq 0 \\ -x, & x < 0 \end{cases}$$

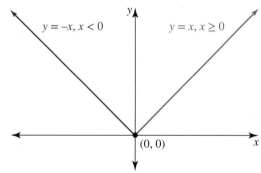

$y = -x, x < 0$ $\quad$ $y = x, x \geq 0$

$(0, 0)$

Graphing this function gives a line with positive gradient to the right of the y-axis and a line with a negative gradient to the left of the y-axis.

The function is said to be continuous at $x = 0$.

As for any function, each x-value can only be paired to exactly one y-value. To calculate the corresponding y-value for a given value of x, the choice of which branch of the rule to use depends to which section of the domain the x-value belongs.

WORKED EXAMPLE 9 Constructing a continuous piece-wise graph

A continuous piece-wise linear graph is constructed from the following linear graphs.

$$y = 2x + 1, \, x \leq a$$
$$y = 4x - 1, \, x > a$$

a. By solving the equations simultaneously, determine the point of intersection and hence determine the value of a.

b. Sketch the piece-wise linear graph.

THINK

a. 1. Calculate the intersection point of the two graphs by solving the equations simultaneously.

WRITE/DRAW

a. $y = 2x + 1$
$y = 4x - 1$
Solve by substitution:
$2x + 1 = 4x - 1$
$1 = 4x - 2x - 1$
$1 = 2x - 1$
$1 + 1 = 2x$
$2 = 2x$
$x = 1$

Substitute $x = 1$ to calculate y:

$y = 2(1) + 1$

$\quad = 3$

The point of intersection is $(1, 3)$.

$x = 1$ and $y = 3$

2. The x-value of the point of intersection determines the x-values for where the linear graphs meet.

$x = 1$; therefore, $a = 1$.

b. 1. Sketch the two graphs without taking into account the intervals.

b.

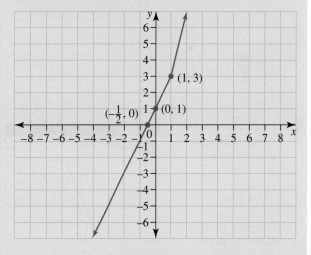

2. Identify which graph exists within the stated x-intervals to sketch the piece-wise linear graph.

$y = 2x + 1$ exists for $x \leq 1$.

$y = 4x - 1$ exists for $x > 1$.

Remove the sections of each graph that do not exist for these values of x.

TI | THINK

WRITE

a. 1. On a Calculator page, press MENU then select:
3: Algebra
7: Solve System of Equations
2: Solve System of Linear Equations
Complete the fields as:
Number of equations: 2
Variables: x, y
then select OK.
Complete the entry line as:
linSolve
$\left(\begin{cases} y = 2x + 1 \\ y = 4x - 1 \end{cases}, \{x, y\} \right)$
then press ENTER.

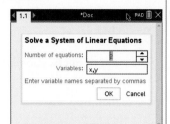

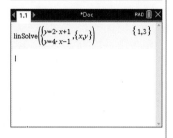

2. The answer appears on the screen.

The point of intersection is $(1, 3)$, so $a = 1$.

CASIO | THINK

WRITE

a. 1. On an Equation screen, select Simultaneous by pressing F1.

Select 2 unknowns by pressing F1.

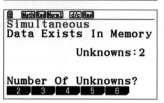

2. Rearrange the given equations into the form $ax + by = c$. Enter the coefficients for x and y, and the constant term, into the matrix on the screen.

$y = 2x + 1 \Rightarrow 2x - y = -1$
$y = 4x - 1 \Rightarrow 4x - y = 1$

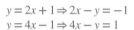

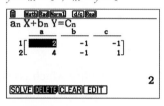

3. Select SOLVE by pressing F1.

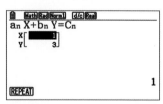

4. The answer appears on the screen.

The point of intersection is $(1, 3)$, so $a = 1$.

b. 1. On a Graphs page, complete the entry line for function 1 as:
$f1(x) = \begin{cases} 2x + 1, x \le 1 \\ 4x - 1, x > 1 \end{cases}$
then press ENTER.
Note: The piecewise function template can be found by pressing the [⫶⫶] button.

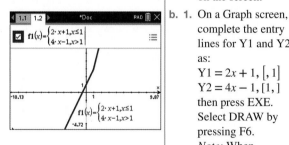

b. 1. On a Graph screen, complete the entry lines for Y1 and Y2 as:
Y1 = 2x + 1, [, 1]
Y2 = 4x − 1, [1,]
then press EXE.
Select DRAW by pressing F6.
Note: When restricting the domain of a function, use interval notation, leaving the upper or lower bound blank to represent ± ∞.

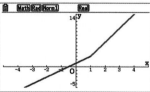

2. To mark the point where the branches join, press MENU, then select
5: Trace
1: Graph Trace.
Type '1' then press ENTER twice.

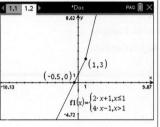

3. To determine the *x*-intercept, press MENU then select:
6: Analyze Graph
1: Zero.
Move the cursor to the left of the *x*-intercept when prompted for the lower bound, then press ENTER.
Move the cursor to the right of the *x*-intercept when prompted for the upper bound, then press ENTER.

4. To determine the *y*-intercept, press MENU then select:
5: Trace
1: Graph Trace.
Type '0' then press ENTER twice.

2. To mark the point where the branches join, select Trace by pressing F1. Type '1 ' then press EXE twice.

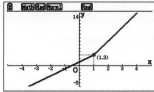

3. To find the *x*-intercept, select G-Solv by pressing F5, then select ROOT by pressing F1.
Use the up/down arrows to select the graph of Y1, then press EXE twice.

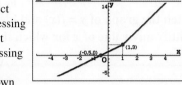

4. To find the *y*-intercept, select G-Solv by pressing F5, then select Y-ICEPT by pressing F4. Use the up/down arrows to select the graph of Y1, then press EXE twice.

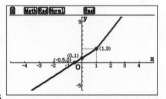

4.4.2 Discontinuous piece-wise functions

Not all branches of piece-wise functions will join. This means the function is discontinuous at that point of its domain. Care needs to be taken at the point of discontinuity, as only one end point can be included for a function. This is indicated with open and closed dots.

An example of a simple discontinuous piece-wise function is given below.

Consider the function defined by the rule:

$$y = \begin{cases} x+1, & x \geq 0 \\ x+2, & x < 0 \end{cases}$$

The first graph is linear with a positive gradient starting at the point (0, 1) and including this point.

The second graph is linear with a positive gradient finishing at the point (0, 2) and not including this point.

The function is said to be discontinuous at $x = 0$.

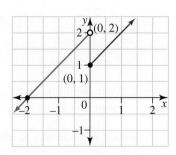

Consider the function:

$$f(x) = \begin{cases} x^2, & x < 1 \\ -x, & x \geq 1 \end{cases}$$

a. **Evaluate:**

   i. $f(-2)$ ii. $f(1)$ iii. $f(2)$.

b. **Sketch the graph of $y = f(x)$ and identify the domain and range.**

c. **Identify any value of x for which the function is not continuous.**

THINK

a. Decide for each x-value which section of the domain it is in and calculate its image using the branch of the piece-wise function's rule that is applicable to that section of the domain.

WRITE

a. $f(x) = \begin{cases} x^2, & x < 1 \\ -x, & x \geq 1 \end{cases}$

   i. $f(-2)$: Since $x = -2$ lies in the domain section $x < 1$, use the rule $f(x) = x^2$.
   $$f(-2) = (-2)^2$$
   $$\therefore f(-2) = 4$$

   ii. $f(1)$: Since $x = 1$ lies in the domain section $x \geq 1$, use the rule $f(x) = -x$.
   $$\therefore f(1) = -1$$

   iii. $f(2)$: Since $x = 2$ lies in the domain section $x \geq 1$, use the rule $f(x) = -x$.
   $$\therefore f(2) = -2$$

b. 1. Sketch each branch over its restricted domain to form the graph of the piece-wise function.

b. Sketch $y = x^2$, $x < 1$.
Parabola, turning point (0, 0) open end point (1, 1).
Sketch $y = -x$, $x \geq 1$.
Line, closed end point (1, −1) Point $x = 2 \Rightarrow (2, -2)$.

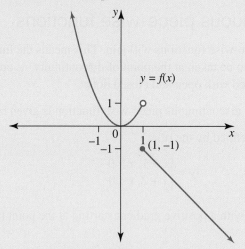

   2. State the domain and range.

The domain is R. The range is $(-\infty, -1] \cup [0, \infty)$.

c. State any value of x where the branches of the graph do not join.

c. The function is not continuous at $x = 1$ because there is a break in the graph.

**TI	THINK**	**WRITE**	**CASIO	THINK**	**WRITE**

a. 1. On a Calculator page, press MENU, then select:
1: Actions
1: Define.
Complete the entry line as:
Define $f1(x) =$
$\begin{cases} x^2, x < 1 \\ -x, x \ge 1 \end{cases}$
then press ENTER.

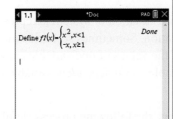

a. 1. On a Graph screen, complete the entry lines for Y1 and Y2 as:
$Y1 = x^2, [\, , 1]$
$Y2 = -x, [1, \,]$
then press EXE. Select DRAW by pressing F6.
Note: When restricting the domain of a function, use interval notation, leaving the upper or lower bound blank to represent $\pm \infty$.

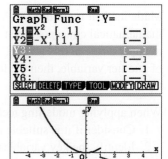

2. Complete the next entry line as:
$f1(-2)$
then press ENTER.

2. Select Trace by pressing F1, then, with the cursor on the graph of Y1, type '−2' and press EXE twice.

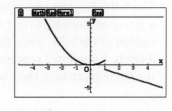

3. Complete the next entry line as:
$f1(1)$
then press ENTER.
Complete the next entry line as:
$f1(2)$
then press ENTER.

3. Select Trace by pressing F1, then use the up/down arrows to move the cursor to the graph of Y2. Type '1' then press EXE twice.
Type '2' then press EXE twice.

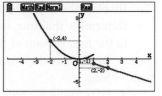

4. The answers appear on the screen.

$f(-2) = 4$, $f(1) = -1$ and $f(2) = -2$.

4. The answers appear on the screen.

$f(-2) = 4$, $f(1) = -1$ and $f(2) = -2$.

b. 1. On a Graphs page, the entry line for function 1 already contains the piece-wise function defined as $f1(x)$ on the Calculator page.
Click the tick box then press ENTER to view the graph.

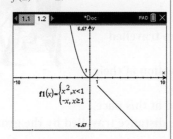

b. 1. See a.1.

2. Draw the graph.

When copying the graph from the screen, remember to use an open circle at the point $(1, 1)$ and a closed circle at $(1, -1)$.

2. Draw the graph.

When copying the graph from the screen, remember to use an open circle at the point $(1, 1)$ and a closed circle at $(1, -1)$.

3. The domain and range can be read from the graph.

The domain is R and the range is $(-\infty, -1] \cup [0, \infty)$.

3. The domain and range can be read from the graph.

The domain is R and the range is $(-\infty, -1] \cup [0, \infty)$.

4.4.3 Modelling with piece-wise functions

Mathematical **modelling** is the process by which a real-life situation or system is represented using mathematical concepts, often in the form of a rule or equation. Sometimes scenarios must be simplified in order to apply a rule at this level of study. In these scenarios, if the values of one variable are influenced by the values of another variable, then the former is the *dependent* variable. An *independent* variable is a factor that influences the dependent variable.

When applying modelling techniques to practical problems, the following process should be considered:
1. Consider if it is suitable to apply a mathematical model to the problem.
2. Identify the key variables and:
    i. identify which is independent and which is dependent
    ii. consider the natural restrictions that are placed on both in the situation given (e.g. time cannot be negative in most cases).
3. Determine the formula or formulas that govern the relationship between key variables.
4. Sketch a graph if possible, considering any natural restrictions on the variables.
5. Use the known information to directly answer the questions being asked. Reflect the language from the question in your worded responses where possible.

When using piece-wise functions to model practical problems, the domain of each function branch must be stated, remembering that each *x*-value may only have one associated *y*-value across the whole piece-wise function.

WORKED EXAMPLE 11 Solving problems involving piece-wise functions

The following two equations represent the distance travelled by a group of students over 5 hours. Equation 1 represents the first section of the hike, when the students are walking at a pace of 4 km/h. Equation 2 represents the second section of the hike, when the students change their walking pace.

$$\text{Equation 1: } d = 4t, \ 0 \le t \le 2$$
$$\text{Equation 2: } d = 2t + 4, \ 2 < t \le 5$$

The variable *d* is the distance in km from the campsite, and *t* is the time in hours.
a. Identify the dependent variable.
b. Determine the time, in hours, for which the group travelled in the first section of the hike.
c. i. Determine their walking pace in the second section of their hike.
   ii. Determine for how long, in hours, they walked at this pace.
d. Sketch a piece-wise linear graph to represent the distance travelled by the group of students over the 5-hour hike.

THINK	WRITE
a. The distance travelled depends on the time.	a. Distance is the dependent variable.
b. 1. Determine which equation the question applies to.	b. This question applies to Equation 1.
2. Look at the time interval for this equation.	$0 \le t \le 2$
3. Interpret the information.	The group travelled for 2 hours.
c. i. 1. Determine which equation the question applies to.	c. i. This question applies to equation 2.

2. Interpret the equation. The walking pace is found by the coefficient of t, as this represents the gradient.

$d = 2t + 4,\ 2 < t \leq 5$
The coefficient of t is 2.

3. Answer the question.

The walking pace is 2 km/h.

ii. 1. Look at the time interval shown.

ii. $2 < t \leq 5$

2. Interpret the information and answer the question.

They walked at this pace for 3 hours.

d. 1. Calculate the distance travelled before the change of pace.

d. Change after $t = 2$ hours:
$d = 4t$
$d = 4 \times 2$
$d = 8\ \text{km}$

2. Sketch the graph $d = 4t$ between $t = 0$ and $t = 2$.

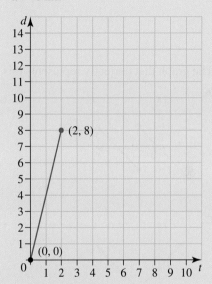

3. Solve the simultaneous equations to calculate the point of intersection.

$4t = 2t + 4$
$4t - 2t = 2t - 2t + 4$
$2t = 4$
$t = 2$
Substitute $t = 2$ into $d = 4t$:
$d = 4 \times 2$
$= 8$

4. Sketch the graph of $d = 2t + 4$ between $t = 2$ and $t = 5$.
For $t = 2$, $d = 2(2) + 4 = 8$.
For $t = 5$, $d = 2(5) + 4 = 14$.
Ensure all points of interest including end points, intercepts and intersections are labelled with their coordinates.

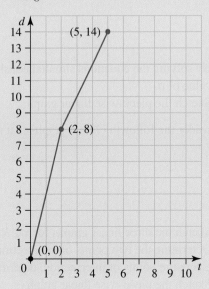

Exercise 4.4 Piece-wise functions

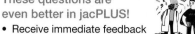

4.4 Exercise	4.4 Exam practice `on`

Simple familiar	Complex familiar	Complex unfamiliar
1, 2, 3, 4, 5, 6, 7, 8, 9, 10, 11, 12, 13, 14	15, 16, 17, 18	19, 20

These questions are even better in jacPLUS!
- Receive immediate feedback
- Access sample responses
- Track results and progress

Find all this and MORE in jacPLUS ▶

Simple familiar

1. **MC** Consider the following piece-wise function:

$$f(x) = \begin{cases} -x & x < 1 \\ x, & x \geq 1 \end{cases}$$

a. The graph that correctly represents this function is:

A.
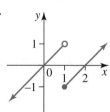

B.
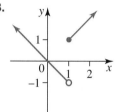

C.
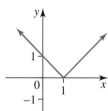

D.
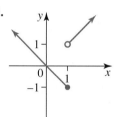

b. The range of this piece-wise function is:

   A. R　　　　　**B.** $[-1, \infty)$　　　　　**C.** $(-1, \infty)$　　　　　**D.** $[0, \infty)$

2. **WE 9** A continuous piece-wise linear graph is constructed from the following linear graphs.

$$y = -3x - 3, \ x \leq a$$
$$y = x + 1, \ x > a$$

a. By solving the equations simultaneously, determine the point of intersection and hence determine the value of a.

b. Sketch the piece-wise linear graph.

3. Consider the following linear graphs that make up a piece-wise linear graph.

$$y = 2x - 3, \ x \leq a$$
$$y = 3x - 4, \ a < x \leq b$$
$$y = 5x - 12, \ x > b$$

a. Sketch the three linear graphs for $x \in R$.
b. Determine the two points of intersection.
c. Using the points of intersection, determine the values of a and b.
d. Sketch the piece-wise linear graph.

4. a. Sketch the graph of the function $g(x) = \begin{cases} x^2 + 1, & x \geq 0 \\ 2 - x, & x < 0 \end{cases}$

   b. Identify the range of g.

   c. Evaluate:

      i. $g(-1)$ ii. $g(0)$ iii. $g(1)$.

5. a. Sketch the graph of the function $f(x) = \begin{cases} x - 2, & x < -2 \\ x^2 - 4, & -2 \leq x \leq 2 \\ x + 2, & x > 2 \end{cases}$

   b. Identify any value of x for which the function is not continuous.

   c. Identify the range of f.

   d. Evaluate:

      i. $f(-3)$ ii. $f(-2)$ iii. $f(1)$ iv. $f(2)$ v. $f(5)$.

6. Consider the function defined by $f(x) = \begin{cases} 4x + a, & x < 1 \\ \dfrac{2}{x}, & 1 \leq x \leq 4 \end{cases}$

   a. Determine the value of a so the function will be continuous at $x = 1$.

   b. Explain whether the function is continuous at $x = 0$.

7. A step graph is a special type of piece-wise function consisting of a series of horizontal line segments. Determine the equations that make up the step graph shown.

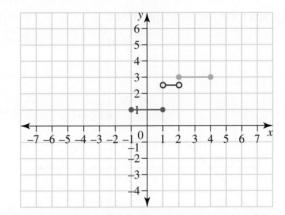

8. **WE 11** The following two equations represent water being added to a water tank over 15 hours, where w is the water in litres and t is the time in hours.

   Equation 1: $w = 25t$, $0 \leq t \leq 5$

   Equation 2: $w = 30t - 25$, $5 < t \leq 15$

   a. Identify the dependent variable.

   b. Determine how many litres of water are in the tank after 5 hours.

   c. i. Determine the rate at which the water is being added to the tank after 5 hours.

      ii. Determine for how long the water is added to the tank at this rate.

   d. Sketch a piece-wise graph to represent the water in the tank at any time, t, over the 15-hour period.

9. Determine the points of intersection for the following four linear graphs, and hence complete the intervals for x by determining the values of a, b and c.

   a. i. $y = x + 4$, $x \leq a$ ii. $y = 2x + 3$, $a \leq x \leq b$
       iii. $y = x + 6$, $b \leq x \leq c$ iv. $y = 3x + 1$, $x \geq c$

   b. Describe the problem that you encounter when trying to sketch a piece-wise linear graph formed by these four linear graphs.

10. **WE10** Consider the function for which $f(x) = \begin{cases} x^3, & x < 1 \\ 2, & x \geq 1 \end{cases}$

   a. Evaluate the following.

      i. $f(-2)$ ii. $f(1)$ iii. $f(2)$

   b. Sketch the graph of $y = f(x)$ and identify the domain and range.

11. a. Sketch the graph of the function $f(x) = \begin{cases} \dfrac{1}{x}, & x < 0 \\ x + 1, & x \geq 0 \end{cases}$.

   b. Identify the range of f.

12. Specify the rule for the function represented by the graph shown.

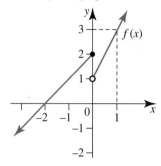

13. Form the rule in piece-wise graph form for the graph shown.

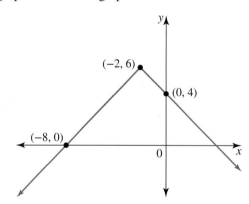

14. The postage costs to send parcels from the Northern Territory to Sydney are shown in the table.

   Where a parcel weight appears next to more than one cost, the higher price is applied.

   a. Represent this information in a graph.
   b. Pammie has two parcels to post to Sydney from the Northen Territory. One parcel weighs 450 g and the other weighs 525 g. Determine whether it is cheaper to send the parcels individually or together. Justify your answer using calculations.

Weight of parcel (kg)	Cost ($)
0 – 0.5	6.60
0.5 – 1	16.15
1 – 2	21.35
2 – 3	26.55
3 – 4	31.75
4 – 5	36.95

15. A car hire company charges a flat rate of $50 plus 75 cents per kilometre up to and including 150 kilometres. An equation to represent this cost, C, in dollars is given as $C = 50 + ak$, $0 \leq k \leq b$, where k is the distance travelled in kilometres.
    The cost charged for distances over 150 kilometres is given by the equation $C = 87.50 + 0.5k$.
    Sketch both equations on the same set of axes.

16. Airline passengers are charged an excess for any luggage that weighs 20 kg or over. The following graph shows these charges for luggage weighing over 20 kg.

    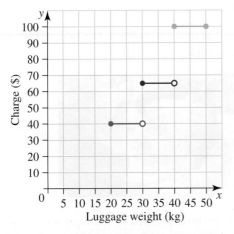

    a. Nerada checks in her luggage and is charged $40. Determine the maximum excess luggage she could have without having to pay any more.
    b. Hilda and Hanz have two pieces of luggage between them. One piece weighs 32 kg and the other piece weighs 25 kg. Explain how they could minimise their excess luggage charges.

17. The growth of a small tree was recorded over 6 months. It was found that the tree's growth could be represented by three linear equations, where h is the height in centimetres and t is the time in months.

    $$\text{Equation 1: } h = 2t + 20, 0 \leq t \leq a$$
    $$\text{Equation 2: } h = t + 22, a < t \leq b$$
    $$\text{Equation 3: } h = 3t + 12, b < t \leq c$$

    Sketch the piece-wise linear graph that shows the height of the tree over the 6-month period.

18. The temperature of a wood-fired oven, $T\,°C$, steadily increases until it reaches 200 °C. Initially the oven has a temperature of 18 °C and it reaches the temperature of 200 °C in 10 minutes.
    Once the oven has heated up for 10 minutes, a loaf of bread is placed in the oven to cook for 20 minutes.
    After 20 minutes of cooking, the oven's temperature is lowered.
    The temperature decreases steadily, and after 30 minutes the oven's temperature reaches 60 °C.
    Sketch the graph that shows the changing temperature of the wood-fired oven during the 60-minute interval.

19. The speed, v km/h, at which a car is driven for a 25-minute time interval is described by the function with the rule

$$v = \begin{cases} 5t^2, & 0 \le t \le 4 \\ a, & 4 < t < 24 \\ b - 80t, & 24 \le t \le 25 \end{cases}$$

The time t is in minutes, and a and b are constants.
Determine at what time the car was being driven at 45 km/h.

20. Stamp duty is a government tax on the purchase of items such as cars and houses. The table shows the range of stamp duty charges for purchasing a used car in South Australia.

Used car price (P)	Stamp duty (S)
0–1000	1%
1000–2000	$10 + 2\%(P - 1000)$
2000–3000	$30 + 3\%(P - 2000)$
3000+	$60 + 4\%(P - 3000)$

The stamp duty charge for a car purchased for $1000 or less can be expressed by the equation $S = 0.01P$, where S is the stamp duty charge and P is the purchase price of the car for $0 \le P \le 1000$.
Similar equations can be used to express the charges for cars with higher prices.

Equation 1: $S = 0.01P$, $0 \le P \le 1000$

Equation 2: $S = 0.02P - 10$, $a < P \le b$

Equation 3: $S = 0.03P - c$, $2000 < P \le d$

Equation 4: $S = fP - e$, $P > 3000$

Suki and Boris purchase a car and pay $45 in stamp duty. Determine the price they paid for their car.

Fully worked solutions for this chapter are available online.

LESSON
4.5 Review

4.5.1 Summary

Hey students! Now that it's time to revise this chapter, go online to:

Access the chapter summary

Review your results

Practise exam questions

Find all this and MORE in jacPLUS

4.5 Exercise

learn on

4.5 Exercise	4.5 Exam practice on

Simple familiar	Complex familiar	Complex unfamiliar
1, 2, 3, 4, 5, 6, 7, 8, 9, 10, 11, 12	13, 14, 15, 16	17, 18, 19, 20

These questions are even better in jacPLUS!
- Receive immediate feedback
- Access sample responses
- Track results and progress

Find all this and MORE in jacPLUS

Simple familiar

1. **MC** The interval shown below is:

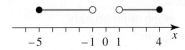

A. $[-5, -1] \cup [0, 4]$ B. $[-5, -1) \cup [0, 4]$ C. $(-5, -1) \cup (1, 4]$ D. $[-5, -1) \cup (1, 4]$

2. **MC** The rule describing the relation shown in the graph is:

A. $y = 2x$
B. $y = 2x, x \in \{1, 2, 3, 4\}$
C. $y = 2x, x \in N$
D. $y = \dfrac{x}{2}$

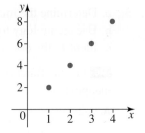

3. Identify whether each of the following relations are continuous or not.

a. $\{(-4, 4), (-3, 2), (-2, 0), (-1, -2), (0, 0), (1, 2), (2, 4)\}$

b.

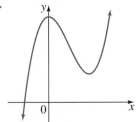

c.

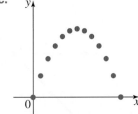

4. Identify which of the following relations are functions. Identify the domain and range for each function. *Hint:* It may be helpful to view the graphs of the relations in parts **e** and **f** using technology.
   a. $\{(0, 2), (0, 3), (1, 3), (2, 4), (3, 5)\}$
   b. $\{(-3, -2), (-1, -1), (0, 1), (1, 3), (2, -2)\}$
   c. $\{(x, y) : y = 2, x \in R\}$
   d. $\{(x, y) : x = -3, y \in J\}$
   e. $y = x^3 + x$
   f. $x = y^2 + 1$

5. Identify the domain and range of each of the following relations.
   a.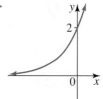
   b.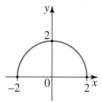
   c.

   d.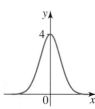
   e.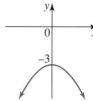
   f.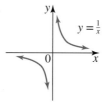

6. Identify the implied domains of the rational functions with the following rules. *Hint:* It may be helpful to view the graphs of these functions using technology.
   a. $y = \dfrac{1}{16 - x^2}$
   b. $y = \dfrac{2 - x}{x^2 + 3}$

7. A function f is defined as follows: $f : [-2, a] \to R$, where $f(x) = (x - 1)^2 - 4$.
   a. Determine $f(-2)$, $f(-1)$, $f(0)$, $f(1)$, $f(3)$.
   b. If $f(a) = 12$, determine the value of a.
   c. Sketch the function f, labelling the graph appropriately.
   d. From the graph or otherwise, identify the:
      i. domain of $f(x)$
      ii. range of $f(x)$.

8. Consider $f : R \to R$, $f(x) = x^3 - x^2$.
   a. Determine the image of 2.
   b. Use technology to sketch the graph of $y = f(x)$.
   c. Identify the domain and range of the function f.

9. **MC** For the function $f : [-2, 4] \to R$, $f(x) = ax + b$, $f(0) = 1$ and $f(1) = 0$. The image of -2 under the mapping is:
   A. -2
   B. -1
   C. 1
   D. 3

10. a. Form a rule for the graph of the piece-wise function shown.

b. Form the rule for the graph of the piece-wise function shown.

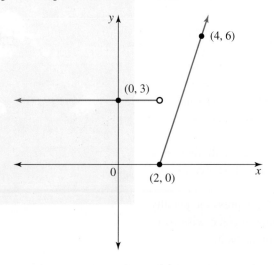

11. Determine the values of a and b so that the function with the rule

$$f(x) = \begin{cases} a, & x \in (-\infty, -3] \\ x + 2, & x \in (-3, 3) \\ b, & x \in [3, \infty) \end{cases}$$

is continuous for $x \in R$. For these values, sketch the graph of $y = f(x)$.

12. Consider $f : R \to R, f(x) = \begin{cases} -x - 1, & x < -1 \\ \sqrt{1 - x^2}, & -1 \leq x \leq 1 \\ x + 1, & x > 1. \end{cases}$

a. Calculate the value of:

   i. $f(0)$ **ii.** $f(3)$ **iii.** $f(-2)$ **iv.** $f(1)$

b. Demonstrate the function is not continuous at $x = 1$.

c. Use technology to sketch the graph of $y = f(x)$.

d. Determine the value of a such that $f(a) = a$.

Complex familiar

13. Define $f(x) = x^3 + lx^2 + mx + n$. Given $f(3) = -25$, $f(5) = 49$ and $f(7) = 243$, use technology to determine the image of 1.2.

14. Determine the piece-wise function for the graph shown and identify its domain and range.

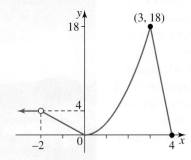

15. In an effort to reduce the time her children spend in the shower, a mother introduced a penalty scheme with fines to be paid from the children's pocket money according to the following:
If someone spends more than 5 minutes in the shower, the fine in dollars is equal to the shower time in minutes; if someone spends up to and including 5 minutes in the shower, there is no fine.
If someone chooses not to shower at all, there is a fine of $2 because that child won't be nice to be near.
Defining appropriate symbols, express the penalty scheme as a mathematical rule in piece-wise form and sketch the graph that represents it.

16. The amount of money in a bank account over 12 months is shown in the piece-wise graph, where A is the amount of money in dollars and t is the time in months.

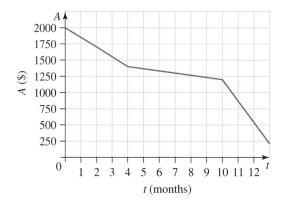

t (months)

Originally, the account balance was $2000, but after 4 months it was $1400, after 10 months it was $1100 and after 12 months the balance was $500.
Determine the rule for this piece-wise function that illustrates the amount of money in the bank account for the 12-month period.

Complex unfamiliar

17. The country bus leaves Alexander at 6:30 a.m. for Collinsville, maintaining an average speed of 80 km/h for 3 hours, stopping at Beauty Point for a rest break of 30 minutes. It then continues travelling the next 3 hours at an average speed of 70 km/h to reach Collinsville on time.
A motorhome leaves Collinsville for Alexander at the same time, travelling at an average speed of 50 km/h. The vehicles pass each other on the road. Determine the time and the distance from Alexander when they pass. Express your answers to the nearest minute and kilometre.

18. Jerri and Samantha have both entered a 10-km fun run for charity.

The distance travelled by Jerri can be modelled by the linear equation $d = 6t - 0.1$, where d is the distance in km from the starting point and t is time in hours.

The distance Samantha is from the starting point at any time, t hours, can be modelled by the piece-wise linear graph:

$$d = \begin{cases} 4t, & 0 \leq t < \dfrac{1}{2} \\ 8t - 2, & \dfrac{1}{2} \leq t \leq a \end{cases},$$

where a is a constant.

Determine when and where Samantha and Jerri will meet on the course, and determine the winner and their winning margin over the other runner.

19. The speed, v km/h, at which a motorbike is ridden for a 50-minute time interval is described by the function with the rule

$$v = \begin{cases} at, & 0 \leq t \leq 5 \\ 60, & 5 < t < 30 \\ b(50 - t)^2, & 30 \leq t \leq 50 \end{cases}$$

The time t is in minutes, and a and b are constants. The function is continuous over the time interval. Determine at what time the motorbike is travelling at 15 km/h.

20. A ray of light comes in along the line $x + y = 2$ above the x-axis and is reflected off the axis so that the angle of departure (the angle of reflection) is equal to the angle of arrival (the angle of incidence).

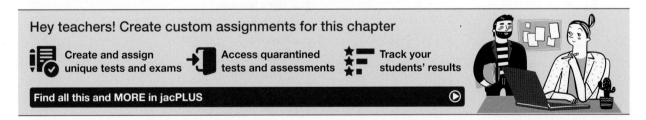

Calculate the magnitude of the angle of departure and express the path of the incoming and departing rays in terms of a piece-wise function.

Fully worked solutions for this chapter are available online.

Answers

Chapter 4 Introduction to functions and relations

4.2 Functions and relations

4.2 Exercise

1. a. $[-2, \infty)$ b. $(-\infty, 5)$
   c. $(-3, 4]$ d. $(-\infty, -1]$
   e. $(-5, -2] \cup [3, \infty)$ f. $(-3, 1) \cup (2, 4]$

2. a. $[-6, 2)$ b. $(-9, -3)$

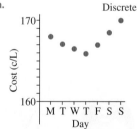

   c. $(-\infty, 2]$ d. $(1, 10]$

   e. $(-\infty, -2) \cup [1, 3)$ f. $[-8, 0) \cup (2, 6]$

3. a. $\{x : -4 \le x < 2\}$
      $= [-4, 2)$

   b. $\left\{ y : -1 < y < \sqrt{3} \right\}$
      $\left(-1, \sqrt{3} \right)$

   c. $\{x : x > 3\}$
      $(3, \infty)$

   d. $\{x : x \le -3\}$
      $(-\infty, -3]$

   e. R or $(-\infty, \infty)$

   f. $(-\infty, 0) \cup (0, \infty)$

4. B
5. A
6. D

7. a.

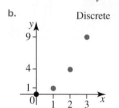

   b.

8. a. Not a function
   b. Function
   c. Function
   d. Function
   e. Function

9. a. Not a function
   b. Function
   c. Function
   d. Function
   e. Not a function
   f. Function

10. The functions are a, b, c, d.

11. The functions are b, c, d, e.

12. C

13. C

14. C

c.

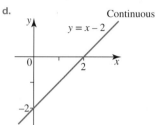

d.

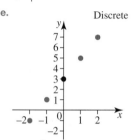

e.

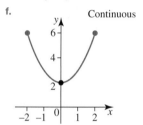

f.

15. a.

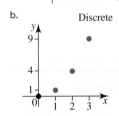

b.

Discrete

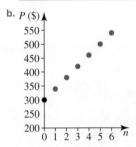

c. Because the variables are continuous, measurements can be taken in between the given values.

d. Half of the initial temperature is 40°C. It takes approximately 11 minutes.

16. a.

n	0	1	2	3	4	5	6
p	300	340	380	420	460	500	540

b.

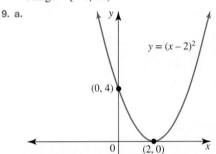

c. The variables are discrete.
Only whole numbers of computers can be sold.

d. $P = 300 + 40n$

17. Function

18. Relation

19. Function

20. Function

4.3 Function notation

4.3 Exercise

1. C

2. B

3. D

4. A

5. a. Domain is $\{0, 2, 3, 4\}$, range is $\{-1, 0, 3, 4\}$. This is a function.

b. Domain is R and range is $(-\infty, 4]$. It is a function.

c. Domain is $[0, 3]$, range is $[0, 4]$. This is a function.

d. Domain is $[-2, \infty)$, range is R. This is not a function.

6. a. Domain $[0, 5]$, range $[0, 15]$

b. Domain $[-4, 2) \cup (2, \infty)$, range $(-\infty, 10)$

c. Domain $[-3, 6]$, range $[0, 8]$

d. Domain $[-2, 2]$, range $[-4, 4]$

e. Domain $\{3\}$, range R

f. Domain R, range R

7. The graphs of **d** and **e** are not of functions.

8. a. Domain $= \{3, 4, 5, 6, 7\}$
Range $= \{8, 10, 12, 14, 16\}$

b. Domain $= \{1.1, 1.3, 1.5, 1.7\}$
Range $= \{2, 1.8, 1.6, 1.4\}$
or $= \{1.4, 1.6, 1.8, 2\}$

c. Domain $= \{3, 4, 5, 6\}$
Range $= \{110, 130, 150, 170\}$

d. Domain $= \{M, Tu, W, Th, Fr\}$
Range $= \{25, 30, 35\}$

e. Domain $= \{3, 4, 5\}$
Range $= \{13, 18, 23\}$

f. Domain $= R$
Range $= [-1, \infty)$

9. a.

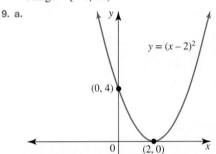

Domain R, range $R^+ \cup \{0\}$

b. An answer is $[2, \infty)$.

10. a. i. $f(0) = 1$ **ii.** $f(2) = 7$
iii. $f(-2) = -5$ **iv.** $f(5) = 16$

b. i. $g(0) = 2$ **ii.** $g(-3) = 1$
iii. $g(5) = 3$ **iv.** $g(-4) = 0$

c. i. $g(1) = 3$ **ii.** $g\left(\dfrac{1}{2}\right) = 2$

iii. $g\left(-\dfrac{1}{2}\right) = 6$ **iv.** $g\left(-\dfrac{1}{5}\right) = 9$

d. i. $f(0) = 9$ **ii.** $f(-2) = 1$
iii. $f(1) = 16$ **iv.** $f(a) = a^2 + 6a + 9$

11. $f(x) = x^2 + 2x - 3$

a. i. $f(-2) = -3$ **ii.** $f(9) = 96$

b. i. $f(2a) = 4a^2 + 4a - 3$ **ii.** $f(1-a) = a^2 - 4a$

c. $f(x+h) - f(x) = 2xh + h^2 + 2h$

d. $\{x : x < -3\} \cup \{x : x > 1\}$

e. $x = -5, x = 3$

f. $x = -4, x = 1$

12. a. $x = 3$ **b.** $x = \pm 3$

c. $x = \dfrac{1}{3}$ **d.** $x = 3$ or $x = 2$

e. $x = -4$ or $x = 1$ **f.** $x = -1$

13. a.
$a = 2$
$b = 3$
$\Rightarrow f(x) = 2x + 3$

b. $x = -\dfrac{3}{2}$

c. $f(-3) = -3$

d. $g : (-\infty, 0] \to R, g(x) = 2x + 3$

14. C

15. $f : [0, 7) \to R, f(x) = x^2 - 6x + 10$
Domain is $[0, 7)$, range is $[1, 17)$.

16. a. $P = 4x + 6$

b. Domain $(1, 6]$, range $(10, 30]$

17. a.

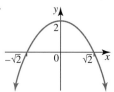

Domain $= (-\infty, \infty)$
Range $= (-\infty, 2]$

b.

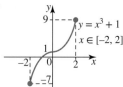

Domain $= [-2, 2]$
Range $= [-7, 9]$

c.

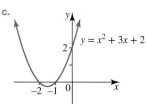

Domain $= (-\infty, \infty)$
Range $= \left[-\dfrac{1}{4}, \infty\right)$

d.

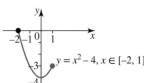

Domain $= [-2, 1]$
Range $= [-4, 0]$

e.

$y = 2x - 5, x \in [-1, 4)$

Domain $= [-1, 4)$
Range $= [-7, 3)$

f.

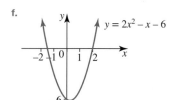

Domain $= (-\infty, \infty)$

Range $= \left[-\dfrac{49}{8}, \infty\right)$

or $= \left[-6\dfrac{1}{8}, \infty\right)$

18. a. Domain $= R$ b. Domain $= [0, \infty)$
c. Domain $= [-4, 4]$ d. Domain $= R$
e. Domain $= R\backslash\{0\}$ f. Domain $= R$

19. a. 47

b. 21

c. 9 weeks

d. As t increases $\dfrac{96}{t+3}$ gets smaller and approaches zero.
$N(t) \to 15$, so no.

20. a. $f(x) = \dfrac{5}{2} + \dfrac{3}{4}x - \dfrac{1}{4}x^2$

b. $g(x) = \dfrac{3}{2} + \dfrac{5}{4}x - \dfrac{1}{4}x^2$

c. $x = 2$

d.

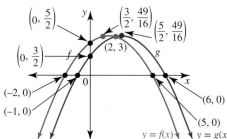

The graph of function g has the same shape as the graph of function f but g has horizontally translated 1 unit to the right.

4.4 Piece-wise functions

4.4 Exercise

1. a. B   b. C

2. a. Point of intersection $= (-1, 0)$; therefore, $a = -1$.

b.

3. a.

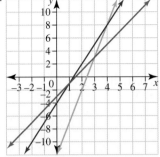

b. $(1, -1)$ and $(4, 8)$

c. $a = 1$ and $b = 4$

d.

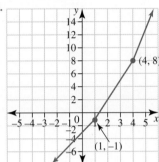

4. a.

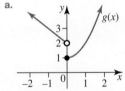

b. Range $[1, \infty)$

c. i. $g(-1) = 3$

ii. $g(0) = 1$

iii. $g(1) = 2$

5. a.

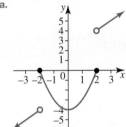

b. $x = -2$ and $x = 2$

c. $(-\infty, 0] \cup (4, \infty)$

d. i. $f(-3) = -5$

ii. $f(-2) = 0$

iii. $f(1) = -3$

iv. $f(2) = 0$

v. $f(5) = 7$

6. a. $a = -2$

b. $x = 0$ lies in the domain for which the rule is the linear function $y = 4x + a$, so the graph will be continuous at this point.

7. $y = 1, \qquad 1 \le x \le 1$
$y = 2.5, \qquad 1 < x < 2$
$y = 3, \qquad 2 \le x \le 4$

8. a. Water

b. 125 L

c. i. 30 L/h ii. 10 hours

d.

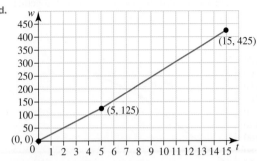

9. a. $(1, 5), (3, 9)$ and $(2.5, 8.5)$
$a = 1, b = 3, c = 2.5$

b. $b > c$, which means that graph iii is not valid and the piece-wise linear graph cannot be sketched.

10. a. i. $f(-2) = -8$ ii. $f(1) = 2$ iii. $f(2) = 2$

b.

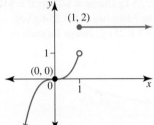

Domain R; range $(-\infty, 1) \cup \{2\}$

11. a.

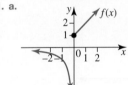

b. Range of $f = (-\infty, 0) \cup [1, \infty)$

12. $f(x) = \begin{cases} x + 2 & x \le 0 \\ 2x + 1 & x > 0 \end{cases}$

13. Samantha and Jerri meet on the course after 57 minutes, 5.6 km from the starting line. Samantha is the winner by 11 minutes.

14. a.

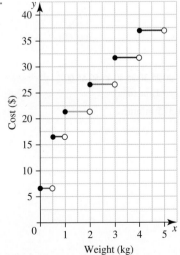

b. Individually: 450 g costs $6.60, 525 g costs $16.15, total cost = $22.75. Together: total weight = 450 + 525 = 975, costs $16.15 to send.

It is cheaper to post them together ($16.15 together versus $22.75 individually).

15.

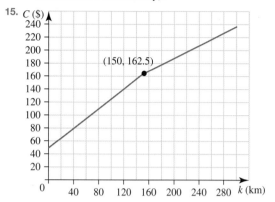

16. a. 10 kg

b. 32-kg charge = $65, 25-kg charge = $40, total = $105. Place 2–3 kg from the 32-kg bag in the 25-kg bag, 32 − 3 = 29 kg. 25 + 3 = 28 kg, charge for each is $40, total = $80.

17.

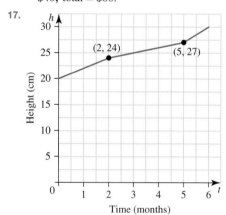

18.

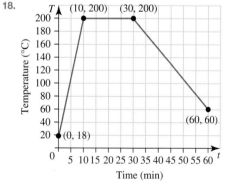

19. 3 minutes and 24.375 minutes

20. $2500

4.5 Review

4.5 Exercise

1. D

2. B

3. a. Not continuous

b. Continuous

c. Not continuous

4. a. Not a function

b. Function
Domain = $\{-3, -1, 0, 1, 2\}$
Range = $\{-2, -1, 1, 3\}$

c. Function
Domain = R
Range = $\{2\}$

d. Not a function

e. Function
Domain = R
Range = R

f. Not a function

5. a. Domain = R
Range = $(0, \infty)$ or $R^{\pm}$

b. Domain = $[-2, 2]$
Range = $[0, 2]$

c. Domain = $[1, \infty)$
Range = R

d. Domain = R
Range = $(0, 4]$

e. Domain = R
Range = $(-\infty, -3]$

f. Domain = $R \backslash \{0\}$
Range = $R \backslash \{0\}$

6. a. $R \backslash \{\pm 4\}$

b. R

7. a. $f(-2) = 5$
$f(-1) = 0$
$f(0) = -3$
$f(1) = -4$
$f(3) = 0$

b. $a = 5$

c.

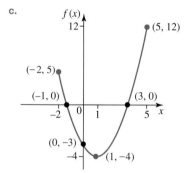

d. i. Domain $[-2, 5]$
ii. Range $[-4, 12]$

8. a. The image of 2 is 4.

b.

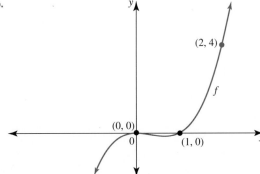

c. Domain R, range R

9. D

10. a. $y = \begin{cases} x+1, & x \le 0 \\ -x+1, & x > 0 \end{cases}$

b. $y = \begin{cases} 3, & x < 2 \\ 3x-6, & x \ge 2 \end{cases}$

11. $a = -1$
$b = 5$

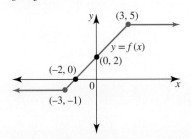

Let the time in the showers be t minutes and the dollar amount of the fine be C.

12. a. i. $f(0) = 1$ **ii.** $f(3) = 4$
iii. $f(-2) = 1$ **iv.** $f(1) = 0$

b. For $y = \sqrt{1-x^2}$, there is a closed end point $(1, 0)$, but for $y = x + 1$, there is an open end point $(1, 2)$. The two branches do not join. Hence, the function is not continuous at $x = 1$ as there will be a break in its graph.

c.

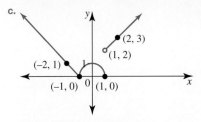

The function is many-to-one.

d. $a = \dfrac{\sqrt{2}}{2}$

13. $f(1.2) = -28.672$

14. Domain $= x \in (-\infty, 4] \setminus \{-2\}$
Range $= [0, 18]$
$$f(x) = \begin{cases} 4, & x \in (-\infty, -2) \\ -2x, & x \in (-2, 0] \\ 2x^2, & x \in [0, 3] \\ -18x + 72, & x \in [3, 4] \end{cases}$$

15. The rule is $C = \begin{cases} 2, & t = 0 \\ 0, & 0 < t \le 5 \\ t, & t > 5 \end{cases}$.

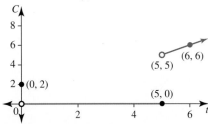

16. $A = \begin{cases} 200 - 150t, & 0 \le t < 4 \\ 1600 - 50t, & 4 \le t < 10 \\ 4100 - 300t, & 10 \le t \le 12 \end{cases}$

17. The vehicles pass each other at 10:18 am, 260 km from Alexander.

18. Samantha and Jerri meet on the course after 57 minutes, 5.6 km from the starting line.

19. The times are 1.25 minutes and 40 minutes.

20. $45°$
$y = x - 2$
$y = \begin{cases} 2 - x, & x < 2 \\ x - 2, & x \ge 2 \end{cases}$

5 Graphs of relations and reciprocal functions

LESSON SEQUENCE

Fully worked solutions for this chapter are available online.

EXAM PREPARATION

Access exam-style questions in every lesson, available online.

on Resources

Solutions	Solutions — Chapter 5 (sol-1272)
Exam questions	Exam question booklet — Chapter 5 (eqb-0293)
Digital documents	Learning matrix — Chapter 5 (doc-41787)
	Chapter summary — Chapter 5 (doc-41800)

LESSON
5.1 Overview

5.1.1 Introduction

The graphs examined in this chapter – the hyperbola, the parabola and the circle – are members of a family of curves known as the **conic sections**. They are named as such because their shapes are the result of the intersection of a plane with a solid cone at different angles.

The motion of spacecraft, satellites and probes as they move through space can be modelled by the equations of conic

Hyperbola   Parabola   Circle

curves. The International Space Station traces out a nearly circular path as it orbits the Earth sixteen times a day at an altitude of 400 kilometres.

5.1.2 Syllabus links

Lesson	Lesson title	Syllabus links
5.2	**The circle**	○ Recognise and determine features of the graphs of $x^2 + y^2 = r^2$ and $(x - h)^2 + (y - k)^2 = r^2$, including their circular shapes, centres and radii.
		○ Sketch the graphs of relations, with or without technology.
		○ Model and solve problems that involve relations, with and without technology.
5.3	**The sideways parabola and square root function**	○ Recognise and determine features of the graph of $y^2 = x$, including its parabolic shape and axis of symmetry.
		○ Recognise and determine features of the graphs of $y = a\sqrt{x - h} + k$, including their shape, intercepts, and behaviour as $x \to \infty$ and $x \to -\infty$.
		○ Sketch the graphs of relations, with or without technology.
		○ Model and solve problems that involve relations, with and without technology.
5.4	**The reciprocal function**	○ Recognise features of the graphs of $y = \dfrac{1}{x}$ and $y = \dfrac{a}{(x - h)} + k$, including their hyperbolic shape, intercepts, asymptotes, and behaviour as $x \to \infty$ and $x \to -\infty$.
		○ Model and solve problems that involve reciprocal functions, with and without technology.
		○ Sketch the graphs of reciprocal functions, with or without technology.

Source: Mathematical Methods Senior Syllabus 2024 © State of Queensland (QCAA) 2024; licensed under CC BY 4.0.

LESSON
5.2 The circle

5.2.1 Equation of a circle

The **circle** is an example of a **relation**. A circle is not a function.

If the centre is at $(0, 0)$ and the radius is r, then the circle has equation $x^2 + y^2 = r^2$, with domain $[-r, r]$ and range $[-r, r]$.

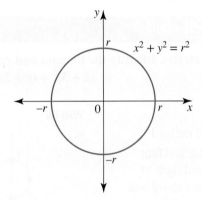

To obtain the equation of a circle, consider a circle of radius r and centre at the point C (h, k).

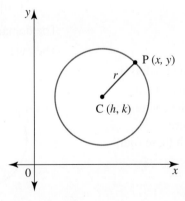

Let P (x, y) be any point on the circumference. CP, of length r, is the radius of the circle.

Using the formula for the distance between two points:

$$\sqrt{(x - h)^2 + (y - k)^2} = CP = r$$
$$(x - h)^2 + (y - k)^2 = r^2$$

The end points of the horizontal diameter are $(h-r, k)$ and $(h+r, k)$; the end points of the vertical diameter are $(h, k-r)$ and $(h, k+r)$. These points, together with the centre point, are usually used to sketch the circle. The intercepts with the coordinate axes are not always calculated.

The domain and range are obtained from the end points of the horizontal and vertical diameters.

The circle with the centre (h, k) and radius r has domain $[h-r, h+r]$ and range $[k-r, k+r]$.

General equation of a circle

The equation of a circle with centre (h, k) and radius r is:

$$(x-h)^2 + (y-k)^2 = r^2$$

where the key features are:
- **centre (h, k)**
- **radius r**
- **domain $[h-r, h+r]$**
- **range $[k-r, k+r]$.**

WORKED EXAMPLE 1 Sketching circles

Sketch the graphs of the following circles. Identify the domain and range of each.
a. $x^2 + (y-3)^2 = 1$ **b.** $(x+3)^2 + (y+2)^2 = 9$

THINK

a. 1. This circle has centre $(0, 3)$ and radius 1.

   2. On a set of axes, mark the centre and four points: 1 unit (the radius) left and right of the centre, and 1 unit (the radius) above and below the centre.

   3. Draw a circle that passes through these four points.

   4. Identify the domain.

   5. Identify the range.

b. 1. This circle has centre $(-3, -2)$ and radius 3.

   2. On a set of axes, mark the centre and four points: 3 units left and right of the centre, and 3 units above and below the centre.

   3. Draw a circle that passes through these four points.

   4. Identify the domain.

   5. Identify the range.

WRITE

a.

(graph of circle $x^2 + (y-3)^2 = 1$ with centre $(0,3)$, points at $y=2$, $y=4$, $x=-1$, $x=1$)

The domain is $[-1, 1]$.

The range is $[2, 4]$.

b.

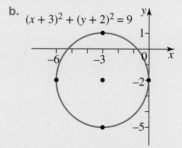

(graph of circle $(x+3)^2 + (y+2)^2 = 9$ with centre $(-3,-2)$)

The domain is $[-6, 0]$.

The range is $[-5, 1]$.

**TI	THINK**	**WRITE**	**CASIO	THINK**	**WRITE**

a. 1. On a Graphs page, press MENU, then select:
3: Graph Entry/Edit
3: Equation Templates
3: Circle
1: Center form
$(x - h)^2 + (y - k)^2 = r^2$
Complete the entry line as:
$(x - (-3))^2 + (y - (-2))^2 = 3^2$
then press ENTER.

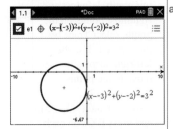

a. 1. On a Conic Graphs screen, use the up/down arrows to select the equation for a circle, then press EXE.

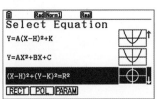

2. To determine the coordinates of the centre of the circle, press MENU, then select:
6: Analyze Graph
8: Analyze Conics
1: Center
Click on the centre of the circle then press ENTER.

2. Replace the values of H, K and R:
$H = -3$
$K = -2$
$R = 3$
Press EXE.

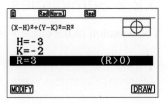

3. To plot maximum, minimum, leftmost and rightmost points on the graph, press MENU, then select:
5: Trace
1: Graph trace
Type '−3' and press ENTER twice.
Use the left/right arrows to move to the opposite side of the circle, then type '−3' again and press ENTER twice. Repeat this process to plot the points $(-6, -2)$ and $(0, -2)$.

3. Select DRAW by pressing F6.

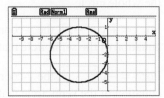

4. The domain and range can be read from the graph.

The domain is $[-6, 0]$ and the range is $[-5, 1]$.

4. To determine the coordinates of the centre of the circle, select G-Solve by pressing F5, then select CENTER by pressing F1. Press EXE.

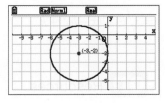

5. To plot the maximum, minimum, leftmost and rightmost points on the graph, select Trace by pressing F1. Use the left/right arrows to move to the points $(0, -2)$, $(-3, 1)$, $(-6, -2)$ and $(-3, -5)$ and press EXE on each one.

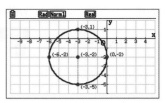

6. The domain and range can be read from the graph.

The domain is $[-6, 0]$ and the range is $[-5, 1]$.

5.2.2 Semicircles

The equation of the circle $x^2 + y^2 = r^2$ can be rearranged to make y the subject.

$$x^2 + y^2 = r^2$$
$$y^2 = r^2 - x^2$$
$$y = \pm\sqrt{r^2 - x^2}$$

The equation of the circle can be expressed as $y = \pm\sqrt{r^2 - x^2}$. This form of the equation indicates two **semicircle** functions that together make up the whole circle.

For $y = +\sqrt{r^2 - x^2}$, the y-coordinates must be positive (or zero), so this is the equation of the semicircle that lies above the x-axis.

For $y = -\sqrt{r^2 - x^2}$, the y-coordinates must be negative (or zero), so this is the equation of the semicircle that lies below the x-axis.

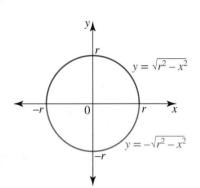

The semicircle $y = \sqrt{r^2 - x^2}$

The semicircle with equation $y = \sqrt{r^2 - x^2}$ is a function. It is the top half of the circle, with centre $(0, 0)$, radius r, domain $[-r, r]$ and range $[0, r]$.

The domain can be deduced algebraically since $\sqrt{r^2 - x^2}$ is only real if $r^2 - x^2 \geq 0$. From this, the domain requirement $-r \leq x \leq r$ can be obtained.

For the circle with centre (h, k) and radius r, rearranging its equation $(x - h)^2 + (y - k)^2 = r^2$ gives the equation of the top, or upper, semicircle as

$$y = \sqrt{r^2 - (x - h)^2} + k.$$

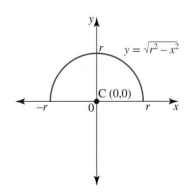

a. Sketch the graph of $y = \sqrt{5 - x^2}$ and identify the domain and range.

b. For the circle with equation $4x^2 + 4y^2 = 1$, determine the equation of its lower semicircle and identify its domain and range.

THINK

a. 1. State the centre and radius of the circle this semicircle is part of.

2. Sketch the graph.

3. Read from the graph its domain and range.

b. 1. Rearrange the equation of the circle to make y the subject and state the equation of the lower semicircle.

2. Identify the domain and range.

WRITE

a. $y = \sqrt{5 - x^2}$ is the equation of a semicircle in the form of $y = \sqrt{r^2 - x^2}$.

Centre: $(0, 0)$

Radius: $r^2 = 5 \Rightarrow r = \sqrt{5}$ since r cannot be negative.

This is an upper semicircle.

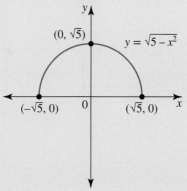

Domain $\left[-\sqrt{5}, \sqrt{5}\right]$; range $\left[0, \sqrt{5}\right]$

b. $4x^2 + 4y^2 = 1$

Rearrange: $4y^2 = 1 - 4x^2$

$$y^2 = \frac{1 - 4x^2}{4}$$

$$y = \pm\sqrt{\frac{1}{4} - x^2}$$

Therefore, the lower semicircle has the equation

$$y = -\sqrt{\frac{1}{4} - x^2}.$$

The domain is $\left[-\dfrac{1}{2}, \dfrac{1}{2}\right]$. This is the lower semicircle,

so the range is $\left[-\dfrac{1}{2}, 0\right]$.

TI | THINK	WRITE	CASIO | THINK	WRITE

a. 1. On a Graphs page, complete the entry line for function 1 as: $f1(x) = \sqrt{5 - x^2}$ then press ENTER.

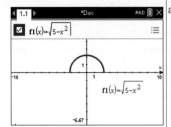

a. 1. On a Graph screen, complete the entry line for Y1 as: $Y1 = \sqrt{5 - x^2}$, $\left[-\sqrt{5}, \sqrt{5}\right]$ then press EXE. Select DRAW by pressing F6.

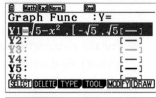

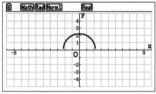

2. To find the x-intercepts, press MENU, then select:
6: Analyze Graph
1: Zero
Move the cursor to the left of the x-intercept when prompted for the lower bound, then press ENTER. Move the cursor to the right of the x-intercept when prompted for upper bound, then press ENTER. Repeat this process to find the other x-intercept.

2. To find the x-intercepts, select G-Solve by pressing F5, then select ROOT by pressing F1. Press EXE then use the left/right arrows to move across to the other x-intercept and press EXE.

3. To find the maximum, press MENU, then select:
6: Analyze Graph
3: Maximum
Move the cursor to the left of the maximum when prompted for the lower bound, then, press ENTER. Move the cursor to the right of the maximum when prompted for the upper bound, then press ENTER.

3. To find the maximum, select G-Solve by pressing F5, then select MAX by pressing F2. Press EXE.

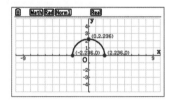

4. State the domain and range.
Note: The calculator only displays approximate values for points of interest on the screen. However, exact values should be used when stating the domain and range.

The domain is $[-\sqrt{5}, \sqrt{5}]$ and the range is $[0, \sqrt{5}]$.

4. State the domain and range.
Note: The calculator only displays approximate values for points of interest on the screen. However, exact values should be used when stating the domain and range.

The domain is $[-\sqrt{5}, \sqrt{5}]$ and the range is $[0, \sqrt{5}]$.

Exercise 5.2 The circle

5.2 Exercise **5.2 Exam practice** on

These questions are
even better in jacPLUS!
• Receive immediate feedback
• Access sample responses
• Track results and progress

Find all this and MORE in jacPLUS ▶

Simple familiar	Complex familiar	Complex unfamiliar
1, 2, 3, 4, 5, 6, 7, 8, 9, 10, 11	12, 13, 14, 15, 16	17, 18, 19, 20

Simple familiar

1. State the equation of each of the circles and semicircles graphed below.

a. 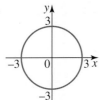   b. 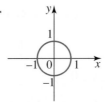   c. 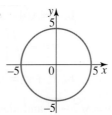   d.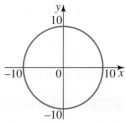

e. 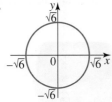   f. 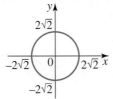   g. 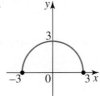   h.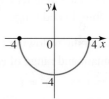

2. Identify the domain and range of each circle and semicircle in question 1.

3. Sketch the graph of each of the following relations.

   a. $x^2 + y^2 = 4$ b. $x^2 + y^2 = 16$ c. $x^2 + y^2 = 49$

   d. $x^2 + y^2 = 7$ e. $x^2 + y^2 = 12$ f. $x^2 + y^2 = \dfrac{1}{4}$

4. **WE1** Sketch the graphs of the following circles. Identify the domain and range of each.

   a. $x^2 + (y+2)^2 = 1$ b. $x^2 + (y-2)^2 = 4$

   c. $(x-4)^2 + y^2 = 9$ d. $(x-2)^2 + (y+1)^2 = 16$

   e. $(x+3)^2 + (y+2)^2 = 25$ f. $(x-3)^2 + (y-2)^2 = 9$

   g. $(x+5)^2 + (y-4)^2 = 36$ h. $\left(x - \dfrac{1}{2}\right)^2 + \left(y + \dfrac{3}{2}\right)^2 = \dfrac{9}{4}$

5. **MC** Consider the circle shown.

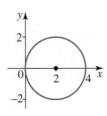

   a. The equation of the circle is:

   A. $x^2 + (y-2)^2 = 4$ **B.** $(x-2)^2 + y^2 = 16$ **C.** $(x+2)^2 + y^2 = 16$ **D.** $(x-2)^2 + y^2 = 4$

   b. The range of the relation is:

   A. R **B.** $[-2, 2]$ **C.** $[0, 4]$ **D.** $[2, 4]$

6. **MC** Consider the equation $(x+3)^2 + (y-1)^2 = 1$.

   a. The graph that represents this relation is:

   A. 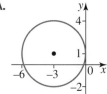 B. 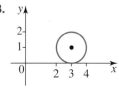 C. 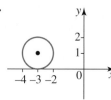 D.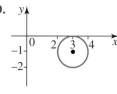

   b. The domain of the relation is:

   A. $[-3.5, -2.5]$ B. $(-4, -2)$ C. R D. $[-4, -2]$

7. Classify each of the following equations as describing either a function (F) or a non-function (N).

   a. $y = \pm\sqrt{81 - x^2}$

   b. $y = \sqrt{4 - x^2}$

   c. $y = -\sqrt{1 - x^2}$

   d. $y = \sqrt{\dfrac{1}{9} - x^2}$

   e. $y = -\sqrt{\dfrac{1}{4} - x^2}$

   f. $y = \sqrt{5 - x^2}$

   g. $y = \pm\sqrt{10 - x^2}$

   h. $x^2 + y^2 = 3, -\sqrt{3} \le x \le 0$

8. **WE2** a. Sketch the graph of $y = \sqrt{7 - x^2}$ and identify the domain and range.

   b. For the circle with equation $9x^2 + 9y^2 = 1$, identify the equation of its upper semicircle and identify its domain and range.

9. Sketch the following circles and state the centre, radius, domain and range of each.

   a. $x^2 + (y-1)^2 = 1$
   b. $(x+2)^2 + (y+4)^2 = 9$
   c. $16x^2 + 16y^2 = 81$

10. Determine the equations of the following circles from the given information.

   a. Centre $(-8, 9)$; radius 6
   b. Centre $(7, 0)$; radius $2\sqrt{2}$
   c. Centre $(1, 6)$ and containing the point $(-5, -4)$

   d. The end points of a diameter are $\left(-\dfrac{4}{3}, 2\right)$ and $\left(\dfrac{4}{3}, 2\right)$.

11. A region is bounded by a circle described by the equation $(x-2)^2 + (y+4)^2 = 25$. Evaluate whether the following points lie **i** inside the circle, **ii** on the circle or **iii** outside the circle.

   a. $(2, 1)$ b. $(0, 0)$ c. $(1, 3)$ d. $(4, -3)$ e. $(5, 3)$

Complex familiar

12. Determine a general rule that can be used to determine if a point (m, n) lies inside the region bounded by the equation $(x-h)^2 + (y-k)^2 = r^2$.

13. Deduce the two values of a so that the point $(a, 2)$ lies on the circle $x^2 + y^2 + 8x - 3y + 2 = 0$.

14. The points $(0, 4)$ and $(4, 0)$ lie on the circle $x^2 + y^2 = 16$. Determine whether it is possible to draw two other circles with different radii to this one that also pass through this pair of points. Explain your decision.

15. Calculate the coordinates of the points of intersection of the line $y = 2x$ and the circle $(x-2)^2 + (y-2)^2 = 1$.

16. Calculate the coordinates of the points of intersection of $y = 7 - x$ with the circle $x^2 + y^2 = 49$. On a diagram, sketch the region $((x, y) : y \ge 7 - x) \cap \{(x, y) : x^2 + y^2 \le 49\}$.

17. Circular ripples are formed when a water drop hits the surface of a pond. If one ripple is represented by the equation $x^2 + y^2 = 4$ and then 3 seconds later by $x^2 + y^2 = 190$, where the length of measurements are in centimetres, calculate how fast the ripple is moving outwards. (State your answers to 1 decimal place.)

18. Determine the shortest and longest distance from the origin to the circle $(x - 2)^2 + (y + 4)^2 = 9$.

19. Consider the circle with equation $(x - 1)^2 + (y - 2)^2 = 25$. Using clearly explained mathematical analysis, calculate the exact distance of the centre of the circle from the chord joining the points $(5, -1)$ and $(4, 6)$.

20. When a drop of water hits the flat surface of a pool, circular ripples are made. One ripple is represented by the equation $x^2 + y^2 = 9$, and 5 seconds later, the ripple is represented by the equation $x^2 + y^2 = 225$, where the lengths of the radii are in centimetres.
If the ripple continues to move at the same rate, determine when it will hit the edge of the pool, which is 2 metres from its centre.

Fully worked solutions for this chapter are available online.

LESSON
5.3 The sideways parabola and square root function

SYLLABUS LINKS
- Recognise and determine features of the graph of $y^2 = x$, including its parabolic shape and axis of symmetry.
- Recognise and determine features of the graphs of $y = a\sqrt{x - h} + k$, including their shape, intercepts, and behaviour as $x \to \infty$ and $x \to -\infty$.
- Sketch the graphs of relations, with or without technology.
- Model and solve problems that involve relations, with and without technology.

Source: Mathematical Methods Senior Syllabus 2024 © State of Queensland (QCAA) 2024; licensed under CC BY 4.0.

5.3.1 The relation $y^2 = x$

The relation $y^2 = x$ cannot be a function since, for example, $x = 1$ is paired with both $y = 1$ and $y = -1$; the graph of $y^2 = x$ therefore fails the vertical line test for a function.

The shape of the graph of $y^2 = x$ could be described as a **sideways parabola** opening to the right, like the reflector in a car's headlight.

The features of the graph of $y^2 = x$ are:
- domain $R^+ \cup \{0\}$ with the graph opening to the right or $[0, \infty)$
- range R
- a turning point, usually called a vertex, at $(0, 0)$
- a horizontal axis of symmetry with equation $y = 0$ (the x-axis).

The graph of $y^2 = -x$ will open to the left, with domain $R^- \cup \{0\}$ or $(-\infty, 0]$.

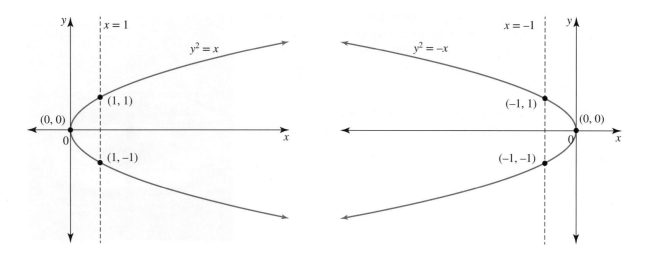

5.3.2 The square root function

The **square root function** is formed from two branches of the relation $y^2 = x$.

Since $y^2 = x \Rightarrow y = \pm \sqrt{x}$, the upper branch has the equation $y = \sqrt{x}$ and the lower branch has the equation $y = -\sqrt{x}$.

These two functions are drawn on the diagram.

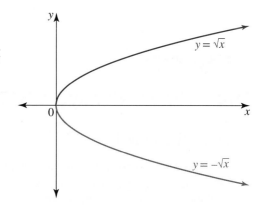

5.3.3 The graph of $y = \sqrt{x}$

The square root function is defined by the equation $y = \sqrt{x}$ or $y = x^{\frac{1}{2}}$. The y-values of this function must be such that $y \geq 0$. No term under a square root symbol can be negative, so this function also requires that $x \geq 0$. The graph of the square root function is shown in the diagram.

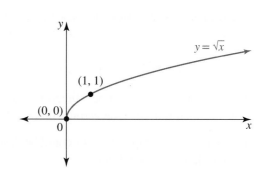

The key features of the graph of $y = \sqrt{x}$ are as follows:
- The end point is $(0, 0)$.
- As $x \to \infty, y \to \infty$.
- The function is defined only for $x \geq 0$, so the domain is $R^+ \cup \{0\}$ or $[0, \infty)$.
- y-values cannot be negative, so the range is $R^+ \cup \{0\}$ or $[0, \infty)$.
- The graph is of a function.
- The shape is the upper half of the sideways parabola $y^2 = x$.

Variations of the basic graph

The term under the square root cannot be negative, so $y = \sqrt{-x}$ has a domain requiring $-x \geq 0 \Rightarrow x \in (-\infty, 0]$. This graph can be obtained by reflecting the graph of $y = \sqrt{x}$ in the y-axis. The four variations in position of the basic graph are shown in the diagram.

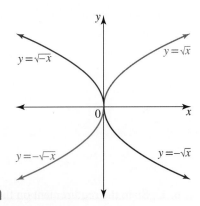

The diagram could also be interpreted as displaying the graphs of the two relations $y^2 = x$ (on the right of the y-axis) and $y^2 = -x$ (on the left of the y-axis). The end points of the square root functions are the vertices of these sideways parabolas, the point $(0, 0)$.

5.3.4 Transformations of the square root function

The general equation of the square root function

$$y = a\sqrt{x - h} + k$$

The key features of the graph of $y = a\sqrt{x - h} + k$ are as follows:

- The end point is (h, k).
- If $a > 0$, the end point is a minimum point.
- If $a < 0$, the end point is a maximum point.
- The domain is $[h, \infty)$, since $x - h \geq 0 \rightarrow x \geq h$.
- If $a > 0$, the range is $[k, \infty)$.
- If $a < 0$, the range is $(-\infty, k]$.
- There are either one or no x-intercepts.
- There are either one or no y-intercepts.

The graph of $y = a\sqrt{-(x - h)} + k$ also has its end point at (h, k), but its domain is $(-\infty, h]$.

WORKED EXAMPLE 3 Sketching square root functions

a. Sketch $y = 2\sqrt{x + 1} - 4$, and identify its domain and range. Include any axis intercepts with their coordinates.

b. For the function $f(x) = \sqrt{2 - x}$:
   i. identify its domain
   ii. sketch the graph of $y = f(x)$.

THINK

a. 1. State the coordinates of the end point.

   2. Calculate any intercepts with the coordinate axes.

WRITE

a. $y = 2\sqrt{x + 1} - 4$

End point $(-1, -4)$

y-intercept: let $x = 0$.

$y = 2\sqrt{1} - 4$

  $= -2$

y-intercept $(0, -2)$

x-intercept: let $y = 0$.

$2\sqrt{x + 1} - 4 = 0$

$\therefore \sqrt{x + 1} = 2$

Square both sides:

$x + 1 = 4$

$\therefore x = 3$

x-intercept $(3, 0)$

3. Sketch the graph and state its domain and range.

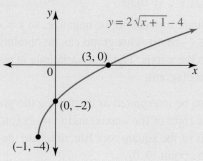

Domain $[-1, \infty)$, range $[-4, \infty)$

b. i. State the requirement on the expression under the square root and then state the domain.

b. i. $f(x) = \sqrt{2-x}$
Domain: require $2 - x \geq 0$.
$-x \geq -2$
$\therefore x \leq 2$
Therefore, the domain is $(-\infty, 2]$.

ii. Identify the key points of the graph and sketch.

ii. $f(x) = \sqrt{2-x}$
$y = \sqrt{2-x}$
$\quad = \sqrt{-(x-2)}$
The end point $(2, 0)$ is also the x-intercept.
y-intercept: let $x = 0$ in $y = \sqrt{2-x}$.
$y = \sqrt{2}$
The y-intercept is $\left(0, \sqrt{2}\right)$.

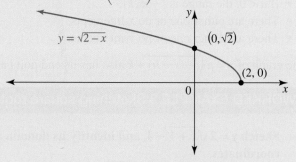

TI	THINK	DISPLAY/WRITE	CASIO	THINK	DISPLAY/WRITE

a. 1. On a Graphs page, complete the entry line as:
$f1(x) = 2\sqrt{x+1} - 4$
Then press ENTER.

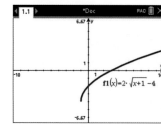

a. 1. On a Graphs screen, complete the entry line as:
$y1 = 2\sqrt{x+1} - 4$
Press EXE and then f6 to DRAW.

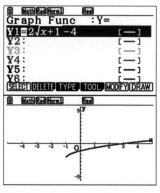

2. State the domain and range.

The domain is $[-1, \infty)$.
The range is $[-4, \infty)$.

2. State the domain and range.

The domain is $[-1, \infty)$.
The range is $[-4, \infty)$.

5.3.5 Determining the equation of a square root function

If a diagram is given, the direction of the graph will indicate whether to use the equation in the form $y = a\sqrt{x - h} + k$ or $y = a\sqrt{-(x - h)} + k$. If a diagram is not given, the domain or a rough sketch of the given information may clarify which form of the equation to use.

WORKED EXAMPLE 4 Determining the equation of a square root graph

Determine a possible equation for the square root function shown.

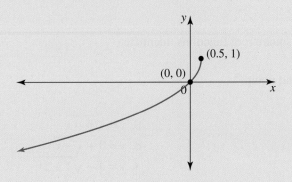

THINK

1. Note the direction of the graph to decide which form of the equation to use.

2. Substitute the coordinates of the end point into the equation.

3. Use a second point on the graph to determine the value of a.

4. Determine the equation of the graph.

5. Express the equation in a simplified form.

WRITE

The graph opens to the left with $x \leq 0.5$.

Let the equation be $y = a\sqrt{-(x - h)} + k$.

End point $(0.5, 1)$
$$y = a\sqrt{-(x - 0.5)} + 1$$
$$= a\sqrt{-x + 0.5} + 1$$

$(0, 0)$ lies on the graph.
$$0 = a\sqrt{0.5} + 1$$
$$-1 = a \times \frac{1}{\sqrt{2}}$$
$$a = -\sqrt{2}$$

The equation is $y = -\sqrt{2}\sqrt{-x + 0.5} + 1$.

$$y = -\sqrt{2}\sqrt{-x + \frac{1}{2}} + 1$$
$$= -\sqrt{2}\sqrt{\frac{-2x + 1}{2}} + 1$$
$$= -\sqrt{2}\frac{\sqrt{-2x + 1}}{\sqrt{2}} + 1$$
$$= -\sqrt{-2x + 1} + 1$$

Therefore, the equation is $y = 1 - \sqrt{1 - 2x}$.

Simple familiar

1. For each of the following square root functions, identify:

   i. its domain
   ii. the coordinates of its end point
   iii. its range.

   a. $y = \sqrt{x-9}$

   b. $y = \sqrt{4-x}$

   c. $y = -\sqrt{x+3}$

   d. $y = 3 + \sqrt{-x}$

   e. $y = \sqrt{3x-6} - 7$

   f. $y = 4 - \sqrt{1-2x}$

2. **WE3** a. Sketch $y = \sqrt{x-1} - 3$, and identify its domain and range. Include any axis intercepts with their coordinates.

   b. For the function $f(x) = -\sqrt{2x+4}$:

   i. identify the domain
   ii. sketch the graph of $y = f(x)$.

3. Sketch the graph of each of the following square root functions, showing end points and any intersections with the coordinate axes.

   a. $y = \sqrt{x+4} - 1$

   b. $y = 3 - \sqrt{3x}$

   c. $y = -\sqrt{6-x}$

   d. $y = \sqrt{9-2x} + 1$

4. Sketch the following relations and identify their domains and ranges.

   a. $y = \sqrt{x+3} - 2$

   b. $y = 5 - \sqrt{5x}$

   c. $y = 2\sqrt{9-x} + 4$

   d. $y = \sqrt{49-7x}$

   e. $y = 2 + \sqrt{x+4}$

   f. $y + 1 + \sqrt{-2x+3} = 0$

5. **WE4** Form a possible equation for the square root function shown.

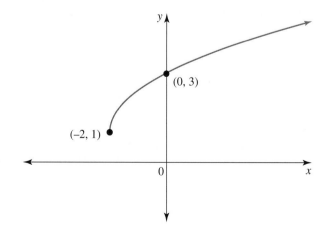

6. Determine the equation for the graph shown, given it represents a square root function with end point $(-2, 2)$.

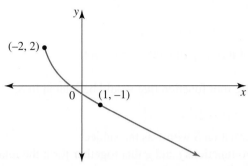

7. Form the equation of the square root function with end point $(4, -1)$ and containing the point $(0, 9)$. Determine the coordinates of the point at which this function cuts the x-axis.

8. Determine the equation of the function that has the same shape as $y = \sqrt{-x}$ and an end point with coordinates $(4, -4)$.

9. A square root function graph has an equation in the form $y = a\sqrt{x - h} + k$. If the end point is $(3, 2)$ and the graph passes through the point $(4, 6)$, determine the equation of the graph.

10. Determine the equation of the square root function shown in the diagram, expressing it in the form $y = a\sqrt{x - h} + k$.

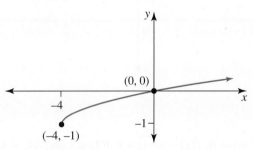

11. **MC** The graph of the function $f(x) = 1 + \sqrt{3 - x}$ could be:

A.

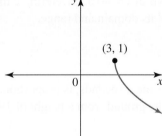

B.

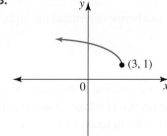

C.

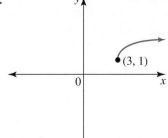

D.

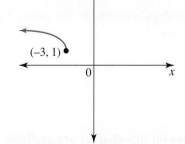

12. **MC** The coordinates of the end point of the square root function with equation $y - 2 = \sqrt{3 - x}$ are:

    A. $(2, 3)$ **B.** $(2, -3)$ **C.** $(3, 2)$ **D.** $(-3, 2)$

13. Consider the function $S = \{(x, y) : y + 2 = 3\sqrt{x - 1}\}$.

    a. Determine the coordinates of its end point and x-intercept.

    b. Sketch the graph of S.

    c. Form the equation of a square root function that would be related to S.

14. Consider the relation $S = \{(x, y) : (y - 2)^2 = 1 - x\}$.

    a. Express the equation of the relation S with y as the subject.

    b. Define, as mappings, the two functions f and g that together form the relation S.

    c. Sketch, on the same diagram, the graphs of $y = f(x)$ and $y = g(x)$.

    d. Determine the image of -8 for each function.

15. The equation of the graph shown has the form $y = \sqrt{a(x - h)} + k$. Determine the values of a, h and k, and state the equation.

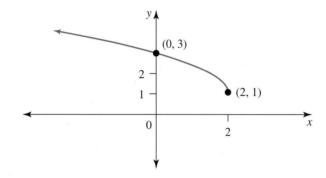

16. A function is defined as $f : [0, \infty) \rightarrow R, f(x) = \sqrt{mx + n}, f(1) = 1$ and $f(4) = 4$.
Sketch the graph of $y = f(x)$ together with the graph of $y = x$ and state $\{x : f(x) > x\}$.

17. The function $y = a - \sqrt{b(x - c)}$ has an end point at $(2, 5)$ and cuts the x-axis at $x = -10.5$. Determine the values of a, b and c, and hence determine the equation of the function and its domain and range.

18. A small object falls to the ground from a vertical height of h metres. The time, t seconds, is proportional to the square root of the height. If it takes 2 seconds for the object to reach the ground from a height of 19.6 metres, obtain the rule for t in terms of h.

19. Consider the curve $y = \sqrt{x} - 1$ and the straight line $y = kx$.
Determine the values of k if the line and the curve intersect at two distinct points.

20. Discuss the relationships between the two curves $y = x^2$ and $y^2 = x$.
Justify your decisions.

Fully worked solutions for this chapter are available online.

LESSON
5.4 The reciprocal function

SYLLABUS LINKS

- Recognise features of the graphs of $y = \dfrac{1}{x}$ and $y = \dfrac{a}{(x-h)} + k$, including their hyperbolic shape, intercepts, asymptotes, and behaviour as $x \to \infty$ and $x \to -\infty$.
- Model and solve problems that involve reciprocal functions, with and without technology.
- Sketch the graphs of reciprocal functions, with or without technology.

Source: Mathematical Methods Senior Syllabus 2024 © State of Queensland (QCAA) 2024; licensed under CC BY 4.0.

5.4.1 The graph of $y = \dfrac{1}{x}$

The rule $y = x^{-1}$ can also be written as $y = \dfrac{1}{x}$. This is the rule for a **rational function** called a **hyperbola**. Two things can be immediately observed from the rule:

- $x = 0$ must be excluded from the domain, since division by zero is not defined.
- $y = 0$ must be excluded from the range, since there is no number whose reciprocal is zero.

The lines $x = 0$ and $y = 0$ are **asymptotes**. An asymptote is a line the graph will approach but never reach. As these two asymptotes $x = 0$ and $y = 0$ are a pair of perpendicular lines, the hyperbola is known as a **rectangular hyperbola**. The asymptotes are a key feature of the graph of a hyperbola.

Completing a table of values can give us a 'feel' for this graph.

x	-10	-4	-2	-1	$-\dfrac{1}{2}$	$-\dfrac{1}{4}$	$-\dfrac{1}{10}$	0	$\dfrac{1}{10}$	$\dfrac{1}{4}$	$\dfrac{1}{2}$	1	2	4	10
y	$-\dfrac{1}{10}$	$-\dfrac{1}{4}$	$-\dfrac{1}{2}$	-1	-2	-4	-10	no value possible	10	4	2	1	$\dfrac{1}{2}$	$\dfrac{1}{4}$	$\dfrac{1}{10}$

The values in the table illustrate that as $x \to \infty$, $y \to 0$, and as $x \to -\infty$, $y \to 0$.

The table also illustrates that as $x \to 0$, either $y \to -\infty$ or $y \to \infty$.

These observations describe the asymptotic behaviour of the graph.

The key features: of the graph $y = \dfrac{1}{x}$ (shown) are as follows:

- The vertical asymptote has equation $x = 0$ (the y-axis).
- The horizontal asymptote has equation $y = 0$ (the x-axis).
- The domain is $R\backslash\{0\}$.
- The range is $R\backslash\{0\}$.
- As $x \to \infty$, $y \to 0$ from above, and as $x \to -\infty$, $y \to 0$ from below. This can be written as: $x \to \infty$, $y \to 0^+$ and as $x \to -\infty, y \to 0^-$.
- As $x \to 0^-$, $y \to -\infty$, and as $x \to 0^+$, $y \to \infty$.
- The graph is that of a one-to-one function.
- The graph has two branches separated by the asymptotes.
- As the two branches do not join at $x = 0$, the function is said to be **discontinuous** at $x = 0$.
- The graph lies in **quadrants** 1 and 3 as defined by the asymptotes.

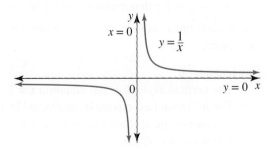

With the basic shape of the hyperbola established, transformations of the graph of $y = \dfrac{1}{x}$ can be studied.

5.4.2 Transformations of the graph $y = \dfrac{1}{x}$

Dilation from the x-axis

The effect of a dilation factor on the graph can be illustrated by comparing $y = \dfrac{1}{x}$ and $y = \dfrac{2}{x}$.

For $x = 1$, the point $(1, 1)$ lies on $y = \dfrac{1}{x}$, whereas the point $(1, 2)$ lies on $y = \dfrac{2}{x}$. The dilation effect on $y = \dfrac{2}{x}$ is to move the graph further out from the x-axis. The graph has a dilation factor of 2 from the x-axis or in the y-direction.

Reflection in the x-axis

The graph of $y = -\dfrac{1}{x}$ illustrates the effect of inverting the graph by reflecting $y = \dfrac{1}{x}$ in the x-axis.

The graph of $y = -\dfrac{1}{x}$ lies in quadrants 2 and 4 as defined by the asymptotes.

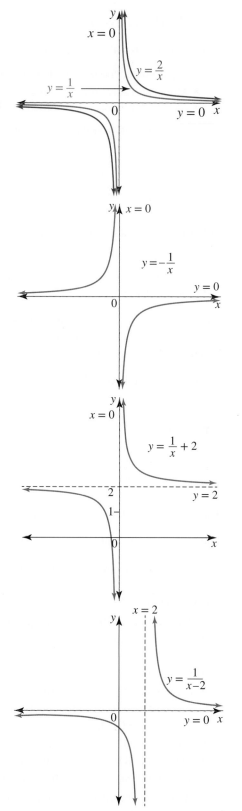

Vertical translation

The graph of $y = \dfrac{1}{x} + 2$ illustrates the effect of a vertical translation of 2 units upwards.

Key features:
- The horizontal asymptote has equation $y = 2$. This means that as $x \to \pm\infty$, $y \to 2$.
- The vertical asymptote is unaffected and remains $x = 0$.
- The domain is $R\backslash\{0\}$.
- The range is $R\backslash\{2\}$.

Horizontal translation

For $y = \dfrac{1}{x - 2}$ as the denominator cannot be zero, $x - 2 \neq 0 \Rightarrow x \neq 2$. The domain must exclude $x = 2$, so the line $x = 2$ is the vertical asymptote.

Key features:
- The vertical asymptote has equation $x = 2$.
- The horizontal asymptote is unaffected by the horizontal translation and still has the equation $y = 0$.
- The domain is $R\backslash\{2\}$.
- The range is $R\backslash\{0\}$.

The graph of $y = \dfrac{1}{x - 2}$ demonstrates the same effect that we have seen with other graphs that are translated 2 units to the right.

5.4.3 The general equation of a hyperbola

The general equation of the hyperbola

$$y = \frac{a}{(x-h)} + k$$

where a, h and k are constants.

The key features of the graph $y = \frac{a}{(x-h)} + k$ are as follows:

- The vertical asymptote has the equation $x = h$.
- The horizontal asymptote has the equation $y = k$.
- The domain is $R \backslash \{h\}$.
- The range is $R \backslash \{k\}$.
- Asymptotic behaviour: as $x \to \pm\infty$, $y \to k$, and as $x \to h$, $y \to \pm\infty$.
- There are two branches to the graph, and the graph is discontinuous at $x = h$.
- If $a > 0$, the graph lies in the asymptote-formed quadrants 1 and 3.
- If $a < 0$, the graph lies in the asymptote-formed quadrants 2 and 4.
- $|a|$ gives the dilation factor from the x-axis.

WORKED EXAMPLE 5 Describing transformations to $y = \dfrac{1}{x}$

Determine the transformations that should be made to the graph of $y = \dfrac{1}{x}$ to obtain the graph of $y = \dfrac{-4}{x+2} - 1$.

THINK	WRITE
1. Write the general equation of the hyperbola.	$y = \dfrac{a}{(x-h)} + k$
2. Identify the value of a.	$a = -4$
3. Describe the transformations to $y = \dfrac{1}{x}$ caused by a.	The graph of $y = \dfrac{1}{x}$ is dilated by the factor of 4 in the y direction or from the x-axis and reflected in the x-axis.
4. Identify the value of h.	$h = -2$
5. Describe the transformations to the graph caused by h.	The graph is translated 2 units to the left.
6. Identify the value of k.	$k = -1$
7. Describe the transformations to the graph caused by k.	The graph is translated 1 unit down.

Sketching the graph of the hyperbola by hand can be easily done by following these steps:

Step 1 Decide whether the hyperbola is positive or negative.

Step 2 Determine the position of the asymptotes.

Step 3 Determine the values of the intercepts with the axes.

Step 4 On the set of axes, draw the asymptotes (using dotted lines) and mark the intercepts with the axes.

Step 5 Treating the asymptotes as the new set of axes, sketch either the positive or the negative hyperbola, making sure it passes through the intercepts that have been previously marked.

WORKED EXAMPLE 6 Sketching hyperbolas

Sketch the graph of $y = \dfrac{2}{x+2} - 4$, clearly identifying the intercepts with the axes and the positions of the asymptotes.

THINK

1. Compare the given equation with $y = \dfrac{a}{x-h} + k$ and state the values of a, h and k.

2. Write a short statement about the effects of a, h and k on the graph of $y = \dfrac{1}{x}$.

3. Write the equations of the asymptotes.

4. Determine the value of the y-intercept by letting $x = 0$.

5. Determine the value of the x-intercept by making $y = 0$.

WRITE

$a = 2, h = -2, k = -4$

The graph of $y = \dfrac{1}{x}$ is dilated by the factor of 2 in the y direction or from the x-axis and translated 2 units to the left and 4 units down.

Asymptotes: $x = -2; \; y = -4$

y-intercept: when $x = 0$, $y = \dfrac{2}{0+2} - 4$
$= 1 - 4$
$= -3$

x-intercept: when $y = 0$, $\quad 0 = \dfrac{2}{x+2} - 4$

$\dfrac{2}{x+2} = 4$

$2 = 4(x+2)$

$2 = 4x + 8$

$4x = 2 - 8$

$4x = -6$

$x = -\dfrac{6}{4}$

$\therefore x = -\dfrac{3}{2}$

6. To sketch the graph, follow these steps:
- Draw the set of axes and label them.
- Use dashed lines to draw the asymptotes and label them.
- Mark the intercepts with the axes.
- Treating the asymptotes as your new set of axes, sketch the graph of the hyperbola (as a is positive, the graph is not reflected). Make sure the upper branch passes through the x- and y-intercepts previously marked.

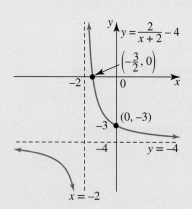

**TI	THINK**	**WRITE**	**CASIO	THINK**	**WRITE**

1. On a Graphs page, complete the entry line for function 1 as:
$$f1(x) = \frac{2}{x+2} - 4$$
then press ENTER.

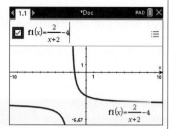

1. On a Graph screen, complete the entry line for Y1 as:
$$Y1 = \frac{2}{x+2} - 4$$
then press EXE. Select DRAW by pressing F6.

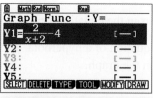

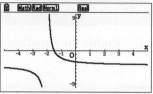

2. To find the x-intercept, press MENU, then select:
6: Analyze Graph
1: Zero
Move the cursor to the left of the x-intercept when prompted for the lower bound, then press ENTER. Move the cursor to the right of the x-intercept when prompted for the upper bound, then press ENTER.

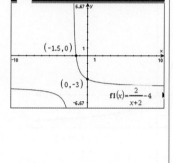

2. To find the x-intercept, select G-Solve by pressing F5, then select ROOT by pressing F1. Press EXE.

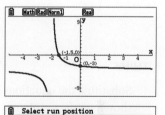

3. To find the y-intercept, press MENU, then select:
5: Trace
1: Graph Trace
Type '0', then press ENTER twice.

3. To find the y-intercept, select G-Solve by pressing F5, then select Y-ICEPT by pressing F4. Press EXE.

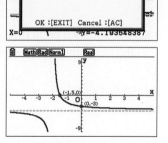

4. To draw the horizontal asymptote on the screen, press MENU, then select:
8: Geometry
4: Construction
1: Perpendicular
Click on the y-axis, then click on the point on the y-axis where $y = -4$.
Press MENU, then select:
1: Actions
4: Attributes
Click on the asymptote, press the down arrow and then press the right arrow twice to change the line style to dashed. Press ENTER.

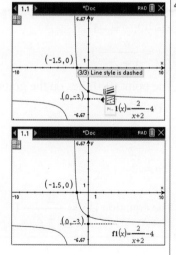

4. To draw the horizontal asymptote on the screen, select Sketch by pressing F4, press F6 to scroll across to more menu options, then select Horz by pressing F5. Use the up/down arrows to position the horizontal line, then select FORMAT by pressing SHIFT 5. Change the Line Style to Broken, then select OK by pressing EXIT. Press EXE.

5.4.4 Determine the equation of a hyperbola

From the equation $y = \dfrac{a}{(x-h)} + k$, it can be seen that three pieces of information will be needed to form the equation of a hyperbola. These are usually the equations of the asymptotes and the coordinates of a point on the graph.

WORKED EXAMPLE 7 Determining the equation of a hyperbola

Determine the equation of the hyperbola shown.

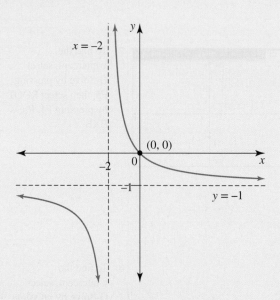

THINK	WRITE
1. Substitute the equations of the asymptotes shown on the graph into the general equation of a hyperbola.	Let the equation of the graph be $y = \dfrac{a}{(x-h)} + k$. From the graph, the asymptotes have equations $x = -2,\ y = -1$. $\therefore y = \dfrac{a}{x+2} - 1$
2. Use a known point on the graph to determine the remaining unknown constant.	The point $(0, 0)$ lies on the graph. $0 = \dfrac{a}{2} - 1$ $\therefore 1 = \dfrac{a}{2}$ $\therefore a = 2$
3. Identify the equation of the hyperbola.	The equation is $y = \dfrac{2}{x+2} - 1$.

5.4.5 Modelling with the hyperbola

Applications involving the use of the hyperbolic function for modelling and predicting data may need domain restrictions. Unlike many polynomial models, the hyperbola has neither maximum nor minimum turning points, so the asymptotes are often where the interest will lie. The horizontal asymptote is often of particular interest as it represents the limiting value of the model.

A relocation plan to reduce the number of bats in a public garden is formed, and t months after the plan is introduced, the number of bats N in the garden is thought to be modelled by $N = 250 + \dfrac{30}{t+1}$.

a. Determine how many bats were removed from the garden in the first 9 months of the relocation plan.

b. Sketch the graph of the bat population over time using the given model and identify its domain and range.

c. Determine the maximum number of bats that will be relocated according to this model.

THINK

a. Find the number of bats at the start of the plan and the number after 9 months, and calculate the difference.

b. 1. Identify the asymptotes and other key features that are appropriate for the restriction $t \geq 0$.

2. Sketch the part of the graph of the hyperbola that is applicable and label axes appropriately.
   Note: The vertical scale is broken in order to better display the graph.

3. Identify the domain and range for this model.

c. 1. Interpret the meaning of the horizontal asymptote.

2. Write the answer.

WRITE

a. $N = 250 + \dfrac{30}{t+1}$

When $t = 0$, $N = 250 + \dfrac{30}{1}$.

Therefore, there were 280 bats when the plan was introduced.

When $t = 9$, $N = 250 + \dfrac{30}{10}$.

Therefore, 9 months later there were 253 bats.

Difference $= 280 - 253$

$\qquad\qquad = 27$

Hence, over the first 9 months, 27 bats were removed.

b. $N = 250 + \dfrac{30}{t+1}, t \geq 0$

Vertical asymptote $t = -1$ (not applicable)

Horizontal asymptote $N = 250$

The initial point is $(0, 280)$.

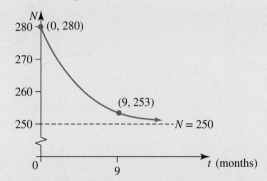

Domain $\{t : t \geq 0\}$ or $t \in [0, \infty)$

Range $(250, 280]$

c. The horizontal asymptote shows that as $t \to \infty, N \to 250$. This means $N = 250$ gives the limiting population of the bats.

Since the population of bats cannot fall below 250 and there were 280 initially, the maximum number of bats that can be relocated is 30.

5.4 Exercise	5.4 Exam practice on

Simple familiar	Complex familiar	Complex unfamiliar
1, 2, 3, 4, 5, 6, 7, 8, 9, 10, 11, 12, 13	14, 15, 16, 17	18, 19, 20

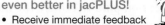

These questions are even better in jacPLUS!
- Receive immediate feedback
- Access sample responses
- Track results and progress

Find all this and MORE in jacPLUS ▶

Simple familiar

1. **WE5** Describe the changes that should be made to the graph of $y = \dfrac{1}{x}$ to obtain the graph of each of the following.

   a. $y = \dfrac{2}{x}$

   b. $y = -\dfrac{3}{x}$

   c. $y = \dfrac{1}{x-6}$

   d. $y = \dfrac{2}{x+4}$

2. **WE5** Describe the changes that should be made to the graph of $y = \dfrac{1}{x}$ to obtain the graph of each of the following.

   a. $y = \dfrac{1}{x} + 7$

   b. $y = \dfrac{2}{x} - 5$

   c. $y = \dfrac{1}{4+x} - 3$

   d. $y = -\dfrac{4}{x-1} - 4$

3. Identify which of the following transformations were applied to the graph of $y = \dfrac{1}{x}$ to obtain each of the graphs shown below.

   i. Translation to the right   **ii.** Translation to the left   **iii.** Translation up

   iv. Translation down   **v.** Reflection in the x-axis

   a.

   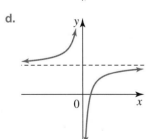

   b.

   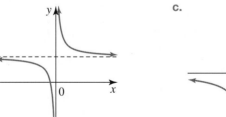

   c.

   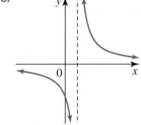

   d.

   e.

   f.

   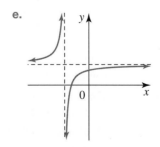

   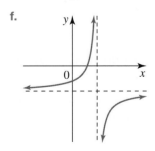

4. For each of the following graphs, identify:
   i. the equations of the asymptotes
   ii. the domain
   iii. the range.

a.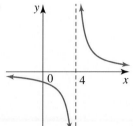

b.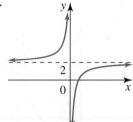

c.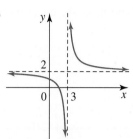

d.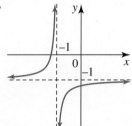

e.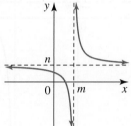

f.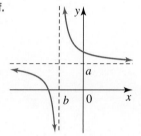

5. **WE6** Sketch each of the following, clearly identifying the position of the asymptotes and the intercepts with the axes.

   a. $y = \dfrac{1}{x+3}$

   b. $y = \dfrac{1}{x+2} - 1$

   c. $y = \dfrac{3}{x-1} - \dfrac{3}{4}$

   d. $y = -\dfrac{2}{x+5}$

6. **WE6** Sketch each of the following, clearly identifying the positions of the asymptotes and the intercepts with the axes.

   a. $y = \dfrac{6}{1-x} - 3$

   b. $y = -\dfrac{3}{x-2} + 6$

   c. $y = 1 - \dfrac{1}{2-x}$

   d. $y = \dfrac{1}{2x+3} + 4$

7. **MC** Identify the equation that represents the graph shown.

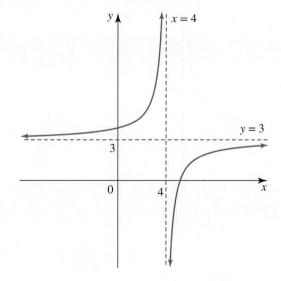

A. $y = 3 + \dfrac{1}{x-4}$

B. $y = 3 - \dfrac{1}{4-x}$

C. $y = \dfrac{1}{4-x} - 3$

D. $y = 3 - \dfrac{1}{x-4}$

8. If a function is given by $f(x) = \dfrac{1}{x}$, sketch each of the following, labelling the asymptotes and the intercepts with the axes.

   a. $f(x+2)$ b. $f(x) - 1$ c. $-f(x) - 2$

9. **WE7** Determine the equation of the hyperbola shown.

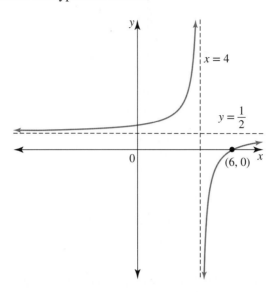

10. Identify the equations of the asymptotes of the following hyperbolas.

   a. $y = \dfrac{1}{x+5} + 2$ b. $y = \dfrac{8}{x} - 3$ c. $y = \dfrac{-3}{4x}$ d. $y = \dfrac{-3}{14+x} - \dfrac{3}{4}$

11. Sketch the graphs of the following functions and identify the domain and range of each function.

   a. $y = \dfrac{1}{x+1} - 3$ b. $y = 4 - \dfrac{3}{x-3}$ c. $y = -\dfrac{5}{3+x}$ d. $y = -\left(1 + \dfrac{5}{2-x}\right)$

12. **MC** A possible equation for the graph shown is:

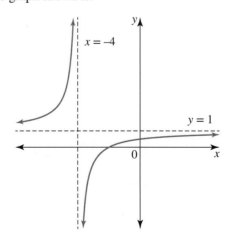

   A. $y = \dfrac{2}{x+4} - 1$ B. $y = \dfrac{2}{x+4} + 1$

   C. $y = \dfrac{-2}{x-4} + 1$ D. $y = \dfrac{-2}{x+4} + 1$

13. **WE8** The number of cattle, P, owned by a farmer at a time t years after purchase is modelled by

$$P = 30 + \frac{100}{2 + t}.$$

a. Determine how many cattle the herd is reduced by after the first 2 years.
b. Sketch the graph of the number of cattle over time using the given model and state its domain and range.
c. Determine the minimum number the herd of cattle is expected to reach according to this model.

Complex familiar

14. Deduce the equations of each of these graphs.

a.

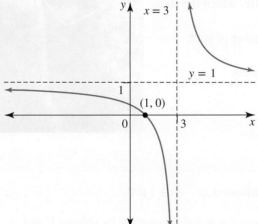

b.

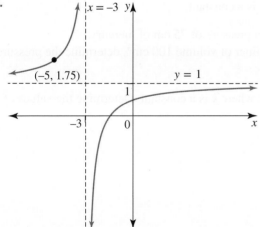

15. A hyperbola is undefined when $x = \frac{1}{4}$. As $x \to -\infty$, its graph approaches the line $y = -\frac{1}{2}$ from below.

The graph cuts the x-axis where $x = 1$.

a. Determine the equation of the hyperbola.
b. Express the function in mapping notation.

16. a. If $\dfrac{11-3x}{4-x} = a - \dfrac{b}{4-x}$, determine the values of a and b.

   b. Hence, sketch the graph of $y = \dfrac{11-3x}{4-x}$.

   c. Determine the values of x for which $\dfrac{11-3x}{4-x} > 0$.

17. In an effort to protect a rare species of stick insect, 20 of the species were captured and relocated to a small island where there were few predators. After 2 years, the population had grown to 240 stick insects.
A model for the size N of the stick insect population after t years on the island is thought to be defined by the function $N: R^+ \cup \{0\} \to R$,

$$N(t) = a + \frac{b}{t+2}.$$

   a. Calculate the values of a and b.

   b. Determine the length of time, to the nearest month, after which the stick insect population reached 400.

   c. Determine when the model would predict that the number of stick insects will reach 500.

   d. Determine how large the stick insect population can grow.

Complex unfamiliar

18. Determine the domain and range of the hyperbola with equation $xy - 4y + 1 = 0$.

19. Boyle's Law says that if the temperature of a given mass of gas remains constant, its volume V and pressure P are connected by the formula $V = \dfrac{k}{P}$, where k is a constant.

A container of volume 50 cm^3 is filled with a gas under a pressure of 75 cm of mercury.
If the container is connected by a hose to an empty container of volume 100 cm^3, determine the pressure in the two containers.

20. Consider the two functions $y = 1 - \dfrac{1}{(x+2)}$ and $y = x + k$, where k is a constant. Determine the values of k if the two functions intersect once, twice or not at all.

Fully worked solutions for this chapter are available online.

LESSON
5.5 Review

5.5.1 Summary

doc-
41800

Hey students! Now that it's time to revise this chapter, go online to:

Access the chapter summary	**Review your** results	**Practise exam** questions

Find all this and MORE in jacPLUS

5.5 Exercise

learn on

5.5 Exercise	5.5 Exam practice on

Simple familiar	Complex familiar	Complex unfamiliar
1, 2, 3, 4, 5, 6, 7, 8, 9, 10, 11, 12	13, 14, 15, 16	17, 18, 19, 20

These questions are even better in jacPLUS!
- Receive immediate feedback
- Access sample responses
- Track results and progress

Find all this and MORE in jacPLUS

Simple familiar

1. **MC** If $f(x) = \dfrac{2}{x} + 1$, then $f(x) + 2$ will have:

   A. the horizontal asymptote $y = 2$
   B. the horizontal asymptote $y = 1$
   C. the horizontal asymptote $y = 3$
   D. the vertical asymptote $x = 2$.

2. **MC** The equation of the graph shown is likely to be:

   A. $y = -\dfrac{2}{x-2} - 1$

   B. $y = 1 - \dfrac{2}{x+2}$

   C. $y = -\dfrac{2}{x+1} - 2$

   D. $y = \dfrac{-2}{x+2} - 1$

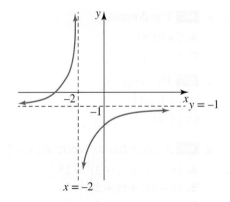

3. The graph of $y = \dfrac{1}{x}$ was dilated by the factor of 4 in the y direction or from the x-axis, reflected in the x-axis and then translated 2 units to the left and 1 unit down.

   a. Identify the equation of the asymptotes.
   b. Identify the domain and range.
   c. Identify the equation of the new graph.
   d. Sketch the graph.

4. Sketch the graph of each of the following, clearly showing the positions of the asymptotes and the intercepts with the axes.

   a. $y = \dfrac{2}{x-2}$

   b. $y = -\dfrac{4}{x} - 1$

   c. $y = \dfrac{2}{x-4} + 2$

5. Determine the equation of the hyperbola shown in the diagram.

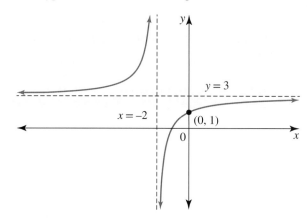

6. **MC** The equation of the circle shown is:

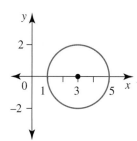

   A. $(x+3)^2 + y^2 = 4$

   B. $(x-3)^2 + y^2 = 2$

   C. $(x+3)^2 + y^2 = 2$

   D. $(x-3)^2 + y^2 = 4$

The circle with equation $(x+1)^2 + (y-4)^2 = 9$ applies to questions **7** and **8**.

7. **MC** The domain is:

   A. $[-10, 8]$

   B. $[-2, 4]$

   C. $[-4, 2]$

   D. $(-2, 4)$

8. **MC** The range is:

   A. $[-7, -1]$

   B. $[-5, 13]$

   C. $[1, 7]$

   D. $[-3, 3]$

9. **MC** A circle has its centre at $(4, -2)$ and a radius of $\sqrt{5}$. The equation of the circle is:

   A. $(x-4)^2 + (y+2)^2 = 25$

   B. $(x-4)^2 + (y+2)^2 = 5$

   C. $(x+4)^2 + (y-2)^2 = 5$

   D. $(x+4)^2 + (y-2)^2 = 25$

10. a. Sketch the graph of the relation $x^2 + y^2 = 100$.

    b. From this relation, form two functions and identify the domain and range of each.

11. Sketch the graph of $y = -\sqrt{4 - x^2}$ and identify its domain and range.

12. **MC** For the graph $y = \dfrac{1}{x} - 3$, identify which of the following statements is *not* true.
   A. The domain is $R\setminus\{0\}$.
   B. The graph passes through the point $(1, -2)$.
   C. There are asymptotes at $x = 3$ and $y = 0$.
   D. The range is $R\setminus\{-3\}$.

Complex familiar

13. The point $(2, -1)$ lies on the graph of a hyperbola. If the equations of the asymptotes are $x = 0$ and $y = 5$, determine the equation of the hyperbola.

14. Part of a hyperbola is shown in the diagram.

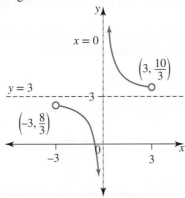

   a. Identify the domain and range of the graph.
   b. Determine the equation of the graph.

15. Determine the equation of the square root function with end point $(3, -4)$ and containing the point $(12, 2)$, and identify the point where the function cuts the x-axis.

16. The end points of a diameter of a circle are $(4, -3)$ and $(-8, 9)$.
   Determine the equation of the circle.

Complex unfamiliar

17. An eagle soars from the top of a cliff that is 48.4 metres above the ground and then descends towards unsuspecting prey below.

   The eagle's height, h metres above the ground, at time t seconds can be modelled by the equation

   $$h = 50 + \frac{a}{t - 25}$$

   where $0 \leq t < 25$ and a is a constant.
   Determine how many seconds the eagle takes to reach the ground.

18. Consider the graph $y = a\sqrt{x - h}$. Determine the equation of this graph if it passes through the points $(3, 5)$ and $(18, 10)$.

19. From Ohm's Law, the electrical resistance, R ohms, of a metal conductor in which the voltage is constant and the current, I amperes, are connected by the following formula where k is a constant.

$$R = \frac{k}{I},$$

If a current of 0.6 amperes produces a resistance of 400 ohms, determine the percentage change in resistance if the current is increased by 20%.

20. A toy train runs around a circular track that can be described by the equation $(x - 1)^2 + \left(y + \frac{3}{2}\right)^2 = \frac{25}{4}$.

There is also a straight track tangential to the circle at point P, which leads to a train depot at point G (vertically above the centre of the track). All units are in metres.

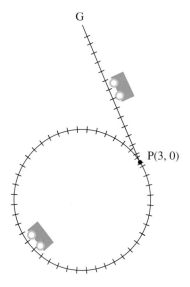

One train engine continues to travel round and round the circular track at a speed of π m/s, while a second train engine continues to shunt backwards and forwards between P and G at a speed of 1 m/s. If the small child playing with the trains releases both engines at the same time from point P, determine how long it takes before they collide at P.

Fully worked solutions for this chapter are available online.

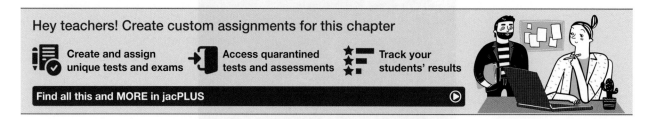

Answers

Chapter 5 Graphs of relations and reciprocal functions

5.2 The circle

5.2 Exercise

1. a. $x^2 + y^2 = 9$
   b. $x^2 + y^2 = 1$
   c. $x^2 + y^2 = 25$
   d. $x^2 + y^2 = 100$
   e. $x^2 + y^2 = 6$
   f. $x^2 + y^2 = 8$
   g. $y = \sqrt{9 - x^2}$
   h. $y = -\sqrt{16 - x^2}$

2. a. Both $[-3, 3]$
   b. Both $[-1, 1]$
   c. Both $[-5, 5]$
   d. Both $[-10, 10]$
   e. Both $[-\sqrt{6}, \sqrt{6}]$
   f. Both $[-2\sqrt{2}, 2\sqrt{2}]$
   g. $[-3, 3], [0, 3]$
   h. $[-4, 4], [-4, 0]$

3. a.
   b.
   c.
   d.
   e.
   f.

4. a.
   $[-1, 1]$ and $[-3, -1]$
   b.
   $[-2, 2]$ and $[0, 4]$
   c.
   $[1, 7]$ and $[-3, 3]$
   d.
   $[-2, 6]$ and $[-5, 3]$
   e.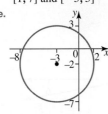
   $[-8, 2]$ and $[-7, 3]$
   f.
   $[0, 6]$ and $[-1, 5]$

g.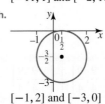

$[-11, 1]$ and $[-2, 10]$

h.

$[-1, 2]$ and $[-3, 0]$

5. a. D
   b. B

6. a. C
   b. D

7. F: b, c, d, e, f
   N: a, g, h

8. a.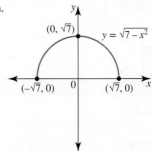

$y = \sqrt{7 - x^2}$

Domain $[-\sqrt{7}, \sqrt{7}]$, range $[0, \sqrt{7}]$

b. $y = \sqrt{\dfrac{1}{9} - x^2}$

Domain $\left[-\dfrac{1}{3}, \dfrac{1}{3}\right]$, range $\left[0, \dfrac{1}{3}\right]$

9.

	Centre	Radius	Domain	Range
a.	$(0, 1)$	1	$[-1, 1]$	$[0, 2]$

	Centre	Radius	Domain	Range
b.	$(-2, -4)$	3	$[-5, 1]$	$[-7, -1]$

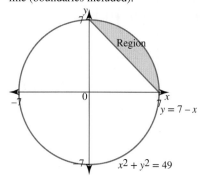

	Centre	Radius	Domain	Range
c.	$(0, 0)$	$\dfrac{9}{4}$	$\left[-\dfrac{9}{4}, \dfrac{9}{4}\right]$	$\left[-\dfrac{9}{4}, \dfrac{9}{4}\right]$

10. a. $(x + 8)^2 + (y - 9)^2 = 36$
    b. $(x - 7)^2 + y^2 = 8$
    c. $(x - 1)^2 + (y - 6)^2 = 136$
    d. $9x^2 + 9(y - 2)^2 = 16$

11. a. On   b. Outside   c. Outside
    d. Inside   e. Outside

12. (m, n) lies inside the region of the circle if $(m - h)^2 + (n - k)^2 < r^2$.

13. $a = 0$ and $a = -8$

14. Yes. Sample responses can be found in the worked solutions in the online resources.

15. $\left(\dfrac{7}{5}, \dfrac{14}{5}\right)$ and $(1, 2)$

16. $(0, 7), (7, 0)$; the region is inside the circle and above the line (boundaries included).

17. 3.9 cm/s

18. Shortest distance $(2\sqrt{5} - 3)$ units
    Longest distance $(2\sqrt{5} + 3)$ units

19. $\dfrac{5\sqrt{2}}{2}$

20. 1 minute 22.1 seconds after it is dropped

5.3 The sideways parabola and square root function

5.3 Exercise

1. a. i. $[9, \infty)$ ii. $(9, 0)$ iii. $[0, \infty)$
   b. i. $(-\infty, 4]$ ii. $(4, 0)$ iii. $[0, \infty)$
   c. i. $[-3, \infty)$ ii. $(-3, 0)$ iii. $(-\infty, 0]$
   d. i. $(-\infty, 0]$ ii. $(0, 3)$ iii. $[3, \infty)$
   e. i. $[2, \infty)$ ii. $(2, -7)$ iii. $[-7, \infty)$
   f. i. $\left(-\infty, \dfrac{1}{2}\right]$ ii. $\left(\dfrac{1}{2}, 4\right)$ iii. $(-\infty, 4]$

2. a. Domain $[1, \infty)$; range $[-3, \infty)$

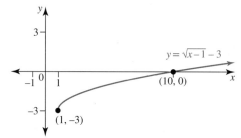

   b. i. $[-2, \infty)$
      ii.

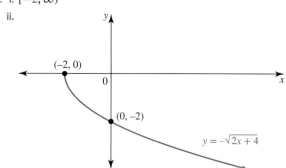

3. a.

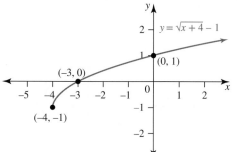

   b.

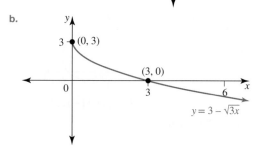

c.

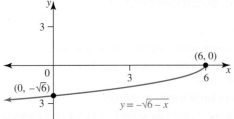

d.

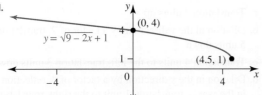

4. a. Domain $[-3, \infty)$, range $[-2, \infty)$

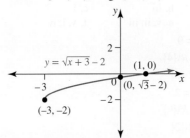

b. Domain $[0, \infty)$, range $(-\infty, 5]$

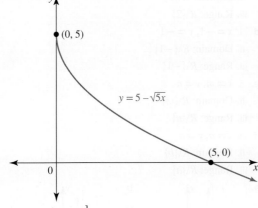

c. Domain $(-\infty, 9]$, range $[4, \infty)$

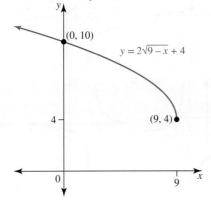

d. Domain $(-\infty, 7]$, range $[0, \infty)$

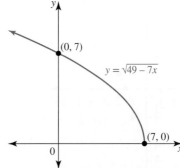

e. Domain $[-4, \infty)$, range $[2, \infty)$

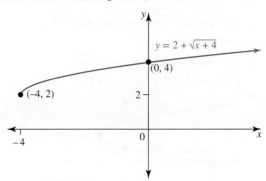

f. Domain $\left(-\infty, \dfrac{3}{2}\right]$, range $(-\infty, -1]$

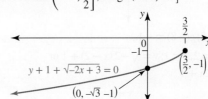

5. $y = \sqrt{2}\sqrt{x+2} + 1 \Rightarrow y = \sqrt{2(x+2)} + 1$

6. $y = -\sqrt{3(x+2)} + 2$

7. $y = 5\sqrt{4-x} - 1; \left(\dfrac{99}{25}, 0\right)$

8. $y = \sqrt{4-x} - 4$

9. $y = 4\sqrt{x-3} + 2$

10. $y = \dfrac{1}{2}\sqrt{x+4} - 1$

11. B

12. C

13. a. $(1, -2), \left(\dfrac{13}{9}, 0\right)$

b.

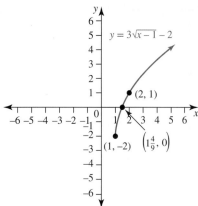

$y = 3\sqrt{x-1} - 2$

(2, 1)

(1, −2) $\left(1\tfrac{4}{9}, 0\right)$

c. $y = -2 - 3\sqrt{x-1}$ or $y + 2 = -3\sqrt{x-1}$

14. a. $y = 2 \pm \sqrt{1-x}$

b. $f : (-\infty, 1] \rightarrow R, f(x) = 2 + \sqrt{1-x}$
$g : (-\infty, 1] \rightarrow R, g(x) = 2 - \sqrt{1-x}$

c.

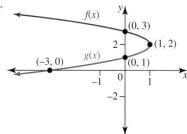

$f(x)$

(0, 3)

(1, 2)

(−3, 0) $g(x)$ (0, 1)

d. $f(-8) = 5; g(-8) = -1.$

15. $a = -2, h = 2, k = 1; y = \sqrt{-2(x-2)} + 1$ or
$y = \sqrt{4 - 2x} + 1$

16.

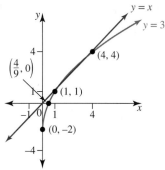

$y = x$

$y = 3\sqrt{x} - 2$

$\left(\tfrac{4}{9}, 0\right)$

(4, 4)

(1, 1)

(0, −2)

$\{x : 1 < x < 4\}$

17. $a = 5, b = -2, c = 2$

Equation: $y = 5 - \sqrt{-2(x-2)}$ or $y = 5 - \sqrt{4 - 2x}$

Domain: $(-\infty, 2]$

Range: $(-\infty, 5]$

18. $t = \sqrt{\dfrac{h}{4.9}}$

19. $0 < k < \dfrac{1}{4}$

20. They are both parabolas. See the worked solution in the online resources for further discussion.

5.4 The reciprocal function

5.4 Exercise

1. a. Dilation in the y direction by a factor of 2

b. Dilation in the y direction by a factor of 3, reflection in the x-axis

c. Translation 6 units to the right

d. Dilation in the y direction by a factor of 2, translation 4 units to the left

2. a. Translation 7 units up

b. Dilation in the y direction by a factor of 2, translation 5 units down

c. Translation 4 units to the left, translation 3 units down

d. Dilation in the y direction by a factor of 4, reflection in the x-axis, translation 1 unit to the right, translation 4 units down

3. a. v   **b.** iii   **c.** i

    d. v, iii   **e.** v, ii, iii   **f.** v, i, iv

4. a. **i.** $x = 4, y = 0$

    ii. Domain: $R\backslash\{4\}$

    iii. Range: $R\backslash\{0\}$

b. **i.** $x = 0, y = 2$

    ii. Domain: $R\backslash\{0\}$

    iii. Range: $R\backslash\{2\}$

c. **i.** $x = 3, y = 2$

    ii. Domain: $R\backslash\{3\}$

    iii. Range: $R\backslash\{2\}$

d. **i.** $x = -1, y = -1$

    ii. Domain: $R\backslash\{-1\}$

    iii. Range: $R\backslash\{-1\}$

e. **i.** $x = m, y = n$

    ii. Domain: $R\backslash\{m\}$

    iii. Range: $R\backslash\{n\}$

f. **i.** $x = b, y = a$

    ii. Domain: $R\backslash\{b\}$

    iii. Range: $R\backslash\{a\}$

5. a.

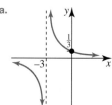

$\tfrac{1}{3}$

−3

b.

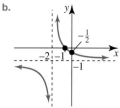

$-\tfrac{1}{2}$

−2 −1

−1

c.

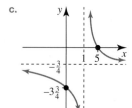

1 5

$-\tfrac{3}{4}$

$-3\tfrac{3}{4}$

d.

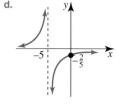

−5

$-\tfrac{2}{5}$

6. a.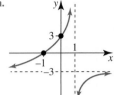

b.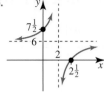

c.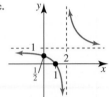

d.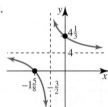

8. a.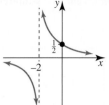

b.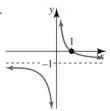

c.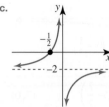

7. D

9. $y = \dfrac{-1}{x-4} + \dfrac{1}{2}$

10. a. $x = -5, y = 2$ **b.** $x = 0, y = -3$

     c. $x = 0, y = 0$ **d.** $x = -14, y = -\dfrac{3}{4}$

11.

	Asymptotes	y-intercept	x-intercept	Domain	Range	Point
a.	$x = -1, y = -3$	$(0, -2)$	$\left(-\dfrac{2}{3}, 0\right)$	$R\backslash\{-1\}$	$R\backslash\{-3\}$	$(-2, -4)$

$$y = \frac{1}{x+1} - 3$$

$x = -1$ $\left(-\dfrac{2}{3}, 0\right)$ $(0, -2)$ $y = -3$ $(-2, -4)$

	Asymptotes	y-intercept	x-intercept	Domain	Range	Point
b.	$x = 3, y = 4$	$(0, 5)$	$\left(\dfrac{15}{4}, 0\right)$	$R\backslash\{3\}$	$R\backslash\{4\}$	

$$y = 4 - \frac{3}{x-3}$$

$x = 3$ $y = 4$ $(0, 5)$ $\left(\dfrac{15}{4}, 0\right)$

	Asymptotes	y-intercept	x-intercept	Domain	Range	Point
c.	$x=-3, y=0$	$\left(0, -\dfrac{5}{3}\right)$	None	$R\backslash\{-3\}$	$R\backslash\{0\}$	$(-4, 5)$

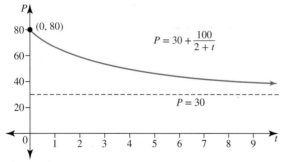

	Asymptotes	y-intercept	x-intercept	Domain	Range	Point
d.	$x=2, y=-1$	$\left(0, -\dfrac{7}{2}\right)$	$(7, 0)$	$R\backslash\{2\}$	$R\backslash\{-1\}$	

12. D

13. a. Reduced by 25 cattle

b. Domain $\{t : t \geq 0\}$, range $(30, 80)$

c. The number of cattle will never go below 30.

14. a. $y = \dfrac{2}{x-3} + 1$

b. $y = -\dfrac{1.5}{x+3} + 1$

15. a. $y = \dfrac{\frac{3}{8}}{x - \frac{1}{4}} - \dfrac{1}{2} = \dfrac{3}{8x-2} - \dfrac{1}{2}$

b. $f : R\backslash\left\{\dfrac{1}{4}\right\} \to R, f(x) = \dfrac{3}{8x-2} - \dfrac{1}{2}$

16. a. $a = 3; b = 1$

b.

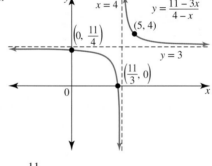

c. $x < \dfrac{11}{3}$ or $x > 4$

17. a. $a = 460; b = -880$

b. 12 years 8 months

c. Never reaches 500 insects

d. Cannot be larger than 460

18. Domain $R\backslash\{4\}$, range $R\backslash\{0\}$

19. 25 cm of mercury

20. One intersection: $k = 1$ or $k = 5$
Two intersections: $k > 1$ or $k < 5$
No intersection: $1 < k < 5$

5.5 Review

5.5 Exercise

1. C

2. D

3. a. $x = -2, y = -1$

   b. Domain $R\backslash\{-2\}$, range $R\backslash\{-1\}$

   c. $y = \dfrac{-4}{x+2} - 1$

   d.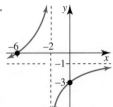

4. a.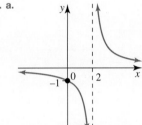

   b.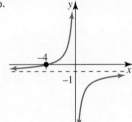

   c.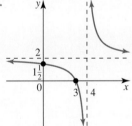

5. $y = 3 - \dfrac{4}{x+2}$

6. D

7. C

8. C

9. B

10. a.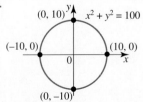

b. $f_1: [-10, 10] \to R, \ f(x) = \sqrt{(100 - x^2)}$ with domain
   $f = [-10, 10]$, range $f = [0, 10]$ and
   $f_2: [-10, 10] \to R, \ f(x) = -\sqrt{(100 - x^2)}$ with domain
   $f = [-10, 10]$, range $f = [-10, 0]$

11.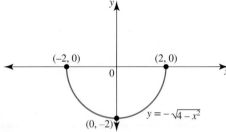

12. C

13. $y = -\dfrac{12}{x} + 5$

14. a. Domain: $(-3, 3)\backslash\{0\}$

   Range: $\left(-\infty, \dfrac{8}{3}\right) \cup \left(\dfrac{10}{3}, \infty\right)$

   b. $y = \dfrac{1}{x} + 3, \ -3 < x < 3, \ x \neq 0$

15. $y = 2\sqrt{x - 3} - 4$; x-intercept $(7, 0)$

16. $(x + 2)^2 + (y - 3)^2 = 72$

17. 24.2 seconds

18. $y = \sqrt{5}\sqrt{x + 2}$ or $y = \sqrt{5(x + 2)}$

19. Decreased by $16\dfrac{2}{3}\%$

20. 20 seconds

6 Trigonometric functions

LESSON SEQUENCE

Fully worked solutions for this chapter are available online.

EXAM PREPARATION

Access exam-style questions in every lesson, available online.

on Resources

 Solutions   Solutions — Chapter 6 (sol-1273)

 Exam questions   Exam question booklet — Chapter 6 (eqb-0294)

 Digital documents   Learning matrix — Chapter 6 (doc-41788)
                                  Chapter summary — Chapter 6 (doc-41801)

LESSON
6.1 Overview

6.1.1 Introduction

Periodicity forms a natural part of our lives. Our heart rates, blood pressure and other vital statistics fluctuate over a 24-hour period, our long-term sleep and wakefulness patterns are periodic, and our mood-affecting biorhythms and ovulation cycles are periodic. Even in the world of fashion, such measures as dress length or width of trouser leg can be cyclical. What is old-fashioned today will, in the fullness of time, often regain its popularity.

The length of time for one repetition of a cycle is known as the period. The rotation of the Earth creates a day–night cycle with a period of 24 hours; the revolution of the Earth about the sun creates seasonal cycles with a period of 12 months.

The trigonometric sine and cosine functions are the most important examples of periodic functions. These form models for many periodic phenomena, smoothing out random fluctuations to show the overall oscillatory nature of the phenomena about an equilibrium position. The functions model the wave form of alternating current in most electrical power circuits; they model the depth of water at a pier from low to high tide levels, they model sound and light waves, weather patterns and temperature fluctuations; and they model the height above ground of a person on a Ferris wheel; all of these are examples of their applicability.

The simple harmonic motion of a pendulum swinging under gravity is modelled by a sine function. Such pendulums have provided a means of timekeeping ever since their invention in 1656. London's Big Ben is one such example. Despite quartz clocks being more accurate, grandfather clocks are still found today in private homes and antique shops.

Radian measure makes trigonometric functions mathematically simpler. The concept of a radian was first recognised in 1714 by the English mathematician Roger Cotes, a colleague of Isaac Newton. The actual term 'radian', however, was introduced in 1873 by James Thomson, mathematics professor at Queen's University, Belfast. Cotes died from illness at a young age, with Newton quoted as saying, 'If he had lived we would have known something.'

6.1.2 Syllabus links

Lesson	Lesson title	Syllabus links
6.2	**Trigonometry review**	○ Understand and use the exact values of $\cos(\theta)$, $\sin(\theta)$ and $\tan(\theta)$ at integer multiples of $\dfrac{\pi}{6}$ and $\dfrac{\pi}{4}$.
6.3	**Radian measure**	○ Define and use radian measure and understand its relationship with degree measure. ○ Calculate lengths of arcs and areas of sectors in circles.
6.4	**Unit circle definitions**	○ Understand the unit circle definition of $\cos(\theta)$, $\sin(\theta)$ and $\tan(\theta)$ and periodicity using radians.
6.5	**Exact values and symmetry properties**	○ Understand and use the exact values of $\cos(\theta)$, $\sin(\theta)$ and $\tan(\theta)$ at integer multiples of $\dfrac{\pi}{6}$ and $\dfrac{\pi}{4}$. ○ Understand the unit circle definition of $\cos(\theta)$, $\sin(\theta)$ and $\tan(\theta)$ and periodicity using radians.
6.6	**Graphs of sine, cosine and tangent functions**	○ Sketch the graphs of $y = \sin(x)$, $y = \cos(x)$ and $y = \tan(x)$ on extended domains.
6.7	**Solving trigonometric equations**	○ Solve trigonometric equations, with and without technology, including the use of the Pythagorean identity $\sin^2(A) + \cos^2(A) = 1$.
6.8	**Transformations of the sine and cosine functions**	○ Recognise and determine the effect of the parameters a, b, h and k, on the graphs of $y = a \sin(b(x - h)) + k$ and $y = a \cos(b(x - h)) + k$, with and without technology. ○ Sketch the graphs of $y = a \sin(b(x - h)) + k$, $y = a \cos(b(x - h)) + k$, with and without technology.
6.9	**Modelling with trigonometric functions**	○ Model and solve problems that involve trigonometric functions, with and without technology.

Source: Mathematical Methods Senior Syllabus 2024 © State of Queensland (QCAA) 2024; licensed under CC BY 4.0.

LESSON
6.2 Trigonometry review

SYLLABUS LINKS

- Understand and use the exact values of $\cos(\theta)$, $\sin(\theta)$ and $\tan(\theta)$ at integer multiples of $\dfrac{\pi}{6}$ and $\dfrac{\pi}{4}$.

Source: Mathematical Methods Senior Syllabus 2024 © State of Queensland (QCAA) 2024; licensed under CC BY 4.0.

The process of calculating all side lengths and all angle magnitudes of a triangle is called **solving the triangle**. Here we review the use of trigonometry to solve right-angled triangles.

6.2.1 Right-angled triangles

The hypotenuse is the longest side of a right-angled triangle and it lies opposite the 90° angle, the largest angle in the triangle. The other two sides are labelled relative to one of the other angles in the triangle, an example of which is shown in the diagram.

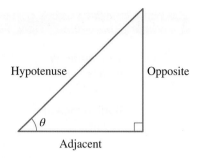

It is likely that the trigonometric ratios of sine, cosine and tangent, possibly together with Pythagoras' theorem $(a^2 + b^2 = c^2)$, will be required to solve a right-angled triangle.

Trigonometric ratios

$$\sin(\theta) = \frac{\text{opposite}}{\text{hypotenuse}}$$

$$\cos(\theta) = \frac{\text{adjacent}}{\text{hypotenuse}}$$

$$\tan(\theta) = \frac{\text{opposite}}{\text{adjacent}}$$

The trigonometric ratios, usually remembered as SOH, CAH, TOA, cannot be applied to triangles that do not have a right angle. However, isosceles and equilateral triangles can easily be divided into two right-angled triangles by dropping a perpendicular from the vertex between a pair of equal sides to the midpoint of the opposite side.

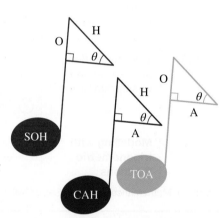

By and large, as first recommended by the French mathematician François Viète in the sixteenth century, decimal notation has been adopted for magnitudes of angles rather than the sexagesimal system of degrees and minutes, although even today we still may see written, for example, either 15°24′ or 15.4° for the magnitude of an angle.

WORKED EXAMPLE 1 Calculating sides and angles in right-angled triangles

Calculate, to 2 decimal places, the value of the pronumeral shown in each diagram.

a.
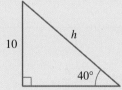

b.
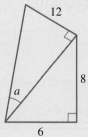

THINK

a. 1. Choose the appropriate trigonometric ratio.

WRITE

a. Relative to the angle, the sides marked are the opposite and the hypotenuse.

$$\sin(\theta) = \frac{\text{opposite}}{\text{hypotenuse}}$$

$$\sin(40°) = \frac{10}{h}$$

2. Rearrange to make the required side the subject and evaluate, checking the calculator is in degree mode.

$$h = \frac{10}{\sin(40°)}$$
$$= 15.56 \text{ to 2 decimal places}$$

b. 1. Obtain the hypotenuse length of the lower triangle.

2. In the upper triangle choose the appropriate trigonometric ratio.

b. From Pythagoras' theorem the sides 6, 8, 10 form a Pythagorean triple, so the hypotenuse is 10.

The opposite and adjacent sides to the angle $a°$ are now known.

$$\tan(a) = \frac{12}{10}$$

3. Rearrange to make the required angle the subject and evaluate.

$$\tan(a) = 1.2$$
$$\therefore a = \tan^{-1}(1.2)$$
$$= 50.19° \text{ to 2 decimal places}$$

TI	THINK	WRITE	CASIO	THINK	WRITE
a. 1. Write a trigonometric ratio that can be solved for h using the information given.	$\sin(40) = \dfrac{10}{h}$	a. 1. Write a trigonometric ratio that can be solved for h using the information given.	$\sin(40) = \dfrac{10}{h}$
2. Put the Calculator in DEGREE mode. On a Calculator page, press MENU, then select: 3: Algebra 1: Numerical Solve Complete the entry line as: $\text{nSolve}\left(\sin(40) = \dfrac{10}{h}, h\right)$ then press ENTER.	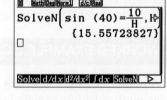	2. On a Run-Matrix screen, press OPTN and select CALC by pressing F4, then select SolveN by pressing F5. Complete the entry line as: SolveN $\left(\sin(40) = \dfrac{10}{H}, H\right)$ then press EXE.	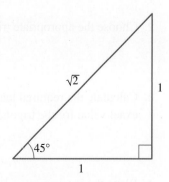
3. The answer appears on the screen.	$h = 15.56$ (2 decimal places)	3. The answer appears on the screen.	$h = 15.56$ (2 decimal places)

6.2.2 Exact values for trigonometric ratios of 30°, 45° and 60°

By considering the isosceles right-angled triangle with equal sides of 1 unit, the trigonometric ratios for 45° can be obtained. Using Pythagoras' theorem, the hypotenuse of this triangle will be $\sqrt{2}$ units.

The equilateral triangle with the side length of 2 units can be divided in half to form a right-angled triangle containing 60° and 30°. This right-angled triangle has a hypotenuse of 2 units and the side divided in half has length 1 unit. Using Pythagoras' theorem, the third side will be $\sqrt{3}$ units.

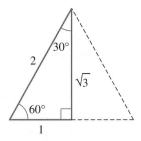

The exact values for trigonometric ratios of 30°, 45° and 60° can be calculated from these triangles using SOH, CAH, TOA. Alternatively, these values can be displayed in a table and committed to memory.

θ	30°	45°	60°
$\sin(\theta)$	$\dfrac{1}{2}$	$\dfrac{1}{\sqrt{2}} = \dfrac{\sqrt{2}}{2}$	$\dfrac{\sqrt{3}}{2}$
$\cos(\theta)$	$\dfrac{\sqrt{3}}{2}$	$\dfrac{1}{\sqrt{2}} = \dfrac{\sqrt{2}}{2}$	$\dfrac{1}{2}$
$\tan(\theta)$	$\dfrac{1}{\sqrt{3}} = \dfrac{\sqrt{3}}{3}$	1	$\sqrt{3}$

As a memory aid, notice the sine values in the table are in the order $\dfrac{\sqrt{1}}{2}, \dfrac{\sqrt{2}}{2}, \dfrac{\sqrt{3}}{2}$. The cosine values reverse this order, while the tangent values are the sine values divided by the cosine values.

For other angles, a calculator or other technology is required. It is essential to set the calculator mode to degree in order to evaluate a trigonometric ratio involving angles in degree measure.

WORKED EXAMPLE 2 Calculating exact length

A ladder of length 4 metres leans against a fence. If the ladder is inclined (precariously) at 30° to the ground, calculate exactly how far the foot of the ladder is from the fence.

THINK

1. Draw a diagram showing the given information.

WRITE

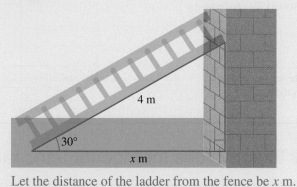

Let the distance of the ladder from the fence be x m.

2. Choose the appropriate trigonometric ratio.

Relative to the angle, the sides marked are the adjacent and the hypotenuse.
$$\cos(30°) = \frac{x}{4}$$

3. Calculate the required length using the exact value for the trigonometric ratio.

$$x = 4\cos(30°)$$
$$= 4 \times \frac{\sqrt{3}}{2}$$
$$= 2\sqrt{3}$$

4. Write the answer.

The foot of the ladder is $2\sqrt{3}$ metres from the fence.

6.2.3 Deducing one trigonometric ratio from another

Given the sine, cosine or tangent value of some unspecified angle, it is possible to obtain the exact value of the other trigonometric ratios of that angle using Pythagoras' theorem.

One common example is that given $\tan(\theta) = \dfrac{4}{3}$ it is possible to deduce that $\sin(\theta) = \dfrac{4}{5}$ and $\cos(\theta) = \dfrac{3}{5}$ without evaluating θ. The reason for this is that $\tan(\theta) = \dfrac{4}{3}$ means that the sides opposite and adjacent to the angle θ in a right-angled triangle are in the ratio 4 : 3.

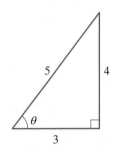

Labelling these sides 4 and 3, respectively, and using Pythagoras' theorem (or recognising the Pythagorean triad '3, 4, 5') leads to the hypotenuse being 5, and hence the ratios $\sin(\theta) = \dfrac{4}{5}$ and $\cos(\theta) = \dfrac{3}{5}$ are obtained.

WORKED EXAMPLE 3 Determining exact trigonometry ratios

A line segment AB is inclined at a degrees to the horizontal, where $\tan(a) = \dfrac{1}{3}$.

a. Deduce the exact value of $\sin(a)$.

b. Calculate the vertical height of B above the horizontal through A if the length of AB is $\sqrt{5}$ cm.

THINK

a. 1. Draw a right-angled triangle with two sides in the given ratio and calculate the third side.

2. State the required trigonometric ratio.

b. 1. Draw the diagram showing the given information.

WRITE

a. $\tan(a) = \dfrac{1}{3} \Rightarrow$ sides opposite and adjacent to angle a are in the ratio 1 : 3.

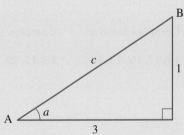

Using Pythagoras' theorem:
$$c^2 = 1^2 + 3^2$$
$$\therefore c = \sqrt{10}$$

$$\sin(a) = \dfrac{1}{\sqrt{10}}$$
$$= \dfrac{\sqrt{10}}{10}$$

b. Let the vertical height be y cm.

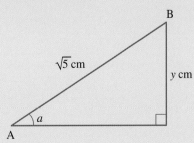

2. Choose the appropriate trigonometric ratio and calculate the required length.

$$\sin(a) = \frac{y}{\sqrt{5}}$$

$$y = \sqrt{5}\,\sin(a)$$

$$= \sqrt{5} \times \frac{1}{\sqrt{10}} \text{ as } \sin(a) = \frac{1}{\sqrt{10}}$$

$$= \frac{1}{\sqrt{2}} \text{ or } \frac{\sqrt{2}}{2}$$

3. Write the answer.

The vertical height of B above the horizontal through A is $\dfrac{\sqrt{2}}{2}$ cm.

 Resources

 Interactivity Trigonometric ratios (int-2577)

Exercise 6.2 Trigonometry review

learn on

6.2 Exercise	**6.2 Exam practice** on

Simple familiar	Complex familiar	Complex unfamiliar
1, 2, 3, 4, 5, 6, 7, 8, 9, 10, 11, 12	13, 14, 15, 16, 17	18, 19, 20

Simple familiar

1. Determine the exact values of the following.
   a. $\sin(45°)$
   b. $\tan(30°)$
   c. $\cos(60°)$
   d. $\tan(45°) + \cos(30°) - \sin(60°)$

2. **WE1** Calculate, to 2 decimal places, the value of the pronumeral shown in each diagram.

   a.
   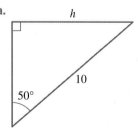

   b.
   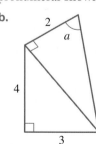

3. Calculate the values of the unknown marked sides, correct to 2 decimal places.

a.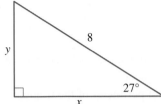

b.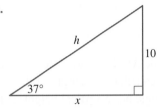

4. Calculate the value of angle θ, correct to 2 decimal places.

a.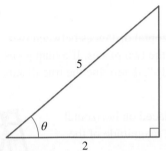

b.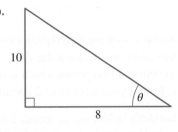

5. **WE2** A ladder of length 4 metres leans against a fence. If the ladder is inclined at 45° to the horizontal ground, calculate exactly how far the foot of the ladder is from the fence.

6. A 1-metre-long broom leaning against a wall makes an angle of 30° with the wall. Calculate exactly how far up the wall the broom reaches.

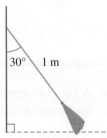

7. Evaluate $\dfrac{\cos(30°)\sin(45°)}{\tan(45°)+\tan(60°)}$, expressing the answer in exact form with a rational denominator.

8. Evaluate $\dfrac{\sin(30°)\cos(45°)}{\tan(60°)}$, expressing the answer in exact form with a rational denominator.

9. **WE3** A line segment AB is inclined at a degrees to the horizontal, where $\tan(a)=\dfrac{2}{3}$.

   a. Deduce the exact value of $\cos(a)$.
   b. Calculate the run of AB along the horizontal through A if the length of AB is 26 cm.

10. For an acute angle θ, obtain the following trigonometric ratios without evaluating θ.

   a. Given $\tan(\theta)=\dfrac{\sqrt{3}}{2}$, form the exact value of $\sin(\theta)$.

   b. Given $\cos(\theta)=\dfrac{5}{6}$, form the exact value of $\tan(\theta)$.

   c. Given $\sin(\theta)=\dfrac{\sqrt{5}}{3}$, form the exact value of $\cos(\theta)$.

11. In order to check the electricity supply, a technician uses a ladder to reach the top of an electricity pole. The ladder reaches 5 metres up the pole and its inclination to the horizontal ground is 54°.

   a. Calculate the length of the ladder to 2 decimal places.
   b. If the foot of the ladder is moved 0.5 metres closer to the pole, calculate its new inclination to the ground and the new vertical height it reaches up the electricity pole, both to 1 decimal place.

12. An 800-metre-long taut chairlift cable enables the chairlift to rise 300 metres vertically. Calculate the angle of elevation (to the nearest degree) of the cable.

Complex familiar

13. a. An isosceles triangle ABC has sides BC and AC of equal length 5 cm. If the angle enclosed between the equal sides is 20°, calculate the length of the third side AB to 3 decimal places.
    b. An equilateral triangle has a vertical height of 10 cm. Calculate the exact perimeter of the triangle.

14. A right-angled triangle contains an angle θ where $\sin(\theta) = \dfrac{3}{5}$. If the longest side of the triangle is 60 cm, calculate the exact length of the shortest side.

15. The distances shown on a map are called *projections*. They give the horizontal distances between two places without taking into consideration the slope of the line connecting the two places. If a map gives the projection as 25 km between two points which actually lie on a slope of 16°, determine the true distance between the points. Round your answer to 2 decimal places.

16. The two legs of a builder's ladder are of length 2 metres. The ladder is placed on horizontal ground with a distance between its two feet of 0.75 metres. Calculate the magnitude of the angle between the legs of the ladder.

17. Triangle ABC has angles such that $\angle CAB = 60°$ and $\angle ABC = 45°$. The perpendicular distance from C to AB is 18 cm. Calculate the exact lengths of each of its sides.

Complex unfamiliar

18. A cube of edge length a units rests on a horizontal table. Calculate the inclination of the diagonal to the horizontal, to 2 decimal places.

19. A lookout tower is 100 metres in height. From the top of this tower, the angle of depression of the top of a second tower standing on the same level ground is 30°; from the bottom of the lookout tower, the angle of elevation to the top of the second tower is 45°. Calculate the height of the second tower and its horizontal distance from the lookout tower, expressing both measurements to 1 decimal place.

20. In the diagram, angles ABC and ACD are right angles and DE is parallel to CA. Angle BAC is a degrees and the length measures of AC and BD are m and n, respectively.
    a. Show that $n = m\sin^2(a)$, where $\sin^2(a)$ is the notation for the square of $\sin(a)$.
    b. If angle EBA is 60° and CD has length measure of $4\sqrt{3}$, calculate the values of a, m and n.

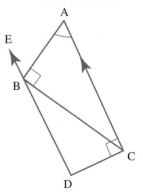

Fully worked solutions for this chapter are available online.

LESSON
6.3 Radian measure

SYLLABUS LINKS

- Define and use radian measure and understand its relationship with degree measure.
- Calculate lengths of arcs and areas of sectors in circles.

Source: Mathematical Methods Senior Syllabus 2024 © State of Queensland (QCAA) 2024; licensed under CC BY 4.0.

Measurements of angles up to now have been given in degree measure. An alternative to degree measure is **radian measure**. This alternative can be more efficient for certain calculations that involve circles, and it is essential for the study of trigonometric functions.

6.3.1 Definition of radian measure

Radian measure is defined in relation to the length of an **arc** of a circle. An arc is a part of the circumference of a circle.

> **Definition of radian measure**
>
> **One radian is the measure of the angle subtended at the centre of a circle by an arc equal in length to the radius of the circle.**

In particular, an arc of 1 unit subtends an angle of 1 radian at the centre of a **unit circle**, a circle with radius 1 unit and, conventionally, a centre at the origin.

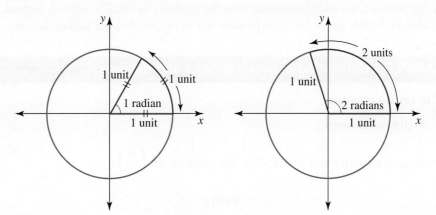

Doubling the arc length to 2 units doubles the angle to 2 radians. This illustrates the direct proportionality between the arc length and the angle in a circle of fixed radius.

The diagram suggests an angle of 1 radian will be a little less than 60° since the sector containing the angle has one 'edge' curved and therefore is not a true equilateral triangle.

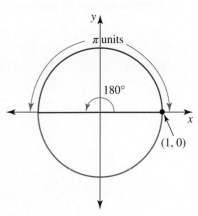

The degree equivalent for 1 radian can be found by considering the angle subtended by an arc that is half the circumference. The circumference of a circle is given by $2\pi r$, so the circumference of a unit circle is 2π.

In a unit circle, an arc of π units subtends an angle of π radians at the centre. But we know this angle to be 180°.

This gives the relationship between radian and degree measure.

Radians to degrees

$$\pi \text{ radians} = 180°$$

Hence, 1 radian equals $\dfrac{180°}{\pi}$, which is approximately 57.3°; 1° equals $\dfrac{\pi}{180}$ radians, which is approximately 0.0175 radians.

Using these relationships, it is possible to convert from radians to degrees and vice versa.

Convert between radians and degrees

To convert radians to degrees, multiply by $\dfrac{180}{\pi}$.

To convert degrees to radians, multiply by $\dfrac{\pi}{180}$.

Radians are often expressed in terms of π, perhaps not surprisingly, since a radian is a circular measure and π is so closely related to the circle.

Notation

π radians can be written as π^c, where c stands for circular measure. However, linking radian measure with the length of an arc, a real number, has such importance that the symbol c is usually omitted. Instead, the onus is on degree measure to always include the degree sign in order not to be mistaken for radian measure.

WORKED EXAMPLE 4 Converting between radians and degrees

a. Convert 30° to radian measure.

b. Convert $\dfrac{4\pi^c}{3}$ to degree measure.

c. Convert $\dfrac{\pi}{4}$ to degree measure and hence state the value of $\sin\left(\dfrac{\pi}{4}\right)$.

THINK

a. Convert degrees to radians.

WRITE

a. To convert degrees to radians, multiply by $\dfrac{\pi}{180}$.

$$30° = 30 \times \dfrac{\pi}{180}$$

$$= \cancel{30} \times \dfrac{\pi}{\cancel{180}_6}$$

$$= \dfrac{\pi}{6}$$

---

b. Convert radians to degrees.
   Note: The degree sign must be used.

b. To convert radians to degrees, multiply by $\dfrac{180}{\pi}$.

$$\frac{4\pi^c}{3} = \left(\frac{4\pi}{3} \times \frac{180}{\pi}\right)^{\circ}$$

$$= \left(\frac{4\cancel{\pi}}{\cancel{3}} \times \frac{\cancel{180}^{60}}{\cancel{\pi}}\right)^{\circ}$$

$$= 240^{\circ}$$

c. 1. Convert radians to degrees.

c. $\dfrac{\pi}{4} = \dfrac{\cancel{\pi}}{\cancel{4}} \times \dfrac{\cancel{180}^{\circ 45}}{\cancel{\pi}}$

$$= 45^{\circ}$$

2. Calculate the trigonometric value.

$$\sin\left(\frac{\pi}{4}\right) = \sin(45^{\circ})$$

$$= \frac{\sqrt{2}}{2}$$

TI | THINK	WRITE	CASIO | THINK	WRITE
a. 1. On a Calculator page in RAD mode, complete the entry line as 30° then press ENTER. *Note:* The degree symbol can be found by pressing the π button.		**a. 1.** Put the Calculator in Radian mode. On the Run-Matrix screen, complete the entry line as 30° then press EXE. *Note:* The degree symbol can be found by pressing OPTN, pressing F6 to scroll across to more menu options, then selecting ANGLE by pressing F5.	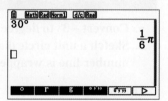
2. The answer appears on the screen.	$30^{\circ} = \dfrac{\pi}{6}^{c}$	**2.** The answer appears on the screen.	$30^{\circ} = \dfrac{\pi}{6}^{c}$
b. 1. On a Calculator page in DEG mode, complete the entry line as $\dfrac{4\pi}{3}r$ then press ENTER. *Note:* The radian symbol can be found by pressing the π button.	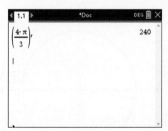	**b. 1.** Put the Calculator in Degree mode. On the Run-Matrix screen, complete the entry line as $\dfrac{4\pi}{3}r$ then press EXE. *Note:* The radian symbol can be found by pressing OPTN, pressing F6 to scroll across to more menu options, then selecting ANGLE by pressing F5.	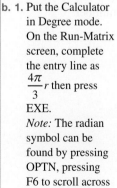
2. The answer appears on the screen.	$\dfrac{4\pi}{3}^{c} = 240^{\circ}$	**2.** The answer appears on the screen.	$\dfrac{4\pi}{3}^{c} = 240^{\circ}$

6.3.2 Extended angle measure and direction

By continuing to rotate around the circumference of the unit circle, larger angles are formed from arcs that are multiples of the circumference. For instance, an angle of 3π radians is formed from an arc of length 3π units created by one and a half revolutions of the unit circle: $3\pi = 2\pi + \pi$. This angle, in degrees, equals $360° + 180° = 540°$, and its end point on the circumference of the circle is in the same position as that of $180°$ or π^c; this is the case with any other angle that is a multiple of 2π added to π^c.

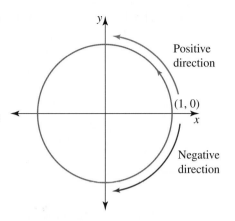

By convention, the positive angles move **anticlockwise** around the unit circle from the point $(1, 0)$. Negative angles move **clockwise** around the unit circle from the point $(1, 0)$.

All trigonometric functions have **periodicity**; that is, at established intervals the function repeats itself in a regular pattern.

WORKED EXAMPLE 5 Converting radians to degrees

a. Convert -3^c to degree measure.
b. Sketch a unit circle diagram to show the position the real number -3 is mapped to when the real number line is wrapped around the circumference of the unit circle.

THINK

a. Convert radians to degrees.

WRITE

a. $-3^c = -\left(3 \times \dfrac{180}{\pi}\right)^°$
As the π can't be cancelled, a calculator is used to evaluate.
$-3^c \approx -171.9°$

b. 1. State how the wrapping of the number line is made.

b. The number zero is placed at the point $(1, 0)$ and the negative number line is wrapped clockwise around the circumference of the unit circle through an angle of $171.9°$ so that the number -3 is its end point.

2. Sketch the unit circle diagram and mark the position of the number.

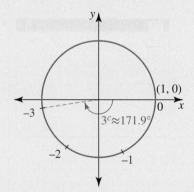

6.3.3 Using radians in calculations

From the definition of a radian, for any circle of radius r, an angle of 1^c is subtended at the centre of the circle by an arc of length r. So, if the angle at the centre of this circle is θ^c, then the length of the arc subtending this angle must be $\theta \times r$.

This gives a formula for calculating the length of an arc.

> **Length of an arc**
>
> $$l = r\theta$$

In the formula l is the arc length and θ is the angle, in radians, subtended by the arc at the centre of the circle of radius r.

Any angles given in degree measure will need to be converted to radian measure to use the arc length formulas.

Some calculations may require recall of the geometry properties of the angles in a circle, such as the angle at the centre of a circle is twice the angle at the circumference subtended by the same arc.

Major and minor arcs

For a minor arc, $\theta < \pi$, and for a major arc, $\theta > \pi$, with the sum of the minor and major arc angles totalling 2π if the major and minor arcs have their end points on the same chord.

To calculate the length of the major arc, the reflex angle with $\theta > \pi$ should be used in the arc length formula. Alternatively, the sum of the minor and major arc lengths gives the circumference of the circle, so the length of the major arc can be calculated as the difference between the lengths of the circumference and the minor arc.

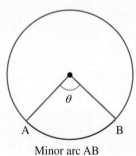

Minor arc AB

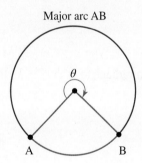

Major arc AB

Trigonometric ratios of angles expressed in radians

Problems in trigonometry may be encountered where angles are given in radian mode and their sine, cosine or tangent value is required to solve the problem. A calculator or other technology can be set on radian or 'rad' mode and the required trigonometric ratio evaluated directly without the need to convert the angle to degrees.

TIP

Care must be taken to ensure the calculator is set to the appropriate degree or radian mode to match the measure in which the angle is expressed. Care is also needed with written presentation: if the angle is measured in degrees, the degree symbol must be given; if there is no degree sign, then it is assumed the measurement is in radians.

a. An arc subtends an angle of 56° at the centre of a circle of radius 10 cm. Calculate the length of the arc to 2 decimal places.

b. Calculate, in degrees, the magnitude of the angle that an arc of length 20π cm subtends at the centre of a circle of radius 15 cm.

THINK	WRITE
a. 1. The angle is given in degrees, so convert it to radian measure.	a. $\theta° = 56°$ $\theta^c = 56 \times \dfrac{\pi}{180}$ $= \dfrac{14\pi}{45}$
2. Calculate the arc length.	$l = r\theta,\ r = 10,\ \theta = \dfrac{14\pi}{45}$ $l = 10 \times \dfrac{14\pi}{45}$ $= \dfrac{28\pi}{9}$ ≈ 9.77 The arc length is 9.77 cm (to 2 decimal places).
b. 1. Calculate the angle at the centre of the circle subtended by the arc.	b. $l = r\theta,\ r = 15,\ l = 20\pi$ $15\theta = 20\pi$ $\therefore \theta = \dfrac{20\pi}{15}$ $= \dfrac{4\pi}{3}$ The angle is $\dfrac{4\pi}{3}$ radians.
2. Convert the angle from radians to degrees.	In degree measure: $\theta° = \dfrac{4\pi}{3} \times \dfrac{180°}{\pi}$ $= 240°$ The magnitude of the angle is 240°.

Sectors

A **sector** is part of the area of a circle. In the figure the shaded area is a sector.

The area of the sector is proportional to the angle θ.

$$A_{\text{sector}} = \pi r^2 \times \dfrac{\theta}{360°}$$

or

$$\pi r^2 \times \dfrac{\theta}{2\pi}$$

$$= \dfrac{r^2\theta}{2}$$

$$A_{\text{sector}} = \dfrac{r^2\theta}{2} \text{ if } \theta \text{ is in radians.}$$

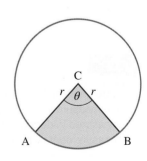

Area of a sector

$$A_{sector} = \pi r^2 \times \frac{\theta}{360°} \text{ if } \theta \text{ is in degrees,}$$

or

$$A_{sector} = \frac{1}{2}r^2\theta \text{ if } \theta \text{ is in radians.}$$

WORKED EXAMPLE 7 Calculating areas of sectors

For the circle in this figure, which has a radius of **6.5 cm**, calculate the area of the sector. Your answer should be rounded to 2 decimal places.

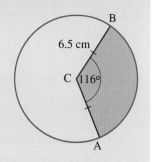

THINK	WRITE
1. Since the angle is measured in degrees, use the formula for the area of a sector in degrees. Determine the quantities r and θ.	$A_{sector} = \pi r^2 \times \dfrac{\theta}{360°}$ $r = 6.5$ $\theta = 116°$
2. Substitute the quantities into the formula and simplicity.	$A_{sector} = \pi(6.5)^2 \times \dfrac{116°}{360°}$ $= 42.7748$
3. Write the answer rounded to 2 decimal places.	$A_{sector} = 42.77 \text{ cm}^2$

Exercise 6.3 Radian measure

learn on

6.3 Exercise	6.3 Exam practice on	These questions are even better in jacPLUS!

Simple familiar	Complex familiar	Complex unfamiliar
1, 2, 3, 4, 5, 6, 7, 8, 9, 10, 11, 12, 13, 14, 15, 16, 17	18, 19, 20, 21	22

These questions are even better in jacPLUS!
- Receive immediate feedback
- Access sample responses
- Track results and progress

Find all this and MORE in jacPLUS

Simple familiar

1. Complete the following tables using technology.

a.

Degrees	30°	45°	60°
Radians			

b.

Degrees	0°	90°	180°	270°	360°
Radians					

2. **WE4** a. Convert $60°$ to radian measure.

   b. Convert $\dfrac{3\pi^c}{4}$ to degree measure.

   c. Convert $\dfrac{\pi}{6}$ to degree measure and hence state the value of $\tan\left(\dfrac{\pi}{6}\right)$.

3. Convert the following to degree measure.

   a. $\dfrac{\pi^c}{5}$ b. $\dfrac{2\pi^c}{3}$ c. $\dfrac{5\pi}{12}$ d. $\dfrac{11\pi}{6}$ e. $\dfrac{7\pi}{9}$ f. $\dfrac{9\pi}{2}$

4. Convert the following to radian measure.

   a. $40°$ b. $150°$ c. $225°$ d. $300°$ e. $315°$ f. $720°$

5. Evaluate the following to 3 decimal places.

   a. $\tan(1.2)$
   b. $\tan(1.2°)$

6. a. Calculate the following to 3 decimal places.

   i. $\tan(1^c)$ ii. $\cos\left(\dfrac{2\pi}{7}\right)$ iii. $\sin(1.46°)$

   b. Complete the following table with exact values.

θ	$\dfrac{\pi}{6}$	$\dfrac{\pi}{4}$	$\dfrac{\pi}{3}$
$\sin(\theta)$			
$\cos(\theta)$			
$\tan(\theta)$			

7. The real number line is wrapped around the circumference of the unit circle. Give two positive and two negative real numbers that lie in the same position as the following numbers.

   a. 0 b. -1

8. **WE5** a. Convert 1.8^c to degree measure.

   b. Draw a unit circle diagram to show the position the real number 1.8 is mapped to when the real number line is wrapped around the circumference of the unit circle.

9. a. For each of the following, draw a unit circle diagram to show the position of the angle and the arc that subtends the angle.

   i. An angle of 2 radians   ii. An angle of -2 radians   iii. An angle of $-\dfrac{\pi}{2}$

   b. For each of the following, draw a unit circle diagram with the real number line wrapped around its circumference to show the position of the number and the associated angle subtended at the centre of the circle.

   i. The number 4   ii. The number -1 iii. The number $\dfrac{7\pi}{3}$

10. **WE6** a. An arc subtends an angle of $75°$ at the centre of a circle of radius 8 cm. Calculate the length of the arc, to 2 decimal places.

    b. Calculate, in degrees, the magnitude of the angle that an arc of length 12π cm subtends at the centre of a circle of radius 10 cm.

11. Express $145.2°$ in radian measure, correct to 2 decimal places.

12. **a.** Express in radian measure, to 3 decimal places.

   i. 3°   **ii.** 112.25°   **iii.** 215.36°

   b. Express in degree measure to 3 decimal places.

   i. 3^c **ii.** 2.3π

   c. Rewrite $\left\{1.5^c, 50°, \dfrac{\pi^c}{7}\right\}$ with the magnitudes of the angles ordered from smallest to largest.

13. Calculate the exact lengths of the following arcs.

   a. The arc that subtends an angle of 150° at the centre of a circle of radius 12 cm

   b. The major arc with end points on a chord that subtends an angle of 60° at the centre of a circle of radius 3 cm

14. An arc of length 6 mm subtends an angle of 0.5^c at the centre of a circle. Calculate the radius of the circle.

15. A circle has a radius of 6 cm. An arc of length $\dfrac{3\pi}{4}$ cm forms part of the circumference of this circle. Determine the angle the arc subtends at the centre of the circle:

   a. in radian measure

   b. in degree measure.

16. **WE7** For the circle in the figure shown, calculate the area of the sector. Your answer should be rounded to 3 decimal places.

17. A circle has a radius of 5.4 metres. A sector with an angle of 1.6 radians at the centre of the circle is created. Calculate the area of the sector.

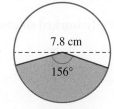

Complex familiar

18. A ball on the end of a rope of length 2.5 metres swings through an arc of 75 cm. Calculate the angle, to the nearest tenth of a degree, through which the ball swings.

19. A fixed point on the rim of a wheel of radius 3 metres rolls along horizontal ground at a speed of 2 m/s. After 5 seconds, calculate the angle the point has rotated through and express the answer in both degrees and radians.

20. An analogue wristwatch has a minute hand of length 11 mm. Calculate, to 2 decimal places, the arc length the minute hand traverses between 9:45 am and 9:50 am.

21. An arc of length 4 cm subtends an angle of 22.5° at the circumference of a circle. Calculate the area of the circle correct to 1 decimal place.

Complex unfamiliar

22. **a.** The Western Australian towns of Broome (B) and Karonie (K) both lie on approximately the same longitude. Broome is approximately 1490 km due north of Karonie (the distance being measured along the meridian). When the sun is directly over Karonie, it is 13.4° south of Broome. Use this information to estimate the radius of the Earth.
   This method dates back to Eratosthenes in 250 BCE, although he certainly didn't use these Australian towns to calculate his results.

   b. A ship sailing due east along the equator from the Galapagos Islands to Ecuador travels a distance of 600 nautical miles. If the ship's longitude changes from 90°W to 80°W during this journey, estimate the radius of the Earth, given that 1 nautical mile is approximately 1.85 km.

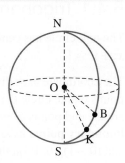

c. Taking the radius of the earth as 6370 km, calculate the distance, to the nearest kilometre, along the meridian between place A, located 20° S, 110° E, and place B, located 34° N, 110° E.

Fully worked solutions for this chapter are available online.

LESSON
6.4 Unit circle definitions

SYLLABUS LINKS

- Understand the unit circle definition of $\cos(\theta)$, $\sin(\theta)$ and $\tan(\theta)$ and periodicity using radians.

Source: Mathematical Methods Senior Syllabus 2024 © State of Queensland (QCAA) 2024; licensed under CC BY 4.0.

With the introduction of radian measure, we encountered positive and negative angles of any size and then associated them with the wrapping of the real number line around the circumference of a unit circle. Before applying this wrapping to define the sine, cosine and tangent functions, we first consider the conventions for angle rotations and the positions of the end points of these rotations.

6.4.1 Trigonometric points

The unit circle has centre $(0, 0)$ and radius 1 unit. Its Cartesian equation is $x^2 + y^2 = 1$.

The coordinate axes divide the Cartesian plane into four quadrants. The points on the circle that lie on the boundaries between the quadrants are the end points of the horizontal and vertical diameters. These **boundary points** have coordinates $(-1, 0), (1, 0)$ on the horizontal axis and $(0, -1), (0, 1)$ on the vertical axis.

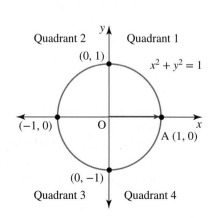

A rotation starts with an initial ray OA, where A is the point $(1, 0)$ and O $(0, 0)$. Angles are created by rotating the initial ray anticlockwise for positive angles and clockwise for negative angles. If the point on the circumference the ray reaches after a rotation of θ is P, then $\angle AOP = \theta$ and P is called the **trigonometric point** $[\theta]$. The angle of rotation θ may be measured in radian or degree measure. In radian measure, the value of θ corresponds to the length of the arc AP of the unit circle the rotation cuts off.

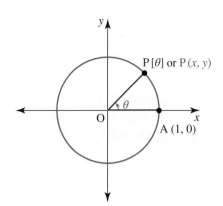

The point P $[\theta]$ has Cartesian coordinates (x, y) where:

- $x > 0$, $y > 0$ if P is in quadrant 1, $0 < \theta < \dfrac{\pi}{2}$

- $x < 0$, $y > 0$ if P is in quadrant 2, $\dfrac{\pi}{2} < \theta < \pi$

- $x < 0$, $y < 0$ if P is in quadrant 3, $\pi < \theta < \dfrac{3\pi}{2}$

- $x > 0$, $y < 0$ if P is in quadrant 4, $\dfrac{3\pi}{2} < \theta < 2\pi$.

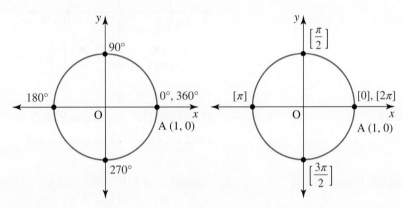

Continued rotation, anticlockwise or clockwise, can be used to form other values for θ greater than 2π or values less than 0, respectively. No trigonometric point has a unique θ value.

The angle θ is said to lie in the quadrant in which its end point P lies.

WORKED EXAMPLE 8 Identifying trigonometric points

a. **Give a trigonometric value, using radian measure, of the point P on the unit circle that lies on the boundary between the quadrants 2 and 3.**

b. **Identify the quadrants the following angles would lie in: $250°, 400°, \dfrac{2\pi^c}{3}, -\dfrac{\pi^c}{6}$.**

c. **Give two other trigonometric points, Q and R, one with a negative angle and one with a positive angle, respectively, that would have the same position as the point P $[250°]$.**

THINK	WRITE
a. 1. State the Cartesian coordinates of the required point.	**a.** The point $(-1, 0)$ lies on the boundary of quadrants 2 and 3.
2. Give a trigonometric value of this point. *Note:* Other values are possible.	An anticlockwise rotation of $180°$ or π^c from the point $(1, 0)$ would have its end point at $(-1, 0)$.
	The point P has the trigonometric value $[\pi]$.

▶

b. 1. Explain how the quadrant is determined.

b. For positive angles, rotate anticlockwise from $(1, 0)$; for negative angles rotate clockwise from $(1, 0)$. The position of the end point of the rotation determines the quadrant.

2. Identify the quadrant the end point of the rotation would lie in for each of the given angles.

Rotating anticlockwise $250°$ from $(1, 0)$ ends in quadrant 3; rotating anticlockwise from $(1, 0)$ through $400°$ would end in quadrant 1; rotating anticlockwise from $(1, 0)$ by $\dfrac{2\pi}{3}$ would end in quadrant 2; rotating clockwise from $(1, 0)$ by $\dfrac{\pi}{6}$ would end in quadrant 4.

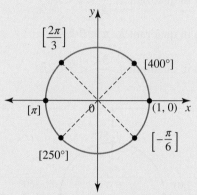

3. State the answer.

The angle $250°$ lies in quadrant 3, $400°$ in quadrant 1, $\dfrac{2\pi^c}{3}$ in quadrant 2 and $-\dfrac{\pi^c}{6}$ in quadrant 4.

c. 1. Identify a possible trigonometric point Q.

c. A rotation of $110°$ in the clockwise direction from $(1, 0)$ would end in the same position as $P[250°]$. Therefore, the trigonometric point could be $Q[-110°]$.

2. Identify a possible trigonometric point R.

A full anticlockwise revolution of $360°$ plus another anticlockwise rotation of $250°$ would end in the same position as $P[250°]$. Therefore, the trigonometric point could be $R[610°]$.

6.4.2 Unit circle definitions of the sine and cosine functions

Consider the unit circle and trigonometric point $P[\theta]$ with Cartesian coordinates (x, y) on its circumference. In the triangle ONP, $\angle NOP = \theta = \angle AOP$, $ON = x$ and $NP = y$.

As the triangle ONP is right-angled, $\cos(\theta) = \dfrac{x}{1} = x$ and $\sin(\theta) = \dfrac{y}{1} = y$. This enables the following definitions to be given.

For a rotation from the point $(1, 0)$ of any angle θ with end point P $[\theta]$ on the unit circle:

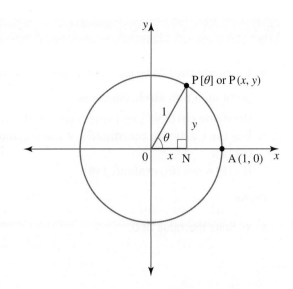

Coordinates of a trigonometric point

$\cos(\theta)$ is the x-coordinate of the trigonometric point P $[\theta]$.

$\sin(\theta)$ is the y-coordinate of the trigonometric point P $[\theta]$.

Substituting these coordinate values into the equation of the circle gives:

$$x^2 + y^2 = 1$$
$$(\cos(\theta))^2 + (\sin(\theta))^2 = 1$$
$$\cos^2(\theta) + \sin^2(\theta) = 1$$

This relationship, $\cos^2(\theta) + \sin^2(\theta) = 1$, is known as the **Pythagorean identity**. It is a true statement for any value of θ.

The Pythagorean identity may be rearranged to give either $\cos^2(\theta) = 1 - \sin^2(\theta)$ or $\sin^2(\theta) = 1 - \cos^2(\theta)$.

The Pythagorean identity

$$\cos^2(\theta) + \sin^2(\theta) = 1$$

The importance of these definitions is that they enable sine and cosine functions to be defined for any real number θ. With θ measured in radians, the trigonometric point $[\theta]$ also marks the position the real number θ is mapped to when the number line is wrapped around the circumference of the unit circle, with zero placed at the point $(1, 0)$. This relationship enables the sine or cosine of a real number θ to be evaluated as the sine or cosine of the angle of rotation of θ radians in a unit circle: $\sin(\theta) = \sin(\theta^c)$ and $\cos(\theta) = \cos(\theta^c)$.

The sine and cosine functions are $f: R \to R$, $f(\theta) = \sin(\theta)$ and $f: R \to R$, $f(\theta) = \cos(\theta)$.

They are **trigonometric functions**, also referred to as **circular functions**. The use of parentheses in writing $\sin(\theta)$ or $\cos(\theta)$ emphasises their functionality.

The mapping has a many-to-one correspondence as many values of θ are mapped to the one trigonometric point. The functions have a **period** of 2π, since rotations of θ and of $2\pi + \theta$ have the same end point on the circumference of the unit circle. The cosine and sine values repeat after each complete revolution around the unit circle.

For $f(\theta) = \sin(\theta)$, the image of a number such as 4 is $f(4) = \sin(4) = \sin(4^c)$. This is evaluated as the y-coordinate of the trigonometric point $[4]$ on the unit circle.

The values of a function for which $f(t) = \cos(t)$, where t is a real number, can be evaluated through the relation $\cos(t) = \cos(t^c)$ as t will be mapped to the trigonometric point $[t]$ on the unit circle.

The sine and cosine functions are periodic functions that have applications in contexts which may have nothing to do with angles, as we shall study in later chapters.

a. **Calculate the Cartesian coordinates of the trigonometric point P $\left[\dfrac{\pi}{3}\right]$ and show the position of this point on a unit circle diagram.**

b. **Illustrate $\cos{(330°)}$ and $\sin{(2)}$ on a unit circle diagram.**

c. **Use the Cartesian coordinates of the trigonometric point $[\pi]$ to obtain the values of $\sin{(\pi)}$ and $\cos{(\pi)}$.**

d. **If $f(\theta) = \cos{(\theta)}$, evaluate $f(0)$.**

THINK	WRITE
a. 1. State the value of θ.	a. P $\left[\dfrac{\pi}{3}\right]$ This is the trigonometric point with $\theta = \dfrac{\pi}{3}$.
2. Calculate the exact Cartesian coordinates. *Note:* The exact values for sine and cosine of $\dfrac{\pi^c}{3}$, or 60°, need to be known.	The Cartesian coordinates are: $x = \cos{(\theta)}$ $y = \sin{(\theta)}$ $= \cos\left(\dfrac{\pi}{3}\right)$ $= \sin\left(\dfrac{\pi}{3}\right)$ $= \dfrac{1}{2}$ $= \dfrac{\sqrt{3}}{2}$ Therefore, P has coordinates $\left(\dfrac{1}{2}, \dfrac{\sqrt{3}}{2}\right)$.
3. Show the position of the given point on a unit circle diagram.	P $\left[\dfrac{\pi}{3}\right]$ or P $\left(\dfrac{1}{2}, \dfrac{\sqrt{3}}{2}\right)$ lies in quadrant 1 on the circumference of the unit circle.

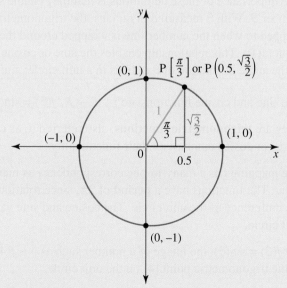

b. 1. Identify the trigonometric point and which of its Cartesian coordinates gives the first value.	b. $\cos{(330°)}$: The value of $\cos{(330°)}$ is given by the x-coordinate of the trigonometric point [330°].
2. State the quadrant in which the trigonometric point lies.	The trigonometric point [330°] lies in quadrant 4.

3. Identify the trigonometric point and which of its Cartesian coordinates gives the second value.

4. State the quadrant in which the trigonometric point lies.

5. Draw a unit circle showing the two trigonometric points and construct the line segments that illustrate the x- and y-coordinates of each point.

$\sin(2)$:

The value of $\sin(2)$ is given by the y-coordinate of the trigonometric point [2].

As $\dfrac{\pi}{2} \approx 1.57 < 2 < \pi \approx 3.14$, the trigonometric point [2] lies in quadrant 2.

For each of the points on the unit circle diagram, the horizontal line segment gives the x-coordinate and the vertical line segment gives the y-coordinate.

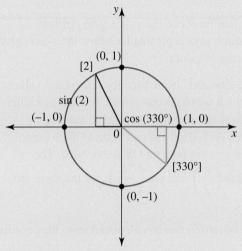

6. Label the line segments that represent the appropriate coordinate for each point.

The value of $\cos(330°)$ is the length measure of the horizontal line segment.
The value of $\sin(2)$ is the length measure of the vertical line segment.
The line segments illustrating these values are highlighted in orange on the diagram.

c. 1. State the Cartesian coordinates of the given point.

c. An anticlockwise rotation of π from $(1, 0)$ gives the end point $(-1, 0)$.
The trigonometric point $[\pi]$ is the Cartesian point $(-1, 0)$.

2. State the required values.

The point $(-1, 0)$ has $x = -1, y = 0$.
Since $x = \cos(\theta)$,
$\cos(\pi) = x$
$\qquad\quad = -1$
Since $y = \sin(\theta)$,
$\sin(\pi) = y$
$\qquad\quad = 0$

d. 1. Substitute the given value in the function rule.

d. $f(\theta) = \cos(\theta)$
$\therefore f(0) = \cos(0)$

2. Identify the trigonometric point and state its Cartesian coordinates.

The trigonometric point [0] has Cartesian coordinates $(1, 0)$.

3. Evaluate the required value of the function.

The value of $\cos(0)$ is given by the x-coordinate of the point $(1, 0)$.
$\therefore \cos(0) = 1$
$\qquad \therefore f(0) = 1$

6.4.3 Unit circle definition of the tangent function

Consider again the unit circle with centre O $(0,0)$ containing the points A $(1,0)$ and the trigonometric point P $[\theta]$ on its circumference. A tangent line to the circle is drawn at point A. The radius OP is extended to intersect the tangent line at point T.

For any point P $[\theta]$ on the unit circle, tan (θ) is defined as the length of the intercept AT that the extended ray OP cuts off on the tangent drawn to the unit circle at the point A $(1,0)$.

Intercepts that lie above the x-axis give positive tangent values; intercepts that lie below the x-axis give negative tangent values.

Unlike the sine and cosine functions, there are values of θ for which tan (θ) is undefined. These occur when OP is vertical and therefore parallel to the tangent line through A $(1,0)$; these two vertical lines cannot intersect, no matter how far OP is extended. The values of $\tan\left(\dfrac{\pi}{2}\right)$ and $\tan\left(\dfrac{3\pi}{2}\right)$, for instance, are undefined.

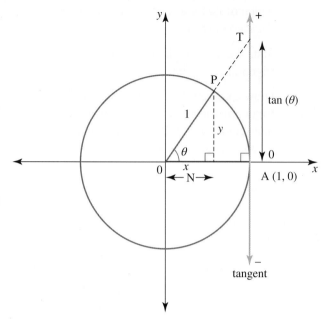

The value of tan (θ) can be calculated from the coordinates (x, y) of the point P $[\theta]$, provided the x-coordinate is not zero.

Using the ratio of sides of the similar triangles ONP and OAT:

$$\frac{AT}{OA} = \frac{NP}{ON}$$

$$\frac{\tan(\theta)}{1} = \frac{y}{x}$$

Hence:

$\tan(\theta) = \dfrac{y}{x}, x \neq 0$, where (x, y) are the coordinates of the trigonometric point P $[\theta]$.

Since $x = \cos(\theta), y = \sin(\theta)$, this can be expressed as the following relationship.

Unit circle definition of tan (θ)

$$\tan(\theta) = \frac{\sin(\theta)}{\cos(\theta)}$$

WORKED EXAMPLE 10 Evaluating tan (θ)

a. **Illustrate tan $(130°)$ on a unit circle diagram and use a calculator to evaluate tan $(130°)$ to 3 decimal places.**

b. **Use the Cartesian coordinates of the trigonometric point P $[\pi]$ to determine the value of tan (π).**

THINK

a. 1. State the quadrant in which the angle lies.

2. Draw the unit circle with the tangent at the point A (1, 0).
Note: The tangent line is always drawn at this point (1, 0).

WRITE

a. 130° lies in the second quadrant.

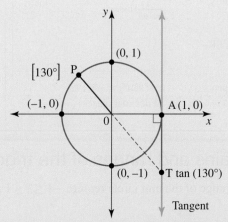

3. Extend PO until it reaches the tangent line.

Let T be the point where the extended radius PO intersects the tangent drawn at A.
The intercept AT is tan (130°).

4. State whether the required value is positive, zero or negative.

The intercept lies below the x-axis, which shows that tan (130°) is negative.

5. Calculate the required value.

The value of tan (130°) = −1.192, correct to 3 decimal places.

b. 1. Identify the trigonometric point and state its Cartesian coordinates.

b. The trigonometric point P [π] is the end point of a rotation of π^c or 180°. It is the Cartesian point P (−1, 0).

2. Calculate the required value.

The point (−1, 0) has $x = -1, y = 0$.
Since $\tan(\theta) = \dfrac{y}{x}$,

$$\tan(\pi) = \frac{0}{-1}$$
$$= 0$$

3. Check the answer using the unit circle diagram.

Check:
PO is horizontal and runs along the x-axis. Extending PO, it intersects the tangent at the point A. This means the intercept is 0, which means tan (π) = 0.

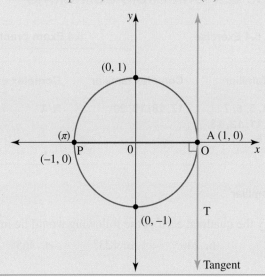

TI	THINK	WRITE	CASIO	THINK	WRITE
a. 1. Put the Calculator in Degree mode. On a Calculator page, complete the entry line as: tan (130) then press ENTER.	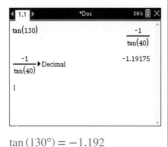	**a. 1.** Put the Calculator in Degree mode. On a Run-Matrix screen, complete the entry line as: tan 130 then press EXE.	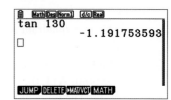
2. The answer appears on the screen.	$\tan(130°) = -1.192$ (3 decimal places)	**2.** The answer appears on the screen.	$\tan(130°) = -1.192$ (3 decimal places)

6.4.4 Domains and ranges of the trigonometric functions

The domain and range of the unit circle require $-1 \le x \le 1$ and $-1 \le y \le 1$, so $-1 \le \cos(\theta) \le 1$ and $-1 \le \sin(\theta) \le 1$.

Since θ can be any real number, this means that the function f where f is either sine or cosine has domain R and range $[-1, 1]$.

Unlike the sine and cosine functions, the domain of the tangent function is not the set of real numbers R, since $\tan(\theta)$ is not defined for any value of θ that is an odd multiple of $\dfrac{\pi}{2}$. Excluding these values, intercepts of any size may be cut off on the tangent line, so $\tan(\theta) \in R$.

This means that the function f where f is tangent has domain $R \setminus \left\{ \pm \dfrac{\pi}{2}, \pm \dfrac{3\pi}{2}, ... \right\}$ and range R.

The domain of the tangent function can be written as $R \setminus \left\{ (2n + 1)\dfrac{\pi}{2}, n \in Z \right\}$ and the tangent function as the mapping $f: R \setminus \left\{ (2n + 1)\dfrac{\pi}{2}, n \in Z \right\} \to R, f(\theta) = \tan(\theta)$.

 Resources

 Interactivity The unit circle (int-2582)

Exercise 6.4 Unit circle definitions

learn

6.4 Exercise	**6.4 Exam practice**

These questions are even better in jacPLUS!
- Receive immediate feedback
- Access sample responses
- Track results and progress

Find all this and MORE in jacPLUS ▶

Simple familiar	Complex familiar	Complex unfamiliar
1, 2, 3, 4, 5, 6, 7, 8, 9, 10, 11, 12, 13, 14, 15, 16	17, 18, 19, 20	N/A

Simple familiar

1. Identify the quadrant each of the following would lie in.
   a. 24°   b. 240°   c. 123°   d. 365°   e. −50°   f. −120°

2. Identify the quadrant or boundary in which each of the following lies.

   a. $585°$ b. $\dfrac{11\pi}{12}$ c. -18π d. $\dfrac{7\pi}{4}$

3. a. The trigonometric point $[\theta]$ lies on the boundary between quadrants 1 and 2.

   i. State the Cartesian coordinates of the trigonometric point $[\theta]$.
   ii. State the value of $\sin(\theta)$.

   b. The trigonometric point $[\alpha]$ lies on the boundary between quadrants 2 and 3.

   i. State the Cartesian coordinates of the trigonometric point $[\alpha]$.
   ii. State the value of $\cos(\alpha)$.

   c. The trigonometric point $[\beta]$ lies on the boundary between quadrants 3 and 4.

   i. State the Cartesian coordinates of the trigonometric point $[\beta]$.
   ii. State the value of $\tan(\beta)$.

   d. The trigonometric point $[v]$ lies on the boundary between quadrants 4 and 1.

   i. State the Cartesian coordinates of the trigonometric point $[v]$.
   ii. State the value of $\sin(v), \cos(v)$ and $\tan(v)$.

4. **WE8** a. Give a trigonometric value, using radian measure, of the point P on the unit circle that lies on the boundary between the quadrants 1 and 2.

   b. Identify the quadrants the following angles would lie in: $120°, -400°, \dfrac{4\pi^{\,c}}{3}, \dfrac{\pi^{\,c}}{4}$.

   c. Give two other trigonometric points, Q with a negative angle and R with a positive angle, that would have the same position as the point P $[120°]$.

5. **WE9** a. Calculate the Cartesian coordinates of the trigonometric point P $\left[\dfrac{\pi}{6}\right]$ and show the position of this point on a unit circle diagram.

   b. Illustrate $\sin(225°)$ and $\cos(1)$ on a unit circle diagram.

   c. Use the Cartesian coordinates of the trigonometric point $\left[-\dfrac{\pi}{2}\right]$ to obtain the values of $\sin\left(-\dfrac{\pi}{2}\right)$ and $\cos\left(-\dfrac{\pi}{2}\right)$.

   d. If $f(\theta) = \sin(\theta)$, evaluate $f(0)$.

6. a. For the function $f(t) = \sin(t)$, evaluate $f\left(\dfrac{3\pi}{2}\right)$.

   b. For the function $g(t) = \cos(t)$, evaluate $g(4\pi)$.
   c. For the function $h(t) = \tan(t)$, evaluate $h(-\pi)$.
   d. For the function $k(t) = \sin(t) + \cos(t)$, evaluate $k(6.5\pi)$.

7. Identify the quadrants where:

   a. $\sin(\theta)$ is always positive
   b. $\cos(\theta)$ is always positive.

8. a. Calculate the Cartesian coordinates of the trigonometric point P $\left[\dfrac{\pi}{4}\right]$.

   b. Express the Cartesian point P $(0, -1)$ as two different trigonometric points, one with a positive value for θ and one with a negative value for θ.

9. Illustrate the following on a unit circle diagram.

   a. $\cos(40°)$ b. $\sin(165°)$ c. $\cos(-60°)$ d. $\sin(-90°)$

10. Illustrate the following on a unit circle diagram.

    a. $\sin\left(\dfrac{5\pi}{3}\right)$

    b. $\cos\left(\dfrac{3\pi}{5}\right)$

    c. $\cos(5\pi)$

    d. $\sin\left(-\dfrac{2\pi}{3}\right)$

11. Illustrate the following on a unit circle diagram.

    a. $\tan(315°)$

    b. $\tan\left(\dfrac{5\pi}{6}\right)$

    c. $\tan\left(\dfrac{4\pi}{3}\right)$

    d. $\tan(-300°)$

12. a. The trigonometric point $P[\theta]$ has Cartesian coordinates $(-0.8, 0.6)$. Identify the quadrant in which P lies and state the values of $\sin(\theta)$, $\cos(\theta)$ and $\tan(\theta)$.

    b. The trigonometric point $Q[\theta]$ has Cartesian coordinates $\left(\dfrac{\sqrt{2}}{2}, -\dfrac{\sqrt{2}}{2}\right)$. Identify the quadrant in which Q lies and state the values of $\sin(\theta)$, $\cos(\theta)$ and $\tan(\theta)$.

    c. For the trigonometric point $R[\theta]$ with Cartesian coordinates $\left(\dfrac{2}{\sqrt{5}}, \dfrac{1}{\sqrt{5}}\right)$, identify the quadrant in which R lies and state the values of $\sin(\theta)$, $\cos(\theta)$ and $\tan(\theta)$.

    d. The Cartesian coordinates of the trigonometric point $S[\theta]$ are $(0, 1)$. Describe the position of S and state the values of $\sin(\theta)$, $\cos(\theta)$ and $\tan(\theta)$.

13. By locating the appropriate trigonometric point and its corresponding Cartesian coordinates, obtain the exact values of the following.

    a. $\cos(0)$

    b. $\sin\left(\dfrac{\pi}{2}\right)$

    c. $\tan(\pi)$

    d. $\cos\left(\dfrac{3\pi}{2}\right)$

    e. $\sin(2\pi)$

    f. $\cos\left(\dfrac{17\pi}{2}\right) + \tan(-11\pi) + \sin\left(\dfrac{11\pi}{2}\right)$

14. Consider $\left\{\tan(-3\pi), \tan\left(\dfrac{5\pi}{2}\right), \tan(-90°), \tan\left(\dfrac{3\pi}{4}\right), \tan(780°)\right\}$.

    a. Identify the elements in the set that are not defined.
    b. Identify the elements that have negative values.

15. **WE10** a. Illustrate $\tan(230°)$ on a unit circle diagram and use a calculator to evaluate $\tan(230°)$ to 3 decimal places.

    b. Use the Cartesian coordinates of the trigonometric point $P[2\pi]$ to determine the value of $\tan(2\pi)$.

16. a. On a unit circle diagram show the trigonometric point $P[2]$ and the line segments $\sin(2)$, $\cos(2)$ and $\tan(2)$. Label them with their length measures expressed to 2 decimal places.

    b. State the Cartesian coordinates of P to 2 decimal places.

Complex familiar

17. Consider O, the centre of the unit circle, and the trigonometric points $P\left[\dfrac{3\pi}{10}\right]$ and $Q\left[\dfrac{2\pi}{5}\right]$ on its circumference.

    a. Sketch the unit circle showing these points.
    b. Determine how many radians are contained in the angle $\angle QOP$.
    c. Express each of the trigonometric points P and Q with a negative θ value.
    d. Express each of the trigonometric points P and Q with a larger positive value for θ than the given values $P\left[\dfrac{3\pi}{10}\right]$ and $Q\left[\dfrac{2\pi}{5}\right]$.

18. On a unit circle diagram, show the trigonometric points A [0] and P [θ] where θ is acute, and show the line segments sin (θ) and tan (θ). By comparing the lengths of the line segments with the length of the arc AP, explain why sin (θ) < θ < tan (θ) for acute θ.

19. a. Use technology to obtain the exact Cartesian coordinates of the trigonometric points $P\left[\frac{7\pi}{4}\right]$ and $Q\left[\frac{\pi}{4}\right]$, and describe the relative position of the points P and Q on the unit circle.

    b. Use technology to obtain the Cartesian coordinates of the trigonometric points $R\left[\frac{4\pi}{5}\right]$ and $S\left[\frac{\pi}{5}\right]$. Describe the relative position of these points on the unit circle.

    c. Give the sine, cosine and tangent values of:

       i. $\frac{7\pi}{4}$ and $\frac{\pi}{4}$, and compare the values

       ii. $\frac{4\pi}{5}$ and $\frac{\pi}{5}$, and compare the values.

20. Use technology to calculate the exact values of the following.

    a. $\cos^2\left(\frac{7\pi}{6}\right) + \sin^2\left(\frac{7\pi}{6}\right)$

    b. $\cos\left(\frac{7\pi}{6}\right) + \sin\left(\frac{7\pi}{6}\right)$

    c. $\sin^2\left(\frac{7}{6}\right) + \cos^2\left(\frac{7}{6}\right)$

    d. $\sin^2(76°) + \cos^2(76°)$

    e. $\sin^2(t) + \cos^2(t)$; explain the result with reference to the unit circle

Fully worked solutions for this chapter are available online.

LESSON
6.5 Exact values and symmetry properties

SYLLABUS LINKS

- Understand and use the exact values of $\cos(\theta)$, $\sin(\theta)$ and $\tan(\theta)$ at integer multiples of $\frac{\pi}{6}$ and $\frac{\pi}{4}$.
- Understand the unit circle definition of $\cos(\theta)$, $\sin(\theta)$ and $\tan(\theta)$ and periodicity using radians.

Source: Mathematical Methods Senior Syllabus 2024 © State of Queensland (QCAA) 2024; licensed under CC BY 4.0.

There are relationships between the coordinates and associated trigonometric values of trigonometric points placed in symmetric positions in each of the four quadrants. These will now be investigated.

6.5.1 The signs of the sine, cosine and tangent values in the four quadrants

The definitions $\cos(\theta) = x$, $\sin(\theta) = y$, $\tan(\theta) = \frac{y}{x}$ where (x, y) are the Cartesian coordinates of the trigonometric point $[\theta]$ have been established.

If θ lies in the first quadrant, **All** of the trigonometric values will be positive, since $x > 0$, $y > 0$.

If θ lies in the second quadrant, only the **Sine** value will be positive, since $x < 0$, $y > 0$.

If θ lies in the third quadrant, only the **Tangent** value will be positive, since $x < 0$, $y < 0$.

If θ lies in the fourth quadrant, only the **Cosine** value will be positive, since $x > 0$, $y < 0$.

This is illustrated in the diagram shown.

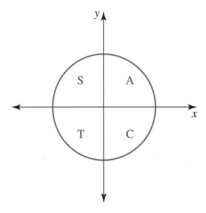

There are several mnemonics for remembering the allocation of signs in this diagram: we shall use 'ASTC' and refer to the diagram as the ASTC diagram, starting in the first quadrant.

A common saying is '**All Stations To Central**'.

 Resources

Interactivity All Sin Cos Tan (int-2583)

6.5.2 The sine, cosine and tangent values at the boundaries of the quadrants

The points that do not lie within a quadrant are the four coordinate axes intercepts of the unit circle. These are called the boundary points. Since the Cartesian coordinates and the trigonometric positions of these points are known, the boundary values can be summarised by the following table.

Boundary point	$(1, 0)$	$(0, 1)$	$(-1, 0)$	$(0, -1)$	$(1, 0)$
θ radians	0	$\dfrac{\pi}{2}$	π	$\dfrac{3\pi}{2}$	2π
θ degrees	0°	90°	180°	270°	360°
$\sin(\theta)$	0	1	0	-1	0
$\cos(\theta)$	1	0	-1	0	1
$\tan(\theta)$	0	Undefined	0	Undefined	0

Other values of θ could be used for the boundary points, including negative values.

a. **Identify the quadrant(s) where both cos (θ) and sin (θ) are negative.**

b. **If $f(\theta) = \cos(\theta)$, evaluate $f(-6\pi)$.**

THINK	WRITE
a. 1. Refer to the ASTC diagram.	a. $\cos(\theta) = x$, $\sin(\theta) = y$ The quadrant where both x and y are negative is quadrant 3.
b. 1. Substitute the given value in the function rule.	b. $f(\theta) = \cos(\theta)$ $\therefore f(-6\pi) = \cos(-6\pi)$
2. Identify the Cartesian coordinates of the trigonometric point.	A clockwise rotation of 6π from $(1, 0)$ shows that the trigonometric point $[-6\pi]$ is the boundary point with coordinates $(1, 0)$.
3. Evaluate the required value of the function.	The x-coordinate of the boundary point gives the cosine value. $\cos(-6\pi) = 1$ $\therefore f(-6\pi) = 1$

Exact trigonometric values of $\dfrac{\pi}{6}$, $\dfrac{\pi}{4}$ and $\dfrac{\pi}{3}$

As the exact trigonometric ratios are known for angles of 30°, 45° and 60°, these give the trigonometric ratios for $\dfrac{\pi}{6}$, $\dfrac{\pi}{4}$ and $\dfrac{\pi}{3}$, respectively. A summary of these is given with the angles in each triangle expressed in radian measure. The values should be memorised.

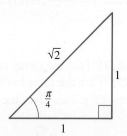 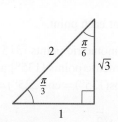

θ	$\dfrac{\pi}{6}$ or 30°	$\dfrac{\pi}{4}$ or 45°	$\dfrac{\pi}{3}$ or 60°
$\sin(\theta)$	$\dfrac{1}{2}$	$\dfrac{1}{\sqrt{2}} = \dfrac{\sqrt{2}}{2}$	$\dfrac{\sqrt{3}}{2}$
$\cos(\theta)$	$\dfrac{\sqrt{3}}{2}$	$\dfrac{1}{\sqrt{2}} = \dfrac{\sqrt{2}}{2}$	$\dfrac{1}{2}$
$\tan(\theta)$	$\dfrac{1}{\sqrt{3}} = \dfrac{\sqrt{3}}{3}$	1	$\sqrt{3}$

These values can be used to calculate the exact trigonometric values for other angles that lie in positions symmetric to these first-quadrant angles.

6.5.3 Trigonometric points symmetric to [θ] where $\theta \in \left\{ 30°, 45°, 60°, \dfrac{\pi}{6}, \dfrac{\pi}{4}, \dfrac{\pi}{3} \right\}$

The symmetrical points to [45°] are shown in the diagram.

Each radius of the circle drawn to each of the points makes an acute angle of 45° with either the positive or the negative x-axis. The symmetric points to [45°] are the end points of a rotation that is 45° short of, or 45° beyond, the horizontal x-axis. The calculations $180° - 45°$, $180° + 45°$ and $360° - 45°$ give the symmetric trigonometric points [135°], [225°] and [315°], respectively.

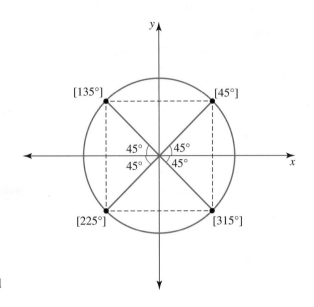

Comparisons between the coordinates of these trigonometric points with those of the first quadrant point [45°] enable the trigonometric values of these non-acute angles to be calculated from those of the acute angle 45°.

Consider the y-coordinate of each point.

As the y-coordinates of the trigonometric points [135°] and [45°] are the same, $\sin(135°) = \sin(45°)$. Similarly, the y-coordinates of the trigonometric points [225°] and [315°] are the same, but both are the negative of the y-coordinate of [45°]. Hence, $\sin(225°) = \sin(315°) = -\sin(45°)$. This gives the following exact sine values.

$$\sin(45°) = \frac{\sqrt{2}}{2}; \sin(135°) = \frac{\sqrt{2}}{2}; \sin(225°) = -\frac{\sqrt{2}}{2}; \sin(315°) = -\frac{\sqrt{2}}{2}$$

Now consider the x-coordinate of each point.

As the x-coordinates of the trigonometric points [315°] and [45°] are the same, $\cos(315°) = \cos(45°)$. Similarly, the x-coordinates of the trigonometric points [135°] and [225°] are the same but both are the negative of the x-coordinate of [45°]. Hence, $\cos(135°) = \cos(225°) = -\cos(45°)$. This gives the following exact cosine values.

$$\cos(45°) = \frac{\sqrt{2}}{2}; \cos(135°) = -\frac{\sqrt{2}}{2}; \cos(225°) = -\frac{\sqrt{2}}{2}; \cos(315°) = \frac{\sqrt{2}}{2}$$

Either by considering the intercepts cut off on the vertical tangent drawn at $(1, 0)$ or by using $\tan(\theta) = \dfrac{y}{x} = \dfrac{\sin(\theta)}{\cos(\theta)}$, you will find that the corresponding relationships for the four points are $\tan(225°) = \tan(45°)$ and $\tan(135°) = \tan(315°) = -\tan(45°)$. Hence, the exact tangent values are:

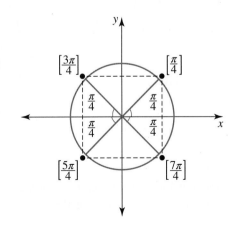

$$\tan(45°) = 1; \tan(135°) = -1; \tan(225°) = 1; \tan(315°) = -1.$$

The relationships between the Cartesian coordinates of [45°] and each of [135°], [225°] and [315°] enable the trigonometric values of 135°, 225° and 315° to be calculated from those of 45°.

If, instead of degree measure, the radian measure of $\dfrac{\pi}{4}$ is used, the symmetric points to $\left[\dfrac{\pi}{4}\right]$ are the end points

of rotations that lie $\dfrac{\pi}{4}$ short of, or $\dfrac{\pi}{4}$ beyond, the horizontal x-axis. The positions of the symmetric points

are calculated as $\pi - \dfrac{\pi}{4}, \pi + \dfrac{\pi}{4}, 2\pi - \dfrac{\pi}{4}$, giving the symmetric trigonometric points $\left[\dfrac{3\pi}{4}\right], \left[\dfrac{5\pi}{4}\right], \left[\dfrac{7\pi}{4}\right]$,
respectively.

By comparing the Cartesian coordinates of the symmetric points with those of the first quadrant point $\left[\dfrac{\pi}{4}\right]$, it is
possible to obtain results such as the following selection.

Second quadrant

$$\cos\left(\dfrac{3\pi}{4}\right) = -\cos\left(\dfrac{\pi}{4}\right)$$
$$= -\dfrac{\sqrt{2}}{2}$$

Third quadrant

$$\tan\left(\dfrac{5\pi}{4}\right) = \tan\left(\dfrac{\pi}{4}\right)$$
$$= 1$$

Fourth quadrant

$$\sin\left(\dfrac{7\pi}{4}\right) = -\sin\left(\dfrac{\pi}{4}\right)$$
$$= -\dfrac{\sqrt{2}}{2}$$

A similar approach is used to generate symmetric points to the first quadrant points $\left[\dfrac{\pi}{6}\right]$ and $\left[\dfrac{\pi}{3}\right]$.

WORKED EXAMPLE 12 Calculating exact values

Calculate the exact values of the following.

a. $\cos\left(\dfrac{5\pi}{3}\right)$

b. $\sin\left(\dfrac{7\pi}{6}\right)$

c. $\tan(-30°)$

THINK

a. 1. State the quadrant in which the trigonometric point lies.

2. Identify the first-quadrant symmetric point.

WRITE

a. $\cos\left(\dfrac{5\pi}{3}\right)$

As $\dfrac{5\pi}{3} = \dfrac{5}{3}\pi = 1\dfrac{2}{3}\pi$,

the point $\left[\dfrac{5\pi}{3}\right]$ lies in quadrant 4.

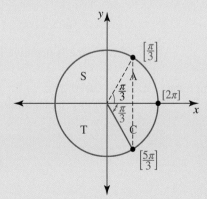

Since $\dfrac{5\pi}{3} = 2\pi - \dfrac{\pi}{3}$, the rotation of $\dfrac{5\pi}{3}$ stops short

of the x-axis by $\dfrac{\pi}{3}$. The points $\left[\dfrac{\pi}{3}\right]$ and $\left[\dfrac{5\pi}{3}\right]$ are

symmetric.

3. Compare the coordinates of the symmetric points and obtain the required value.
Note: Check the +/− sign follows the ASTC diagram rule.

The *x*-coordinates of the symmetric points are equal.

$$\cos\left(\frac{5\pi}{3}\right) = +\cos\left(\frac{\pi}{3}\right)$$

$$= \frac{1}{2}$$

Check: cosine is positive in quadrant 4.

b. 1. State the quadrant in which the trigonometric point lies.

b. $\sin\left(\frac{7\pi}{6}\right)$

$$\frac{7\pi}{6} = 1\frac{1}{6}\pi$$

The point lies in quadrant 3.

2. Identify the first-quadrant symmetric point.

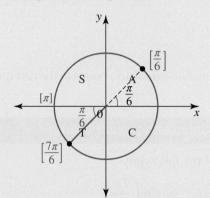

As $\frac{7\pi}{6} = \pi + \frac{\pi}{6}$, the rotation of $\frac{7\pi}{6}$ goes beyond the

x-axis by $\frac{\pi}{6}$. The points $\left[\frac{\pi}{6}\right]$ and $\left[\frac{7\pi}{6}\right]$ are symmetric.

3. Compare the coordinates of the symmetric points and obtain the required value.

The *y*-coordinate of $\left[\frac{7\pi}{6}\right]$ is the negative of that of $\left[\frac{\pi}{6}\right]$ in the first quadrant.

$$\sin\left(\frac{7\pi}{6}\right) = -\sin\left(\frac{\pi}{6}\right)$$

$$= -\frac{1}{2}$$

Check: sine is negative in quadrant 3.

c. 1. State the quadrant in which the trigonometric point lies.

c. $\tan(-30°)$
$[-30°]$ lies in quadrant 4.

2. Identify the first-quadrant symmetric point.

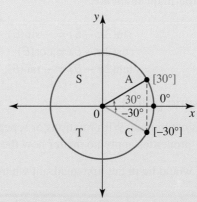

−30° is a clockwise rotation of 30° from the horizontal, so the symmetric point in the first quadrant is [30°].

3. Compare the coordinates of the symmetric points and obtain the required value.
Note: Alternatively, consider the intercepts that would be cut off on the vertical tangent at $(1, 0)$.

The points [30°] and [−30°] have the same x-coordinates but opposite y-coordinates. The tangent value is negative in quadrant 4.

$$\tan(-30°) = -\tan(30°)$$
$$= -\frac{\sqrt{3}}{3}$$

6.5.4 Symmetry properties

The **symmetry properties** give the relationships between the trigonometric values in quadrants $2, 3, 4$ and that of the first quadrant value, called the base, with which they are symmetric. The symmetry properties are simply a generalisation of what was covered for the bases $\frac{\pi}{6}, \frac{\pi}{4}, \frac{\pi}{3}$.

For any real number θ where $0 < \theta < \frac{\pi}{2}$, the trigonometric point $[\theta]$ lies in the first quadrant. The other quadrant values can be expressed in terms of the base θ, since the symmetric values will either be θ short of, or θ beyond, the horizontal x-axis.

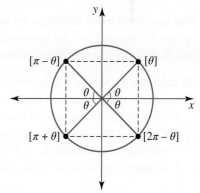

The symmetric points to $[\theta]$ are:
- second quadrant $[\pi - \theta]$
- third quadrant $[\pi + \theta]$
- fourth quadrant $[2\pi - \theta]$.

Comparing the Cartesian coordinates with those of the first-quadrant base leads to the following general statements.

The symmetry properties for the second quadrant are:

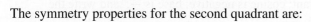

$$\sin(\pi - \theta) = \sin(\theta)$$
$$\cos(\pi - \theta) = -\cos(\theta)$$
$$\tan(\pi - \theta) = -\tan(\theta).$$

The symmetry properties for the third quadrant are:

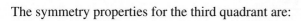

$$\sin(\pi + \theta) = -\sin(\theta)$$
$$\cos(\pi + \theta) = -\cos(\theta)$$
$$\tan(\pi + \theta) = \tan(\theta).$$

The symmetry properties for the fourth quadrant are:

$$\sin(2\pi - \theta) = -\sin(\theta)$$
$$\cos(2\pi - \theta) = \cos(\theta)$$
$$\tan(2\pi - \theta) = -\tan(\theta).$$

Other forms for the symmetric points

The rotation assigned to a point is not unique. With rotations or repeated revolutions, other values are always possible. However, the symmetry properties apply no matter how the points are described.

The trigonometric point $[2\pi + \theta]$ would lie in the first quadrant where all ratios are positive. Hence:

$$\sin(2\pi + \theta) = \sin(\theta)$$
$$\cos(2\pi + \theta) = \cos(\theta)$$
$$\tan(2\pi + \theta) = \tan(\theta).$$

The trigonometric point $[-\theta]$ would lie in the fourth quadrant where only cosine is positive. Hence:

$$\sin(-\theta) = -\sin(\theta)$$
$$\cos(-\theta) = \cos(\theta)$$
$$\tan(-\theta) = -\tan(\theta).$$

For negative rotations, the points symmetric to $[\theta]$ could be given as:
- fourth quadrant $[-\theta]$
- third quadrant $[-\pi + \theta]$
- second quadrant $[-\pi - \theta]$
- first quadrant $[-2\pi + \theta]$.

Using symmetry properties to calculate values of trigonometric functions

Trigonometric values are either the same as or the negative of the associated trigonometric values of the first-quadrant base; the sign is determined by the ASTC diagram.

The base involved is identified by noting the rotation needed to reach the x-axis and determining how far short of or how far beyond this the symmetric point is. It is important to emphasise that for the points to be symmetric, this is always measured from the horizontal and not the vertical axis.

To calculate a value of a trigonometric function, follow these steps.
- Locate the quadrant in which the trigonometric point lies.
- Identify the first-quadrant base with which the trigonometric point is symmetric.
- Compare the coordinates of the trigonometric point with the coordinates of the base point or use the ASTC diagram rule to form the sign in the first instance.
- Evaluate the required value exactly if there is a known exact value involving the base.

For example, to evaluate $\cos\left(\dfrac{3\pi}{4}\right)$, think: 'Second quadrant; cosine is negative; base is $\dfrac{\pi}{4}$,' and write the following.

$$\cos\left(\frac{3\pi}{4}\right) = -\cos\left(\frac{\pi}{4}\right)$$
$$= -\frac{\sqrt{2}}{2}$$

a. **Identify the symmetric points to [20°]. Of these points, identify the point at which the tangent value is the same as tan(20°).**

b. **Express $\sin\left(\dfrac{6\pi}{5}\right)$ in terms of a first-quadrant value.**

c. **If $\cos(\theta) = 0.6$, give the values of $\cos(\pi - \theta)$ and $\cos(2\pi - \theta)$.**

d. **Calculate the exact values of the following.**

    i. $\tan\left(\dfrac{7\pi}{6}\right)$ ii. $\sin\left(\dfrac{11\pi}{3}\right)$

THINK	WRITE
a. 1. Calculate the symmetric points to the given point.	a. Symmetric points to [20°] will be $\pm 20°$ from the x-axis. The points are: second quadrant $[180° - 20°] = [160°]$ third quadrant $[180° + 20°] = [200°]$ fourth quadrant $[360° - 20°] = [340°]$.
2. Identify the quadrant.	The point [20°] is in the first quadrant, so $\tan(20°)$ is positive. As tangent is also positive in the third quadrant, $\tan(200°) = \tan(20°)$.
3. State the required point.	The tangent value at the trigonometric point [200°] has the same value as $\tan(20°)$.
b. 1. Express the trigonometric value in the appropriate quadrant form.	b. $\dfrac{6\pi}{5}$ is in the third quadrant. $$\sin\left(\dfrac{6\pi}{5}\right) = \sin\left(\pi + \dfrac{\pi}{5}\right)$$
2. Apply the symmetry property for that quadrant.	$\therefore \sin\left(\dfrac{6\pi}{5}\right) = -\sin\left(\dfrac{\pi}{5}\right)$
c. 1. Use the symmetry property for the appropriate quadrant.	c. $(\pi - \theta)$ is second quadrant form. $\therefore \cos(\pi - \theta) = -\cos(\theta)$
2. State the answer.	Since $\cos(\theta) = 0.6$, $\cos(\pi - \theta) = -0.6$.
3. Use the symmetry property for the appropriate quadrant.	$2\pi - \theta$ is fourth quadrant form. $\therefore \cos(2\pi - \theta) = \cos(\theta)$
4. State the answer.	Since $\cos(\theta) = 0.6$, $\cos(2\pi - \theta) = 0.6$.
d. i. 1. Express the trigonometric value in an appropriate quadrant form. Apply the symmetry property and evaluate.	d. i. $\dfrac{7\pi}{6}$ is in the third quadrant, so $\tan\left(\dfrac{7\pi}{6}\right)$ is positive. $$\tan\left(\dfrac{7\pi}{6}\right) = \tan\left(\pi + \dfrac{\pi}{6}\right)$$ $$= \tan\left(\dfrac{\pi}{6}\right)$$ $$= \dfrac{\sqrt{3}}{3}$$ $$\therefore \tan\left(\dfrac{7\pi}{6}\right) = \dfrac{\sqrt{3}}{3}$$

ii. 1. Express the trigonometric value in an appropriate quadrant form.

ii. $\frac{11\pi}{3}$ is in quadrant 4.

2. Apply the symmetry property and evaluate.

$$\sin\left(\frac{11\pi}{3}\right) = \sin\left(4\pi - \frac{\pi}{3}\right)$$

$$= \sin\left(2\pi - \frac{\pi}{3}\right)$$

$$= -\sin\left(\frac{\pi}{3}\right)$$

$$= -\frac{\sqrt{3}}{2}$$

$$\therefore \sin\left(\frac{11\pi}{3}\right) = -\frac{\sqrt{3}}{2}$$

 Resources

 Interactivity Symmetry points and quadrants (int-2584)

Exercise 6.5 Exact values and symmetry properties

 learn

6.5 Exercise	**6.5 Exam practice** on

Simple familiar	Complex familiar	Complex unfamiliar
1, 2, 3, 4, 5, 6, 7, 8, 9, 10, 11, 12, 13, 14, 15, 16, 17	18, 19, 20	N/A

These questions are even better in jacPLUS!
- Receive immediate feedback
- Access sample responses
- Track results and progress

Find all this and MORE in jacPLUS ▶

Simple familiar

1. State the value of each of the following. (It may help to show the boundary points on a diagram.)

   a. $\cos(4\pi)$
   b. $\tan(9\pi)$
   c. $\sin(7\pi)$
   d. $\sin\left(\frac{13\pi}{2}\right)$
   e. $\cos\left(-\frac{9\pi}{2}\right)$
   f. $\tan(-20\pi)$

2. Identify the quadrant(s) or boundaries for which the following apply.

   a. $\cos(\theta) > 0, \sin(\theta) < 0$
   b. $\tan(\theta) > 0, \cos(\theta) > 0$
   c. $\sin(\theta) > 0, \cos(\theta) < 0$
   d. $\cos(\theta) = 0$
   e. $\cos(\theta) = 0, \sin(\theta) > 0$
   f. $\sin(\theta) = 0, \cos(\theta) < 0$

---

3. Determine positions for the points in quadrants 2, 3 and 4 that are symmetric to the trigonometric point [θ] for which the value of θ is:

    a. $\dfrac{\pi}{3}$ b. $\dfrac{\pi}{6}$ c. $\dfrac{\pi}{4}$

    d. $\dfrac{\pi}{5}$ e. $\dfrac{3\pi}{8}$ f. 1.

4. Calculate the exact values of the following.

    a. $\cos(120°)$ b. $\tan(225°)$ c. $\sin(330°)$

    d. $\tan(-60°)$ e. $\cos(-315°)$ f. $\sin(510°)$

5. Calculate the exact values of the following.

    a. $\sin\left(\dfrac{3\pi}{4}\right)$ b. $\tan\left(\dfrac{2\pi}{3}\right)$ c. $\cos\left(\dfrac{5\pi}{6}\right)$

    d. $\cos\left(\dfrac{4\pi}{3}\right)$ e. $\tan\left(\dfrac{7\pi}{6}\right)$ f. $\sin\left(\dfrac{11\pi}{6}\right)$

6. Calculate the exact values of the following.

    a. $\cos\left(-\dfrac{\pi}{4}\right)$ b. $\sin\left(-\dfrac{\pi}{3}\right)$ c. $\tan\left(-\dfrac{5\pi}{6}\right)$

    d. $\sin\left(\dfrac{8\pi}{3}\right)$ e. $\cos\left(\dfrac{9\pi}{4}\right)$ f. $\tan\left(\dfrac{23\pi}{6}\right)$

7. **WE11** a. Identify the quadrant(s) where $\cos(\theta)$ is negative and $\tan(\theta)$ is positive.

    b. If $f(\theta) = \tan(\theta)$, evaluate $f(4\pi)$.

8. If $f(t) = \sin(\pi t)$, evaluate $f(2.5)$.

9. **WE12** Calculate the exact values of the following.

    a. $\sin\left(\dfrac{4\pi}{3}\right)$ b. $\tan\left(\dfrac{5\pi}{6}\right)$ c. $\cos(-30°)$

10. Calculate the exact values of $\sin\left(-\dfrac{5\pi}{4}\right)$, $\cos\left(-\dfrac{5\pi}{4}\right)$ and $\tan\left(-\dfrac{5\pi}{4}\right)$.

11. If $\cos(\theta) = 0.2$, use the symmetry properties to identify the values of the following.

    a. $\cos(\pi - \theta)$ b. $\cos(\pi + \theta)$

    c. $\cos(-\theta)$ d. $\cos(2\pi + \theta)$

12. If $\sin(t) = 0.9$ and $\tan(x) = 4$, calculate the value of the following.

    a. $\tan(-x)$ b. $\sin(\pi - t)$

    c. $\tan(2\pi - t)$ d. $\sin(-t) + \tan(\pi + x)$

13. Given $\cos(\theta) = 0.91$, $\sin(t) = 0.43$ and $\tan(x) = 0.47$, use the symmetry properties to identify the values of the following.

    a. $\cos(\pi + \theta)$ b. $\sin(\pi - t)$

    c. $\tan(2\pi - x)$ d. $\cos(-\theta)$

    e. $\sin(-t)$ f. $\tan(2\pi + x)$

14. If $\sin(\theta) = p$, express the following in terms of p.

    a. $\sin(2\pi - \theta)$ b. $\sin(3\pi - \theta)$

    c. $\sin(-\pi + \theta)$ d. $\sin(\theta + 4\pi)$

15. Calculate the exact values of the following.

a. $\cos\left(\dfrac{7\pi}{6}\right) + \cos\left(\dfrac{2\pi}{3}\right)$

b. $2\sin\left(\dfrac{7\pi}{4}\right) + 4\sin\left(\dfrac{5\pi}{6}\right)$

c. $\sqrt{3}\tan\left(\dfrac{5\pi}{4}\right) - \tan\left(\dfrac{5\pi}{3}\right)$

d. $\sin\left(\dfrac{8\pi}{9}\right) + \sin\left(\dfrac{10\pi}{9}\right)$

e. $2\cos^2\left(-\dfrac{5\pi}{4}\right) - 1$

f. $\dfrac{\tan\left(\frac{17\pi}{4}\right)\cos(-7\pi)}{\sin\left(-\frac{11\pi}{6}\right)}$

16. **WE13** a. Identify the symmetric points to [75°]. Of these points, identify the point at which the cosine value is the same as $\cos(75°)$.

b. Express $\tan\left(\dfrac{6\pi}{7}\right)$ in terms of a first quadrant value.

c. If $\sin(\theta) = 0.8$, give the values of $\sin(\pi - \theta)$ and $\sin(2\pi - \theta)$.

d. Calculate the exact value of the following.

i. $\cos\left(\dfrac{5\pi}{4}\right)$

ii. $\sin\left(\dfrac{25\pi}{6}\right)$

17. Given $\cos(\theta) = p$, express the following in terms of p.

a. $\cos(-\theta)$

b. $\cos(5\pi + \theta)$

Complex familiar

18. a. Verify that $\sin^2\left(\dfrac{5\pi}{4}\right) + \cos^2\left(\dfrac{5\pi}{4}\right) = 1$.

b. Explain, with the aid of a unit circle diagram, why $\cos(-\theta) = \cos(\theta)$ is true for $\theta = \dfrac{5\pi}{6}$.

c. The point [ϕ] lies in the second quadrant and has Cartesian coordinates $(-0.5, 0.87)$. Show this on a diagram and give the values of $\sin(\pi + \phi)$, $\cos(\pi + \phi)$ and $\tan(\pi + \phi)$.

d. Simplify $\sin(-\pi + t) + \sin(-3\pi - t) + \sin(t + 6\pi)$.

e. Use the unit circle to give two values of an angle A for which $\sin(A) = \sin(144°)$.

f. With the aid of the unit circle, give three values of B for which $\sin(B) = -\sin\left(\dfrac{2\pi}{11}\right)$.

19. a. Identify the quadrant in which the point P [4.2] lies.
    b. Calculate the Cartesian coordinates of point P [4.2] to 2 decimal places.
    c. Identify the trigonometric positions, to 4 decimal places, of the points in the other three quadrants that are symmetric to the point P [4.2].

20. Consider the point Q [θ], $\tan(\theta) = 5$.

    a. Identify the two quadrants in which Q could lie.
    b. Determine, to 4 decimal places, the value of θ for each of the two points.
    c. Calculate the Cartesian coordinates of the two points

Fully worked solutions for this chapter are available online.

LESSON
6.6 Graphs of the sine, cosine and tangent functions

As the two functions sine and cosine are closely related, we shall initially consider their graphs together.

6.6.1 The graphs of $y = \sin(x)$ and $y = \cos(x)$

The graphs of $y = \sin(x)$ and $y = \cos(x)$ can be plotted using the boundary values from continued rotations, clockwise and anticlockwise, around the unit circle.

x	0	$\dfrac{\pi}{2}$	π	$\dfrac{3\pi}{2}$	2π
$\sin(x)$	0	1	0	-1	0
$\cos(x)$	1	0	-1	0	1

The diagram shows four cycles of the graphs drawn on the domain $[-4\pi, 4\pi]$. The graphs continue to repeat their wavelike pattern over their maximal domain R; the interval, or **period**, of each repetition is 2π.

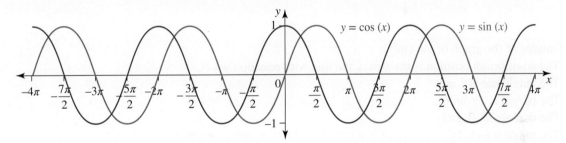

The first observation that strikes us about these graphs is how remarkably similar they are: a horizontal translation of $\dfrac{\pi}{2}$ to the right will transform the graph of $y = \cos(x)$ into the graph of $y = \sin(x)$, and a horizontal translation of $\dfrac{\pi}{2}$ to the left transforms the graph of $y = \sin(x)$ into the graph of $y = \cos(x)$.

Recalling our knowledge of transformations of graphs, this observation can be expressed as follows.

$$\cos\left(x - \frac{\pi}{2}\right) = \sin(x)$$

$$\sin\left(x + \frac{\pi}{2}\right) = \cos(x)$$

The two functions are said to be '**out of phase**' by $\dfrac{\pi}{2}$ or to have a **phase difference** or **phase shift** of $\dfrac{\pi}{2}$.

Both graphs oscillate up and down one unit from the x-axis. The x-axis is the **equilibrium** or **mean position**, and the distance the graphs oscillate up and down from this mean position to a maximum or minimum point is called the **amplitude**.

The graphs keep repeating this cycle of oscillations up and down from the equilibrium position, with the amplitude measuring half the vertical distance between maximum and minimum points and the period measuring the horizontal distance between successive maximum points or between successive minimum points.

6.6.2 One cycle of the graph of $y = \sin(x)$

The basic graph of $y = \sin(x)$ has the domain $[0, 2\pi]$, which restricts the graph to one cycle.

The graph of the function $f: [0, 2\pi] \to R, f(x) = \sin(x)$ is shown.

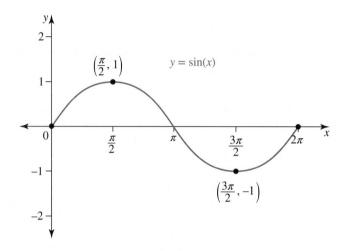

Key features of the graph of $y = \sin(x)$:
- The equilibrium position is the x-axis, the line with equation $y = 0$.
- The amplitude is 1 unit.
- The period is 2π units.
- The domain is $[0, 2\pi]$.
- The range is $[-1, 1]$.
- The x-intercepts occur at $x = 0, \pi, 2\pi$.
- The type of correspondence is many-to-one.

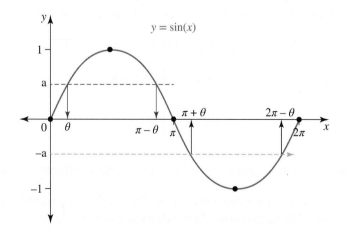

The graph lies above the x-axis for $x \in (0, \pi)$ and below for $x \in (\pi, 2\pi)$, which matches the quadrant signs of sine given in the ASTC diagram. The symmetry properties of sine are displayed in its graph as $\sin(\pi - \theta) = \sin(\theta)$ and $\sin(\pi + \theta) = \sin(2\pi - \theta) = -\sin(\theta)$.

6.6.3 One cycle of the graph of $y = \cos(x)$

The basic graph of $y = \cos(x)$ has the domain $[0, 2\pi]$, which restricts the graph to one cycle.

The graph of the function $f:\ [0, 2\pi] \to R, f(x) = \cos(x)$ is shown.

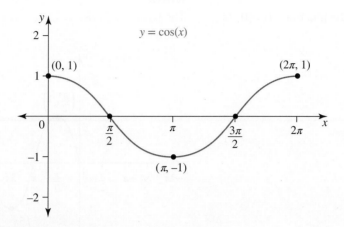

Key features of the graph of $y = \cos \pi(x)$:
- The equilibrium position is the x-axis, the line with equation $y = 0$.
- The amplitude is 1 unit.
- The period is 2π units.
- The domain is $[0, 2\pi]$.
- The range is $[-1, 1]$.
- The x-intercepts occur at $x = \dfrac{\pi}{2}, \dfrac{3\pi}{2}$.
- The type of correspondence is many-to-one.

The graph of $y = \cos(x)$ has the same amplitude, period, equilibrium (or mean) position, domain, range and type of correspondence as the graph of $y = \sin(x)$.

6.6.4 Guide to sketching the graphs on extended domains

There is a pattern of 5 points to the shape of the basic sine and cosine graphs created by the division of the period into four equal intervals.

For $y = \sin(x)$: the first point starts at the equilibrium; the second point, at $\dfrac{1}{4}$ of the period, reaches up one amplitude to the maximum point; the third point, at $\dfrac{1}{2}$ of the period, is back at equilibrium; the fourth point, at $\dfrac{3}{4}$ of the period, goes down one amplitude to the minimum point; the fifth point, at the end of the period interval, returns back to equilibrium.

In other words:

equilibrium $\to$ range maximum $\to$ equilibrium $\to$ range minimum $\to$ equilibrium.

For $y = \cos(x)$: the pattern for one cycle is summarised as:

range maximum $\to$ equilibrium $\to$ range minimum $\to$ equilibrium $\to$ range maximum.

This pattern only needs to be continued in order to sketch the graph of $y = \sin(x)$ or $y = \cos(x)$ on a domain other than $[0, 2\pi]$.

Sketch the graph of $y = \sin(x)$ over the domain $[-2\pi, 4\pi]$ and state the number of cycles of the sine function drawn.

THINK	WRITE
1. Draw the graph of the function over $[0, 2\pi]$.	The basic graph of $y = \sin(x)$ over the domain $[0, 2\pi]$ is drawn. 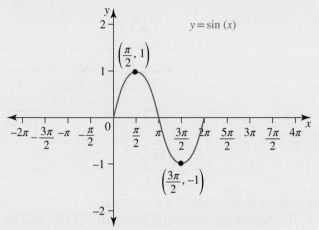
2. Extend the pattern to cover the domain specified.	The pattern is extended for one cycle in the negative direction and one further cycle in the positive direction to cover the domain $[-2\pi, 4\pi]$. 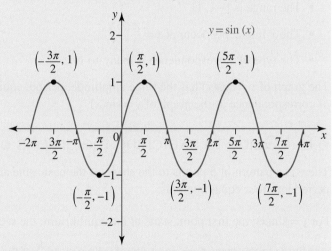
3. State the number of cycles of the function that are shown in the graph.	Altogether, 3 cycles of the sine function have been drawn.

TI	THINK	WRITE	CASIO	THINK	WRITE

TI | THINK

1. On a Graphs page, set the Graphing Angle to Radian.
   Press MENU, then select:
   4: Window/Zoom
   1: Window Settings …
   Complete the fields as:
   XMin: -3π
   XMax: 5π
   XScale: $\pi/2$
   YMin: -2
   YMax: 2
   YScale: 1
   then select OK.
   Note: The calculator will only give decimal approximations for intercepts, minimums and maximums, so it is important to have the x-axis scale as a multiple of π so that important points can be easily read from the graph.

WRITE

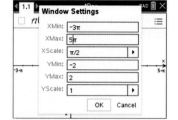

2. Complete the entry line for function 1 as:
   $f1(x) = \sin(x)| -2\pi \le x \le 4\pi$
   then press ENTER.

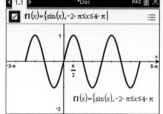

3. Identify the coordinates of the x-intercepts from the graph.

The x-axis scale is $\frac{\pi}{2}$, so it can be seen that the graph has x-intercepts at $(-2\pi, 0), (-\pi, 0), (0, 0), (\pi, 0), (2\pi, 0), (3\pi, 0)$ and $(4\pi, 0)$.
Be sure to mark these points when sketching the graph.

4. Identify the coordinates of the maximums from the graph.

The x-axis scale is $\frac{\pi}{2}$, so it can be seen that the graph has maximums at $\left(-\frac{3\pi}{2}, 1\right), \left(\frac{\pi}{2}, 1\right)$ and $\left(\frac{5\pi}{2}, 1\right)$.
Be sure to mark these points when sketching the graph.

CASIO | THINK

1. Put the Calculator in Radian mode.
   On a Graph screen, press SHIFT then F3 to open the V_WIN.
   Complete the fields as:
   Xmin: -3π
   max: 5π
   scale: $\pi/2$
   Ymin: -2
   max: 2
   scale: 1
   then press EXIT.
   Note: The calculator will only give decimal approximations for intercepts, minimums and maximums, so it is important to have the x-axis scale as a multiple of π so that important points can be easily read from the graph.

WRITE

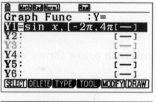

2. Complete the entry line for y1 as:
   $y1 = \sin x, [-2\pi, 4\pi]$
   then press EXE.
   Select DRAW by pressing F6.
   Select DRAW.

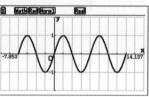

3. Identify the coordinates of the x-intercepts from the graph.

The x-axis scale is $\frac{\pi}{2}$, so it can be seen that the graph has x-intercepts at $(-2\pi, 0), (-\pi, 0), (0, 0), (\pi, 0), (2\pi, 0), (3\pi, 0)$ and $(4\pi, 0)$.
Be sure to mark these points when sketching the graph.

4. Identify the coordinates of the maximums from the graph.

The x-axis scale is $\frac{\pi}{2}$, so it can be seen that the graph has maximums at $\left(-\frac{3\pi}{2}, 1\right), \left(\frac{\pi}{2}, 1\right)$ and $\left(\frac{5\pi}{2}, 1\right)$.
Be sure to mark these points when sketching the graph.

| 5. Identify the coordinates of the minimums from the graph. | The x-axis scale is $\dfrac{\pi}{2}$, so it can be seen that the graph has minimums at $\left(-\dfrac{\pi}{2}, -1\right)$, $\left(\dfrac{3\pi}{2}, -1\right)$ and $\left(\dfrac{7\pi}{2}, -1\right)$. Be sure to mark these points when sketching the graph. | 5. Identify the coordinates of the minimums from the graph. | The x-axis scale is $\dfrac{\pi}{2}$, so it can be seen that the graph has minimums at $\left(-\dfrac{\pi}{2}, -1\right)$, $\left(\dfrac{3\pi}{2}, -1\right)$ and $\left(\dfrac{7\pi}{2}, -1\right)$. Be sure to mark these points when sketching the graph. |
| 6. Count the number of cycles of the sine function drawn. | 3 cycles are drawn. | 6. Count the number of cycles of the sine function drawn. | 3 cycles are drawn. |

6.6.5 The graph of $y = \tan(x)$

The graph of $y = \tan(x)$ has a distinct shape that is quite different to those of the sine and cosine graphs. As values such as $\tan\left(\dfrac{\pi}{2}\right)$ and $\tan\left(\dfrac{3\pi}{2}\right)$ are undefined, a key feature of the graph of $y = \tan(x)$ is the presence of vertical asymptotes at odd multiples of $\dfrac{\pi}{2}$.

Features of tan(x)

The relationship $\tan(x) = \dfrac{\sin(x)}{\cos(x)}$ shows that:
- **$\tan(x)$ will be undefined whenever $\cos(x) = 0$**
- **$\tan(x) = 0$ whenever $\sin(x) = 0$.**

The graph below shows $y = \tan(x)$ over the domain $[0, 2\pi]$.

The key features of the graph of $y = \tan(x)$ are as follows:
- There are vertical asymptotes at $x = \dfrac{\pi}{2}, x = \dfrac{3\pi}{2}$ for the domain $[0, 2\pi]$. For extended domains this pattern continues with asymptotes at $x = (2n + 1)\dfrac{\pi}{2}, n \in Z$.

- The period is π. Two cycles are completed over the domain $[0, 2\pi]$.
- The range is R.
- x-intercepts occur at $x = 0, \pi, 2\pi$ for the domain $[0, 2\pi]$. For extended domains this pattern continues with x-intercepts at $x = n\pi, n \in Z$.
- The mean position is $y = 0$.
- The asymptotes are one period apart.
- The x-intercepts are one period apart.

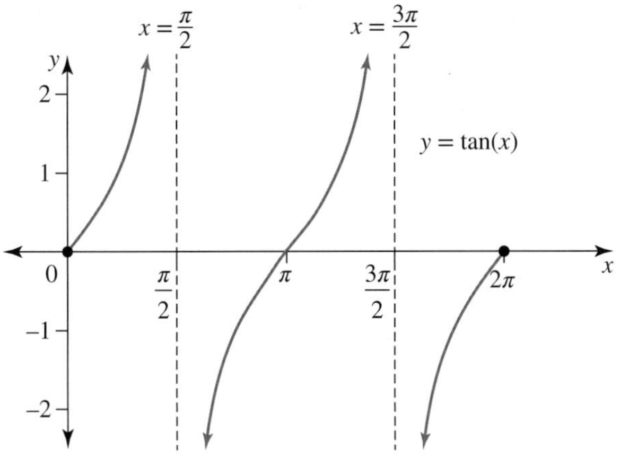

Unlike the sine and cosine graphs, 'amplitude' has no meaning for the tangent graph. As for any graph, the x-intercepts of the tangent graph are the solutions to the equation formed when $y = 0$.

Sketch the graph of $y = \tan(x)$ for $x \in \left(-\dfrac{3\pi}{2}, \dfrac{\pi}{2} \right)$.

THINK

1. State the equations of the asymptotes.

WRITE

Period: the period of $y = \tan(x)$ is π.

Asymptotes: the graph has an asymptote at $x = \dfrac{\pi}{2}$.

As the asymptotes are a period apart, for the

domain $\left(-\dfrac{3\pi}{2}, \dfrac{\pi}{2} \right)$ there is an asymptote at $x = \dfrac{\pi}{2} -$

$\pi \Rightarrow x = -\dfrac{\pi}{2}$ and another at $x = -\dfrac{\pi}{2} - \pi = -\dfrac{3\pi}{2}$.

2. State where the graph cuts the x-axis.
   Note: The x-intercepts can be found by
   solving the equation
   $\tan(x) = 0$, $x \in \left(-\dfrac{3\pi}{2}, \dfrac{\pi}{2} \right)$.

The x-intercepts occur midway between the
asymptotes. $(-\pi, 0)$ and $(0, 0)$ are the x-intercepts.

3. Sketch the graph.

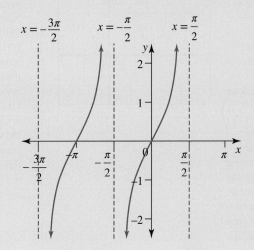

TI| THINK	WRITE	CASIO| THINK	WRITE

1. On a Graphs page, set the Graphing Angle to Radian. Press MENU, then select
4: Window/Zoom
1: Window
Settings …
Complete the fields as:
XMin: -2π
XMax: π
XScale: $\pi/2$
YMin: -5
YMax: 5
YScale: 1
then select OK.

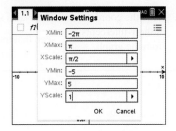

1. Put the Calculator in Radian mode.
   On a Graph screen, press SHIFT then F3 to open the V_WIN. Complete the fields as:
   Xmin: -2π
   max: π
   scale: $\pi/2$
   Ymin: -5
   max: 5
   scale: 1
   then press EXE.

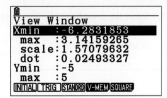

2. Complete the entry line for function 1 as: $f1(x) = \tan(x)|-\dfrac{3\pi}{2} \le x \le \dfrac{\pi}{2}$ then press ENTER.

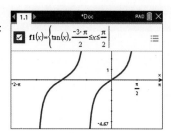

2. Complete the entry line for y1 as: $y1 = \tan x$, $[-\dfrac{3\pi}{2}, \dfrac{\pi}{2}]$ then press EXE. Select DRAW by pressing F6.

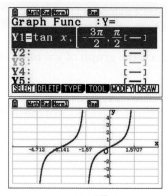

3. Identify the positions of the vertical asymptotes.

The x-axis scale is $\dfrac{\pi}{2}$, so it can be seen that the graph has vertical asymptotes at $x = -\dfrac{3\pi}{2}$, $x = -\dfrac{\pi}{2}$ and $x = \dfrac{\pi}{2}$.

3. Identify the positions of the vertical asymptotes.

The x-axis scale is $\dfrac{\pi}{2}$, so it can be seen that the graph has vertical asymptotes at $x = -\dfrac{3\pi}{2}$, $x = -\dfrac{\pi}{2}$ and $x = \dfrac{\pi}{2}$.

4. To draw the vertical asymptotes, press MENU, then select:
8: Geometry
4: Construction
1: Perpendicular
Click on the
x-axis, then click on the point $\left(-\dfrac{3\pi}{2}, 0\right)$.
Repeat this step to draw the other vertical asymptotes.
Note: To change the style of the vertical lines, place the cursor on the line and press CTRL then MENU, then select:
3: Attributes.

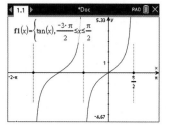

4. To draw the vertical asymptotes, select Sketch by pressing SHIFT then F4. Press F6 to scroll across to more menu options, then select Vertical by pressing F4.
Use the left/right arrows to position the vertical line, then press EXE. Repeat this step to draw the other vertical asymptotes.

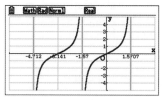

5. Read the coordinates of the x-intercepts from the graph.

The x-axis scale is $\dfrac{\pi}{2}$, so it can be seen that the graph has x-intercepts at $(-\pi, 0)$ and $(0, 0)$.

5. Read the coordinates of the x-intercepts from the graph.

The x-axis scale is $\dfrac{\pi}{2}$, so it can be seen that the graph has x-intercepts at $(-\pi, 0)$ and $(0, 0)$.

 Resources

Interactivity Graph plotter: Tangent (int-2978)

6.6 Exercise	6.6 Exam practice **on**

Simple familiar	Complex familiar	Complex unfamiliar
1, 2, 3, 4, 5, 6, 7, 8, 9, 10	11, 12, 13, 14, 15, 16	17, 18, 19, 20

These questions are even better in jacPLUS!
- Receive immediate feedback
- Access sample responses
- Track results and progress

Find all this and MORE in jacPLUS ▶

Simple familiar

1. Consider the graph of the function $y = f(x)$ shown.

   a. State the domain and range of the graph.
   b. Select the appropriate equation for the function shown
   c. Write down the coordinates of each of the four turning points.
   d. Identify the period and the amplitude of the graph.
   e. Give the equation of the mean (or equilibrium) position.
   f. Identify the values of x for which $f(x) > 0$.

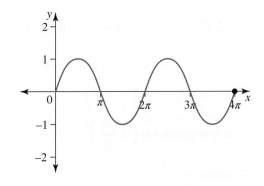

2. Consider the graph of the function $y = g(x)$ shown.

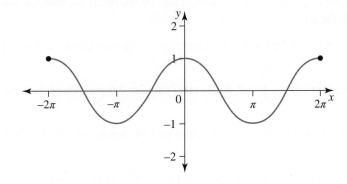

   a. State the domain and range of the graph.
   b. Select the appropriate equation for the function shown
   c. Write down the coordinates of each of the minimum turning points.
   d. Identify the period and the amplitude of the graph and state the equation of its mean (or equilibrium) position.
   e. Write down the coordinates of the x-intercepts of the graph.
   f. Identify the values of x for which $g(x) < 0$.

3. **WE 14** Sketch the graph of $y = \cos(x)$ over the domain $[-2\pi, 4\pi]$ and state the number of cycles of the cosine function drawn.

4. Sketch the graphs of $y = \sin(x)$ and $y = \cos(x)$ over the given domain intervals.

   a. $y = \sin(x), 0 \le x \le 6\pi$
   b. $y = \cos(x), -4\pi \le x \le 2\pi$
   c. $y = \cos(x), -\dfrac{\pi}{2} \le x \le \dfrac{3\pi}{2}$
   d. $y = \sin(x), -\dfrac{3\pi}{2} \le x \le \dfrac{\pi}{2}$

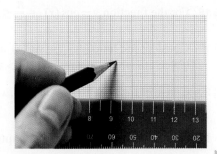

5. **a.** State the number of maximum turning points on the graph of the function $f : [-4\pi, 0] \to R, f(x) = \sin(x)$.
   b. State the number of minimum turning points of the graph of the function $f : [0, 14\pi] \to R, f(x) = \cos(x)$.

6. Sketch the graph of $y = \cos(x), -4\pi \leq x \leq 5\pi$ and state the number of cycles of the cosine function drawn.

7. State the number of intersections that the graphs of the following make with the x-axis.
   a. $y = \cos(x), 0 \leq x \leq \dfrac{7\pi}{2}$
   b. $y = \sin(x), -2\pi \leq x \leq 4\pi$
   c. $y = \sin(x), 0 \leq x \leq 20\pi$
   d. $y = \cos(x), \pi \leq x \leq 4\pi$

8. On the same set of axes, sketch the graphs of $y = \cos(x)$ and $y = \sin(x)$ over the domain $[0, 2\pi]$ and shade the region $\{(x, y) : \sin(x) \geq \cos(x), x \in [0, 2\pi]\}$.

9. **WE15** Sketch the graph of $y = \tan(x)$ for $x \in \left(-\dfrac{\pi}{2}, \dfrac{3\pi}{2}\right)$.

10. Sketch the following over the given interval.
    a. $y = \tan(x), x \in \left(\dfrac{\pi}{2}, \dfrac{3\pi}{2}\right)$
    b. $y = \tan(x), x \in (-\pi, 0)$
    c. $y = \tan(x), x \in \left(0, \dfrac{5\pi}{2}\right)$

Complex familiar

11. The graph of the function $f : [0, a] \to R, f(x) = \cos(x)$ has 10 intersections with the x-axis. Determine the smallest value possible for a.

12. The graph of the function $f : [b, 5\pi] \to R, f(x) = \sin(x)$ has 7 turning points. If $f(b) = 0$, determine the value of b.

13. If the graph of the function $f : [-c, c] \to R, f(x) = \sin(x)$ covers 2.5 periods of the sine function, determine the value of c.

14. Sketch one cycle of the cosine graph over $[0, 2\pi]$ and give the values of x in this interval for which $\cos(x) < 0$.

15. Explain how the graph in question **14** illustrates what the ASTC diagram says about the sign of the cosine function.

16. Consider the function defined by $y = \tan(x), -\pi \leq x \leq \pi$.
    a. Sketch the graph over the given interval.
    b. Obtain the x-coordinates of the points on the graph for which the y-coordinate is $-\sqrt{3}$.

Complex unfamiliar

17. For the function $y = \sin(x), -\dfrac{\pi}{2} \leq x \leq \dfrac{\pi}{2}$, determine where $\sin(x) \geq \dfrac{1}{2}$

18. For the function $y = \cos(x), -\dfrac{\pi}{2} \leq x \leq \dfrac{\pi}{2}$, determine where $2\cos(x) \leq 1$.

19. Consider the functions $y = \sin(x)$ and $y = \cos(x)$, for $0 \leq x \leq \pi$.
    Calculate where $\sin(x) - \cos(x) > 0$.

20. Consider the function $y = \tan(x), x \in [0, \pi]$.
    Calculate where $\tan(x) + 1 \leq 0$.

Fully worked solutions for this chapter are available online.

LESSON
6.7 Solving trigonometric equations

SYLLABUS LINKS

- Solve trigonometric equations, with and without technology, including the use of the Pythagorean identity $\sin^2(A) + \cos^2(A) = 1$.

Source: Mathematical Methods Senior Syllabus 2024 © State of Queensland (QCAA) 2024; licensed under CC BY 4.0.

The symmetry properties of trigonometric functions can be used to obtain solutions to equations of the form $f(x) = a$ where f is sine, cosine or tangent. If f is sine or cosine, then $-1 \leq a \leq 1$, and if f is tangent, then $a \in R$.

Once the appropriate base value of the first quadrant is known, symmetric points in any other quadrant can be obtained. However, there are many values, generated by both positive and negative rotations, that can form these symmetric quadrant points. Consequently, the solution of a trigonometric equation such as $\sin(x) = a, x \in R$ would have infinite solutions. We shall consider trigonometric equations in which a subset of R is specified as the domain in order to have a finite number of solutions.

6.7.1 Solving trigonometric equations on finite domains

To solve the basic type of equation $\sin(x) = a, 0 \leq x \leq 2\pi$, use the following steps:
- Identify the quadrants in which solutions lie from the sign of a.
  - If $a > 0$, x must lie in quadrants 1 and 2 where sine is positive.
  - If $a < 0$, x must be in quadrants 3 and 4 where sine is negative.
- Obtain the base value, or first-quadrant value, by solving $\sin(x) = a$ if $a > 0$ or ignoring the negative sign if $a < 0$ (to ensure the first-quadrant value is obtained).
  - This may require recognition of an exact value ratio, or it may require the use of a calculator.
- Once obtained, use the base value to generate the values for the quadrants required from their symmetric forms.

The basic equations $\cos(x) = a$ or $\tan(x) = a, 0 \leq x \leq 2\pi$ are solved in a similar manner, with the sign of a determining the quadrants in which solutions lie.
- For $\cos(x) = a$:
  - if $a > 0$, x must lie in quadrants 1 and 4 where cosine is positive
  - if $a < 0$, x must lie in quadrants 2 and 3 where cosine is negative.
- For $\tan(x) = a$:
  - if $a > 0$, x must lie in quadrants 1 and 3 where tangent is positive
  - if $a < 0$, x must lie in quadrants 2 and 4 where tangent is negative.

6.7.2 Symmetric forms

For one positive and one negative rotation, the symmetric points to the first-quadrant base are shown in the diagrams.

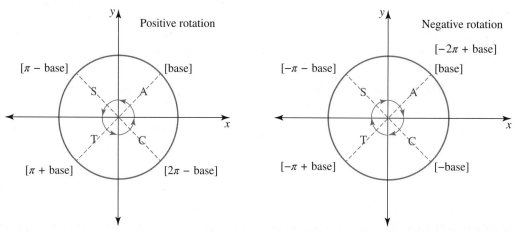

WORKED EXAMPLE 16 Solving equations with exact values

Solve the following equations to obtain exact values for x.

a. $\sin(x) = \dfrac{\sqrt{3}}{2}, 0 \leq x \leq 2\pi$

b. $\sqrt{2}\,\cos(x) + 1 = 0, 0 \leq x \leq 2\pi$

c. $\sqrt{3} - 3\tan(x) = 0, -2\pi \leq x \leq 2\pi$

THINK	WRITE
a. 1. Identify the quadrants in which the solutions lie.	a. $\sin(x) = \dfrac{\sqrt{3}}{2}, 0 \leq x \leq 2\pi$ Sine is positive in quadrants 1 and 2. 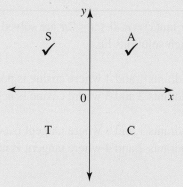
2. Use knowledge of exact values to state the first-quadrant base.	The base is $\dfrac{\pi}{3}$ since $\sin\left(\dfrac{\pi}{3}\right) = \dfrac{\sqrt{3}}{2}$.
3. Generate the solutions using the appropriate quadrant forms.	Since $x \in [0, 2\pi]$, there will be two positive solutions, one from quadrant 1 and one from quadrant 2. $\therefore x = \dfrac{\pi}{3}$ or $x = \pi - \dfrac{\pi}{3}$
4. Calculate the solutions from their quadrant forms.	$\therefore x = \dfrac{\pi}{3}$ or $\dfrac{2\pi}{3}$

b. 1. Rearrange the equation so the trigonometric function is isolated on one side.

b. $\sqrt{2}\cos(x) + 1 = 0, 0 \le x \le 2\pi$

$$\sqrt{2}\cos(x) = -1$$

$$\cos(x) = -\frac{1}{\sqrt{2}}$$

2. Identify the quadrants in which the solutions lie.

Cosine is negative in quadrants 2 and 3.

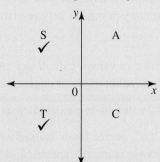

3. Identify the base.
Note: The negative sign is ignored in identifying the base since the base is the first-quadrant value.

Since $\cos\left(\dfrac{\pi}{4}\right) = \dfrac{1}{\sqrt{2}}$, the base is $\dfrac{\pi}{4}$.

4. Generate the solutions using the appropriate quadrant forms.

Since $x \in [0, 2\pi]$, there will be two positive solutions.
$$\therefore x = \pi - \frac{\pi}{4} \text{ or } x = \pi + \frac{\pi}{4}$$

5. Calculate the solutions from their quadrant forms.

$$\therefore x = \frac{3\pi}{4} \text{ or } \frac{5\pi}{4}$$

c. 1. Rearrange the equation so the trigonometric function is isolated on one side.

c. $\sqrt{3} - 3\tan(x) = 0, -2\pi \le x \le 2\pi$
$$\therefore \tan(x) = \frac{\sqrt{3}}{3}$$

2. Identify the quadrants in which the solutions lie.

Tangent is positive in quadrants 1 and 3.

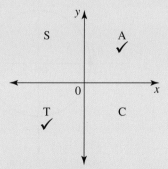

3. Identify the base.

The base is $\dfrac{\pi}{6}$ since $\tan\left(\dfrac{\pi}{6}\right) = \dfrac{\sqrt{3}}{3}$.

4. Generate the solutions using the appropriate quadrant forms.

Since $-2\pi \le x \le 2\pi$, there will be 4 solutions, two from a positive rotation and two from a negative rotation.
$$x = \frac{\pi}{6}, \pi + \frac{\pi}{6} \text{ or } x = -\pi + \frac{\pi}{6}, = -2\pi + \frac{\pi}{6}$$

5. Calculate the solutions from their quadrant forms.

$$\therefore x = \frac{\pi}{6}, \frac{7\pi}{6}, -\frac{5\pi}{6}, -\frac{11\pi}{6}$$

6.7.3 Trigonometric equations with boundary value solutions

Recognition of exact trigonometric values allows us to identify the base for solving trigonometric equations to obtain exact solutions. However, there are also exact trigonometric values for boundary points. These need to be recognised should they appear in an equation. The simplest strategy to solve trigonometric equations involving boundary values is to use a unit circle diagram to generate the solutions. The domain for the equation determines the number of rotations required around the unit circle. It is not appropriate to consider quadrant forms to generate solutions, since boundary points lie between two quadrants.

Using technology

When bases are not recognisable from exact values, calculators are needed to identify the base. Whether the calculator, or other technology, is set on radian mode or degree mode is determined by the given equation.

For example, if $\sin(x) = -0.7, 0 \leq x \leq 2\pi$, the base is calculated as $\sin^{-1}(0.7)$ in radian mode.

However, for $\sin(x) = -0.7, 0° \leq x \leq 360°$, degree mode is used when calculating the base as $\sin^{-1}(0.7)$. The degree of accuracy required for the answer is usually specified in the question; if not, express answers rounded to 2 decimal places.

WORKED EXAMPLE 17 Solving equations with and without technology

a. **Solve for x: $3\cos(x) + 3 = 0, -4\pi \leq x \leq 4\pi$.**
b. **Solve for x to 2 decimal places: $\sin(x) = -0.75, 0 \leq x \leq 4\pi$.**
c. **Solve for x to 1 decimal place: $\tan(x°) + 12.5 = 0, -180° \leq x° \leq 180°$.**

THINK	WRITE
a. 1. Express the equation with the trigonometric function as subject.	a. $3\cos(x) + 3 = 0, -4\pi \leq x \leq 4\pi$ $\therefore \cos(x) = -1$
2. Identify any boundary points.	-1 is a boundary value since $\cos(\pi) = -1$. The boundary point $[\pi]$ has Cartesian coordinates $(-1, 0)$.
3. Use a unit circle to generate the solutions.	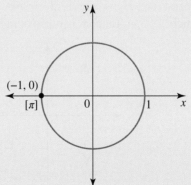

As $-4\pi \leq x \leq 4\pi$, this means 2 anticlockwise revolutions and 2 clockwise revolutions around the circle are required, with each revolution generating one solution.
The solutions are:
$$x = \pi, 3\pi \text{ and } = -\pi, -3\pi$$
$$\therefore x = \pm\pi, \pm 3\pi$$

| b. 1. Identify the quadrants in which the solutions lie. | b. $\sin(x) = -0.75, 0 \leq x \leq 4\pi$
 Sine is negative in quadrants 3 and 4. |
| 2. Calculate the base. | The base is $\sin^{-1}(0.75)$. Using radian mode,
 $\sin^{-1}(0.75) = 0.848$ to 3 decimal places. |

3. Generate the solutions using the appropriate quadrant forms.

Since $x \in [0, 4\pi]$, there will be four positive solutions from two anticlockwise rotations.
$x = \pi + 0.848, 2\pi - 0.848$ or $x = 3\pi + 0.848, 4\pi - 0.848$

4. Calculate the solutions to the required accuracy.
Note: If the base is left as $\sin^{-1}(0.75)$, then the solutions such as $x = \pi + \sin^{-1}(0.75)$ could be calculated on radian mode in one step.

$\therefore x = 3.99, 5.44, 10.27, 11.72$ (correct to 2 decimal places)

c. 1. Identify the quadrants in which the solutions lie.

c. $\tan(x°) + 12.5 = 0, -180° \leq x° \leq 180°$
$\tan(x°) = -12.5$
Tangent is negative in quadrants 2 and 4.

2. Calculate the base.

The base is $\tan^{-1}(12.5)$. Using degree mode,
$\tan^{-1}(12.5) = 85.43°$ to 2 decimal places.

3. Generate the solutions using the appropriate quadrant forms.

Since $-180° \leq x° \leq 180°$, a clockwise rotation of 180° gives one negative solution in quadrant 4 and an anticlockwise rotation of 180° gives one positive solution in quadrant 2.
$x° = -85.43°$ or $x° = 180° - 85.43°$

4. Calculate the solutions to the required accuracy.

$\therefore x = -85.4, 94.6$ (correct to 1 decimal place)

TI	THINK	WRITE	CASIO	THINK	WRITE

b.1. On a Graphs page, set the Graphing Angle to Radian. Press MENU, then select:
4: Window/Zoom
1: Window
Settings …
Complete the fields as:
XMin: 0
XMax: 4π
XScale: $\pi/2$
YMin: -2
YMax: 2
YScale: 1
then select OK.

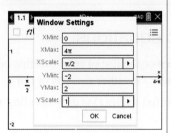

b.1. Put the Calculator in Radian mode.
On a Graph screen, press SHIFT then F3 to open the V_WIN. Complete the fields as:
Xmin: 0
max: 4π
scale: $\pi/2$
Ymin: -2
max: 2
scale: 1
then press EXE.

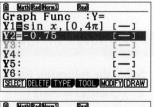

2. Complete the entry line for function 1 as:
$f1(x) = \sin(x)|$
$0 \leq x \leq 4\pi$
then press ENTER.
Complete the entry line for function 2 as:
$f2(x) = -0.75$
then press ENTER.

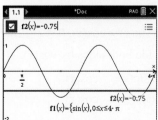

2. Complete the entry line for y1 as:
$y1 = \sin x, [0, 4\pi]$
then press EXE.
Complete the entry line for y2 as:
$y2 = -0.75$
then press EXE.
Select DRAW by pressing F6.

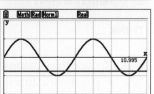

3. To find the points of intersection, press MENU, then select:
6: Analyze Graph
4: Intersection
Move the cursor to the left of the point of intersection when prompted for the lower bound, then press ENTER. Move the cursor to the right of the point of intersection when prompted for the upper bound, then press ENTER. Repeat this step to find the other points of intersection.

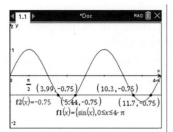

4. The answers appear on the screen.

The x-coordinates of the points of intersection of the graphs $f1(x) = \sin(x)$ and $f2(x) = -0.75$ represent the solutions to the equation $\sin(x) = -0.75$ over the domain $[0, 4\pi]$.
$\therefore x = 3.99, 5.44, 10.27, 11.72$ (correct to 2 decimal places)

3. To find the points of intersection, select Analysis, then G-Solve by pressing SHIFT then F5, then select INTSECT by pressing F5. With the cursor on the first point of intersection, press EXE. Use the left/right arrows to move to the next point of intersection and mark it on the graph by pressing EXE. Repeat this step to find the other points of intersection.

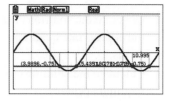

4. The answers appear on the screen.

The x-coordinates of the points of intersection of the graphs $y1 = \sin x$ and $y2 = -0.75$ represent the solutions to the equation $\sin x = -0.75$ over the domain $[0, 4\pi]$.
$\therefore x = 3.99, 5.44, 10.27, 11.72$ (correct to 2 decimal places)

6.7.4 Further types of trigonometric equations

Trigonometric equations may require algebraic techniques or the use of relationships between the functions before they can be reduced to the basic form $f(x) = a$, where f is either sin, cos or tan.

- Equations of the form $\sin(x) = a \cos(x)$ can be converted to $\tan(x) = a$ by dividing both sides of the equation by $\cos(x)$, $\cos(x) \neq 0$.
- Equations of the form $\sin^2(x) = a$ can be converted to $\sin(x) = \pm \sqrt{a}$ by taking the square roots of both sides of the equation.
- Equations of the form $\sin^2(x) + b \sin(x) + c = 0$ can be converted to standard quadratic equations by using the substitution $u = \sin(x)$.

Since $-1 \leq \sin(x) \leq 1$ and $-1 \leq \cos(x) \leq 1$, neither $\sin(x)$ nor $\cos(x)$ can have values greater than 1 or smaller than -1. This may have implications requiring the rejection of some steps when working with sine or cosine trigonometric equations. As $\tan(x) \in R$, there is no restriction on the values the tangent function can take.

WORKED EXAMPLE 18 Solving equations by substitution

Solve the following for x, where $0 \leq x \leq 2\pi$.

a. $\sqrt{3} \sin(x) = \cos(x)$

b. $\cos^2(x) + \cos(x) - 2 = 0$

THINK

WRITE

a. 1. Reduce the equation to one trigonometric function.

a. $\sqrt{3} \sin(x) = \cos(x)$, $0 \leq x \leq 2\pi$

Divide both sides by $\cos(x)$.

$$\frac{\sqrt{3} \sin(x)}{\cos(x)} = \frac{\cos(x)}{\cos(x)}$$

$$\therefore \sqrt{3}\tan(x) = 1$$
$$\therefore \tan(x) = \frac{1}{\sqrt{3}}$$

2. Calculate the solutions.

Tangent is positive in quadrants 1 and 3.

The base is $\dfrac{\pi}{6}$.

$$x = \frac{\pi}{6}, \pi + \frac{\pi}{6}$$

3. State the solutions.

$$x = \frac{\pi}{6}, \frac{7\pi}{6}$$

b. 1. Use substitution to form a quadratic equation.

b. $\cos^2(x) + \cos(x) - 2 = 0, 0 \le x \le 2\pi$

Let $a = \cos(x)$.

The equation becomes $a^2 + a - 2 = 0$.

2. Solve the quadratic equation.

$(a+2)(a-1) = 0$
$$\therefore a = -2 \text{ or } a = 1$$

3. Substitute back for the trigonometric function.

Since $a = \cos(x)$, $\cos(x) = -2$ or $\cos(x) = 1$.

Reject $\cos(x) = -2$ since $-1 \le \cos(x) \le 1$.

$$\therefore \cos(x) = 1$$

4. Solve the remaining trigonometric equation.

$\cos(x) = 1, \quad 0 \le x \le 2\pi$

Boundary value since $\cos(0) = 1$

$$\therefore x = 0, 2\pi$$

5. Write the answer.

$x = 0, 2\pi$

| TI | THINK | WRITE | CASIO | THINK | WRITE |

b.1. On a Graphs page, set the Graphing Angle to Radian.
Press MENU, then select:
4: Window/Zoom
1: Window Settings
…
Complete the fields as:
XMin: 0
XMax: 2π
XScale: $\pi/2$
YMin: -3
YMax: 3
YScale: 1
then select OK.
Note: The calculator will only give decimal approximations for intercepts, minimums and maximums, so it is important to have the x-axis scale as a multiple of π so that important points can be read easily from the graph.

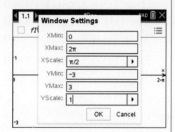

b.1. Put the Calculator in Radian mode.
On a Graph screen, press SHIFT then F3 to open the V_WIN.
Complete the fields as:
Xmin: 0
max: 2π
scale: $\pi/2$
Ymin: -3
max: 3
scale: 1
then press EXE.
Note: The calculator will only give decimal approximations for intercepts, minimums and maximums, so it is important to have the x-axis scale as a multiple of π so that important points can be read easily from the graph.

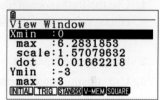

2. Complete the entry
line for function 1 as:
$f1(x) = (\cos(x))^2 + \cos(x) - 2 | 0 \le x \le 2\pi$
then press ENTER.

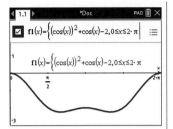

The x-intercepts of the graph
represent the solutions to the
equation
$(\cos(x))^2 + \cos(x) - 2 = 0$.
$\therefore x = 0$ and 2π

3. The answer appears
on the screen.

2. Complete the entry
line for $y1$ as:
$y1 = (\cos x)^2 + \cos x - 2, [0, 2\pi]$
then press EXE.
Select DRAW by
pressing F6.

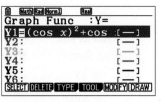

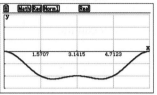

3. The answer appears
on the screen.

The x-intercepts of the graph
represent the solutions to the
equation
$(\cos(x))^2 + \cos(x) - 2 = 0$.
$\therefore x = 0$ and 2π

6.7.5 Using the Pythagorean identity

Trigonometric equations may also require using the Pythagorean identity before they can be reduced to the basic
form $f(x) = a$, where f is either sine or cosine.

WORKED EXAMPLE 19 Solving equations using the Pythagorean identity

Solve $2\sin^2(x) + 3\cos(x) - 3 = 0$, where $0 \le x \le 2\pi$.

THINK

1. Use the Pythagorean identity to express the
   equation in terms of $\cos(x)$.

2. Simplify the equation.

3. Use substitution to form a quadratic
   equation.

4. Solve the quadratic equation.

5. Substitute back for the trigonometric
   function.

6. Solve the trigonometric equations for the
   given domain.

7. Write the answer.

WRITE

$2\sin^2(x) + 3\cos(x) - 3 = 0$
$2(1 - \cos^2(x)) + 3\cos(x) - 3 = 0$
$2 - 2\cos^2(x) + 3\cos(x) - 3 = 0$

$2\cos^2(x) - 3\cos(x) + 1 = 0$

Let $u = \cos(x)$.
The equation becomes: $2u^2 - 3u + 1 = 0$
$(2u - 1)(u - 1) = 0$

$\therefore u = \dfrac{1}{2}$ or $u = 1$

Hence, $\cos(x) = \dfrac{1}{2}$ or $\cos(x) = 1$.

For $\cos(x) = \dfrac{1}{2}, 0 \le x \le 2\pi$.

Cosine is positive in quadrants 1 and 4.

The base angle is $\dfrac{\pi}{3}$ since $\cos\left(\dfrac{\pi}{3}\right) = \dfrac{1}{2}$.

$\therefore x = \dfrac{\pi}{3}$ or $2\pi - \dfrac{\pi}{3}$

$x = \dfrac{\pi}{3}$ or $\dfrac{5\pi}{3}$

For $\cos(x) = 1, 0 \le x \le 2\pi$.
This is a boundary point as $\cos(0) = 1$.
$\therefore x = 0$ or 2π

$x = 0, \dfrac{\pi}{3}, \dfrac{5\pi}{3}, 2\pi$

6.7.6 Solving trigonometric equations that require a change of domain

Equations such as $\sin(2x) = 1, 0 \leq x \leq 2\pi$ can be expressed in the basic form by the substitution of $\theta = 2x$. The accompanying domain must be changed to be the domain for θ. This requires the end points of the domain for x to be multiplied by 2. Hence, $0 \leq x \leq 2\pi \Rightarrow 2 \times 0 \leq 2x \leq 2 \times 2\pi$ gives the domain requirement for θ as $0 \leq \theta \leq 4\pi$.

This allows the equation to be written as $\sin(\theta) = 1, 0 \leq \theta \leq 4\pi$.

This equation can then be solved to give $\theta = \dfrac{\pi}{2}, \dfrac{5\pi}{2}$.

Substituting back for x gives $2x = \dfrac{\pi}{2}, \dfrac{5\pi}{2} \Rightarrow x = \dfrac{\pi}{4}, \dfrac{5\pi}{4}$. The solutions are in the domain specified for x.

The change of domain ensures all possible solutions are obtained.

However, in practice, it is quite common not to formally introduce the pronumeral substitution for equations such as $\sin(2x) = 1, 0 \leq x \leq 2\pi$.

With the domain change, the equation can be written as $\sin(2x) = 1, 0 \leq 2x \leq 4\pi$ and the equation solved for x as follows.

$$\sin(2x) = 1, 0 \leq 2x \leq 4\pi$$
$$\therefore 2x = \frac{\pi}{2}, \frac{5\pi}{2}$$
$$\therefore x = \frac{\pi}{4}, \frac{5\pi}{4}$$

WORKED EXAMPLE 20 Solving equations with changed domains

a. Solve $\cos(3x) = -\dfrac{1}{2}$ for x, $0 \leq x \leq 2\pi$.

b. Use substitution to solve the equation $\tan\left(2x - \dfrac{\pi}{4}\right) = -1, 0 \leq x \leq \pi$.

THINK

a. 1. Change the domain to be that for the given multiple of the variable.

2. Solve the equation for $3x$.
Note: Alternatively, substitute $\theta = 3x$ and solve for θ.

3. Calculate the solutions for x and write the answer.

WRITE

a. $\cos(3x) = -\dfrac{1}{2}, 0 \leq x \leq 2\pi$

Multiply the end points of the domain of x by 3

$\therefore \cos(3x) = -\dfrac{1}{2}, 0 \leq 3x \leq 6\pi$

Cosine is negative in quadrants 2 and 3.

The base is $\dfrac{\pi}{3}$.

As $3x \in [0, 6\pi]$, each of the three revolutions will generate 2 solutions, giving a total of 6 values for $3x$.

$3x = \pi - \dfrac{\pi}{3}, \pi + \dfrac{\pi}{3}, 3\pi - \dfrac{\pi}{3}, 3\pi + \dfrac{\pi}{3}, 5\pi - \dfrac{\pi}{3}, 5\pi + \dfrac{\pi}{3}$

$= \dfrac{2\pi}{3}, \dfrac{4\pi}{3}, \dfrac{8\pi}{3}, \dfrac{10\pi}{3}, \dfrac{14\pi}{3}, \dfrac{16\pi}{3}$

Divide each of the 6 values by 3 to obtain the solutions for x.

$x = \dfrac{2\pi}{9}, \dfrac{4\pi}{9}, \dfrac{8\pi}{9}, \dfrac{10\pi}{9}, \dfrac{14\pi}{9}, \dfrac{16\pi}{9}$

b. 1. State the substitution required to express the equation in basic form.

b. $\tan\left(2x - \dfrac{\pi}{4}\right) = -1,\ 0 \le x \le \pi$

Let $\theta = 2x - \dfrac{\pi}{4}$.

2. Change the domain of the equation to that of the new variable.

For the domain change:
$$0 \le x \le \pi$$
$$\therefore 0 \le 2x \le 2\pi$$
$$\therefore -\dfrac{\pi}{4} \le 2x - \dfrac{\pi}{4} \le 2\pi - \dfrac{\pi}{4}$$
$$\therefore -\dfrac{\pi}{4} \le \theta \le \dfrac{7\pi}{4}$$

3. State the equation in terms of θ.

The equation becomes $\tan(\theta) = -1,\ -\dfrac{\pi}{4} \le \theta \le \dfrac{7\pi}{4}$.

4. Solve the equation for θ.

Tangent is negative in quadrants 2 and 4.

The base is $\dfrac{\pi}{4}$.
$$\theta = -\dfrac{\pi}{4},\ \pi - \dfrac{\pi}{4},\ 2\pi - \dfrac{\pi}{4}$$
$$= -\dfrac{\pi}{4},\ \dfrac{3\pi}{4},\ \dfrac{7\pi}{4}$$

5. Substitute back in terms of x.

$$\therefore 2x - \dfrac{\pi}{4} = -\dfrac{\pi}{4},\ \dfrac{3\pi}{4},\ \dfrac{7\pi}{4}$$

6. Calculate the solutions for x and write the answer.

Add $\dfrac{\pi}{4}$ to each value.
$$\therefore 2x = 0,\ \pi,\ 2\pi$$
Divide by 2.
$$\therefore x = 0,\ \dfrac{\pi}{2},\ \pi$$

Exercise 6.7 Solving trigonometric equations

6.7 Exercise	6.7 Exam practice

These questions are even better in jacPLUS!
- Receive immediate feedback
- Access sample responses
- Track results and progress

Find all this and MORE in jacPLUS ▶

Simple familiar	Complex familiar	Complex unfamiliar
1, 2, 3, 4, 5, 6, 7, 8, 9	10, 11, 12, 13, 14, 15, 16, 17	18, 19, 20

Simple familiar

1. Solve the following for x, given $0 \le x \le 2\pi$.

  a. $\cos(x) = \dfrac{1}{\sqrt{2}}$ **b.** $\sin(x) = -\dfrac{1}{\sqrt{2}}$ **c.** $\tan(x) = -\dfrac{1}{\sqrt{3}}$

  d. $2\sqrt{3}\ \cos(x) + 3 = 0$ **e.** $4 - 8\ \sin(x) = 0$ **f.** $2\sqrt{2}\ \tan(x) = \sqrt{24}$

2. Determine the exact solutions for $\theta \in [-2\pi, 2\pi]$ for which:
   a. $\tan(\theta) = 1$
   b. $\cos(\theta) = -0.5$
   c. $1 + 2\sin(\theta) = 0$.

3. a. Determine the solutions to the equation $3\tan(x) + 3\sqrt{3} = 0$ over the domain $x \in [0, 3\pi]$.
   b. For $0 \le t \le 4\pi$, determine the exact solutions to $10\sin(t) - 3 = 2$.
   c. Calculate the exact values of v that satisfy $4\sqrt{2}\cos(v) = \sqrt{2}\cos(v) + 3, -\pi \le v \le 5\pi$.

4. **WE16** Solve the following equations to obtain exact values for x.
   a. $\sin(x) = \dfrac{1}{2}, 0 \le x \le 2\pi$
   b. $\sqrt{3} - 2\cos(x) = 0, 0 \le x \le 2\pi$
   c. $4 + 4\tan(x) = 0, -2\pi \le x \le 2\pi$

5. Consider the equation $\cos(\theta) = -\dfrac{1}{2}, -180° \le \theta \le 180°$.
   a. Determine how many solutions for θ the equation has.
   b. Calculate the solutions of the equation.

6. Solve for $a°$, given $0° \le a° \le 360°$.
   a. $\sqrt{3} + 2\sin(a°) = 0$
   b. $\tan(a°) = 1$
   c. $6 + 8\cos(a°) = 2$
   d. $4(2 + \sin(a°)) = 11 - 2\sin(a°)$

7. Obtain all values for $t, t \in [-\pi, 4\pi]$, for which:
   a. $\tan(t) = 0$
   b. $\cos(t) = 0$
   c. $\sin(t) = -1$
   d. $\cos(t) = 1$
   e. $\sin(t) = 1$
   f. $\tan(t) = 1$.

8. a. **WE17** Solve $1 - \sin(x) = 0, -4\pi \le x \le 4\pi$ for x.
   b. Solve $\tan(x) = 0.75, 0 \le x \le 4\pi$ for x, to 2 decimal places.
   c. Solve $4\cos(x°) + 1 = 0, -180° \le x° \le 180°$ for x, to 1 decimal place.

9. Solve the following equations, where possible, to obtain the values of the pronumerals.
   a. $4\sin(a) + 3 = 5, -2\pi < a < 0$
   b. $6\tan(b) - 1 = 11, -\dfrac{\pi}{2} < b < 0$
   c. $8\cos(c) - 7 = 1, -\dfrac{9\pi}{2} < c < 0$
   d. $\dfrac{9}{\tan(d)} - 9 = 0, 0 < d \le \dfrac{5\pi}{12}$
   e. $2\cos(e) = 1, -\dfrac{\pi}{6} \le e \le \dfrac{13\pi}{6}$
   f. $\sin(f) = -\cos(150°), -360° \le f \le 360°$

Complex familiar

10. **WE18** Solve the following for x, given $0 \le x \le 2\pi$.
    a. $\sqrt{3}\sin(x) = 3\cos(x)$
    b. $\sin^2(x) - 5\sin(x) + 4 = 0$

11. Solve $\cos^2(x) = \dfrac{3}{4}, 0 \le x \le 2\pi$ for x.

12. Solve the following for x, where $0 \le x \le 2\pi$.
    a. $\sin(x) = \sqrt{3}\cos(x)$
    b. $\sin(x) = -\dfrac{\cos(x)}{\sqrt{3}}$
    c. $\sin(2x) + \cos(2x) = 0$
    d. $\dfrac{3\sin(x)}{8} = \dfrac{\cos(x)}{2}$
    e. $\sin^2(x) = \cos^2(x)$
    f. $\cos(x)(\cos(x) - \sin(x)) = 0$

13. a. **WE20** Solve $\sin(2x) = \dfrac{1}{\sqrt{2}}, 0 \le x \le 2\pi$ for x.

b. Use a substitution to solve the equation $\cos\left(2x + \dfrac{\pi}{6}\right) = 0, 0 \le x \le \dfrac{3\pi}{2}$.

14. Solve the following for $\theta, 0 \le \theta \le 2\pi$.

a. $\sqrt{3} \tan(3\theta) + 1 = 0$

b. $2\sqrt{3} \sin\left(\dfrac{3\theta}{2}\right) - 3 = 0$

c. $4\cos^2(-\theta) = 2$

d. $\sin\left(2\theta + \dfrac{\pi}{4}\right) = 0$

15. **WE19** Solve $2\sin^2(x) - 5\cos(x) + 1 = 0$, where $0 \le x \le 2\pi$.

16. Solve the following for x, where $0 \le x \le 2\pi$.

a. $\sin^2(x) = \dfrac{1}{2}$

b. $2\cos^2(x) + 3\cos(x) = 0$

c. $2\sin^2(x) - \sin(x) - 1 = 0$

d. $\tan^2(x) + 2\tan(x) - 3 = 0$

e. $2\cos^2(x) + \sin(x) = 1$

f. $\cos^2(x) - 9 = 0$

17. Solve $\sin\left(\dfrac{x}{2}\right) = \sqrt{3}\cos\left(\dfrac{x}{2}\right), 0 \le x \le 2\pi$ for x.

Complex unfamiliar

18. Use technology to calculate the values of θ, correct to 2 decimal places, that satisfy the following conditions.

a. $2 + 3\cos(\theta) = 0, 0 \le \theta \le 2\pi$

b. $\tan(\theta) = \dfrac{1}{\sqrt{2}}, -2\pi \le \theta \le 3\pi$

c. $5\sin(\theta°) + 4 = 0, -270° \le \theta° \le 270°$

d. $\cos^2(\theta°) = 0.04, 0° \le \theta \le 360°$

19. Consider the function $f: [0, 2\pi] \to R, f(x) = a\sin(x)$.

a. If $f\left(\dfrac{\pi}{6}\right) = 4$, calculate the value of a.

b. Use the answer to part a to detemine, where possible, any values of x, to 2 decimal places, for which the following apply.

i. $f(x) = 3$ ii. $f(x) = 8$ iii. $f(x) = 10$

20. If $f(t) = 1 - 5\sin\left(\dfrac{\pi t}{6}\right)$ where $t \in [-3, 3]$, determine the value(s) of t if $f(t) + 3 = f(-t) - 2$.

Fully worked solutions for this chapter are available online.

LESSON
6.8 Transformations of the sine and cosine functions

SYLLABUS LINKS

- Recognise and determine the effect of the parameters a, b, h and k, on the graphs of $y = a \sin(b(x - h)) + k$ and $y = a \cos(b(x - h)) + k$, with and without technology.
- Sketch the graphs of $y = a \sin(b(x - h)) + k$, $y = a \cos(b(x - h)) + k$, with and without technology.

Source: Mathematical Methods Senior Syllabus 2024 © State of Queensland (QCAA) 2024; licensed under CC BY 4.0.

The shapes of the graphs of the two functions, sine and cosine, are now familiar. Each graph has a wavelike pattern of period 2π, and the pair are 'out of phase' with each other by $\dfrac{\pi}{2}$. We now consider the effect of transformations on the shape and on the properties of these basic sine and cosine graphs.

6.8.1 Transformations of functions

The following transformations are applied to the graph of $y = f(x)$ to form the graph of $y = af(b(x - h)) + k$.
- dilation of factor a (assuming $a > 0$) from the x-axis parallel to the y-axis
- reflection in the x-axis if $a < 0$
- dilation of factor $\dfrac{1}{b}$ (assuming $b > 0$) from the y-axis parallel to the x-axis
- reflection in the y-axis if $b < 0$
- horizontal translation of h units
  - to the left if $h > 0$
  - to the right if $h < 0$
- vertical translation of k units
  - up if $k > 0$
  - down if $k < 0$

We can therefore infer that the graph of $y = a \sin(b(x - h)) + k$ is the image of the basic $y = \sin(x)$ graph after the same set of transformations are applied.

However, the period, amplitude and equilibrium, or mean, position of a trigonometric graph are such key features that we shall consider each transformation in order to interpret its effect on each of these features.

6.8.2 Amplitude changes

Consider the graphs of $y = \sin(x)$ and $y = 2\sin(x)$.

Comparison of the graph of $y = 2\sin(x)$ with the graph of $y = \sin(x)$ shows the dilation of factor 2 from the x-axis affects the amplitude, but neither the period nor the equilibrium position is altered.

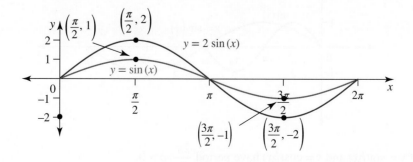

Consider the graphs of $y = \cos(x)$ and $y = -\frac{1}{2}\cos(x)$.

Comparison of the graph of $y = -\frac{1}{2}\cos(x)$ where $a = -\frac{1}{2}$ with the graph of $y = \cos(x)$ shows the dilation factor affecting the amplitude is $\frac{1}{2}$ and the graph of $y = -\frac{1}{2}\cos(x)$ is reflected in the x-axis.

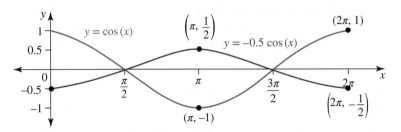

- The graphs of $y = a\sin(x)$ and $y = a\cos(x)$ have amplitude a if $a > 0$.
- If $a < 0$, the graph is reflected in the x-axis, (inverted) and the amplitude is the positive part of a (or $|a|$).

6.8.3 Period changes

Comparison of the graph of $y = \cos(2x)$ with the graph of $y = \cos(x)$ shows the dilation factor of $\frac{1}{2}$ from the y-axis affects the period: it halves the period. The period of $y = \cos(x)$ is 2π, and the period of $y = \cos(2x)$ is $\frac{1}{2}$ of 2π; that is, $y = \cos(2x)$ has a period of $\frac{2\pi}{2} = \pi$. Neither the amplitude nor the equilibrium position has been altered.

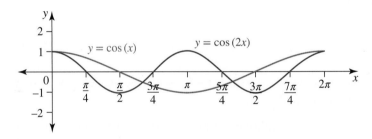

Comparison of one cycle of the graph of $y = \sin\frac{x}{2}$ with one cycle of the graph of $y = \sin(x)$ shows the dilation factor of 2 from the y-axis doubles the period. The period of the graph of $y = \sin\left(\frac{x}{2}\right)$ is $\frac{2\pi}{\frac{1}{2}} = 2 \times 2\pi = 4\pi$.

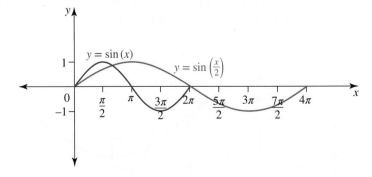

- The graphs of $y = \sin(bx)$ and $y = \cos(bx)$ have period $\frac{2\pi}{b}$, $b > 0$.

6.8.4 Equilibrium (or mean) position changes

Comparison of the graph of $y = \sin(x) - 2$ with the graph of $y = \sin(x)$ shows that vertical translation affects the equilibrium position. The graph of $y = \sin(x) - 2$ oscillates about the line $y = -2$, so its range is $[-3, -1]$. Neither the period nor the amplitude is affected.

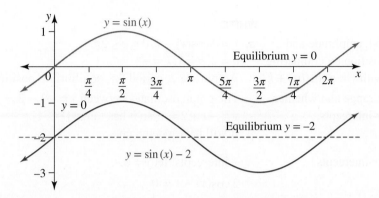

The graphs of $y = \sin(x) + k$ and $y = \cos(x) + k$ both oscillate about the equilibrium (or mean) position $y = k$.

The range of both graphs is $[k-1, k+1]$, since the amplitude is 1.

Key features of $y = a\sin(bx) + k$ and $y = a\cos(bx) + k$

The graphs of $= a\sin(bx) + k$ and $y = a\cos(bx) + k$ have:

- **amplitude a for $a > 0$; the graphs are reflected in the x-axis (inverted) if $a < 0$**
- **period $\dfrac{2\pi}{b}$ (for $b > 0$); the graphs are reflected in the y-axis if $b < 0$**
- **equilibrium or mean position $y = k$**
- **range $[k - a, k + a]$.**

The oscillation about the equilibrium position of the graph of $y = a\sin(bx) + k$ always starts at the equilibrium with the pattern, for each period divided into quarters, of:

- equilibrium → range maximum → equilibrium → range minimum → equilibrium if $a > 0$, or:

  equilibrium → range minimum → equilibrium → range maximum → equilibrium if $a < 0$.

The oscillation about the equilibrium position of the graph of $y = a\cos(bx) + k$ either starts from its maximum or minimum point with the pattern:

- range maximum → equilibrium → range minimum → equilibrium → range maximum if $a > 0$, or: range

  minimum → equilibrium → range maximum → equilibrium → range minimum if $a < 0$.

When sketching the graphs, any intercepts with the x-axis are usually obtained by solving the trigonometric equation $a\sin(bx) + k = 0$ or $a\cos(bx) + k = 0$.

Sketch the graphs of the following functions.

a. $y = 2\cos(x) - 1, 0 \le x \le 2\pi$

b. $y = 4 - 2\sin(3x), -\dfrac{\pi}{2} \le x \le \dfrac{3\pi}{2}$

THINK

WRITE

a. 1. State the period, amplitude and equilibrium position by comparing the equation with $y = a\cos(bx) + k$.

2. Determine the range and whether there will be x-intercepts.

3. Calculate the x-intercepts.

4. Scale the axes by marking $\dfrac{\pi}{2}$-period intervals on the x-axis. Mark the equilibrium position and end points of the range on the y-axis. Then plot the graph using its pattern.

a. $y = 2\cos(x) - 1, 0 \le x \le 2\pi$

$a = 2, b = 1, k = -1$

Amplitude 2, period 2π, equilibrium position $y = -1$

The graph oscillates between $y = -1 - 2 = -3$, and $y = -1 + 2 = 1$, so it has range $[-3, 1]$.

It will have x-intercepts.

x-intercepts: let $y = 0$.

$2\cos(x) - 1 = 0$

$\therefore \cos(x) = \dfrac{1}{2}$

Base $\dfrac{\pi}{3}$, quadrants 1 and 4

$x = \dfrac{\pi}{3}, 2\pi - \dfrac{\pi}{3}$

$= \dfrac{\pi}{3}, \dfrac{5\pi}{3}$

x-intercepts are $\left(\dfrac{\pi}{3}, 0\right), \left(\dfrac{5\pi}{3}, 0\right)$.

The period is 2π, so the scale on the x-axis is in multiples of $\dfrac{\pi}{2}$. Since $a > 0$, graph starts at range maximum at its y-intercept $(0, 1)$.

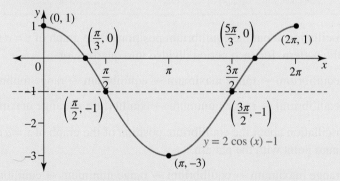

5. Label all key features of the graph including the maximum and minimum points.

The maximum points are $(0, 1)$ and $(2\pi, 1)$. The minimum point is $(\pi, -3)$.

b. 1. State the information the equation provides by comparing the equation with $y = a\sin(bx) + k$.
Note: The amplitude is always a positive value.

b. $y = 4 - 2\sin(3x), -\dfrac{\pi}{2} \le x \le \dfrac{3\pi}{2}$

$y = -2\sin(3x) + 4$, domain $\left[-\dfrac{\pi}{2}, \dfrac{3\pi}{2}\right]$

$a = -2, b = 3, k = 4$

Amplitude 2; graph is inverted; period $\dfrac{2\pi}{3}$; equilibrium $y = 4$

2. Determine the range and whether there will be x-intercepts.

The graph oscillates between $y = 4 - 2 = 2$ and $y = 4 + 2 = 6$, so its range is $[2, 6]$.
There are no x-intercepts.

3. Scale the axes and extend the $\frac{1}{4}$-period intervals on the x-axis to cover the domain. Mark the equilibrium position and end points of the range on the y-axis. Then plot the graph using its pattern and continue the pattern over the given domain.

Dividing the period of $\frac{2\pi}{3}$ into four gives a horizontal scale of $\frac{\pi}{6}$. The first cycle of the graph starts at its equilibrium position at its y-intercept $(0, 4)$ and decreases as $a < 0$.

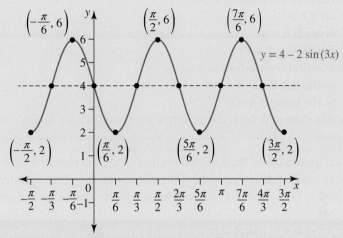

4. Label all key features of the graph including the maximum and minimum points.
Note: Successive maximum points are one period apart, as are the successive minimum points.

The maximum points are $\left(-\frac{\pi}{6}, 6\right)$, $\left(\frac{\pi}{2}, 6\right)$ and $\left(\frac{7\pi}{6}, 6\right)$. The minimum points are $\left(-\frac{\pi}{2}, 2\right)$, $\left(\frac{\pi}{6}, 2\right)$, $\left(\frac{5\pi}{6}, 2\right)$ and $\left(\frac{3\pi}{2}, 2\right)$.

6.8.5 Phase changes

Horizontal translations of the sine and cosine graphs do not affect the period, amplitude or equilibrium, as one cycle of each of the graphs of $y = \sin(x)$ and $y = \sin\left(x - \frac{\pi}{4}\right)$ illustrate.

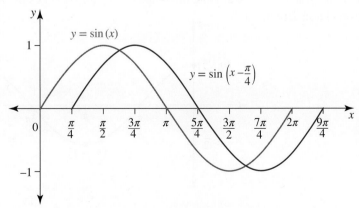

The horizontal translation causes the two graphs to be 'out of phase' by $\frac{\pi}{4}$.

- The graph of $y = \sin(x - h)$ has a phase shift of h from the graph of $y = \sin(x)$.
- The graph of $y = \cos(x - h)$ has a phase shift of h from the graph of $y = \cos(x)$.

6.8.6 The graphs of $y = a\sin(b(x-h)) + k$ and $y = a\cos(b(x-h)) + k$

Key features of $y = a\sin(b(x-h)) + k$ and $y = a\cos(b(x-h)) + k$

The graphs of $= a\sin(b(x-h)) + k$ and $y = a\cos(b(x-h)) + k$ have:

- **amplitude** a for $a > 0$; the graphs are reflected in the x-axis (inverted) if $a < 0$
- **period** $\dfrac{2\pi}{b}$ (for $b > 0$); the graphs are reflected in the y-axis if $b < 0$
- **equilibrium or mean position** $y = k$
- **vertical translation** of k units
  - up if $k > 0$
  - down if $k < 0$
- **range** $[k - a, k + a]$
- **horizontal translation** of h units
  - to the left if $h > 0$
  - to the right if $h < 0$
- **phase shift** of h from the graph of $y = a\sin(bx) + k$ or $y = a\cos(bx) + k$.

Horizontal translation of the 5 key points that create the pattern for the graph of either $y = a\sin(bx)$ or $y = a\cos(bx)$ will enable one cycle of the graph with the phase shift to be sketched. This transformed graph may be extended to fit a given domain, with its rule used to calculate the coordinates of end points.

WORKED EXAMPLE 22 Sketching graphs using transformations

a. Sketch the graph of $y = \sqrt{2}\cos\left(x - \dfrac{\pi}{4}\right), 0 \le x \le 2\pi$.

b. State the period, amplitude, range and phase shift factor for the graph of $y = -2\sin\left(4x - \dfrac{\pi}{3}\right) + 5$.

THINK

a. 1. Identify the key features of the graph from the given equation.

2. Sketch one cycle of the graph without the horizontal translation.

WRITE

a. $y = \sqrt{2}\cos\left(x - \dfrac{\pi}{4}\right), 0 \le x \le 2\pi$

Period 2π; amplitude $\sqrt{2}$; equilibrium position $y = 0$; horizontal translation of $\dfrac{\pi}{4}$ to the right; domain $[0, 2\pi]$.

Sketching the graph of $y = \sqrt{2}\cos(x)$ using the pattern gives:

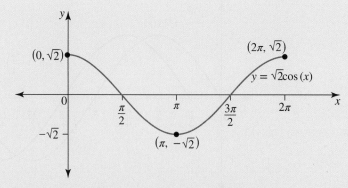

3. Sketch the required graph using horizontal translation.

The key points of the graph of $y = \sqrt{2}\cos(x)$ become, under a horizontal translation of $\dfrac{\pi}{4}$ units to the right:

$$\left(0, \sqrt{2}\right) \quad \rightarrow \quad \left(\dfrac{\pi}{4}, \sqrt{2}\right)$$

$$\left(\dfrac{\pi}{2}, 0\right) \quad \rightarrow \quad \left(\dfrac{3\pi}{4}, 0\right)$$

$$\left(\pi, -\sqrt{2}\right) \quad \rightarrow \quad \left(\dfrac{5\pi}{4}, -\sqrt{2}\right)$$

$$\left(\dfrac{3\pi}{2}, 0\right) \quad \rightarrow \quad \left(\dfrac{7\pi}{4}, 0\right)$$

$$\left(2\pi, \sqrt{2}\right) \quad \rightarrow \quad \left(\dfrac{9\pi}{4}, \sqrt{2}\right)$$

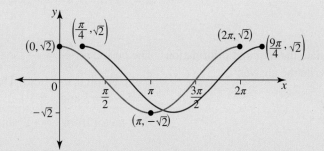

4. Calculate the end points of the domain.

The translated graph is not on the required domain.

End points for the domain $[0, 2\pi]$:

$$y = \sqrt{2}\cos\left(x - \dfrac{\pi}{4}\right)$$

When $x = 0$,

$$y = \sqrt{2}\cos\left(-\dfrac{\pi}{4}\right)$$

$$= \sqrt{2}\cos\left(\dfrac{\pi}{4}\right)$$

$$= \sqrt{2} \times \dfrac{1}{\sqrt{2}}$$

$$= 1$$

When $x = 2\pi$,

$$y = \sqrt{2}\cos\left(2\pi - \dfrac{\pi}{4}\right)$$

$$= \sqrt{2}\cos\left(\dfrac{\pi}{4}\right)$$

$$= 1$$

The end points are $(0, 1)$ and $(2\pi, 1)$.

5. Sketch the graph on the required domain.

Note: As the graph covers one full cycle, the end points should have the same y-coordinates.

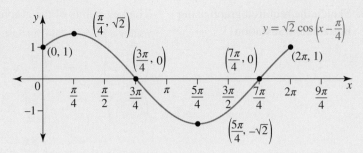

b. 1. Express the equation in the form $y = a \sin(b(x - h)) + k$.

$$y = -2 \sin\left(4x + \frac{\pi}{3}\right) + 5$$

$$= -2 \sin\left(4\left(x + \frac{\pi}{12}\right)\right) + 5$$

$$a = -2, b = 4, h = \frac{-\pi}{12}, k = 5$$

2. Calculate the required information and write the answer.

The period is $\dfrac{2\pi}{n} = \dfrac{2\pi}{4}$, so the period is $\dfrac{\pi}{2}$. The amplitude is 2 (the graph is inverted). The graph oscillates between $y = 5 - 2 = 3$ and $y = 5 + 2 = 7$, so the range is $[3, 7]$. The phase shift factor from $y = -2 \sin(4x)$ is $-\dfrac{\pi}{12}$.

6.8.7 Forming the equation of a sine or cosine graph

The graph of $y = \sin\left(x + \dfrac{\pi}{2}\right)$ is the same as the graph of $y = \cos(x)$, since sine and cosine have a phase difference of $\dfrac{\pi}{2}$. This means that it is possible for the equation of the graph to be expressed in terms of either function. This is true for all sine and cosine graphs, so their equations are not uniquely expressed. Given the choice, it is simpler to choose the form that does not require a phase shift.

WORKED EXAMPLE 23 Determining the equation of a sine or cosine graph

Determine two possible equations for the graph shown.

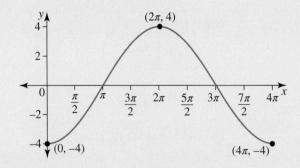

THINK

1. Identify the key features of the given graph.

WRITE

The graph has a period of 4π and amplitude 4, and the equilibrium position is $y = 0$.

2. Form a possible equation for the graph that does not involve any horizontal translation.

The graph could be an inverted cosine graph. A possible cosine equation for the graph could be $y = a \cos(bx) + k$ with $a = -4$ and $k = 0$.
$\therefore y = -4 \cos(bx)$

The period is $\dfrac{2\pi}{b}$.

From the diagram, the period is 4π.

$$\dfrac{2\pi}{b} = 4\pi$$
$$\dfrac{2\pi}{4\pi} = b$$
$$b = \dfrac{1}{2}$$

Therefore, a possible equation is $y = -4 \cos\left(\dfrac{1}{2}x\right)$.

3. Form a possible equation for the graph that does involve a horizontal translation.
Note: Other equations for the graph are possible by considering other phase shifts.

Alternatively, the graph could be a sine function that has been horizontally translated π units to the right. This sine graph is not inverted, so a is positive. A possible sine equation could be:
$y = a \sin(b(x - h)) + k$ with $a = 4$, $h = \pi$, $k = 0$
$\therefore y = 4 \sin(b(x - \pi))$.

The graph has the same period of 4π, so $b = \dfrac{1}{2}$.

Therefore, a possible equation is
$$y = 4 \sin\left(\dfrac{1}{2}(x - \pi)\right).$$

When applying transformations to $y = f(x)$ to form the graph of $y = af(b(x - h)) + k$, the order of operations can be important, so any dilation or reflection should be applied before any translation.

Exercise 6.8 Transformations of the sine and cosine functions **learn** on

6.8 Exercise	6.8 Exam practice on

Simple familiar	Complex familiar	Complex unfamiliar
1, 2, 3, 4, 5, 6, 7, 8, 9, 10, 11	12, 13, 14, 15, 16, 17, 18, 19	20

Simple familiar

1. State the period and amplitude of each of the following.
   a. $y = 6 \cos(2x)$
   b. $y = -7 \cos\left(\dfrac{x}{2}\right)$

c. $y = -\dfrac{3}{5}\sin\left(\dfrac{3x}{5}\right)$

d. $y = \sin\left(\dfrac{6\pi x}{7}\right)$

e.

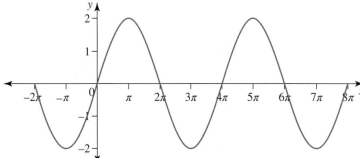

f.

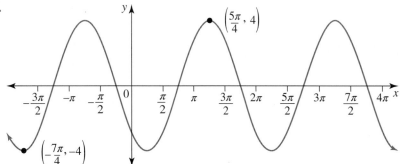

2. State the amplitude, range and period, then sketch the graph of one complete period for each of the following functions.

a. $y = \dfrac{1}{4}\sin(4x)$

b. $y = \dfrac{1}{2}\sin\left(\dfrac{1}{4}x\right)$

c. $y = -\sin(3x)$

d. $y = 3\cos(2x)$

e. $y = -6\cos\left(\dfrac{2x}{5}\right)$

f. $y = \dfrac{1}{3}\cos(5\pi x)$.

3. Sketch the following graphs over the given domains.

a. $y = 3\cos(2x),\ 0 \le x \le 2\pi$

b. $y = 2\sin\left(\dfrac{1}{2}x\right),\ 0 \le x \le 4\pi$

c. $y = -5\sin(4x),\ 0 \le x \le 2\pi$

d. $y = -\cos(\pi x),\ 0 \le x \le 4$

4. Sketch each of the following over the domain specified and state the range.

a. $y = \sin(x) + 3,\ 0 \le x \le 2\pi$

b. $y = \cos(x) - 1,\ 0 \le x \le 2\pi$

c. $y = \cos(x) + 2,\ -\pi \le x \le \pi$

d. $y = 4 - \sin(x),\ -\pi \le x \le 2\pi$

5. **WE21** Sketch the graphs of the following functions.

a. $y = 2\sin(x) + 1,\ 0 \le x \le 2\pi$

b. $y = 4 - 3\cos(2x),\ -\pi \le x \le 2\pi$

6. Sketch the graph of $y = f(x)$ for the function $f : [0, 12] \to R, f(x) = \sin\left(\dfrac{\pi x}{6}\right)$.

7. State the period, mean position, amplitude and range, then sketch the graph showing one complete cycle of each of the following functions.

a. $y = 4\sin(2x) + 5$

b. $y = -2\sin(3x) + 2$

c. $y = \dfrac{3}{2} - \dfrac{1}{2}\cos\left(\dfrac{x}{2}\right)$

d. $y = 2\cos(\pi x) - \sqrt{3}$

8. Sketch the following graphs over the given domains and state the ranges of each.

   a. $y = 2\cos(2x) - 2, 0 \leq x \leq 2\pi$

   b. $y = 2\sin(x) + \sqrt{3}, 0 \leq x \leq 2\pi$

   c. $y = 3\sin\left(\dfrac{x}{2}\right) + 5, -2\pi \leq x \leq 2\pi$

   d. $y = -4 - \cos(3x), 0 \leq x \leq 2\pi$

   e. $y = 1 - 2\sin(2x) - \pi \leq x \leq 2\pi$

   f. $y = 2[1 - 3\cos(x)], 0° \leq x \leq 360°$

9. a. Give the range of $f : R \to R, f(x) = 3 + 2\sin(5x)$.

   b. Determine the minimum value of the function $f : [0, 2\pi] \to f(x) = 10\cos(2x) - 4$.

   c. Determine the maximum value of the function $f : [0, 2\pi] \to R, f(x) = 56 - 12\sin(x)$ and the value of x for which the maximum occurs.

   d. Describe the sequence of transformations that must be applied for the following.

      i. $\sin(x) \to 3 + 2\sin(5x)$

      ii. $\cos(x) \to 10\cos(2x) - 4$

      iii. $\sin(x) \to 56 - 12\sin(x)$

10. **WE22** a. Sketch the graph of $y = \sqrt{2}\sin\left(x + \dfrac{\pi}{4}\right), 0 \leq x \leq 2\pi$.

    b. State the period, amplitude, range and phase shift factor for the graph of $y = -3\cos\left(2\pi + \dfrac{\pi}{4}\right) + 1$.

11. a. i. Sketch one cycle of each of $y = \cos(x)$ and $y = \cos\left(x + \dfrac{\pi}{6}\right)$ on the same axes.

       ii. Sketch one cycle of each of $y = \cos(x)$ and $y = \cos(x - \pi)$ on a second set of axes.

    b. i. Sketch one cycle of each of $y = \sin(x)$ and $y = \sin\left(x + \dfrac{3\pi}{4}\right)$ on the same axes.

       ii. Sketch one cycle of each of $y = \sin(x)$ and $y = \sin\left(x + \dfrac{3\pi}{2}\right)$ on a second set of axes.

Complex familiar

12. Sketch the following graphs for $0 \leq x \leq 2\pi$.

    a. $y = 2\sin\left(x - \dfrac{\pi}{4}\right)$

    b. $y = -4\sin\left(x + \dfrac{\pi}{6}\right)$

    c. $y = \cos\left(2\left(x + \dfrac{\pi}{3}\right)\right)$

    d. $y = \cos\left(2x - \dfrac{\pi}{2}\right)$

    e. $y = \cos\left(x + \dfrac{\pi}{2}\right) + 2$

    f. $y = 3 - 3\sin(2x - 4\pi)$

13. Sketch the graph of $y = \sin\left(2x - \dfrac{\pi}{3}\right), 0 \leq x \leq \pi$.

14. a. The equation of the graph shown is of the form $y = a\sin(bx)$. Determine the values of a and b, and hence state the equation of the graph.

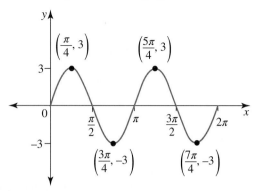

b. The equation of the graph shown is of the form $y = a\cos(bx)$. Determine the values of a and b, and hence state the equation of the graph.

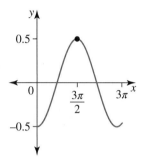

c. The equation of the graph shown is of the form $y = a\cos(bx)$. Determine the values of a and b, and hence state the equation of the graph.

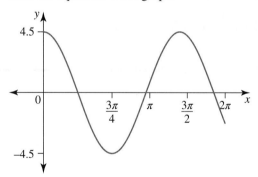

15. Determine the equation of each of the following graphs, given that the equation of each is either of the form $y = a\cos(bx)$ or $y = a\sin(bx)$.

a.

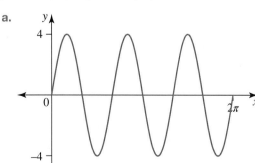

b.

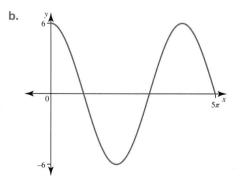

c.

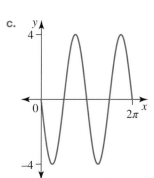

16. The following graphs are either of the form $y = a\cos(bx) + k$ or the form $y = a\sin(bx) + k$. Determine the equation for each graph. Use technology to verify your answer.

a.

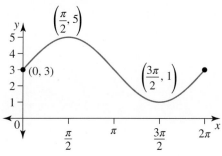

b.

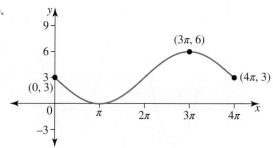

c.

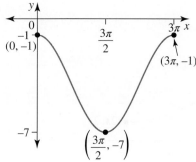

d.

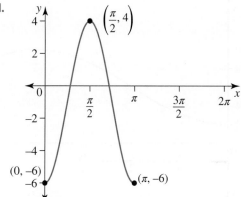

e.

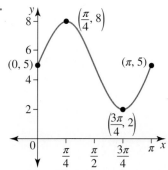

17. In parts **a**–**d**, obtain a possible equation for each of the given graphs.

a.

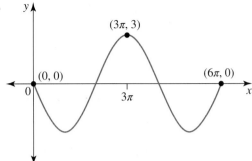

b.

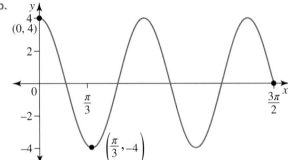

c.

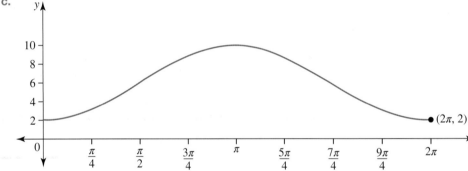

d.

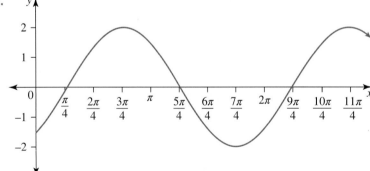

e. Give an alternative equation for the graph shown in part **d**.

f. Use the symmetry properties to give an alternative equation for $y = \cos(-x)$ and for $y = \sin(-x)$.

18. **WE23** Determine two possible equations for the graph shown.

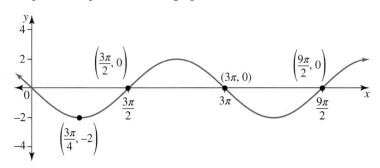

19. A function has the rule $f(x) = a \sin(bx) + k$ and a range of $[5, 9]$.

    $f(x) = f\left(x + \dfrac{2\pi}{3}\right)$ and $\dfrac{2\pi}{3}$ is the smallest positive value for which this relationship holds.

    a. State the period of the function.
    b. Obtain possible values for the positive constants a, b and k.
    c. Sketch one cycle of the graph of $y = f(x)$, stating its domain, D.
    d. A second function has the rule $g(x) = a \cos(bx) + k$, where a, b and k have the same values as those of $y = f(x)$. Sketch one cycle of the graph of $y = g(x)$, $x \in D$, on the same axes as the graph of $y = f(x)$.
    e. Obtain the coordinates of any points of intersection of the graphs of $y = f(x)$ and $y = g(x)$.
    f. Give the values of x that are solutions to the inequation $f(x) \geq g(x)$ where $x \in D$.

Complex unfamiliar

20. a. Sketch the graph of $y = 3 \sin\left(x - \dfrac{\pi}{4}\right) + 1$ for $0 \leq x \leq 2\pi$.

       Justify your procedures and decisions.
    b. Sketch the graph of $y = 1 - 2 \cos(2x - \pi)$ for $x \in [-\pi, \pi]$.
       State the procedures clearly and justify your decisions.

Fully worked solutions for this chapter are available online.

LESSON
6.9 Modelling with trigonometric functions

SYLLABUS LINKS

• Model and solve problems that involve trigonometric functions, with and without technology.

Source: Mathematical Methods Senior Syllabus 2024 © State of Queensland (QCAA) 2024; licensed under CC BY 4.0.

Phenomena that are cyclical in nature can often be modelled by a sine or cosine function.

Examples of periodic phenomena include sound waves, ocean tides and ovulation cycles. Trigonometric models may be able to approximate things such as the movement in the value of the All Ordinaries Index of the stock market, fluctuations in temperature, or the vibrations of violin strings about a mean position.

6.9.1 Maximum and minimum values

As $-1 \le \sin(x) \le 1$ and $-1 \le \cos(x) \le 1$, the maximum value of both $\sin(x)$ and $\cos(x)$ is 1 and the minimum value of both functions is -1. This can be used to calculate, for example, the maximum value of $y = 2\sin(x) + 4$ by substituting 1 for $\sin(x)$:

$$y_{max} = 2 \times 1 + 4 \Rightarrow y_{max} = 6$$

The minimum value can be calculated as:

$$y_{min} = 2 \times (-1) + 4 \Rightarrow y_{min} = 2$$

To calculate the maximum value of $y = 5 - 3\cos(2x)$, the largest negative value of $\cos(2x)$ would be substituted for $\cos(2x)$. Thus:

$$y_{max} = 5 - 3 \times (-1) \Rightarrow y_{max} = 8$$

The minimum value can be calculated by substituting the largest positive value of $\cos(2x)$:

$$y_{min} = 5 - 3 \times 1 \Rightarrow y_{min} = 2$$

Alternatively, identifying the equilibrium position and amplitude enables the range to be calculated.

For $y = 5 - 3\cos(2x)$, with amplitude 3 and equilibrium at $y = 5$, the range is calculated from $y = 5 - 3 = 2$ to $y = 5 + 3 = 8$, giving a range of $[2, 8]$. This also shows the maximum and minimum values.

WORKED EXAMPLE 24 Determining maximum and minimum values

The temperature, $T\,^\circ$C, during one day in April is given by $T = 17 - 4\sin\left(\dfrac{\pi}{12}t\right)$, where t is the time in hours after midnight.

a. Determine the temperature at midnight.
b. Determine the minimum temperature during the day and the time at which it occurred.
c. Identify the interval over which the temperature varied that day.
d. State the period and sketch the graph of the temperature for $t \in [0, 24]$.
e. If the temperature was below c degrees for 2.4 hours, obtain the value of c to 1 decimal place.

THINK	WRITE
a. State the value of t and use it to calculate the required temperature.	a. At midnight, $t = 0$. Substitute $t = 0$ into $T = 17 - 4\sin\left(\dfrac{\pi}{12}t\right)$. $T = 17 - 4\sin(0)$ $\quad = 17$ The temperature at midnight was 17°.
b. 1. State the condition on the value of the trigonometric function for the minimum value of T to occur.	b. $T = 17 - 4\sin\left(\dfrac{\pi}{12}t\right)$ The minimum value occurs when $\sin\left(\dfrac{\pi}{12}t\right) = 1$.
2. Calculate the minimum temperature.	$T_{min} = 17 - 4 \times 1$ $\qquad = 13$ The minimum temperature was 13°.

3. Calculate the time when the minimum temperature occurred.

The minimum temperature occurs when

$$\sin\left(\frac{\pi}{12}t\right) = 1.$$

Solving this equation, $\dfrac{\pi}{12}t = \dfrac{\pi}{2}$

$$t = 6$$

The minimum temperature occurred at 6 am.

c. Use the equilibrium position and amplitude to calculate the range of temperatures.

c. $T = 17 - 4\sin\left(\dfrac{\pi}{12}t\right)$

Amplitude 4, equilibrium $T = 17$
The range is $[17 - 4, 17 + 4] = [13, 21]$.
Therefore, the temperature varied between 13 °C and 21 °C.

d. 1. Calculate the period.

d. $2\pi \div \dfrac{\pi}{12} = 2\pi \times \dfrac{12}{\pi}$

$$= 24$$

The period is 24 hours.

2. Sketch the graph.

Dividing the period into quarters gives a horizontal scale of 6.

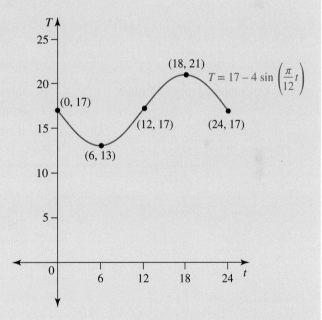

e. 1. Use the symmetry of the curve to deduce the end points of the interval involved.

e. $T < c$ for an interval of 2.4 hours.

For T to be less than the value for a time interval of 2.4 hours, the 2.4-hour interval is symmetric about the minimum point.

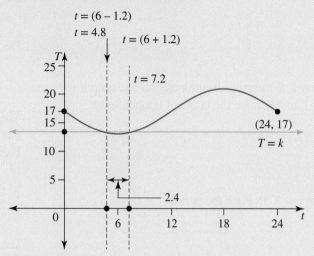

The end points of the t interval must each be $\frac{1}{2} \times 2.4$ from the minimum point where $t = 6$. The end points of the interval occur at $t = 6 \pm 1.2$.

$\therefore T = c$ when $t = 4.8$ or 7.2.

2. Calculate the required value.

Substituting either end point into the temperature model will give the value of c.

If $t = 4.8$:

$$T = 17 - 4\sin\left(\frac{\pi}{12} \times 4.8\right)$$

$$= 17 - 4\sin(0.4\pi)$$

$$= 13.2$$

Therefore, $c = 13.2$ and the temperature is below 13.2 °C for 2.4 hours.

---

on Resources

Interactivity Oscillation (int-2977)

| 6.9 Exercise | 6.9 Exam practice | on |

Simple familiar	Complex familiar	Complex unfamiliar
1, 2, 3, 4, 5, 6, 7, 8, 9, 10	11, 12, 13, 14, 15, 16	17, 18, 19, 20

Simple familiar

1. A child plays with a yo-yo attached to the end of an elastic string. The yo-yo rises and falls about its rest position so that its vertical distance, y cm, above its rest position at time t seconds is given by $y = -40 \cos(t)$.

   a. Sketch the graph of $y = -40 \cos(t)$ showing two complete cycles.
   b. Determine the greatest distance the yo-yo falls below its rest position.
   c. Determine the times at which the yo-yo returns to its rest position during the two cycles.
   d. Determine how many seconds it takes for the yo-yo to first reach a height of 20 cm above its rest position.

2. The temperature from 8 am to 10 pm on a day in February is shown.

   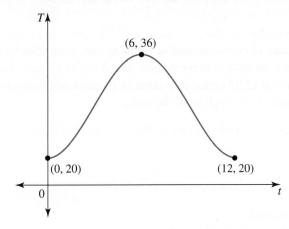

   If T is the temperature in degrees Celsius t hours from 8 am, form the equation of the temperature model and use this to calculate the times during the day when the temperature exceeded 30 degrees.

3. Emotional ups and downs are measured by a wellbeing index that ranges from 0 to 10 in increasing levels of happiness. A graph of this index over a four-week cycle is shown.

   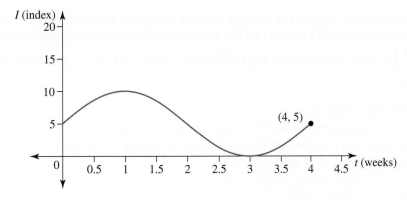

a. Express the relationship between the wellbeing index I and the time t in terms of a trigonometric equation.

b. A person with a wellbeing index of 6 or higher is considered to experience a high level of happiness. Determine the percentage of the four-week cycle for which the model predicts this feeling would last.

4. **WE24** During one day in October the temperature $T\,°C$ is given by $T = 19 - 3\sin\left(\dfrac{\pi}{12}t\right)$, where t is the time in hours after midnight.

a. Determine the temperature at midnight.

b. Determine the maximum temperature during the day and the time at which it occurred.

c. Identify the interval over which the temperature varied that day.

d. State the period and sketch the graph of the temperature for $t \in [0, 24]$.

e. If the temperature was below k degrees for 3 hours, obtain, to 1 decimal place, the value of k.

5. John is a keen amateur share trader who keeps careful records showing the fluctuation in share prices. One share in particular, Zentium, appears to have been following a sinusoidal shape with a period of two weeks over the last five weeks. Its share price has fluctuated between 12 and 15 cents, with its initial price at the start of the observations at its peak.

a. Given the Zentium share price can be modelled by $p = a\cos(nt) + b$, where p is the price in cents of the share t weeks after the start of the recorded observations, determine the values of a, n and b.

b. Sketch the graph of the share price over the last five weeks and state the price of the shares at the end of the five weeks.

c. When John decides to purchase the share, its price is 12.75 cents and rising. He plans to sell it immediately once it reaches 15 cents. According to the model, determine how many days it will be before he sells the share. Round your answer to the nearest day. Assume a 7-day trading week applies.

d. If John buys 10 000 shares at 12.75 cents, sells them at 15 cents and incurs brokerage costs of 1% when buying and selling, calculate how much profit he makes.

6. The depth of water, d metres, at the end of a pier at time t hours after 6 am is modelled by

$$d(t) = \sin\left(\dfrac{\pi t}{6}\right) + 2.5$$

a. State the period of the function.

b. State the maximum and minimum depths of the water.

c. Calculate is the depth at 7 am.

d. Sketch one cycle of the graph of the function.

e. The local council is interested in when the depth of water is at least 3.1 m. Determine for how long, correct to the nearest minute, during a 12-hour period the water is at least 3.1 m deep.

7. James is in a boat out at sea fishing. The weather makes a change for the worse and the water becomes very choppy. The depth of water above the sea bed can be modelled by the function with equation $d = 1.5\sin\left(\dfrac{\pi t}{12}\right) + 12.5$, where d is the depth of water in metres and t is the time in hours since the change of weather began.

a. State the period of the function.

b. State the maximum and minimum heights of the boat above the sea bed.

c. Calculate how far from the sea bed the boat was when the change of weather began.

d. Sketch one cycle of the graph of the function.

e. If the boat is h metres above the seabed for a continuous interval of 4 hours, calculate h correct to 1 decimal place.

f. James has heard on the radio that the cycle of weather should have passed within 12 hours, and when the height of water above the sea bed is at a minimum after that, it will be safe to return to shore. If the weather change occurred at 9:30 am, determine when James will be able to return to shore.

8. The temperature in degrees Celsius at Thredbo on a day in the middle of winter can be modelled by the equation

$$T = 2 - 6\cos\left(\frac{\pi t}{12}\right)$$

where t is the number of hours after 4:00 am.

a. Calculate the minimum and maximum temperatures.
b. Calculate the time(s) of the day at which the temperature is $0\,°C$. Give your answer(s) to the nearest minute.
c. Determine when is the temperature at its maximum.
d. Calculate the temperature at 8:00 am.

9. The water level in a harbour, h metres below a level jetty, at time t hours after 7 am is given by

$$h = 3 - 2.5\sin\left(\frac{1}{2}(t-1)\right)$$

a. Calculate how far below the jetty the water level in the harbour is at 7:30 am. Give your answer correct to 3 decimal places.
b. Determine the greatest and least distances below the jetty.
c. Sketch the graph of h versus t and hence determine the values of t at which the low and high tides first occur. Give your answers correct to 2 decimal places.
d. A boat ties up to the jetty at high tide. State how much extra rope will have to be left so that the boat is still afloat at low tide.

10. The height, h metres, of the tide above mean sea level is given by $h = 4\sin\left(\frac{\pi(t-2)}{6}\right)$, where t is the time in hours since midnight.

a. Determine how far below mean sea level the tide was at 1 am.
b. State the high tide level and show that this first occurs at 5 am.
c. Determine how many hours there are between high tide and the following low tide.
d. Sketch the graph of h versus t for $t \in [0, 12]$.
e. Determine what the height of the tide is predicted to be at 2 pm.
f. Determine how much higher than low tide level the tide is at 11:30 am. Give your answer to 2 decimal places.

11. The diagram shows the cross-section of a sticky tape holder with a circular hole through its side. It is bounded by a curve with the equation $h = a \sin\left(\dfrac{\pi}{5} x\right) + b$, where h cm is the height of the curve above the base of the holder at distance x cm along the base.

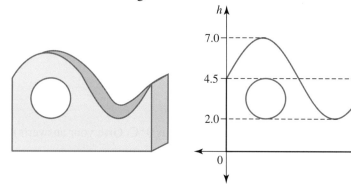

a. Use the measurements given on the graph to state the values of a and b and hence write down the equation of the bounding curve.
b. Calculate the length of the base of the holder.
c. If the centre of the circular hole lies directly below the highest point of the curve, determine the coordinates of the centre.
d. Using the symmetry of the curve, calculate the cross-sectional area lightly shaded in the diagram, to 1 decimal place.

12. In a laboratory, an organism is being grown in a test tube. To help increase the rate at which the organism grows, the test tube is placed in an incubator where the temperature $T\,°C$ varies according to the rule $T = 30 - \cos\left(\dfrac{\pi}{12} t\right)$, where t is the time in minutes since the test tube has been placed in the incubator.

a. State the range of temperature in the incubator.
b. Determine how many minutes after the test tube is placed in the incubator the temperature reaches its greatest value.
c. Sketch the graph of the temperature against time for the first half hour.
d. Determine how many cycles are completed in 1 hour.
e. If the organism is kept in the incubator for 2.5 hours, determine the temperature at the end of this time.
f. Express the rule for the temperature in terms of a sine function.

13. During a particular day in a Mediterranean city, the temperature inside an office building between 10 am and 7:30 pm fluctuates so that t hours after 10 am, the temperature $T\,°C$ is given by $T = 19 + 6\sin\left(\dfrac{\pi t}{6}\right)$.

a. i. State the maximum temperature and the time it occurs.
   ii. State the minimum temperature and the time it occurs.
b. i. Determine the temperature in the building at 11:30 am. Answer to 1 decimal place.
   ii. Determine the temperature in the building at 7:30 pm. Answer to 1 decimal place.
c. Sketch the graph of the temperature against time from 10 am to 7:30 pm.
d. When the temperature reaches 24°, an air conditioner in the boardroom is switched on and it is switched off when the temperature in the rest of the building falls below 24°. Determine for how long the air conditioner is on in the boardroom.
e. The office workers who work the shift between 11:30 am and 7:30 pm complain that the temperature becomes too cool towards the end of their shift. If management agrees that heating can be used for the

coldest two-hour period of their shift, identify the time and temperature at which the heating would be switched on. Express the temperature in both exact form and to 1 decimal place.

14. The height above ground level of a particular carriage on a Ferris wheel is given by $h = 10 - 8.5 \cos\left(\dfrac{\pi}{60}t\right)$ where h is the height in metres above ground level after t seconds.

   a. Calculate the initial height of the carriage.
   b. Calculate how high the carriage will be after one minute.
   c. Determine how many revolutions the Ferris wheel will complete in a four-minute time interval.
   d. Sketch the graph of h against t for the first four minutes.
   e. Determine for how long, to the nearest second, in one revolution, the carriage is higher than 12 metres above the ground.
   f. The carriage is attached by strong wire radial spokes to the centre of the circular wheel. Calculate the length of a radial spoke.

15. A person sunbathing at a position P on a beachfront observes the waves wash onto the beach in such a way that after t minutes, the distance p metres of the end of the water's wave from P is given by $p = 3 \sin(n\pi t) + 5$.

   a. Determine the closest distance the water reaches to the sunbather at P.
   b. Over a one-hour interval, the sunbather counts 40 complete waves that have washed onto the beach. Calculate the value of n.
   c. At some time later in the day, the distance p metres of the end of the water's wave from P is given by $p = a \sin(4\pi t) + 5$. If the water just reaches the sunbather who is still at P, deduce the value of a and determine how many times in half an hour the water reaches the sunbather at P.
   d. Determine in which of the two models of the wave motion, $p = 3 \sin(n\pi t) + 5$ or $p = a \sin(4\pi t) + 5$, the number of waves per minute is greater.

16. A discarded drink can floating in the waters of a creek oscillates vertically up and down 20 cm about its equilibrium position. Its vertical displacement, x metres, from its equilibrium position after t seconds is given by $x = a \sin(bt)$. Initially the can moved vertically downwards from its mean position, and it took 1.5 seconds to rise from its lowest point to its highest point.

   a. Determine the values of a and b and state the equation for the vertical displacement.
   b. Sketch one cycle of the motion.
   c. Calculate the shortest time, T seconds, for the can's displacement to be one-half the value of its amplitude.
   d. Determine the total distance that the can moved in one cycle of the motion.

17. At a suburban shopping centre, one of the stores sells electronic goods such as digital cameras, laptop computers and printers. The store had a one-day sale towards the end of the financial year. The doors opened at 7:55 am and the cash registers opened at 8:00 am. The store closed its doors at 11:00 pm. The total number of people queuing at the six cash registers at any time during the day once the cash registers opened could be modelled by the equation

$$N(t) = 45 \sin\left(\frac{\pi t}{5}\right) - 35 \cos\left(\frac{\pi t}{3}\right) + 68, 0 \le t \le 15$$

where $N(t)$ is the total number of people queuing t hours after the cash registers opened at 8:00 am. Sketch the graph representing the total number of people queuing and determine the maximum number of people in the queue between 3:00 pm and 7:00 pm.

18. For a particular incubated animal cage, the temperature can be modelled by a positive cosine curve, $T = a \cos(b(t-h)) + k$, where $T\,°C$ is the temperature t hours after 8 am. The maximum daily temperature was 28 °C and occurred at 2 pm, and the minimum was 22 °C, occurring at 8 pm. The creature that lives in the incubated cage sleeps when the temperature falls below 24 °C. Determine for how many hours a day, correct to 1 decimal place, the creature sleeps.

19. The intensity, I, of sound emitted by a device is given by $I = 4 \sin(t) - 3 \cos(t)$, where t is the number of hours the device has been operating.
Using technology to sketch the graph of the intensity of sound, determine the maximum intensity the device produces and identify, correct to 2 decimal places, the first value of t for which $I = 0$.

20. The teeth of a tree saw can be approximated by the function $y = x + 4 + 4 \cos(6x)$, $0 \le x \le 4\pi$, where y cm is the vertical height of the teeth at a horizontal distance x cm from the end of the saw.

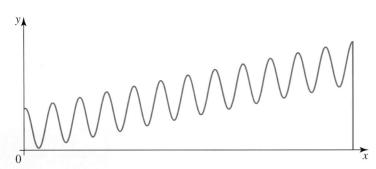

Each peak of the graph represents one of the teeth.
Determine exactly how far apart the successive peaks of the teeth of the saw are and obtain the equation of the linear function that will touch each of the teeth.

Fully worked solutions for this chapter are available online.

LESSON
6.10 Review

6.10.1 Summary

doc-41801

Hey students! Now that it's time to revise this chapter, go online to:

Access the chapter summary

Review your results

Practise exam questions

Find all this and MORE in jacPLUS

6.10 Exercise

learn**on**

| 6.10 Exercise | 6.10 Exam practice **on** |

Simple familiar	Complex familiar	Complex unfamiliar
1, 2, 3, 4, 5, 6, 7, 8, 9, 10, 11, 12	13, 14, 15, 16	17, 18, 19, 20

These questions are even better in jacPLUS!
- Receive immediate feedback
- Access sample responses
- Track results and progress

Find all this and MORE in jacPLUS

Simple familiar

1. **MC** In a rectangle ABCD, the angle CAD is 27° and the side AD is 3 cm. The length of the diagonal AC in cm is closest to:
   - **A.** 6.61
   - **B.** 5.89
   - **C.** 3.37
   - **D.** 2.67

2. **MC** The exact value of $\sin(45°) + \tan(30°) \times \cos(60°)$ is:
   - **A.** $\dfrac{3}{2}$
   - **B.** $\dfrac{\sqrt{2} + 1}{2}$
   - **C.** $\dfrac{3\sqrt{2} + \sqrt{3}}{6}$
   - **D.** $\dfrac{3\sqrt{2} + 2\sqrt{3}}{12}$

3. Express the following in degree measure.
   - **a.** $\dfrac{11\pi^c}{9}$
   - **b.** $-3.5\pi^c$

4. An arc subtends an angle of 114° at the centre of a circle of radius 6.2 cm.
   Find:
   - **a.** the length of the arc
   - **b.** the area of the sector with subtended angle 114°.

5. **MC** Consider the unit circle diagram shown.

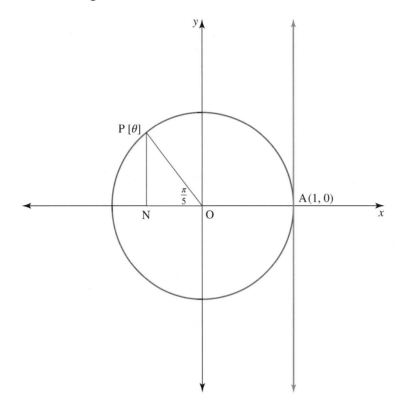

a. A possible value of θ for the trigonometric point P $[\theta]$ is:

A. $\dfrac{\pi}{5}$ **B.** $\dfrac{4\pi}{5}$ **C.** $\dfrac{6\pi}{5}$ **D.** $\dfrac{7\pi}{10}$

b. The value of $\sin(\theta)$ is given by the length of the line segment:

A. OP   **B.** PA   **C.** ON   **D.** NP

6. **MC** Identify the quadrant(s) in which $\sin(\theta) < 0$ and $\cos(\theta) > 0$.

A. First quadrant   **B.** Second quadrant   **C.** Third quadrant   **D.** Fourth quadrant

7. If $\cos(t) = 0.6$, evaluate the following.

a. $\cos(-t)$ b. $\cos(\pi + t)$ c. $\cos(3\pi - t)$ d. $\cos(-2\pi + t)$

8. **MC** The range of the graph of $y = 2 + 4\cos\left(\dfrac{3x}{2}\right)$ is:

A. R

B. $[2, 6]$

C. $[-2, 6]$

D. $[-6, 2]$

9. **MC** A possible equation for the graph shown could be:

A. $y = -3\cos(2x)$

B. $y = 3\cos\left(\dfrac{x}{2}\right)$

C. $y = -3\sin(2x)$

D. $y = -3\sin\left(\dfrac{x}{2}\right)$

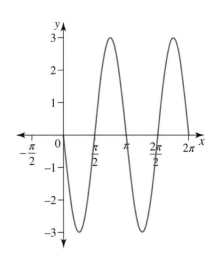

10. Sketch the following graphs over the given domain.

a. $y = \sin(x), 0 \leq x \leq \dfrac{7\pi}{2}$

b. $y = \cos(x), -\pi \leq x < 3\pi$

c. $y = \tan(\theta)$ for the domain $-2\pi \leq \theta \leq 2\pi$

Sketch the following graphs, stating the range and identifying the key features.

d. $y = 5 - 3\cos(x), 0 \leq x \leq 2\pi$

e. $y = 5\sin(3x), 0 \leq x \leq 2\pi$

f. $y = 1 + 2\cos\left(\dfrac{x}{2}\right), -2\pi \leq x \leq 3\pi$

11. **MC** The number of solutions to the equation $\cos(x) = 0, -\dfrac{\pi}{2} \leq x \leq \dfrac{5\pi}{2}$ is:

A. 1 **B.** 2 **C.** 3 **D.** 4

12. **MC** The solutions to the equation $\sin(x) = \dfrac{1}{2}, 0 \leq x \leq 2\pi$ are:

A. $x = \dfrac{\pi}{6}, \dfrac{5\pi}{6}$ **B.** $x = \dfrac{2\pi}{3}, \dfrac{4\pi}{3}$ **C.** $x = \dfrac{5\pi}{6}, \dfrac{7\pi}{6}$ **D.** $x = \dfrac{4\pi}{3}, \dfrac{5\pi}{3}$

Complex familiar

13. Evaluate $\cos\left(\dfrac{11\pi}{6}\right) - \tan\left(\dfrac{11\pi}{3}\right) + \sin\left(-\dfrac{11\pi}{4}\right)$ exactly.

14. a. Given $\tan(x) = \dfrac{5}{2}, x \in \left(\pi, \dfrac{3\pi}{2}\right)$, obtain the exact values for $\cos(x)$ and $\sin(x)$.

b. Given $\sin(y) = -\dfrac{\sqrt{5}}{3}, y \in \left(\dfrac{3\pi}{2}, 2\pi\right)$, obtain the exact values for $\cos(y)$ and $\tan(y)$.

15. If $\cos(\theta) = p$, express the following in terms of p.

a. $\sin\left(\dfrac{\pi}{2} - \theta\right)$ b. $\sin\left(\dfrac{3\pi}{2} + \theta\right)$ c. $\sin^2(\theta)$ d. $\cos(5\pi - \theta)$

16. Solve the following equations for x.

a. $\sqrt{6}\cos(x) = -\sqrt{3}, 0 \leq x \leq 2\pi$

b. $2 - 2\cos(x) = 0, 0 \leq x \leq 4\pi$

c. $2\sin(x) = \sqrt{3}, -2\pi \leq x \leq 2\pi$

d. $\sqrt{5}\sin(x) = \sqrt{5}\cos(x), 0 \leq x \leq 2\pi$

e. $8\sin(3x) + 4\sqrt{2} = 0, -\pi \leq x \leq \pi$

f. $\tan(2x°) = -2, 0° \leq x \leq 270°$

g. $2\sin^2(x) - \sin(x) - 3 = 0, 0 \leq x \leq 2\pi$

h. $2\sin^2(x) - 3\cos(x) - 3 = 0, 0 \leq x \leq 2\pi$

Complex unfamiliar

17. P is the trigonometric point $\left[\dfrac{3\pi}{4}\right]$ and Q is the trigonometric point $[-2]$. Determine the value for the angle POQ where O is the centre of the unit circle, expressing your answer in degrees, correct to 2 decimal places.

18. The depth of water, d metres, at the end of a pier at time t hours after 10 am is modelled by

$$d(t) = 2 \sin\left(\frac{\pi t}{6} - \frac{\pi}{3}\right) + 2.5$$

A particular boat needs a depth of at least 4.2 metres in order to dock. Determine the maximum and minimum depths of the water and the times, to the nearest minute, during which the boat can dock.

19. Two circular pulleys with radii 3 cm and 8 cm have their centres 13 cm apart, as shown in the diagram.

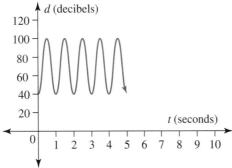

Determine the exact length of DC and the length of the belt required to pass tightly around the pulleys. Give your answer for the length of the belt to 1 decimal place.

20. Sound waves emitted from an amplifier during a band practice session are illustrated over a 5-second interval, with the vertical axis measuring the sound in decibels.

The graph can be expressed in the form $d = a\cos(bt) + c$, where d is the number of decibels of the sound after t seconds. Sound under 80 decibels is considered to be a safe level.

Determine the percentage of time, to 1 decimal place, during 10 minutes of a band practice session that the level of the sound was potentially damaging to anyone present with unprotected hearing.

Fully worked solutions for this chapter are available online.

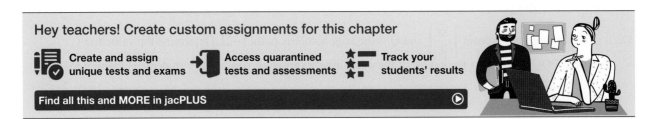

Answers

Chapter 6 Trigonometric functions

6.2 Trigonometry review

6.2 Exercise

1. a. $\dfrac{\sqrt{2}}{2}$ b. $\dfrac{\sqrt{3}}{3}$ c. $\dfrac{1}{2}$ d. 1

2. a. $h = 7.66$ b. $a \approx 68.20°$

3. a. $x = 7.13, y = 3.63$ b. $x = 13.27, h = 16.62$

4. a. $66.42°$ b. $51.34°$

5. $2\sqrt{2}$ m

6. $\dfrac{\sqrt{3}}{2}$ m

7. $\dfrac{3\sqrt{2} - \sqrt{6}}{8}$

8. $\dfrac{\sqrt{6}}{12}$

9. a. $\cos(a) = \dfrac{3}{\sqrt{13}}$

   b. $6\sqrt{13}$ cm

10. a. $\dfrac{\sqrt{21}}{7}$ b. $\dfrac{\sqrt{11}}{5}$ c. $\dfrac{2}{3}$

11. a. 6.18 m b. 59.5°, 5.3 m

12. 22°

13. a. 1.736 cm

    b. $20\sqrt{3}$ cm $= \dfrac{100\sqrt{3}}{3}$ cm^2

14. 36 cm

15. 26.01 km

16. 21.6°

17. $AC = 12\sqrt{3}$ cm; $BC = 18\sqrt{2}$ cm; $AB = 18 + 6\sqrt{3}$ cm

18. 35.26°

19. Height 63.4 m; distance 63.4 m

20. a. Sample responses can be found in the worked solutions in the online resources.

    b. $a = 60°$; $m = 16$; $n = 12$

6.3 Radian measure

6.3 Exercise

1. a.

Degrees	30°	45°	60°
Radians	$\dfrac{\pi}{6}$	$\dfrac{\pi}{4}$	$\dfrac{\pi}{3}$

b.

Degrees	0°	90°	180°	270°	360°
Radians	0	$\dfrac{\pi}{2}$	π	$\dfrac{3\pi}{2}$	2π

2. a. $\dfrac{\pi}{3}$ b. 135° c. 30°; $\dfrac{\sqrt{3}}{3}$

3. a. 36° b. 120° c. 75° d. 330° e. 140° f. 810°

4. a. $\dfrac{2\pi}{9}$ b. $\dfrac{5\pi}{6}$ c. $\dfrac{5\pi}{4}$ d. $\dfrac{5\pi}{3}$ e. $\dfrac{7\pi}{4}$ f. 4π

5. a. 2.572 b. 0.021

6. a. i. 1.557 ii. 0.623 iii. 0.025

b.

θ	$\dfrac{\pi}{6}$	$\dfrac{\pi}{4}$	$\dfrac{\pi}{3}$
$\sin(\theta)$	$\dfrac{1}{2}$	$\dfrac{\sqrt{2}}{2}$	$\dfrac{\sqrt{3}}{2}$
$\cos(\theta)$	$\dfrac{\sqrt{3}}{2}$	$\dfrac{\sqrt{2}}{2}$	$\dfrac{1}{2}$
$\tan(\theta)$	$\dfrac{\sqrt{3}}{3}$	1	$\sqrt{3}$

7. a. $-4\pi, -2\pi, 0, 2\pi, 4\pi$

   b. $-1 - 4\pi, -1 - 2\pi, -1, -1 + 2\pi, -1 + 4\pi$

8. a. $\approx 103.1° \left(\dfrac{324°}{\pi} \right)$

   b.

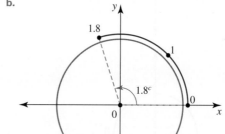

9. a. i.

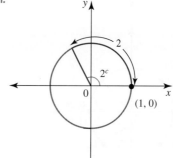

ii.

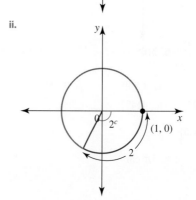

iii.

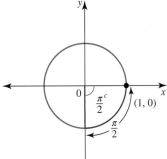

b. i.

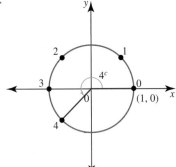

ii.

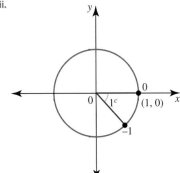

iii.

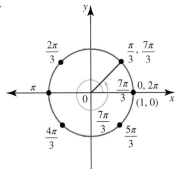

10. a. 10.47 cm   b. 216°
11. 2.53
12. a. i. 0.052   ii. 1.959   iii. 3.759
    b. i. 171.887°   ii. 414°
    c. $\left\{\dfrac{\pi}{7}, 50°, 1.5\right\}$
13. a. 10π cm   b. 5π cm
14. 12 mm

15. a. $\dfrac{\pi}{8}$ b. 22.5°
16. 20.706 cm^2
17. 23.33 m^2
18. 17.2°
19. 191° or $\dfrac{10}{3}$
20. 5.76 mm
21. 81.5 cm^2
22. a. 6371 km   b. 6360 km   c. 6004 km

6.4 Unit circle definitions

6.4 Exercise

1. a. Quadrant 1   b. Quadrant 3   c. Quadrant 2
   d. Quadrant 1   e. Quadrant 4   f. Quadrant 3
2. a. Quadrant 3
   b. Quadrant 2
   c. Boundary of quadrant 1 and quadrant 4
   d. Quadrant 4
3. a. i. $(0, 1)$ ii. $\sin(\theta) = 1$
   b. i. $(-1, 0)$ ii. $\cos(\alpha) = -1$
   c. i. $(0, -1)$ ii. $\tan(\beta)$ is undefined.
   d. i. $(1, 0)$
      ii. $\sin(v) = 0, \cos(v) = 1, \tan(v) = 0$
4. a. $P\left[\dfrac{\pi}{2}\right]$
   b.

120°	$-400° =$ $-360° - 40°$	$\dfrac{4\pi}{3} = \pi + \dfrac{1}{3}\pi$	$\dfrac{\pi}{4}$
Quadrant 2	Quadrant 4	Quadrant 3	Quadrant 1

   c. $Q[-240°]$; $R[480°]$ or other possible answers if keep rotating by 360°.
5. a. $\left(\dfrac{\sqrt{3}}{2}, \dfrac{1}{2}\right)$

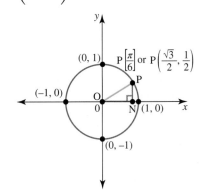

b.

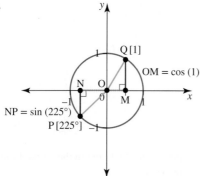

c. $\cos\left(-\dfrac{\pi}{2}\right) = 0; \sin\left(-\dfrac{\pi}{2}\right) = -1$

d. $f(0) = 0$

6. a. $f\left(\dfrac{3\pi}{2}\right) = \sin\left(\dfrac{3\pi}{2}\right) = -1$

b. $g(4\pi) = \cos(4\pi) = 1$

c. $h(-\pi) = \tan(-\pi) = 0$

d. $k(6.5\pi) = \sin\left(\dfrac{13\pi}{2}\right) + \cos\left(\dfrac{13\pi}{2}\right) = 1.$

7. a. First and second quadrants

b. First and fourth quadrants

8. a. $\left(\dfrac{\sqrt{2}}{2}, \dfrac{\sqrt{2}}{2}\right)$ **b.** $\left[\dfrac{3\pi}{2}\right]$ or $\left[-\dfrac{\pi}{2}\right]$

9. a–d.

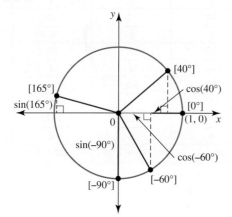

10. a–d.

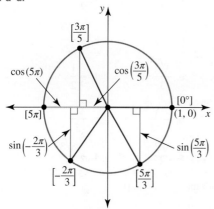

11. a–d.

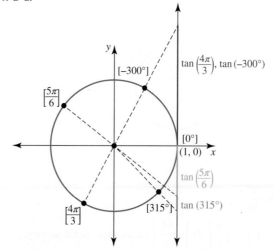

12. a. Quadrant 2; $\sin(\theta) = 0.6, \cos(\theta) = -0.8$, $\tan(\theta) = -0.75$

b. Quadrant 4; $\sin(\theta) = -\dfrac{\sqrt{2}}{2}, \cos(\theta) = \dfrac{\sqrt{2}}{2}$, $\tan(\theta) = -1$

c. Quadrant 1; $\sin(\theta) = \dfrac{1}{\sqrt{5}}, \cos(\theta) = \dfrac{2}{\sqrt{5}}, \tan(\theta) = \dfrac{1}{2}$

d. Boundary between quadrant 1 and quadrant 2; $\sin(\theta) = 1, \cos(\theta) = 0, \tan(\theta) = $ undefined

13. a. 1 **b.** 1 **c.** 0 **d.** 0 **e.** 0 **f.** −1

14. a. $\tan\left(\dfrac{5\pi}{2}\right); \tan(-90°)$ **b.** $\tan\left(\dfrac{3\pi}{4}\right)$

15. a. $\tan(230°) \approx 1.192$

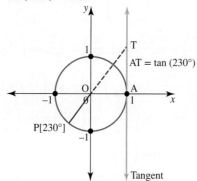

b. $\tan(2\pi) = 0$

16. a.

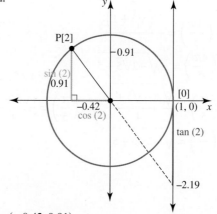

b. $(-0.42, 0.91)$

17. a.

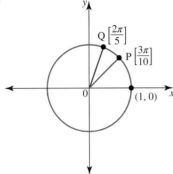

b. $\dfrac{\pi^c}{10}$

c. $P\left[-\dfrac{17\pi}{10}\right]$; $Q\left[-\dfrac{8\pi}{5}\right]$ (other answers are possible)

d. $P\left[\dfrac{23\pi}{10}\right]$; $Q\left[\dfrac{12\pi}{5}\right]$ (other answers are possible)

18. Sample responses can be found in the worked solutions in the online resources.

19. a. $P\left(\dfrac{\sqrt{2}}{2}, \dfrac{-\sqrt{2}}{2}\right)$ and $Q\left(\dfrac{\sqrt{2}}{2}, \dfrac{\sqrt{2}}{2}\right)$ are reflections in the x-axis.

b. $R\left(\dfrac{-(\sqrt{5}+1)}{4}, \dfrac{\sqrt{2(-\sqrt{5}+5)}}{4}\right)$ and

$S\left(\dfrac{\sqrt{5}+1}{4}, \dfrac{\sqrt{2(-\sqrt{5}+5)}}{4}\right)$ are reflections in the y-axis.

c. i. $\sin\left(\dfrac{7\pi}{4}\right) = \dfrac{-\sqrt{2}}{2}, \cos\left(\dfrac{7\pi}{4}\right) = \dfrac{\sqrt{2}}{2}, \tan\left(\dfrac{7\pi}{4}\right) = -1$

$\sin\left(\dfrac{\pi}{4}\right) = \dfrac{\sqrt{2}}{2}, \cos\left(\dfrac{\pi}{4}\right) = \dfrac{\sqrt{2}}{2}, \tan\left(\dfrac{\pi}{4}\right) = 1$

$\sin\left(\dfrac{7\pi}{4}\right) = -\sin\left(\dfrac{\pi}{4}\right), \cos\left(\dfrac{7\pi}{4}\right) = \cos\left(\dfrac{\pi}{4}\right),$

$\tan\left(\dfrac{7\pi}{4}\right) = -\tan\left(\dfrac{\pi}{4}\right)$

ii. $\sin\left(\dfrac{4\pi}{5}\right) = \dfrac{\sqrt{2(-\sqrt{5}+5)}}{4},$

$\cos\left(\dfrac{4\pi}{5}\right) = -\dfrac{\left(\sqrt{5}+1\right)}{4},$

$\tan\left(\dfrac{4\pi}{5}\right) = -\sqrt{-2\sqrt{5}+5}$

$\sin\left(\dfrac{\pi}{5}\right) = \dfrac{\sqrt{2(-\sqrt{5}+5)}}{4}, \cos\left(\dfrac{\pi}{5}\right) = \dfrac{(\sqrt{5}+1)}{4},$

$\tan\left(\dfrac{\pi}{5}\right) = \sqrt{-2\sqrt{5}+5}$

$\sin\left(\dfrac{4\pi}{5}\right) = \sin\left(\dfrac{\pi}{5}\right), \cos\left(\dfrac{4\pi}{5}\right) = -\cos\left(\dfrac{\pi}{5}\right),$

$\tan\left(\dfrac{4\pi}{5}\right) = -\tan\left(\dfrac{\pi}{5}\right)$

20. a. 1

b. $-\dfrac{\sqrt{3}}{2} - \dfrac{1}{2}$

c. 1

d. 1

e. 1; sample responses can be found in the worked solutions in the online resources.

6.5 Exact values and symmetry properties

6.5 Exercise

1. a. 1   **b.** 0   **c.** 0
   d. 1   **e.** 0   **f.** 0

2. a. Fourth

b. First

c. Second

d. Boundary between quadrants 1 and 2, and boundary between quadrants 3 and 4

e. Boundary between quadrants 1 and 2

f. Boundary between quadrants 2 and 3

3.

	Quadrant 2	Quadrant 3	Quadrant 4
a.	$\dfrac{2\pi}{3}$	$\dfrac{4\pi}{3}$	$\dfrac{5\pi}{3}$
b.	$\dfrac{5\pi}{6}$	$\dfrac{7\pi}{6}$	$\dfrac{11\pi}{6}$
c.	$\dfrac{3\pi}{4}$	$\dfrac{5\pi}{4}$	$\dfrac{7\pi}{4}$
d.	$\dfrac{4\pi}{5}$	$\dfrac{6\pi}{5}$	$\dfrac{9\pi}{5}$
e.	$\dfrac{5\pi}{8}$	$\dfrac{11\pi}{8}$	$\dfrac{13\pi}{8}$
f.	$\pi - 1$	$\pi + 1$	$2\pi - 1$

4. a. $-\dfrac{1}{2}$ **b.** 1   **c.** $-\dfrac{1}{2}$

d. $-\sqrt{3}$ **e.** $\dfrac{\sqrt{2}}{2}$ **f.** $\dfrac{1}{2}$

5. a. $\dfrac{\sqrt{2}}{2}$ **b.** $-\sqrt{3}$ **c.** $-\dfrac{\sqrt{3}}{2}$

d. $-\dfrac{1}{2}$ **e.** $\dfrac{\sqrt{3}}{3}$ **f.** $-\dfrac{1}{2}$

6. a. $\dfrac{\sqrt{2}}{2}$ **b.** $-\dfrac{\sqrt{3}}{2}$ **c.** $\dfrac{\sqrt{3}}{3}$

d. $\dfrac{\sqrt{3}}{2}$ **e.** $\dfrac{\sqrt{2}}{2}$ **f.** $-\dfrac{\sqrt{3}}{3}$

7. a. Third quadrant   **b.** 0

8. 1

9. a. $-\dfrac{\sqrt{3}}{2}$　**b.** $-\dfrac{\sqrt{3}}{3}$　**c.** $\dfrac{\sqrt{3}}{2}$

10. $\sin\left(-\dfrac{5\pi}{4}\right) = \dfrac{\sqrt{2}}{2}; \cos\left(-\dfrac{5\pi}{4}\right) = -\dfrac{\sqrt{2}}{2};$

$\tan\left(-\dfrac{5\pi}{4}\right) = -1$

11. a. -0.2　**b.** -0.2　**c.** 0.2　**d.** 0.2

12. a. -4　**b.** 0.9　**c.** -4　**d.** 3.1

13. a. -0.91　**b.** 0.43　**c.** -0.47
d. 0.91　**e.** -0.43　**f.** 0.47

14. a. $-p$　**b.** p　**c.** $-p$　**d.** p

15. a. $\dfrac{-\left(\sqrt{3}+1\right)}{2}$　**b.** $2 - \sqrt{2}$　**c.** $2\sqrt{3}$

d. 0　**e.** 0　**f.** -2

16. a. $[105°]; [255°]; [285°]; \cos(285°) = \cos(75°); [285°]$

b. $-\tan\left(\dfrac{\pi}{7}\right)$

c. $\sin(\pi - \theta) = 0.8, \sin(2\pi - \theta) = -0.8$

d. i. $\cos\left(\dfrac{5\pi}{4}\right) = -\dfrac{\sqrt{2}}{2}$

ii. $\sin\left(\dfrac{25\pi}{6}\right) = \dfrac{1}{2}$

17. a. $\cos(-\theta) = p$
b. $\cos(5\pi + \theta) = -p$

18. a. Sample responses can be found in the worked solutions in the online resources.
b. Same x-coordinates
c. $\sin(\pi + \phi) = -0.87, \cos(\pi + \phi) = 0.5,$
$\tan(\pi + \phi) = -1.74$
d. $\sin(t)$
e. $A = 36°$ or $-216°$ (other answers are possible)
f. $B = \dfrac{13\pi}{11}$ or $\dfrac{20\pi}{11}$ or $-\dfrac{2\pi}{11}$ (other answers are possible)

19. a. Third quadrant
b. $(-0.49, -0.87)$
c. Quadrant 1, $\theta = 1.0587$; quadrant 2, $\theta = 2.0829$; quadrant 4, $\theta = 5.2244$

20. a. Quadrant 1 or quadrant 3
b. Quadrant 1, $\theta = 1.3734$; quadrant 3, $\theta = 4.5150$

c. Quadrant 1: $\left(\dfrac{1}{\sqrt{26}}, \dfrac{5}{\sqrt{26}}\right)$; quadrant 3:

$\left(-\dfrac{1}{\sqrt{26}}, -\dfrac{5}{\sqrt{26}}\right)$

6.6 Graphs of the sine, cosine and tangent functions

6.6 Exercise

1. a. Domain $[0, 4\pi]$, range $[-1, 1]$
b. $y = \sin(x)$
c. $\left(\dfrac{\pi}{2}, 1\right), \left(\dfrac{3\pi}{2}, -1\right), \left(\dfrac{5\pi}{2}, 1\right), \left(\dfrac{7\pi}{2}, -1\right).$
d. Period 2π, amplitude 1
e. $y = 0$ (the x-axis)
f. $x \in (0, \pi) \cup (2\pi, 3\pi)$.

2. a. Domain $[-2\pi, 2\pi]$, range $[-1, 1]$
b. $y = \cos(x)$
c. $(-\pi, -1), (\pi, -1)$
d. Period 2π, amplitude 1, mean position $y = 0$ (the x-axis)
e. $\left(-\dfrac{3\pi}{2}, 0\right), \left(-\dfrac{\pi}{2}, 0\right), \left(\dfrac{\pi}{2}, 0\right), \left(\dfrac{3\pi}{2}, 0\right)$
f. $x \in \left(-\dfrac{3\pi}{2}, -\dfrac{\pi}{2}\right) \cup \left(\dfrac{\pi}{2}, \dfrac{3\pi}{2}\right).$

3. Three cycles

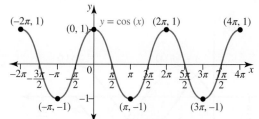

4. a.

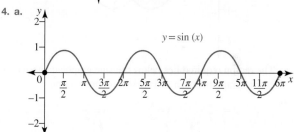

b.

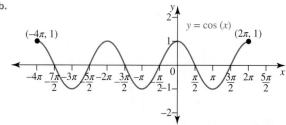

c.

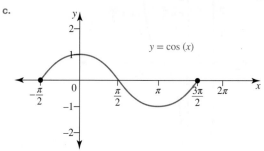

d.

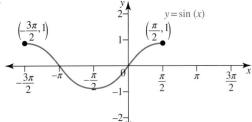

5. a. 2 maximum turning points

   b. 7 minimum turning points

6. $4\frac{1}{2}$ cycles

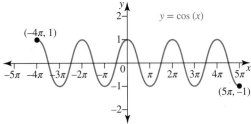

7. a. 4   **b.** 7   **c.** 21   **d.** 3

8.

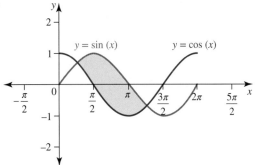

The region required lies below the sine graph and above the cosine graph between their points of intersection.

9.

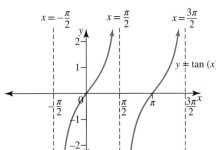

10. a.

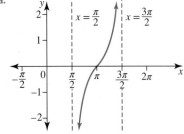

b.

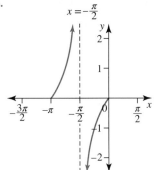

c.

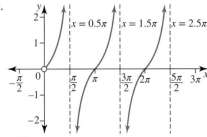

11. $a = \dfrac{19\pi}{4}$

12. $b = -2\pi$

13. $c = \dfrac{5\pi}{2}$

14. $\dfrac{\pi}{2} < x < \dfrac{3\pi}{2}$

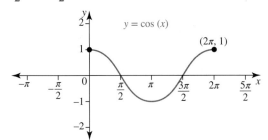

15. Sample responses can be found in the worked solutions in the online resources.

16. a.

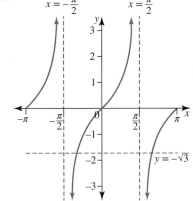

   b. $x = -\dfrac{\pi}{3}, \dfrac{2\pi}{3}$

17. $\dfrac{\pi}{6} \le x \le \dfrac{\pi}{2}$

18. $-\dfrac{\pi}{2} \le x \le -\dfrac{\pi}{3}$ or $\dfrac{\pi}{3} \le x \le \dfrac{\pi}{2}$

19. $\dfrac{\pi}{4} < x \le \pi$

20. $\dfrac{\pi}{2} < x \le \dfrac{3\pi}{4}$

6.7 Solving trigonometric equations

6.7 Exercise

1. a. $\dfrac{\pi}{4}, \dfrac{7\pi}{4}$ **b.** $\dfrac{5\pi}{4}, \dfrac{7\pi}{4}$ **c.** $\dfrac{5\pi}{6}, \dfrac{11\pi}{6}$

   d. $\dfrac{5\pi}{6}, \dfrac{7\pi}{6}$ **e.** $\dfrac{\pi}{6}, \dfrac{5\pi}{6}$ **f.** $\dfrac{\pi}{3}, \dfrac{4\pi}{3}$

2. a. $\theta = -\dfrac{7\pi}{4}, -\dfrac{3\pi}{4}, \dfrac{\pi}{4}, \dfrac{5\pi}{4}$

   b. $\theta = -\dfrac{4\pi}{3}, -\dfrac{2\pi}{3}, \dfrac{2\pi}{3}, \dfrac{4\pi}{3}$

   c. $\theta = -\dfrac{5\pi}{6}, -\dfrac{\pi}{6}, \dfrac{7\pi}{6}, \dfrac{11\pi}{6}$

3. a. $x = \dfrac{2\pi}{3}, \dfrac{5\pi}{3}, \dfrac{8\pi}{3}$

   b. $t = \dfrac{\pi}{6}, \dfrac{5\pi}{6}, \dfrac{13\pi}{6}, \dfrac{17\pi}{6}$

   c. $v = -\dfrac{\pi}{4}, \dfrac{\pi}{4}, \dfrac{7\pi}{4}, \dfrac{9\pi}{4}, \dfrac{15\pi}{4}, \dfrac{17\pi}{4}$

4. a. $\dfrac{\pi}{6}, \dfrac{5\pi}{6}$ **b.** $\dfrac{\pi}{6}, \dfrac{11\pi}{6}$

   c. $-\dfrac{5\pi}{4}, -\dfrac{\pi}{4}, \dfrac{3\pi}{4}, \dfrac{7\pi}{4}$

5. a. 2 solutions   **b.** $\pm 120°$

6. a. $240°, 300°$ **b.** $45°, 225°$

   c. $120°, 240°$ **d.** $30°, 150°$

7. a. $-\pi, 0, \pi, 2\pi, 3\pi, 4\pi$ **b.** $-\dfrac{\pi}{2}, \dfrac{\pi}{2}, \dfrac{3\pi}{2}, \dfrac{5\pi}{2}, \dfrac{7\pi}{2}$

   c. $-\dfrac{\pi}{2}, \dfrac{3\pi}{2}, \dfrac{7\pi}{2}$ **d.** $0, 2\pi, 4\pi$

   e. $\dfrac{\pi}{2}, \dfrac{5\pi}{2}$ **f.** $-\dfrac{3\pi}{4}, \dfrac{\pi}{4}, \dfrac{5\pi}{4}, \dfrac{9\pi}{4}, \dfrac{13\pi}{4}$

8. a. $-\dfrac{7\pi}{2}, -\dfrac{3\pi}{2}, \dfrac{\pi}{2}, \dfrac{5\pi}{2}$ **b.** $0.64, 3.79, 6.93, 10.07$

   c. $x° = \pm 104.5°$

9. a. $-\dfrac{11\pi}{6}, -\dfrac{7\pi}{6}$ **b.** No solution

   c. $-4\pi, -2\pi$ **d.** $\dfrac{\pi}{4}$

   e. $\dfrac{\pi}{3}, \dfrac{5\pi}{3}$ **f.** $-300, -240, 60, 120$

10. a. $\dfrac{\pi}{3}, \dfrac{4\pi}{3}$ **b.** $\dfrac{\pi}{2}$

11. $\dfrac{\pi}{6}, \dfrac{5\pi}{6}, \dfrac{7\pi}{6}, \dfrac{11\pi}{6}$

12. a. $\dfrac{\pi}{3}, \dfrac{4\pi}{3}$ **b.** $\dfrac{5\pi}{6}, \dfrac{11\pi}{6}$

   c. $\dfrac{3\pi}{8}, \dfrac{7\pi}{8}, \dfrac{11\pi}{8}, \dfrac{15\pi}{8}$ **d.** $0.93, 4.07$

   e. $\dfrac{\pi}{4}, \dfrac{3\pi}{4}, \dfrac{5\pi}{4}, \dfrac{7\pi}{4}$ **f.** $\dfrac{\pi}{4}, \dfrac{5\pi}{4}, \dfrac{\pi}{2}, \dfrac{3\pi}{2}$

13. a. $x = \dfrac{\pi}{8}, \dfrac{3\pi}{8}, \dfrac{9\pi}{8}, \dfrac{11\pi}{8}$ **b.** $x = \dfrac{\pi}{6}, \dfrac{2\pi}{3}, \dfrac{7\pi}{6}$

14. a. $\dfrac{5\pi}{18}, \dfrac{11\pi}{18}, \dfrac{17\pi}{18}, \dfrac{23\pi}{18}, \dfrac{29\pi}{18}, \dfrac{35\pi}{18}$

   b. $\dfrac{2\pi}{9}, \dfrac{4\pi}{9}, \dfrac{14\pi}{9}, \dfrac{16\pi}{9}$

   c. $\dfrac{\pi}{4}, \dfrac{3\pi}{4}, \dfrac{5\pi}{4}, \dfrac{7\pi}{4}$

   d. $\dfrac{3\pi}{8}, \dfrac{7\pi}{8}, \dfrac{11\pi}{8}, \dfrac{15\pi}{8}$

15. $x = \dfrac{\pi}{3}, \dfrac{5\pi}{3}$

16. a. $\dfrac{\pi}{4}, \dfrac{3\pi}{4}, \dfrac{5\pi}{4}, \dfrac{7\pi}{4}$ **b.** $\dfrac{\pi}{2}, \dfrac{3\pi}{2}$

   c. $\dfrac{\pi}{2}, \dfrac{7\pi}{6}, \dfrac{11\pi}{6}$ **d.** $\dfrac{\pi}{4}, \dfrac{5\pi}{4}, 1.89, 5.03$

   e. $\dfrac{\pi}{2}, \dfrac{7\pi}{6}, \dfrac{11\pi}{6}$ **f.** No solution

17. $x = \dfrac{2\pi}{3}$

18. a. $2.30, 3.98$

   b. $-5.67, -2.53, 0.62, 3.76, 6.90$

   c. $-53.13, -126.87, 233.13$

   d. $78.46, 101.54, 258.46, 281.54$

19. a. $a = 8$

   b. i. $0.38, 2.76$

     ii. 1.57

     iii. No solution

20. $t = 1$

6.8 Transformations of the sine and cosine functions

6.8 Exercise

1.

	Amplitude	Period
a.	6	π
b.	7	4π
c.	$\dfrac{3}{5}$	$\dfrac{10\pi}{3}$
d.	1	$\dfrac{7}{3}$
e.	2	4π
f.	4	2π

2. a. Amplitude $\frac{1}{4}$, range $\left[-\frac{1}{4}, \frac{1}{4}\right]$, period $\frac{\pi}{2}$

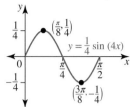

b. Amplitude $\frac{1}{2}$, range $\left[-\frac{1}{2}, \frac{1}{2}\right]$, period 8π

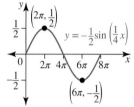

c. Amplitude 1, range $[-1, 1]$, period $\frac{2\pi}{3}$.

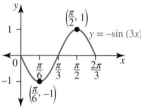

d. Amplitude 3, range $[-3, 3]$, period π

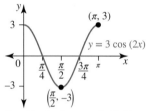

e. Amplitude 6, range $[-6, 6]$, period 5π

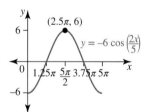

f. Amplitude $\frac{1}{3}$, range $\left[-\frac{1}{3}, \frac{1}{3}\right]$, period 0.4

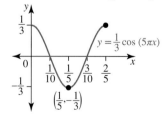

3. a.

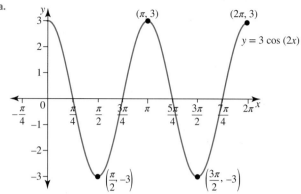

b.

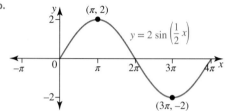

c.

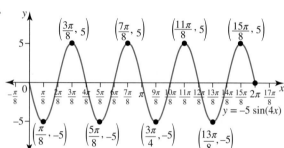

d.

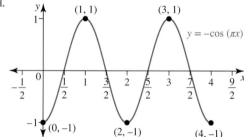

4. a. $[2, 4]$

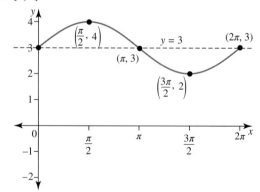

b. $[-2, 0]$

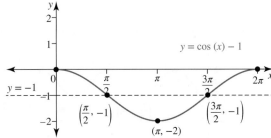

$y = \cos(x) - 1$

$y = -1$

$\left(\dfrac{\pi}{2}, -1\right)$ $\left(\dfrac{3\pi}{2}, -1\right)$

$(\pi, -2)$

c. $[1, 3]$

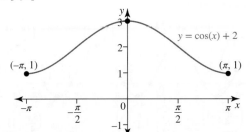

$y = \cos(x) + 2$

$(-\pi, 1)$ $(\pi, 1)$

d. $[3, 5]$

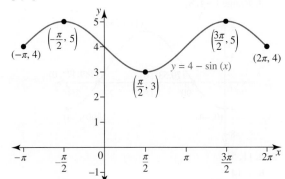

$\left(-\dfrac{\pi}{2}, 5\right)$ $\left(\dfrac{3\pi}{2}, 5\right)$

$(-\pi, 4)$ $(2\pi, 4)$

$y = 4 - \sin(x)$

$\left(\dfrac{\pi}{2}, 3\right)$

5. a.

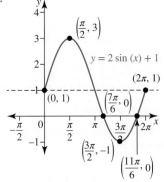

$\left(\dfrac{\pi}{2}, 3\right)$

$y = 2\sin(x) + 1$

$(2\pi, 1)$

$(0, 1)$ $\left(\dfrac{7\pi}{6}, 0\right)$

$\left(\dfrac{3\pi}{2}, -1\right)$

$\left(\dfrac{11\pi}{6}, 0\right)$

Period 2π; amplitude 1; equilibrium position $y = 1$;

range $[-1, 3]$; x-intercepts at $x = \dfrac{7\pi}{6}, \ \dfrac{11\pi}{6}$

b.

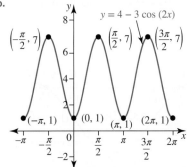

$y = 4 - 3\cos(2x)$

$\left(-\dfrac{\pi}{2}, 7\right)$ $\left(\dfrac{\pi}{2}, 7\right)$ $\left(\dfrac{3\pi}{2}, 7\right)$

$(-\pi, 1)$ $(0, 1)$ $(\pi, 1)$ $(2\pi, 1)$

Period π; amplitude 3; inverted; equilibrium $y = 4$;
range $[1, 7]$

6.

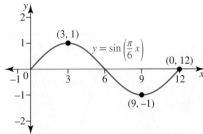

$(3, 1)$ $y = \sin\left(\dfrac{\pi}{6}x\right)$ $(0, 12)$

$(9, -1)$

Period 12; amplitude 1; equilibrium $y = 0$; range $[-1, 1]$;
domain $[0, 12]$

7. a. Period π, mean $y = 5$, amplitude 4, range $[1, 9]$

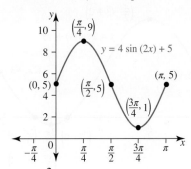

$\left(\dfrac{\pi}{4}, 9\right)$

$y = 4\sin(2x) + 5$

$(0, 5)$ $\left(\dfrac{\pi}{2}, 5\right)$ $(\pi, 5)$

$\left(\dfrac{3\pi}{4}, 1\right)$

b. Period $\dfrac{2\pi}{3}$, mean $y = 2$, amplitude 2, range $[0, 4]$

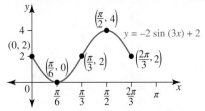

$\left(\dfrac{\pi}{2}, 4\right)$

$(0, 2)$ $y = -2\sin(3x) + 2$

$\left(\dfrac{\pi}{6}, 0\right)$ $\left(\dfrac{\pi}{3}, 2\right)$ $\left(\dfrac{2\pi}{3}, 2\right)$

c. Period 4π, mean $y = \dfrac{3}{2}$, amplitude $\dfrac{1}{2}$, range $[1, 2]$

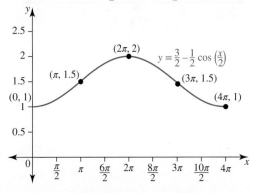

$(2\pi, 2)$ $y = \dfrac{3}{2} - \dfrac{1}{2}\cos\left(\dfrac{x}{2}\right)$

$(\pi, 1.5)$ $(3\pi, 1.5)$

$(0, 1)$ $(4\pi, 1)$

d. Period 2, mean $y = -\sqrt{3}$, amplitude 2, range $[-\sqrt{3} - 2, \ -\sqrt{3} + 2]$

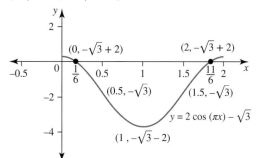

8. a. Range $[-4, 0]$

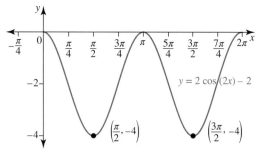

b. Range $[\sqrt{3} - 2, \sqrt{3} + 2]$

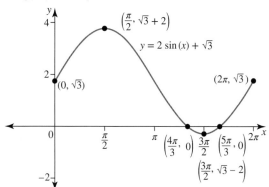

c. Range $[2, 8]$

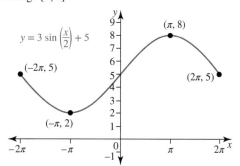

d. Range $[-5, \ -3]$

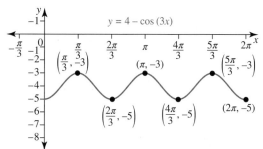

e. Range $[-1, 3]$

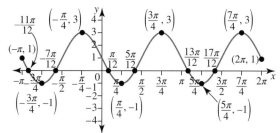

f. Range $[-4, 8]$

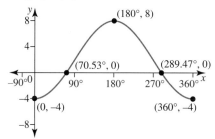

9. a. $[1, 5]$

b. -14

c. $68; x = \dfrac{3\pi}{2}$

d. i. Dilation factor $\dfrac{1}{5}$ from y-axis; dilation factor 2 from x-axis; vertical translation up 3

ii. Dilation factor $\dfrac{1}{2}$ from y-axis; dilation factor 10 from x-axis; vertical translation down 4

iii. Dilation factor 12 from x-axis; reflection in x-axis; vertical translation up 56

10. a.

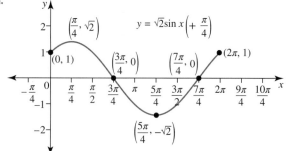

Period 2π; amplitude $\sqrt{2}$; phase shift $-\dfrac{\pi}{4}$ from $y = \sqrt{2}\sin(x)$; end points $(0, 1)$, $(2\pi, 1)$; range $[-\sqrt{2}, \sqrt{2}]$

b. Period π; amplitude 3; phase shift factor $-\dfrac{\pi}{8}$; range $[-2, 4]$

11. a. i.

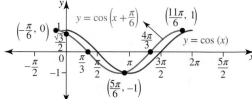

ii.

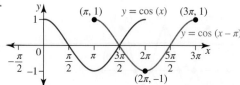

b. i.

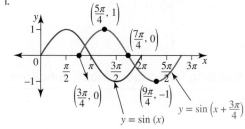

ii.

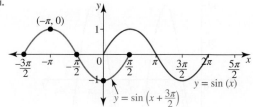

12. a.

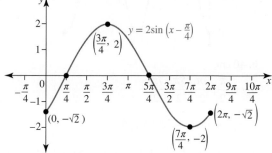

b.

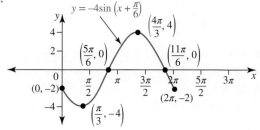

c.

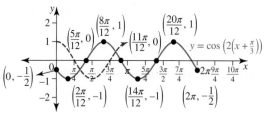

d.

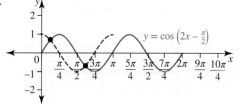

e.

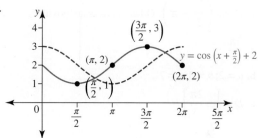

f.

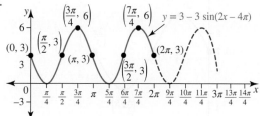

13.

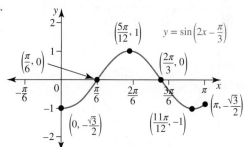

Period π; amplitude 1; phase $\dfrac{\pi}{6}$; end points

$$\left(0, -\dfrac{\sqrt{3}}{2}\right), \left(\pi, -\dfrac{\sqrt{3}}{2}\right)$$

14. a. $a = 3, b = 2, \; y = 3\sin(2x)$

b. $a = -0.5, b = \dfrac{2}{3}, y = -0.5\cos\left(\dfrac{2x}{3}\right)$

c. $a = 4.5, b = \dfrac{4}{3}, y = 4.5\cos\left(\dfrac{4}{3}x\right)$

15. a. $y = 4\sin(3x)$ **b.** $y = 6\cos\left(\dfrac{x}{2}\right)$

c. $y = -4\sin(2x)$

16. a. $y = 2\sin(x) + 3$ **b.** $y = -3\sin\left(\dfrac{x}{2}\right) + 3$

c. $y = 3\cos\left(\dfrac{2x}{3}\right) - 4$ **d.** $y = -5\cos(2x) - 1$

e. $y = 3\sin(2x) + 5$ (other answers are possible)

17. a. $y = -3\sin\left(\dfrac{x}{2}\right)$ **b.** $y = 4\cos(3x)$

c. $y = -4\cos(x) + 6$ **d.** $y = 2\sin\left(x - \dfrac{\pi}{4}\right)$

e. $y = 2\cos\left(x - \dfrac{3\pi}{4}\right)$ **f.** $y = \cos(x);\ y = -\sin(x)$

18. $y = -2\sin\left(\dfrac{2x}{3}\right)$ or $y = -2\cos\left(\dfrac{2x}{3} - \dfrac{\pi}{2}\right)$ or

$y = 2\sin\left(\dfrac{2x}{3} - \pi\right)$. Other answers are possible.

19. a. $\dfrac{2\pi}{3}$

b. $a = 2; b = 3; k = 7$

c. $D = \left[0, \dfrac{2\pi}{3}\right]$

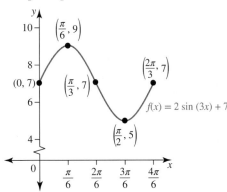

d.

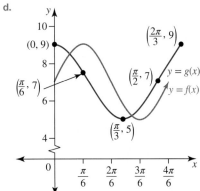

e. $\left(\dfrac{\pi}{12}, 7 + \sqrt{2}\right), \left(\dfrac{5\pi}{12}, 7 - \sqrt{2}\right)$

f. $\dfrac{\pi}{12} \le x \le \dfrac{5\pi}{12}$

20. a.

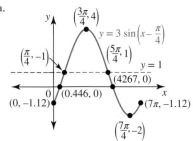

b.

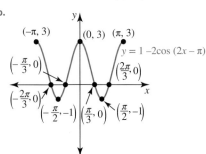

6.9 Modelling with trigonometric functions

6.9 Exercise

1. a.

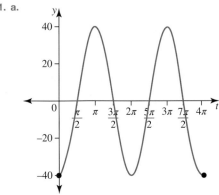

b. 40 cm

c. $t = \dfrac{\pi}{2}, \dfrac{3\pi}{2}, \dfrac{5\pi}{2}, \dfrac{7\pi}{2}$

d. $\dfrac{2\pi}{3}$ seconds ≈ 2.1 seconds

2. $T = 28 - 8\cos\left(\dfrac{\pi}{6}\,t\right)$; between 11.29 am and 4.31 pm

3. a. $I = 5\sin\left(\dfrac{\pi}{2}t\right) + 5$

b. 44%

4. a. 19°

b. 22° at 6 pm

c. Between 16° and 22°

d. The period is 24 hours.

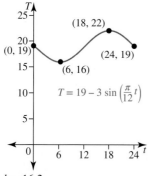

e. $k = 16.2$

5. a. $a = 1.5$; $n = \pi$; $b = 13.5$

b. 12 cents

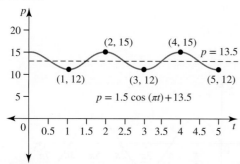

c. 5 days

d. $197.25

6. a. 12 hours

b. Maximum depth: 3.5 metres Minimum depth: 1.5 metres

c. 3 metres

d.

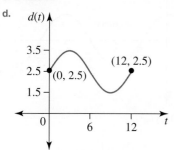

e. The depth will be at least 3.1 metres for 3 hours 33 minutes.

7. a. 24 hours

b. Maximum 14 metres, minimum 11 metres

c. 12.5 metres

d.

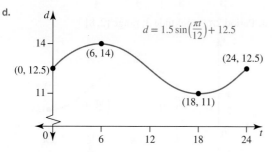

e. $h = \dfrac{3\sqrt{3} + 50}{4} \approx 13.8\,\text{m}$

f. 3:30 am the following day

8. a. Maximum temperature 8°C, minimum temperature −4°C

b. 8:42 am and 11:18 pm

c. 4:00 pm

d. −1°C

9. a. 3.619 m below the jetty

b. 5.5 m and 0.5 m

c.

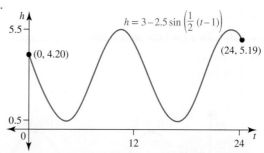

First maximum (high tide) at $t \approx 10.42$, first minimum (low tide) at $t \approx 4.14$

d. 5 m of extra rope

10. a. 2 metres below mean sea level

b. 4 metres above mean sea level; sample responses can be found in the worked solutions in the online resources.

c. 6 hours

d.

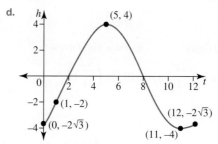

e. At mean sea level

f. Risen by 0.14 metres

11. a. $a = 2.5; b = 4.5; h = 2.5 \sin\left(\dfrac{\pi}{5}x\right) + 4.5$

b. 10 cm

c. $(2.5, \ 3.25)$

d. 40.1 cm^2

12. a. Between 29° and 31°

b. 12 minutes

c.

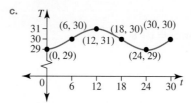

d. 2.5 cycles

e. 30 °C

f. $T = \sin\left(\dfrac{\pi}{12}(t - 6)\right) + 30$

13. a. i. 25° at 1 pm **ii.** 13° at 7 pm

b. i. 23.2 °C **ii.** 13.2 °C

c.

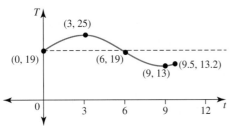

d. 2.24 hours

e. $\left(19 - 3\sqrt{2}\right)° \approx 14.8°$ at 5:30 pm

14. a. 1.5 metres

b. 18.5 metres

c. 2

d.

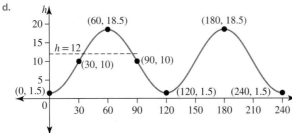

e. 51 seconds

f. 8.5 metres

15. a. 2 metres **b.** $n = \dfrac{4}{3}$

c. $a = 5$; 60 times **d.** $p = a\ \sin\left(4\pi t\right) + 5$

16. a. $a = -20,\ b = \dfrac{2\pi}{3},\ x = -20\ \sin\left(\dfrac{2\pi}{3}t\right)$

b.

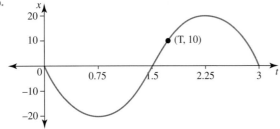

c. $T = 1.75$ seconds

d. 80 cm

17. 86 people

18. 9.4 hours

19. 5 units, 0.64 hours

20. $\dfrac{\pi}{3}$ cm $(\approx 1.047$ cm$); y = x + 8$

6.10 Review

6.10 Exercise

1. C

2. C

3. a. 220°

b. −630°

4. a. 12.34 cm

b. 38.24 cm²

5. a. B

b. D

6. D

7. a. 0.6

b. −0.6

c. −0.6

d. 0.6

8. C

9. C

10. a.

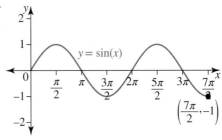

b.

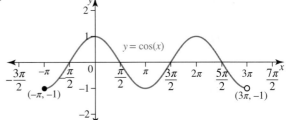

c.

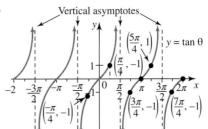

d. Period 2π; amplitude 3; range [2, 8]

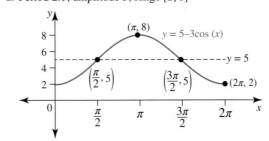

e. Period $\dfrac{2\pi}{3}$; amplitude 5; range $[-5, 5]$

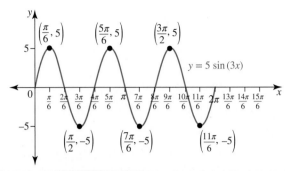

f. Period 4π; amplitude 2; range $[-1, 3]$

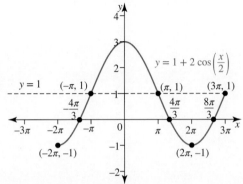

11. D

12. A

13. $\dfrac{3\sqrt{3} - \sqrt{2}}{2}$

14. a. $\cos(x) = -\dfrac{2}{\sqrt{29}}$; $\sin(x) = -\dfrac{5}{\sqrt{29}}$

b. $\cos(y) = \dfrac{2}{3}$; $\tan(y) = -\dfrac{\sqrt{5}}{2}$

15. a. p

b. $-p$

c. $1 - p^2$

d. $-p$

16. a. $\dfrac{3\pi}{4}, \dfrac{5\pi}{4}$

b. $0, 2\pi, 4\pi$

c. $\dfrac{-5\pi}{3}, \dfrac{-4\pi}{3}, \dfrac{\pi}{3}, \dfrac{2\pi}{3}$

d. $\dfrac{\pi}{4}, \dfrac{5\pi}{4}$

e. $-\dfrac{11\pi}{12}, -\dfrac{3\pi}{4}, -\dfrac{\pi}{4}, -\dfrac{\pi}{12}, \dfrac{5\pi}{12}, \dfrac{7\pi}{12}$

f. $58.28, 148.28, 238.28$

g. $\dfrac{3\pi}{2}$

h. $\dfrac{2\pi}{3}, \pi, \dfrac{4\pi}{3}$

17. $110.41°$

18. Maximum depth $= 4.5$
Minimum depth $= 0.5$
The boat can dock from 1:56 pm to 4:04 pm and 1:56 am to 4:04 am.

19. DC: 12 cm; belt: 62.5 cm

20. 39.2%

7 Probability

LESSON SEQUENCE

Fully worked solutions for this chapter are available online.

EXAM PREPARATION

Access exam-style questions in every lesson, available online.

on Resources

Solutions	Solutions — Chapter 7 (sol-1274)
Exam questions	Exam question booklet — Chapter 7 (eqb-0295)
Digital documents	Learning matrix — Chapter 7 (doc-41789)
	Chapter summary — Chapter 7 (doc-41802)

LESSON
7.1 Overview

7.1.1 Introduction

From the earliest times, humans have understood that some events were more likely to happen than others. The ancient Romans invented the insurance policy on this basis, and gamblers since the dawn of time have made and lost money according to their sense of how likely it was they would get the right card, roll the right die or pick the right horse.

Although the popularity of the study of probability for hundreds of years was chiefly due to its usefulness in games of chance, it was eventually recognised that probability could be used to model the future behaviour of many phenomena by analysing how they had behaved in the past.

7.1.2 Syllabus links

Lesson	Lesson title	Syllabus links
7.2	Concepts of probability	○ Use the concepts and language of outcomes, sample spaces and events as sets of outcomes.
		○ Use set language and notation for events, including $\overline{A}$ or A' for the complement of an event, A, $A \cap B$ for the intersection of events A and B, and $A \cup B$ for the union, and recognise mutually exclusive events.
		○ Use everyday occurrences to illustrate set descriptions and representations of events, and set operations, including the use of Venn diagrams.
		○ Use the rules $P(\overline{A}) = 1 - P(A)$ and $P(A \cup B) = P(A) + P(B) - P(A \cap B)$.
		○ Model and solve problems that involve probability, with and without technology.
7.3	Relative frequency	○ Use relative frequencies obtained from data as point estimates of conditional probabilities and as indications of possible independence of events.
		○ Model and solve problems that involve probability, with and without technology.
7.4	Conditional probability	○ Understand the notion of a conditional probability and recognise and use language that indicates conditionality.
		○ Use the notation $P(A|B)$ and the formula $P(A \cap B) = P(A|B)\,P(B)$ to solve problems.
		○ Use relative frequencies obtained from data as point estimates of conditional probabilities and as indications of possible independence of events.
		○ Model and solve problems that involve probability, with and without technology.
7.5	Independence	○ Understand and use the notion of independence of an event A from an event B, as defined by $P(A|B) = P(A)$.
		○ Use the formula $P(A \cap B) = P(A)\,P(B)$ for independent events A and B.
		○ Use relative frequencies obtained from data as point estimates of conditional probabilities and as indications of possible independence of events.
		○ Model and solve problems that involve probability, with and without technology.

Source: Mathematical Methods Senior Syllabus 2024 © State of Queensland (QCAA) 2024; licensed under CC BY 4.0.

LESSON
7.2 Concepts of probability

7.2.1 Notation and language: outcomes, sample spaces and events

Consider the experiment or trial of spinning a wheel that is divided into eight equal sectors, with each sector marked with one of the numbers 1 to 8. If the wheel is unbiased, each of these numbers is equally likely to occur.

The **outcome** of each trial is one of the eight numbers.

The **sample space**, S, is the set of all possible outcomes: $S = \{1, 2, 3, 4, 5, 6, 7, 8\}$.

An **event** is a particular set of outcomes that is a subset of the sample space.

For example, if M is the event of obtaining a number that is a multiple of 3, then $M = \{3, 6\}$.

The **number of elements** contained in set A is denoted by $n(A)$. In the example, the set M contains two outcomes or elements, giving $n(M) = 2$.

The **probability** of an event is the long-term proportion, or relative frequency, of its occurrence.

> ## Probability of an event
>
> For any event A, the probability of its occurrence is $P(A) = \dfrac{n(A)}{n(S)}$,
>
> where $n(A)$ is the number of elements in set A and $n(S)$ is the number of possible outcomes.

Hence, for the event M:

$$P(M) = \frac{n(M)}{n(S)}$$
$$= \frac{2}{8}$$
$$= \frac{1}{4}$$

This value does not mean that a multiple of 3 is obtained once in every four spins of the wheel. However, it does mean that after a very large number of spins of the wheel, the proportion of times that a multiple of 3 will be obtained approaches $\dfrac{1}{4}$. The closeness of this proportion to $\dfrac{1}{4}$ will improve in the long term as the number of spins is further increased.

Probability of an event

For any event A, $0 \leq P(A) \leq 1$.

- If $P(A) = 0$, then it is not possible for A to occur. For example, the chance that the spinner lands on a negative number would be zero.
- If $P(A) = 1$, then the event A is certain to occur. For example, it is 100% certain that the number the spinner lands on will be smaller than 9.

The probability of each outcome is $P(1) = P(2) = P(3) = \ldots = P(8) = \dfrac{1}{8}$ for this spinning wheel. As each outcome is equally likely to occur, the outcomes are **equiprobable**. In other situations, some outcomes may be more likely than others.

For any sample space, the sum of the probabilities of each of the outcomes must total 1.

That is, $P(S) = \dfrac{n(S)}{n(S)} = 1$.

WORKED EXAMPLE 1 Determining sample spaces and outcomes

a. List the sample space of set A, the aces in the deck of cards.
b. Determine $n(B)$, where B is the set of clubs in a deck of cards.

THINK	WRITE
a. There are 4 aces in a deck of cards.	$A = \{$ace of hearts, ace of diamonds, ace of clubs, ace of spades$\}$
b. $n(B)$ is the number of outcomes in set B, the set of clubs in a deck of cards.	$n(B) = 13$

Complementary events

For the spinner example, the event that the number is **not** a multiple of 3 is the **complement** of the event M. The complementary event is written as $\overline{M}$ or M'.

$$\begin{aligned}
P(\overline{M}) &= 1 - P(M) \\
&= 1 - \frac{1}{4} \\
&= \frac{3}{4}
\end{aligned}$$

For any complementary events, $P(A) + P(\overline{A}) = 1$ and therefore $P(\overline{A}) = 1 - P(A)$.

Complementary events

For any complementary events:

$$P(\overline{A}) = 1 - P(A)$$

or

$$P(A) + P(\overline{A}) = 1$$

A spinning wheel is divided into eight sectors, each of which is marked with one of the numbers 1 to 8.
This wheel is biased so that P(8) = 0.3, and the other numbers are equiprobable.
a. Calculate the probability of obtaining the number 4.
b. If A is the event the number obtained is even, calculate P(A) and P($\overline{A}$).

THINK	WRITE
a. 1. State the complement of obtaining the number 8 and the probability of this.	a. The sample space contains the numbers 1 to 8, so the complement of obtaining 8 is obtaining one of the numbers 1 to 7. As P(8) = 0.3, the probability of not obtaining 8 is $1 - 0.3 = 0.7$.
2. Calculate the required probability.	Since each of the numbers 1 to 7 are equiprobable, the probability of each number is $\dfrac{0.7}{7} = 0.1$. Hence, P(4) = 0.1.
b. 1. Identify the elements of the event.	b. $A = \{2, 4, 6, 8\}$
2. Calculate the probability of the event.	$\begin{aligned} P(A) &= P(2 \text{ or } 4 \text{ or } 6 \text{ or } 8) \\ &= P(2) + P(4) + P(6) + P(8) \\ &= 0.1 + 0.1 + 0.1 + 0.3 \\ &= 0.6 \end{aligned}$
3. State the complementary probability.	$\begin{aligned} P(\overline{A}) &= 1 - P(A) \\ &= 1 - 0.6 \\ &= 0.4 \end{aligned}$

7.2.2 Venn diagrams

A **Venn diagram** can be used to graphically represent the relationships between different sets.

The union of two sets is written as $A \cup B$, the outcomes in set A **or** in set B **or** in both sets. The intersection of two sets is written as $A \cap B$, the outcomes in set A **and** set B, or the common outcomes.

A Venn diagram may be helpful in displaying compound events in probability, as illustrated for the sets or events A and B.

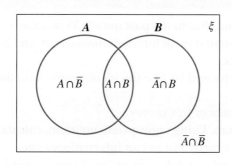

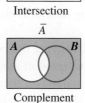

The information shown in the Venn diagram may be the actual outcomes for each event, or it may only show numbers that represent the number of outcomes for each event. Alternatively, the Venn diagram may show the probability of each event.

The total probability is 1; that is, P(S) = 1.

The addition formula

The number of elements contained in set A is denoted by $n(A)$.

The Venn diagram illustrates that $n(A \cup B) = n(A) + n(B) - n(A \cap B)$.

Hence, dividing by the number of elements in the sample space gives:

$$\frac{n(A \cup B)}{n(S)} = \frac{n(A)}{n(S)} + \frac{n(B)}{n(S)} - \frac{n(A \cap B)}{n(S)}$$
$$\therefore P(A \cup B) = P(A) + P(B) - P(A \cap B)$$

The result is known as the addition formula.

The addition formula

$$P(A \cup B) = P(A) + P(B) - P(A \cap B)$$

If the events A and B are **mutually exclusive**, then they cannot occur simultaneously. For mutually exclusive events, $n(A \cap B) = 0$ and therefore $P(A \cap B) = 0$.

For example, event A where a thrown coin lands Heads up and event B where the thrown coin lands Tails up are mutually exclusive events, as the two events can never happen at the same time.

The total of all probabilities in the Venn diagram is 1. That is, $P(S) = 1$.

Mutually exclusive events

For mutually exclusive events:

$$P(A \cup B) = P(A) + P(B)$$

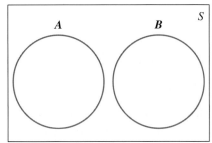

Venn diagram for mutually exclusive events

WORKED EXAMPLE 3 Solving problems with Venn diagrams

From a survey of a group of 50 people, it was found that in the past month 30 members of the group had made a donation to a local charity, 25 had donated to an international charity and 20 had made donations to both local and international charities.

Let L be the set of people donating to a local charity and let I be the set of people donating to an international charity.

a. Construct a Venn diagram to illustrate the results of this survey.

b. One person from the group is selected at random. Using appropriate notation, calculate the probability that this person donated to a local charity but not an international one.

c. Calculate the probability that the person selected at random from the group did not make a donation to either type of charity.

d. Calculate the probability that the person selected at random from the group donated to at least one of the two types of charity.

THINK	**WRITE**
a. Show the given information on a Venn diagram and complete the remaining sections using arithmetic.	**a.** Given: $n(S) = 50$, $n(L) = 30$, $n(I) = 25$ and $n(L \cap I) = 20$ 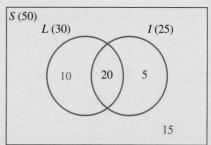
b. 1. State the required probability using set notation.	**b.** $P(L \cap \bar{I}) = \dfrac{n(L \cap \bar{I})}{n(S)}$
2. Identify the value of the numerator from the Venn diagram and calculate the probability.	$P(L \cap \bar{I}) = \dfrac{10}{50}$ $= \dfrac{1}{5}$ $= 0.2$
3. Write the answer in context.	The probability that the randomly chosen person donated to a local charity but not an international one is 0.2.
c. 1. State the required probability using set notation.	**c.** $P(\bar{L} \cap \bar{I}) = \dfrac{n(\bar{L} \cap \bar{I})}{n(S)}$
2. Identify the value of the numerator from the Venn diagram and calculate the probability.	$P(\bar{L} \cap \bar{I}) = \dfrac{15}{50}$ $= \dfrac{3}{10}$ $= 0.3$
3. Write the answer in context.	The probability that the randomly chosen person did not donate is 0.3.
d. 1. State the required probability using set notation.	**d.** $P(L \cup I) = \dfrac{n(L \cup I)}{n(S)}$
2. Identify the value of the numerator from the Venn diagram and calculate the probability.	$P(L \cup I) = \dfrac{10 + 20 + 5}{50}$ $= \dfrac{35}{50}$ $= \dfrac{7}{10}$ $= 0.7$
3. Write the answer in context.	The probability that the randomly chosen person donated to at least one type of charity is 0.7.

7.2.3 Probability tables

For situations involving two events, a **probability table** can provide an alternative to a Venn diagram. Consider the Venn diagram shown.

A probability table presents any known probabilities of the four compound events $A \cap B$, $A \cap \bar{B}$, $\bar{A} \cap B$ and $\bar{A} \cap \bar{B}$ in rows and columns.

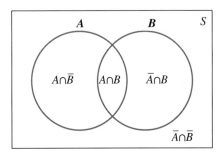

	B	$\bar{B}$	
A	$P(A \cap B)$	$P(A \cap \bar{B})$	$P(A)$
$\bar{A}$	$P(\bar{A} \cap B)$	$P(\bar{A} \cap \bar{B})$	$P(\bar{A})$
	$P(B)$	$P(\bar{B})$	$P(\xi) = 1$

This allows the table to be completed using arithmetic calculations since, for example, $P(A) = P(A \cap B) + P(A \cap \bar{B})$ and $P(B) = P(A \cap B) + P(\bar{A} \cap B)$.

The probabilities of complementary events can be calculated using the formula $P(\bar{A}) = 1 - P(A)$.

To obtain $P(A \cup B)$, the addition formula $P(A \cup B) = P(A) + P(B) - P(A \cap B)$ can be used.

Probability tables are also known as **Karnaugh maps**.

WORKED EXAMPLE 4 Constructing probability tables

$P(A) = 0.4$, $P(B) = 0.7$ and $P(A \cap B) = 0.2$
a. **Construct a probability table for the events A and B.**
b. **Calculate $P(\bar{A} \cup B)$.**

THINK

a. 1. Enter the given information in a probability table.

2. Add in the complementary probabilities.

3. Complete the remaining sections using arithmetic.

WRITE

a. Given: $P(A) = 0.4$, $P(B) = 0.7$, $P(A \cap B) = 0.2$ and also $P(S) = 1$

	B	$\bar{B}$	
A	0.2		0.4
$\bar{A}$			
	0.7		1

$P(\bar{A}) = 1 - 0.4 = 0.6$ and $P(\bar{B}) = 1 - 0.7 = 0.3$.

	B	$\bar{B}$	
A	0.2		0.4
$\bar{A}$			0.6
	0.7	0.3	1

For the first row, $0.2 + 0.2 = 0.4$.
For the first column, $0.2 + 0.5 = 0.7$.
For the second column, $0.2 + 0.1 = 0.3$.

	B	$\bar{B}$	
A	0.2	0.2	0.4
$\bar{A}$	0.5	0.1	0.6
	0.7	0.3	1

b. **1.** State the addition formula.

b. $P(\overline{A} \cup B) = P(\overline{A}) + P(B) - P(\overline{A} \cap B)$

2. Use the values in the probability table to carry out the calculation.

From the probability table, $P(\overline{A} \cap B) = 0.5$.
$\therefore P(\overline{A} \cup B) = 0.6 + 0.7 - 0.5$
$\qquad\qquad\quad = 0.8$

Probability tables are also useful in solving worded problems, as illustrated in the following worked example.

WORKED EXAMPLE 5 Constructing probability tables from worded problems

In a recent survey, students were asked if they liked pizzas or hamburgers. 70% of the students liked pizzas, 40% of the students liked hamburgers and 15% of the students liked neither pizzas nor hamburgers.

a. Determine the probability that a student, chosen at random, liked both pizzas and hamburgers.

b. Determine the probability that a student, chosen at random, liked pizza only.

THINK

a. **1.** Let the set of students who like pizza be Z.

2. Let the set of students who like hamburgers be H.

3. Enter the given information into a probability table.

WRITE

a. Given 70% of students like pizzas:
$P(Z) = 0.7$

Given 40% of students like hamburgers:
$P(H) = 0.4$

Given that 15% of students liked neither:
$P(\overline{Z} \cap \overline{H}) = 0.15$

	Z	$\overline{Z}$	
H			0.4
$\overline{H}$		0.15	
	0.7		1

4. Add the complementary probabilities.

$P(\overline{H}) = 1 - P(H) = 1 - 0.4 = 0.6$
$P(\overline{Z}) = 1 - P(Z) = 1 - 0.7 = 0.3$

	Z	$\overline{Z}$	
H			0.4
$\overline{H}$		0.15	0.6
	0.7	0.3	1

5. Complete the remaining sections using arithmetic.

For the second row, $0.15 + 0.45 = 0.6$.
For the first column, $0.45 + 0.25 = 0.7$.
For the second column, $0.15 + 0.15 = 0.3$.

	Z	$\overline{Z}$	
H	0.25	0.15	0.4
$\overline{H}$	0.45	0.15	0.6
	0.7	0.3	1

6. Use the table to answer the question.

$$P(Z \cup H) = P(Z \cap \overline{H}) + P(\overline{Z} \cap H) + P(Z \cap H)$$

$$= 0.45 + 0.15 + 0.25$$

$$= 0.85$$

or

$$P(Z \cup H) = 1 - P(\overline{Z} \cap \overline{H}) = 1 - 0.15 = 0.85$$

b. The probability that a student likes pizza only is the probability that a student likes pizza and not hamburgers.

b. $P(Z \cap \overline{H}) = 0.45$

The addition formula can be applied to determine probabilities without necessarily using a Venn diagram or constructing a probability table.

For example, consider drawing a card from a well-shuffled deck of 52 playing cards.

The probability of drawing a heart, H, or a queen, Q, can be found by either counting the number of cards that are either hearts or queens, or using the addition formula. If counting the number of cards, the queen of hearts can only be included once.

In words:

The probability or drawing a heart or a queen is the probability of drawing a heart plus the probability of drawing a queen minus the probability of drawing the queen of hearts.

$$P(H \cup Q) = P(H) + P(Q) - P(H \cap Q)$$

$$= \frac{13}{52} + \frac{13}{52} - \frac{1}{52} = \frac{25}{52}$$

7.2.4 Tree diagrams

Tree diagrams are useful displays of two- or three-stage events.

Simple tree diagram

The outcome of each toss of a coin is either Heads (H) or Tails (T). For two tosses, the outcomes are illustrated by the tree diagram shown.

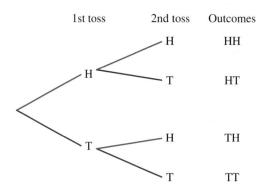

The sample space consists of the four equally likely outcomes HH, HT, TH and TT. This means the probability of obtaining two Heads (HH) in two tosses of a coin would be $\frac{1}{4}$.

The tree diagram could be extended to illustrate repeated tosses of the coin.

A coin is biased so that the chance of it falling as a Head when flipped is 0.75.
a. **Construct a tree diagram to represent the coin being flipped three times.**
b. **Calculate the following probabilities, correct to 3 decimal places:**
   i. **P(HTT)**
   ii. **P(1 H and 2 T)**
   iii. **P(at least 2 Tails).**

THINK	WRITE
a. 1. Tossing a coin has two outcomes. Draw 2 branches from the starting point to show the first toss, 2 branches off each of these to show the second toss and then 2 branches off each of these to show the third toss.	a.
2. Write probabilities on the branches to show the individual probabilities of tossing a Head (0.75) and a Tail. Because tossing a Head and tossing a Tail are mutually exclusive, $P(T) = 1 - P(H) = 0.25$.	

b. i. 1. P(HTT) implies the order: H(0.75), T(0.25), T(0.25).

   2. Multiply the probabilities and round.

b. i. $P(HTT) = P(H) \times P(T) \times P(T)$
   $= (0.75) \times (0.25)^2$
   $= 0.046\,875$
   ≈ 0.047

ii. 1. P(1 H and 2 T) implies: P(HTT), P(THT), P(TTH).

   2. Multiply the probabilities and round.

ii. $P(1 \text{ H and } 2 \text{ T}) = P(HTT) + P(THT) + P(TTH)$
   $= 3(0.75 \times 0.25^2)$
   $= 0.140\,625$
   ≈ 0.141

iii. 1. P(at least 2 Tails) implies: P(HTT), P(THT), P(TTH) and P(TTT).

   2. Add these probabilities and round.

iii. $P(\text{at least } 2 \text{ T}) = P(HTT) + P(THT) + P(TTH) + P(TTT)$
   $= 3(0.75 \times 0.25^2) + 0.25^3$
   $= 0.156\,25$
   ≈ 0.156

| 7.2 Exercise | 7.2 Exam practice |

Simple familiar	Complex familiar	Complex unfamiliar
1, 2, 3, 4, 5, 6, 7, 8, 9, 10, 11, 12, 13	14, 15, 16, 17, 18	19, 20

These questions are even better in jacPLUS!
- Receive immediate feedback
- Access sample responses
- Track results and progress

Find all this and MORE in jacPLUS ▶

Simple familiar

1. **WE1** **a.** List the sample space of set A, the kings in a deck of cards.

   b. Determine $n(B)$, where B is the set of diamonds in a deck of cards.

2. **WE2** A bag contains 5 green, 6 pink, 4 orange and 8 blue counters. A counter is selected at random. Determine the probability that the counter is:

   a. green
   b. orange or blue
   c. not blue
   d. black.

3. An unbiased six-sided die is thrown onto a table. Determine the probability that the number is:

   a. 4
   b. not 4
   c. even
   d. smaller than 5
   e. at least 5
   f. greater than 12.

4. One letter from the alphabet, A through to Z, is chosen at random. Calculate the probability that the letter chosen is:

   a. Q
   b. a vowel
   c. either X or Y or Z
   d. not D
   e. either a consonant or a vowel
   f. one of the letters in the word PROBABILITY.

5. A card is drawn randomly from a standard pack of 52 cards. Calculate the probability that the card is:

   a. not green
   b. from a red suit
   c. a heart
   d. a 10 or from a red suit
   e. not an ace.

6. A spinner is divided into 4 sections coloured red, blue, green and yellow. Each section is equally likely to occur. The spinner is spun twice. List all the possible outcomes and hence determine the probability of obtaining:

   a. the same colour
   b. a red and a yellow
   c. not a green.

   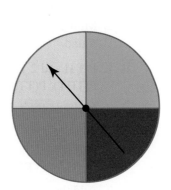

7. **WE3** From a group of 42 students, it was found that 30 students studied Mathematical Methods and 15 studied Geography. Ten of the Geography students did not study Mathematical Methods.

   Let M be the set of students studying Mathematical Methods and let G be the set of students studying Geography.

   a. Construct a Venn diagram to illustrate this situation.

   One student from the group is selected at random.

   b. Using appropriate notation, calculate the probability that this student studies Mathematical Methods but not Geography.

   c. Calculate the probability that this student studies neither Mathematical Methods nor Geography.

   d. Calculate the probability that this student studies only one of Mathematical Methods or Geography.

8. **WE4** Given $P(A) = 0.65$, $P(B) = 0.5$ and $P(\overline{A} \cap \overline{B}) = 0.2$:
    a. construct a probability table for the events A and B
    b. calculate $P(\overline{B} \cup A)$.

9. **WE5** A survey conducted by a streaming network asked viewers in the 16- to 20-year-old demographic which of its two leading programs the viewers preferred to watch. Of the 150 viewers surveyed, 30% watched *Robots in Love*, 60% watched *Talentless* and 20% watched neither show.
    a. Determine the probability that a viewer, chosen at random, watched both shows.
    b. Determine the probability that a viewer, chosen at random, watched *Talentless* but not *Robots in Love*.

10. A coin is tossed three times.
    a. Construct a simple tree diagram to show the possible outcomes.
    b. Determine the probability of obtaining at least one Head.
    c. Calculate the probability of obtaining either exactly two Heads or two Tails.

11. A coin is tossed three times. Show the sample space on a tree diagram and hence calculate the probability of getting:
    a. 2 Heads and 1 Tail
    c. a Head on the first toss of the coin
    e. no more than 1 Tail.
    b. either 3 Heads or 3 Tails
    d. at least 1 Head

12. A survey was carried out to find the type of occupation of 800 adults in a small suburb. There were 128 executives, 180 professionals, 261 trades workers, 178 labourers and 53 unemployed people.
    A person from this group is chosen at random. Calculate the probability that the person chosen is:
    a. a labourer
    b. not employed
    c. not an executive
    d. either a tradesperson or a labourer.

13. Two hundred people applied to do their driving test in October. The results are shown below.

Gender	Passed	Failed
Male	73	26
Female	81	20

    a. Calculate the probability that a person selected at random has failed the test.
    b. Calculate the probability that a person selected at random is a female who passed the test.

Complex familiar

14. **WE6** From a set of 18 cards numbered $1, 2, 3, \dots, 18$, one card is drawn at random.
    Let A be the event of obtaining a multiple of 3, let B be the event of obtaining a multiple of 4 and let C be the event of obtaining a multiple of 5.
    Calculate the following.
    a. $P(A \cup C)$
    b. $P(A \cap \overline{B})$
    c. $P(\overline{A \cap B \cap C})$

15. For two events A and B, it is known that $P(A \cup B) = 0.75$, $P(\overline{A}) = 0.42$ and $P(B) = 0.55$.

 a. Show that $P(\overline{A \cup B}) = P(\overline{A} \cap \overline{B})$.

 b. Construct a Venn diagram for the events A and B.

16. Two unbiased dice are rolled and the larger of the two numbers is noted. If the two dice show the same number, then the sum of the two numbers is recorded.

 a. Construct a table to show all the possible outcomes.

 b. Calculate the probability that the result is:

   i. a number greater than 5

   ii. either a two-digit number or a number greater than 6

   iii. not 9.

17. The 3:38 train to the city is late on average 1 day out of 3.

 a. Construct a probability tree to show the outcomes on three consecutive days.

 b. Calculate the probability that the train is:

   i. late on at least 2 days

   ii. on time on the last day

   iii. on time on all 3 days.

18. A sample of 100 first-year university science students were asked if they studied physics or chemistry. It was found that 63 study physics, 57 study chemistry and 4 study neither. In total, there are 1200 first-year university science students.

 a. Estimate the number of students who are likely to study both physics and chemistry.

 b. Two students are chosen at random from the total number of students. Determine the probability that one of the two students studies neither physics nor chemistry.

Complex unfamiliar

19. A coin is thrown onto a rectangular sheet of cardboard with dimensions 13 cm by 10 cm. The coin lands with its centre inside or on the boundary of the cardboard. Given the radius of the circular coin is 1.5 cm, calculate the probability that the coin lands completely inside the area covered by the rectangular piece of cardboard.

20. In a table tennis competition, each team must play every other team twice.
A regional competition consists of 16 teams, labelled
$A, B, C, \ldots, N, O, P$.
Determine the total number of games played.

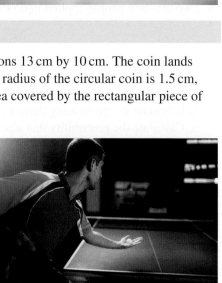

Fully worked solutions for this chapter are available online.

LESSON
7.3 Relative frequency

SYLLABUS LINKS

- Use relative frequencies obtained from data as point estimates of conditional probabilities and as indications of possible independence of events.
- Model and solve problems that involve probability, with and without technology.

Source: Mathematical Methods Senior Syllabus 2024 © State of Queensland (QCAA) 2024; licensed under CC BY 4.0.

7.3.1 Relative frequency

The relative frequency is usually expressed as a decimal or percentage and is calculated using the formula:

$$\text{relative frequency} = \frac{\text{number of times an event has occurred}}{\text{number of trials}}$$

In this formula, a *trial* is the number of times the probability experiment has been conducted.

The formula for relative frequency is similar to that for probability.

The term *relative frequency* refers to actual data obtained, but the term *probability* generally refers to theoretical data unless experimental probability is specifically stated.

WORKED EXAMPLE 7 Determining relative frequency

The weather has been fine on Christmas Day in Brisbane for 32 of the past 40 Christmas Days. Calculate the relative frequency of fine weather on Christmas Day.

THINK	WRITE
1. Write the formula.	$\text{Relative frequency} = \dfrac{\text{number of times an event has occurred}}{\text{number of trials}}$
2. Substitute the number of fine Christmas Days (32) and the number of trials (40).	$\text{Relative frequency} = \dfrac{32}{40}$
3. Calculate the relative frequency as a decimal.	$= 0.8$

The relative frequency can be used to assess the quality of products. This is done by finding the relative frequency of defective products.

WORKED EXAMPLE 8 Determining relative frequency as a percentage

A tyre company tests its tyres and finds that 144 out of a batch of 150 tyres will withstand 20 000 km of normal wear. Determine the relative frequency of tyres that will last 20 000 km. Give the answer as a percentage.

THINK	WRITE
1. Write the formula.	$\text{Relative frequency} = \dfrac{\text{number of times an event has occurred}}{\text{number of trials}}$

2. Substitute 144 (the number of times the event occurred) and 150 (number of trials).	Relative frequency $= \dfrac{144}{150}$	
3. Calculate the relative frequency.	$= 0.96$	
4. Convert the relative frequency to a percentage.	$= 96\%$	

Relative frequencies can be used to solve many practical problems.

WORKED EXAMPLE 9 Solving practical problems using relative frequency

A batch of 200 light globes was tested. The batch is considered unsatisfactory if more than 15% of globes burn for less than 1000 hours. The results of the test are in the table below.

Number of hours, h	Number of globes
$h < 500$	4
$500 \le h < 750$	12
$750 \le h < 1000$	15
$1000 \le h < 1250$	102
$1250 \le h < 1500$	32
$1500 \le h$	35

Determine if the batch is unsatisfactory. Justify your answer.

THINK	WRITE
1. Count the number of light globes that burn for less than 1000 hours.	$4 + 12 + 15 = 31$; thus, 31 light globes burn for less than 1000 hours.
2. Write the formula.	Relative frequency $= \dfrac{\text{number of times an event has occurred}}{\text{number of trials}}$
3. Substitute 31 (number of times the event occurs) and 200 (number of trials).	Relative frequency $= \dfrac{31}{200}$
4. Calculate the relative frequency.	$= 0.155$
5. Convert the relative frequency to a percentage.	$= 15.5\%$
6. Write your conclusion about the quality of the batch of light globes.	The batch is unsatisfactory as more than 15% of the light globes burn for less than 1000 hours.

7.3.2 Using technology to perform simulations

Simulating experiments using manual devices such as dice and spinners can take a lot of time. Simulations are used to reproduce real-life situations when it is not feasible to implement the actual test. A more efficient method of collecting results is to use technology to simulate a real-life scenario.

Random number generators can generate a series of numbers between two given values.

TI | THINK	WRITE	CASIO | THINK	WRITE
1. On a Calculator page, press MENU, 5 Probability, 4 Random for the random number generator menu.	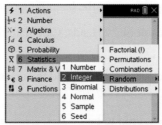	1. On a Run-Matrix page, press OPTN then F6 for more options, then press: F3 for PROB F4 for RAND.	
2. For a random number between 0 and 1, use option 1 Number.		2. For a random number between 0 and 1, use Ran# by pressing F1.	
3. For a random integer between an interval, use option 2 Integer.	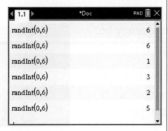	3. For a random integer between an interval, use Int by pressing F2.	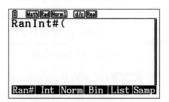
4. Input the range inside the brackets and press ENTER.		4. Input the range inside the brackets and press EXE.	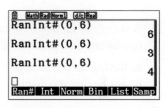
5. Repeatedly pressing ENTER will continue to generate random numbers in the interval.		5. Repeatedly pressing EXE will continue to generate random numbers in the interval.	

Exercise 7.3 Relative frequency

learn on

7.3 Exercise	7.3 Exam practice on

Simple familiar	Complex familiar	Complex unfamiliar
1, 2, 3, 4, 5, 6, 7, 8, 9, 10	11, 12, 13, 14, 15, 16	17, 18, 19, 20

These questions are even better in jacPLUS!
- Receive immediate feedback
- Access sample responses
- Track results and progress

Find all this and MORE in jacPLUS ⊙

Simple familiar

1. **MC** A study of cricket players found that of 150 players, 36 batted left handed. Select the relative frequency of left-handed batsmen.

   A. 0.24   **B.** 0.36   **C.** 0.54   **D.** 0.64

2. **MC** Four surveys were conducted and the following results were obtained. Select the result that has the highest relative frequency.

   A. Of 1500 P-plate drivers, 75 had been involved in an accident.
   B. Of 1200 patients examined by a doctor, 48 had to be hospitalised.
   C. Of 20 000 people at a football match, 950 were attending their first match.
   D. Of 300 drivers breath tested, 170 were found to be over the legal limit.

3. **WE7** At the opening of the ski season, there has been sufficient snow for skiing for 37 out of the past 50 years. Calculate the relative frequency of sufficient snow at the beginning of the ski season.

4. A biased coin has been tossed 100 times with the result of 79 Heads. Calculate the relative frequency of the coin landing Heads.

5. Out of eight Maths tests done by a class during a year, Peter has topped the class three times. Calculate the relative frequency of Peter topping the class.

6. Farmer Jones has planted a wheat crop. For the wheat crop to be successful, Farmer Jones needs 500 mm of rain to fall over the spring months. Past weather records show that this has occurred on 27 of the past 60 years. Determine the relative frequency of:

   a. sufficient rainfall   b. insufficient rainfall.

7. **WE8** Of 300 cars coming off an assembly line, 12 are found to have defective brakes. Calculate the relative frequency of a car having defective brakes. Give the answer as a percentage.

8. A survey of 25 000 new car buyers found that 750 had a major mechanical problem in the first year of operation. Calculate the relative frequency of:

   a. having mechanical problems in the first year
   b. not having mechanical problems in the first year.

9. In an electronics factory, 15 out of every 1400 graphics cards is faulty. Calculate the relative frequency of getting a fully functional graphics card. Express your answer as a percentage.

10. During an election campaign, 2000 people were asked for their voting preferences. One thousand and fifty said they would vote for the government, 875 said they would vote for the opposition, and the remainder were undecided. Calculate the relative frequency of:

    a. government voters
    b. opposition voters
    c. undecided voters.

Complex familiar

11. Research over the past 25 years shows that each November there is an average of two wet days on Sunnybank Island. Travelaround Tours offer one-day tours to Sunnybank Island at a cost of $150 each, with a money back guarantee against rain.
    If Travelaround Tours take 1200 bookings for tours in November, determine how many refunds they would expect to give.

12. An average of 200 robberies takes place each year in the town of Amiak. There are 10 000 homes in this town.
    Each robbery results in an average insurance claim of $20 000. Calculate the minimum premium per home the insurance company would need to charge to cover these claims.

13. **WE9** A car maker recorded the first time that its cars came in for mechanical repairs. The results are in the table below.

Time taken	Number of cars
0–<3 months	5
3–<6 months	12
6–<12 months	37
1–<2 years	49
2–<3 years	62
3 years or more	35

a. The assembly line will need to be upgraded if the relative frequency of cars needing mechanical repair in the first year is greater than 25%. Determine if this will be necessary.

b. Determine, as a percentage, the relative frequency of:
   i. a car needing mechanical repair in the first 3 months
   ii. a car needing mechanical repair in the first 2 years
   iii. a car not needing mechanical repair in the first 3 years.

14. A manufacturer of shock absorbers measures the distance that its shock absorbers can travel before they must be replaced. The results are in the table below.

Number of kilometres	Number of shock absorbers
0–<20 000	1
20 000–<40 000	2
40 000–<60 000	46
60 000–<80 000	61
80 000–<100 000	90

Determine the maximum distance the manufacturer will guarantee so that the relative frequency of the shock absorbers lasting is greater than 0.985.

15. a. Roll a die 120 times and record each result in the table below.

Number	Occurrences	Percentage of throws
1		
2		
3		
4		
5		
6		

b. Describe how close the experimental results are to the theoretical results that were expected. State whether you would expect the difference between the experimental results and the theoretical results to increase or decrease if the die were rolled 1200 times. Explain your answer.

16. The gender of babies in a set of triplets is simulated by flipping 3 coins. If a coin lands Tails up, the baby is a boy. If a coin lands Heads up, the baby is a girl. In the simulation, the trial is repeated 40 times and the following results show the number of Heads obtained in each trial:

0, 3, 2, 1, 1, 0, 1, 2, 1, 0, 1, 0, 2, 0, 1, 0, 1, 2, 3, 2,
1, 3, 0, 2, 1, 2, 0, 3, 1, 3, 0, 1, 0, 1, 3, 2, 2, 1, 2, 1

a. Calculate the probability that exactly one of the babies in a set of triplets is female.
b. Calculate the probability that more than one of the babies in the set of triplets is female.

Complex unfamiliar

17. A die is rolled 20 times. Each roll results in one of the outcomes $\{1, 2, 3, 4, 5, 6\}$.
Explain how you could simulate this experiment to estimate the chance of obtaining a six, and evaluate the reasonableness of your estimation.

18. A random number generator was used to select how many students will receive a free lunch on any given day. The maximum number of students who can receive a free lunch is 10 and the minimum is 0.
Explain how you could simulate the free lunch program over 30 days, and evaluate the reasonableness of your estimation.

19. A mini-lottery game may be simulated as follows. Each game consists of choosing two numbers from the whole numbers 1 to 6. The cost to play one game is $1. A prize of $10 is paid for both numbers correct. No other prizes are awarded.
A particular player, Ethan, always chooses the numbers 1 and 2. Use technology to simulate how much Ethan would win or lose overall if he plays 40 times.

20. Choose one of the topics below (or another of your choice) and calculate the relative frequency of the event. Most of the information needed can be found from books or the internet.
  - Topic A: Examine weather records and determine the relative frequency of rain on New Year's Eve in Brisbane.
  - Topic B: Choose your favourite sporting team. Determine the relative frequency of them winning over the past three seasons.
  - Topic C: Determine the relative frequency of the stock market rising for three consecutive days.
  - Topic D: Check the NRL or AFL competitions and determine the relative frequencies of win, loss and draw for each team.

Fully worked solutions for this chapter are available online.

LESSON
7.4 Conditional probability

7.4.1 Recognising conditional probability

Conditional probability is when the probability of an event is conditional (depends) on another event occurring first. The effect of conditional probability is to reduce the sample space and thus increase the probability of the desired outcome.

For example, if a coin is tossed twice, the sample space is $\{HH, HT, TH, TT\}$. If it is known that at least one Head is obtained, then the sample space reduces to $\{HH, HT, TH\}$ and the probability of obtaining two Heads increases from $\dfrac{1}{4}$ to $\dfrac{1}{3}$.

For two events, A and B, the conditional probability of event A occurring, given that event B has occurred, is written as $P(A|B)$.

Conditional probability can be expressed using a variety of language. Some examples of conditional probability statements follow. The key words to look for in a conditional probability statement have been highlighted in each instance.

- **If** a student buys an ice-cream today, **then** the chance of buying an ice-cream tomorrow is 70%.
- **Given that** a red marble was picked out of the bag with the first pick, the probability of a blue marble being picked out of the bag with the second pick is 0.35.

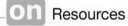 **Resources**

Interactivity Conditional probability and independence (int-6292)

7.4.2 Formula for conditional probability

Consider the events A and B:

$$P(A) = \frac{n(A)}{n(S)}, \; P(B) = \frac{n(B)}{n(S)} \; \text{and} \; P(A \cap B) = \frac{n(A \cap B)}{n(S)}$$

For the conditional probability $P(A|B)$, the sample space is reduced to $n(B)$.

$$
\begin{aligned}
P(A|B) &= \frac{n(A \cap B)}{n(B)} \\
&= n(A \cap B) \div n(B) \\
&= \frac{n(A \cap B)}{n(S)} \div \frac{n(B)}{n(S)} \\
&= P(A \cap B) \div P(B) \\
&= \frac{P(A \cap B)}{P(B)}
\end{aligned}
$$

Hence, the conditional probability formula is as follows.

Conditional probability formula

$$P(A|B) = \frac{P(A \cap B)}{P(B)}$$

The conditional probability formula can be used to determine the probability of event A occurring, given that event B has already occurred.

This formula illustrates that if the events A and B are mutually exclusive so that $P(A \cap B) = 0$, then $P(A|B) = 0$. That is, if B occurs, then it is impossible for A to occur.

However, if B is a subset of A so that $P(A \cap B) = P(B)$, then $P(A|B) = 1$. That is, if B occurs, then it is certain that A will occur.

WORKED EXAMPLE 10 Calculating conditional probability from a given table

The table shows the results of a survey of 100 people aged between 16 and 29 about their preferred choice of food when eating at a café.

	Vegetarian (V)	Non-vegetarian ($\overline{V}$)	
Male (M)	18	38	56
Female (F)	25	19	44
	43	57	100

One person is selected at random from those surveyed. Identify the event and use the table to calculate the following probabilities.
a. $P(M \cap V)$ **b.** $P(M|V)$ **c.** $P(\overline{V}|F)$ **d.** $P(V)$

THINK

a. 1. Describe the event $M \cap V$.

2. Calculate the probability.

b. 1. Describe the conditional probability.

2. State the number of elements in the reduced sample space.

3. Calculate the required probability.

WRITE

a. The event $M \cap V$ is the event the selected person is both male and preferred vegetarian.

$$P(M \cap V) = \frac{n(M \cap V)}{n(S)}$$

$$= \frac{18}{100}$$

$$\therefore P(M \cap V) = 0.18$$

b. The event $M|V$ is the event of a person being male given that the person prefers vegetarian.

Since the person is known to prefer vegetarian, the sample space is reduced to $n(V) = 43$ people.

Of the 43 vegetarians, 18 are male.

$$\therefore P(M|V) = \frac{n(M \cap V)}{n(V)}$$

$$= \frac{18}{43}$$

c. 1. Identify the event.	**c.** The event $\overline{V}\vert F$ is the event of a person preferring non-vegetarian given the person is female.
2. State the number of elements in the reduced sample space.	Since the person is known to be female, the sample space is reduced to $n(F) = 44$ people.
3. Calculate the probability.	Of the 44 females, 19 prefer non-vegetarian. $$\therefore P(\overline{V}\vert F) = \frac{n(\overline{V} \cap F)}{n(F)}$$ $$= \frac{19}{44}$$
d. 1. State the event.	**d.** The event V is the event the selected person prefers vegetarian.
2. Calculate the required probability. *Note:* This is not a conditional probability.	$$P(V) = \frac{n(V)}{n(S)}$$ $$= \frac{43}{100}$$ $$\therefore P(V) = 0.43$$

7.4.3 Multiplication of probabilities

Consider the conditional probability formulas for $P(A\vert B)$ and for $P(B\vert A)$:

$$P(A\vert B) = \frac{P(A \cap B)}{P(B)}$$

$$P(B\vert A) = \frac{P(B \cap A)}{P(A)}$$

Since $A \cap B$ is the same as $B \cap A$, $P(A \cap B) = P(B \cap A)$.

By rearranging the conditional probability formula for $P(A\vert B)$ or for $P(B\vert A)$, the formula for multiplication of probabilities is formed.

Multiplication of probabilities

$$P(A \cap B) = P(A\vert B)\,P(B)$$

$$= P(B\vert A)\,P(A)$$

For example, when selecting two marbles without replacement from a bag containing 6 aqua (A) marbles and 10 black (B) marbles, the probability of obtaining first an aqua and then a black marble would be:

$$P(A \cap B) = P(A)P(B\vert A)$$
$$= \frac{6}{16} \times \frac{10}{15}$$

The multiplication formula can be extended. For example, the probability of obtaining 3 black marbles when selecting three marbles without replacement from the bag of 16 marbles would be:

$$\frac{10}{16} \times \frac{9}{15} \times \frac{8}{14}$$

Notice in the examples above, after each selection from the bag, the number of marbles is decreasing as the marbles are not being replaced.

WORKED EXAMPLE 11 Determining conditional probabilities

If $P(A) = 0.6$, $P(A|B) = 0.6125$ and $P(\overline{B}) = 0.2$, calculate $P(A \cap B)$ and $P(B|A)$.

THINK	WRITE
1. State the conditional probability formula for $P(A|B)$.	**a.** $P(A|B) = \dfrac{P(A \cap B)}{P(B)}$
2. Obtain the value of $P(B)$.	For complementary events: $P(B) = 1 - P(\overline{B})$ $= 1 - 0.2$ $= 0.8$
3. Use the formula to calculate $P(A \cap B)$.	$P(A|B) = \dfrac{P(A \cap B)}{P(B)}$ $0.6125 = \dfrac{P(A \cap B)}{0.8}$ $P(A \cap B) = 0.6125 \times 0.8$ $= 0.49$
4. State the conditional probability formula for $P(B|A)$.	$P(B|A) = \dfrac{P(B \cap A)}{P(A)}$
5. Calculate the required probability.	$P(B \cap A) = P(A \cap B)$ $P(B|A) = \dfrac{P(A \cap B)}{P(A)}$ $= \dfrac{0.49}{0.6}$ $= \dfrac{49}{60}$ ≈ 0.82

7.4.4 Probability tree diagrams

The sample space of a two-stage trial where the outcomes of the second stage are dependent on the outcomes of the first stage can be illustrated with a probability tree diagram.

Each branch is labelled with its probability; conditional probabilities are required for the second-stage branches. Calculations are performed according to the addition and multiplication laws of probability.

The formula for multiplication of probabilities is applied by multiplying the probabilities that lie along the respective branches to calculate the probability of an outcome. For example, to obtain the probability of A and B occurring, we need to multiply the probabilities along the branches A to B, since $P(A \cap B) = P(A) \times P(B|A)$.

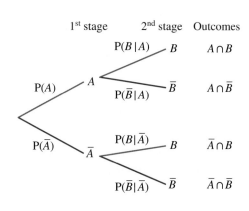

The addition formula for mutually exclusive events is applied by adding the results from separate outcome branches to calculate the union of any of the four outcomes. For example, to obtain the probability that A occurs, add together the results from the two branches where A occurs. This gives $P(A) = P(A)\,P(B|A) + P(A)\,P(\overline{B}|A)$.

Probability tree diagrams

- **Multiply along the branch.**
- **Add the results from each complete branch.**

WORKED EXAMPLE 12 Determining conditional probability using a tree diagram

A box of chocolates contains 6 soft-centre and 4 hard-centre chocolates. A chocolate is selected at random and once eaten, a second chocolate is chosen.
Let S_i be the event a soft-centre chocolate is chosen on the ith selection and H_i be the event that a hard-centre chocolate is chosen on the ith selection, $i = 1, 2$.

a. Deduce the value of $P(H_2|S_1)$.
b. Construct a probability tree diagram to illustrate the possible outcomes.
c. Calculate the probability that the first chocolate has a hard centre and the second a soft centre.
d. Calculate the probability that either both chocolates have soft centres or both have hard centres.

THINK	WRITE
a. 1. Identify the meaning of $P(H_2|S_1)$.	a. $P(H_2|S_1)$ is the probability that the second chocolate has a hard centre given that the first has a soft centre.
2. State the required probability.	If a soft centre has been chosen first, there remain in the box 5 soft- and 4 hard-centre chocolates. $\therefore P(H_2|S_1) = \dfrac{4}{9}$
b. Construct the two-stage probability tree diagram.	b.

1st choice   2nd choice

$\dfrac{6}{10}$ S_1 $\dfrac{5}{9}$ S_2 $S_1 \cap S_2$

  $\dfrac{4}{9}$ H_2 $S_1 \cap H_2$

$\dfrac{4}{10}$ H_1 $\dfrac{6}{9}$ S_2 $H_1 \cap S_2$

  $\dfrac{3}{9}$ H_2 $H_1 \cap H_2$

| c. Identify the appropriate branch and multiply along it to obtain the required probability. *Note:* The multiplication law for probability is $P(H_1 \cap S_2) = P(H_1) \times P(S_2|H_1)$. | c. The required outcome is $H_1 \cap S_2$. $$P(H_1 \cap S_2) = \dfrac{4}{10} \times \dfrac{6}{9}$$ $$= \dfrac{2}{5} \times \dfrac{2}{3}$$ $$= \dfrac{4}{15}$$ The probability is $\dfrac{4}{15}$. |

d. 1. Identify the required outcome.

d. The probability that both chocolates have the same type of centre is $P((S_1 \cap S_2) \cup (H_1 \cap H_2))$.

2. Calculate the probabilities along each relevant branch.

$$P(S_1 \cap S_2) = \frac{6}{10} \times \frac{5}{9}$$

$$P(H_1 \cap H_2) = \frac{4}{10} \times \frac{3}{9}$$

3. Use the addition law for mutually exclusive events by adding the probabilities from the separate branches.

The probability that both chocolates have the same type of centre is:

$$\frac{6}{10} \times \frac{5}{9} + \frac{4}{10} \times \frac{3}{9} = \frac{30}{90} + \frac{12}{90}$$

$$= \frac{42}{90}$$

$$= \frac{7}{15}$$

The probability both centres are the same type is $\frac{7}{15}$.

Exercise 7.4 Conditional probability

learn on

7.4 Exercise	7.4 Exam practice **on**

Simple familiar	Complex familiar	Complex unfamiliar
1, 2, 3, 4, 5, 6, 7, 8, 9, 10, 11, 12, 13, 14	15, 16, 17, 18	19, 20

These questions are even better in jacPLUS!
- Receive immediate feedback
- Access sample responses
- Track results and progress

Find all this and MORE in jacPLUS ▶

Simple familiar

1. A bag contains 3 red (R) balls, 4 purple (P) balls and 2 yellow (Y) balls. One ball is chosen at random and removed from the bag. A second ball is then chosen from the bag.

a. Given that the first ball chosen is yellow, calculate the probability that the second ball will be red. Write the symbol for this conditional probability.

b. Given that the first ball chosen is yellow, calculate the probability that the second ball will be yellow. Write the symbol for this conditional probability.

c. Given that the first ball chosen is red, calculate the probability that the second ball will be purple. Write the symbol for this conditional probability.

d. If the first ball chosen is not red, calculate the probability that the second ball will be red. Write the symbol for this conditional probability.

2. Of a group of 20 students who study either Art or Biology or both subjects, 13 study Art, 16 study Biology, and 9 study both Art and Biology.

Let A be the event a student studies Art and B be the event a student studies Biology.

State whether the following probability statements are true (T) or false (F).

a. $Pr(A) = \dfrac{13}{20}$

b. $Pr(A|B) = \dfrac{13}{20}$

c. $Pr(A|B) = \dfrac{13}{16}$

d. $Pr(A|B) = \dfrac{9}{16}$

e. $Pr(B|A) = \dfrac{9}{13}$

f. $Pr(B|A) = \dfrac{7}{13}$

3. **WE11** If $P(\bar{A}) = 0.6$, $P(B|A) = 0.3$ and $P(B) = 0.5$, calculate $P(A \cap B)$ and $P(A|B)$.

4. If $P(A) = 0.61$, $P(B) = 0.56$ and $P(A \cup B) = 0.81$, calculate the following.

   a. $P(A|B)$ b. $P(A|A \cup B)$ c. $P(A|\bar{A} \cup B)$

5. Given $P(A) = 0.7$, $P(B) = 0.3$ and $P(A \cup B) = 0.8$, determine the following.

   a. $P(A \cap B)$ b. $P(A|B)$ c. $P(B|A)$ d. $P(A|\bar{B})$

6. Given $P(A) = 0.6$, $P(B) = 0.5$ and $P(A \cup B) = 0.8$, determine the following.

   a. $P(A \cap B)$ b. $P(A|B)$ c. $P(B|A)$ d. $P(A|\bar{B})$

7. Given $P(A) = 0.6$, $P(B) = 0.7$ and $P(A \cap B) = 0.4$, determine the following.

   a. $P(A \cup B)$ b. $P(A|B)$ c. $P(B|\bar{A})$ d. $P(\bar{A}|\bar{B})$

8. **WE10** The table shows the results of a survey of 100 people aged between 16 and 29 who were asked whether they rode a bike and what drink they preferred.
   One person is selected at random from those surveyed. Identify the event and use the table to calculate the following.

   a. $P(\bar{B} \cap \bar{C})$ b. $P(\bar{B}|\bar{C})$ c. $P(C|B)$ d. $P(B)$

	Drink containing caffeine (C)	Caffeine-free drink ($\bar{C}$)	
Bike rider (B)	28	16	44
Non–bike rider ($\bar{B}$)	36	20	56
	64	36	100

9. Two six-sided dice are rolled. Calculate the probability that:

   a. the sum of 8 is obtained
   b. a sum of 8 is obtained given the numbers are not the same
   c. the sum of 8 is obtained but the numbers are not the same
   d. the numbers are not the same given the sum of 8 is obtained.

10. A group of 400 people were tested for allergic reactions to two new medications. The results are shown in the table below.

	Allergic reaction	No allergic reaction
Medication A	25	143
Medication B	47	185

   If a person is selected at random from the group, calculate the following.

   a. The probability that the person suffers an allergic reaction
   b. The probability that the person was administered medication A
   c. Given there was an allergic reaction, the probability that medication B was administered
   d. Given the person was administered medication A, the probability that the person did not have an allergic reaction

11. **WE12** A box of jubes contains 5 green jubes and 7 red jubes. One jube is selected at random and once eaten, a second jube is chosen.
    Let G_i be the event a green jube is chosen on the ith selection and R_i the event that a red jube is chosen on the ith selection, $i = 1, 2$.

    a. Construct a probability tree diagram to illustrate the possible outcomes.
    b. Calculate the probability that the first jube is green and the second is red.
    c. Calculate the probability that either both jubes are green or both are red.

12. To get to school, Rodney catches a bus and walks the remaining distance.
   If the bus is on time, Rodney has a 98% chance of arriving at school on time. However, if the bus is late, Rodney's chance of arriving at school on time is only 56%. On average the bus is on time 90% of the time.
   a. Construct a probability tree diagram to describe the given information, defining the symbols used.
   b. Calculate the probability that Rodney will arrive at school on time.

13. Two unbiased dice are rolled and the sum of the topmost numbers is noted. Given that the sum is less than 6, calculate the probability that the sum is an even number.

14. Two cards are drawn randomly from a standard pack of 52 cards. Calculate the probability that:

   a. both cards are diamonds
   b. at least 1 card is a diamond
   c. both cards are diamonds, given that at least one card is a diamond
   d. both cards are diamonds, given that the first card drawn is a diamond.

Complex familiar

---

15. A bag contains 5 red marbles and 7 green marbles. Two marbles are drawn from the bag, one at a time, without replacement. Calculate the probability that:
   a. both marbles are green
   b. at least 1 marble is green
   c. both marbles are green given that at least 1 is green
   d. the first marble drawn is green given that the marbles are of different colours.

16. Sarah and Kate sit a Biology exam. The probability that Sarah passes the exam is 0.9 and the probability that Kate passes the exam is 0.8. Calculate the probability that:
   a. both Sarah and Kate pass the exam
   b. at least one of the two girls passes the exam
   c. only 1 girl passes the exam, given that Sarah passes.

17. In a survey designed to check the number of male and female smokers in a population, it was found that there were 32 male smokers, 41 female smokers, 224 female non-smokers and 203 male non-smokers. A person is selected at random from this group of people. Calculate the probability that the person selected is:
   a. a non-smoker
   b. a smoker, given that the person is male
   c. female, given that the person is a non-smoker.

18. In a sample of 1000 people, it is found that:
   - 82 people are overweight and suffer from hypertension
   - 185 are overweight but do not suffer from hypertension
   - 175 are not overweight but suffer from hypertension
   - 558 are not overweight and do not suffer from hypertension.
     A person is selected at random from the sample. Calculate the probability that the person:
   a. is overweight
   b. suffers from hypertension
   c. suffers from hypertension given that the person is overweight
   d. is overweight given that the person does not suffer from hypertension.

19. When walking home from school during the summer months, Harold buys either an ice-cream or a drink from the corner shop. If Harold bought an ice-cream the previous day, there is a 30% chance that he will buy a drink the next day. If he bought a drink the previous day, there is a 40% chance that he will buy an ice-cream the next day. On Monday, Harold bought an ice-cream. Determine the probability that he buys an ice-cream on Wednesday.

20. In tenpin bowling, a game is made up of 10 frames. Each frame represents one turn for the bowler. In each turn, a bowler is allowed up to 2 rolls of the ball to knock down the 10 pins. If the bowler knocks down all 10 pins with the first ball, this is called a strike. If it takes 2 rolls of the ball to knock the 10 pins, this is called a spare. Otherwise it is called an open frame.
    On average, Richard hits a strike 85% of the time. If he needs a second roll of the ball then, on average, he will knock down the remaining pins 97% of the time. While training for the club championship, Richard plays a game.
    Consider the first 2 frames of his game.
    Determine the probability that the first frame is a strike, given that the second is a strike.

Fully worked solutions for this chapter are available online.

LESSON
7.5 Independence

SYLLABUS LINKS

- Understand and use the notion of independence of an event A from an event B, as defined by $P(A|B) = P(A)$.
- Use the formula $P(A \cap B) = P(A)\,P(B)$ for independent events A and B.
- Use relative frequencies obtained from data as point estimates of conditional probabilities and as indications of possible independence of events.
- Model and solve problems that involve probability, with and without technology.

Source: Mathematical Methods Senior Syllabus 2024 © State of Queensland (QCAA) 2024; licensed under CC BY 4.0.

7.5.1 The concept of independent events

If a coin is tossed twice, the chance of obtaining a Head on the coin on its second toss is unaffected by the result of the first toss. The probability of a Head on the second toss given a Head is obtained on the first toss is still $\frac{1}{2}$.

Events that have no effect on each other are called **independent events**. For such events, $P(A|B) = P(A)$. The given information does not affect the chance of event A occurring.

Events that do affect each other are dependent events. For dependent events, $P(A|B) \neq P(A)$ and the conditional probability formula is used to evaluate $P(A|B)$.

7.5.2 Test for mathematical independence

Although it may be obvious that the chance of obtaining a Head on a coin on its second toss is unaffected by the result of the first toss, in more complex situations it can be difficult to intuitively judge whether events are independent or dependent. For such situations, there is a test for mathematical independence that will determine the matter.

The multiplication formula states that $P(A \cap B) = P(A) \times P(B|A)$.

If the events A and B are independent, then $P(B|A) = P(B)$.

Hence, for independent events the following applies.

> **Independent events**
>
> $$P(A \cap B) = P(A)\, P(B)$$

This result is used to test whether events are mathematically independent or not.

Consider the two probability tables shown.

	A	$\bar{A}$	
B	0.48	0.32	0.8
$\bar{B}$	0.12	0.08	0.2
	0.6	0.4	1

	A	$\bar{A}$	
B	0.5	0.3	0.8
$\bar{B}$	0.1	0.1	0.2
	0.6	0.4	1

$P(A \cap B) = 0.48$

$P(A) \times P(B) = 0.6 \times 0.8 = 0.48$

$\therefore P(A \cap B) = P(A) \times P(B)$

Hence, the two events, A and B, are independent events.

$P(A \cap B) = 0.5$

$P(A) \times P(B) = 0.6 \times 0.8 = 0.48$

$\therefore P(A \cap B) \neq P(A) \times P(B)$

Hence, the two events, A and B, are **not** independent events.

WORKED EXAMPLE 13 Testing for independence

Consider the trial of tossing a coin twice.
Let A be the event of at least one Tail, B be the event of either two Heads or two Tails, and C be the event that the first toss is a Head.

a. **List the sample space and the set of outcomes in each of A, B and C.**
b. **Determine whether A and B are independent.**
c. **Determine whether B and C are independent.**
d. **Use the addition formula to calculate $P(B \cup C)$.**

THINK	WRITE
a. 1. List the elements of the sample space.	**a.** The sample space is the set of equiprobable outcomes: $\{HH, HT, TH, TT\}$.
2. List the elements of A, B and C.	$A = \{HT, TH, TT\}$ $B = \{HH, TT\}$ $C = \{HH, HT\}$
b. 1. State the test for independence.	**b.** A and B are independent if $P(A \cap B) = P(A)P(B)$.
2. Calculate the probabilities needed for the test for independence to be applied.	$P(A) = \dfrac{3}{4}$ and $P(B) = \dfrac{2}{4}$ Since $A \cap B = \{TT\}$, $P(A \cap B) = \dfrac{1}{4}$.
3. Determine whether the events are independent.	Substitute values into the formula $P(A \cap B) = P(A)P(B)$. $\text{LHS} = \dfrac{1}{4}$ $\text{RHS} = \dfrac{3}{4} \times \dfrac{2}{4}$ $\qquad = \dfrac{3}{8}$ Since LHS $\neq$ RHS, the events A and B are not independent.
c. 1. State the test for independence.	**c.** B and C are independent if $P(B \cap C) = P(B)P(C)$.
2. Calculate the probabilities needed for the test for independence to be applied.	$P(B) = \dfrac{2}{4}$ and $P(C) = \dfrac{2}{4}$ Since $P(B \cap C) = \{HH\}$, $P(B \cap C) = \dfrac{1}{4}$.
3. Determine whether the events are independent.	Substitute values in $P(B \cap C) = P(B)P(C)$. $\text{LHS} = \dfrac{1}{4}$ $\text{RHS} = \dfrac{2}{4} \times \dfrac{2}{4}$ $\qquad = \dfrac{1}{4}$ Since LHS $=$ RHS, the events B and C are independent.
d. 1. State the addition formula for $P(B \cup C)$.	**d.** $P(B \cup C) = P(B) + P(C) - P(B \cap C)$
2. Replace $P(B \cap C)$.	Since B and C are independent, $P(B \cap C) = P(B)P(C)$. $\therefore \ P(B \cup C) = P(B) + P(C) - P(B) \times P(C)$
3. Complete the calculation.	$P(B \cup C) = \dfrac{1}{2} + \dfrac{1}{2} - \dfrac{1}{2} \times \dfrac{1}{2}$ $\qquad\qquad = \dfrac{3}{4}$

7.5.3 Independent trials

Consider choosing a ball from a bag containing 6 red and 4 green balls, noting its colour, returning the ball to the bag and then choosing a second ball. These trials are independent as the chance of obtaining a red or green ball is unaltered for each draw.

This is an example of **sampling with replacement**.

The probability tree diagram is as shown.

The branch outcomes for the second stage are not dependent on the results of the first stage.

The probability that both balls are red is $P(R_1 R_2) = \dfrac{6}{10} \times \dfrac{6}{10}$.

As the events are independent:

$$P(R_1 \cap R_2) = P(R_1) \times P(R_2|R_1)$$
$$= P(R_1) \times P(R_2)$$

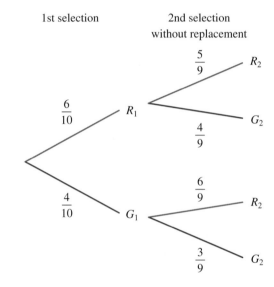

Consider the example above of drawing a ball from the bag, but **sampling without replacement**.

The probability tree diagram is shown.

The branch outcomes for the second stage are **dependent** on the results of the first stage.

Then the probability that both balls are red is

$$P(R_1 \cap R_2) = \dfrac{6}{10} \times \dfrac{5}{9}$$

since the events are not independent.

Sequences of independent events

If the events A, B, C, ... are independent, then $P(A \cap B \cap C \cap ...) = P(A) \times P(B) \times P(C) \times ...$

WORKED EXAMPLE 14 Determining probabilities with independent events

Three overweight people, Ari, Barry and Chris, commence a diet. The chances that each person sticks to the diet are 0.6, 0.8 and 0.7, respectively, independent of each other. Determine the probability that:
a. all three people stick to the diet
b. only Chris sticks to the diet
c. at least one of the three people sticks to the diet.

THINK	WRITE
a. 1. Define the independent events.	a. Let A be the event that Ari sticks to the diet, B be the event that Barry sticks to the diet and C be the event that Chris sticks to the diet. $P(A) = 0.6$, $P(B) = 0.8$ and $P(C) = 0.7$

---

2. State an expression for the required probability.

The probability that all three people stick to the diet is $P(A \cap B \cap C)$.

3. Calculate the required probability.

Since the events are independent,
$$
\begin{aligned}
P(A \cap B \cap C) &= P(A) \times P(B) \times P(C) \\
&= 0.6 \times 0.8 \times 0.7 \\
&= 0.336
\end{aligned}
$$
The probability all three stick to the diet is 0.336.

b. 1. State an expression for the required probability.

b. If only Chris sticks to the diet then neither Ari nor Barry do. The probability is $P(\overline{A} \cap \overline{B} \cap C)$.

2. Calculate the probability.
Note: For complementary events, $P(\overline{A}) = 1 - P(A)$.

$$
\begin{aligned}
P(\overline{A} \cap \overline{B} \cap C) &= P(\overline{A}) \times P(\overline{B}) \times P(C) \\
&= (1 - 0.6) \times (1 - 0.8) \times 0.7 \\
&= 0.4 \times 0.2 \times 0.7 \\
&= 0.056
\end{aligned}
$$
The probability only Chris sticks to the diet is 0.056.

c. 1. Express the required event in terms of its complementary event.

c. The event that at least one of the three people sticks to the diet is the complement of the event that no one sticks to the diet.

2. Calculate the required probability.

$$
\begin{aligned}
&P \text{ (at least one sticks to the diet)} \\
&= 1 - P(\text{no one sticks to the diet}) \\
&= 1 - P(\overline{A} \cap \overline{B} \cap \overline{C}) \\
&= 1 - 0.4 \times 0.2 \times 0.3 \\
&= 1 - 0.024 \\
&= 0.976
\end{aligned}
$$
The probability that at least one person sticks to the diet is 0.976.

Exercise 7.5 Independence

learn on

7.5 Exercise	**7.5 Exam practice** on	These questions are even better in jacPLUS!

Simple familiar	Complex familiar	Complex unfamiliar
1, 2, 3, 4, 5, 6, 7, 8, 9, 10, 11, 12	13, 14, 15, 16	17, 18, 19, 20

These questions are even better in jacPLUS!
- Receive immediate feedback
- Access sample responses
- Track results and progress

Find all this and MORE in jacPLUS ▶

Simple familiar

1. State the condition under which two events A and B are defined as being independent.

2. Two events A and B are such that $P(A) = 0.7$, $P(B) = 0.8$ and $P(A \cup B) = 0.94$. Determine whether A and B are independent.

3. Events A and B are shown in the Venn diagram.
   Determine whether A and B are independent events or not.

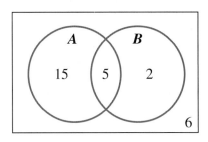

4. a. Given $P(A) = 0.3$, $P(B) = p$ and $P(A \cap B) = 0.12$, determine the value of p if the events A and B are independent.
   b. Given $P(A) = p$, $P(B) = 2p$ and $P(A \cap B) = 0.72$, determine the value of p if the events A and B are independent.
   c. Given $P(A) = q$, $P(B) = \dfrac{1}{5}$ and $P(A \cup B) = \dfrac{4}{5}$, determine the value of q if the events A and B are independent.

5. Events A and B are independent. If $P(B) = \dfrac{2}{3}$ and $P(A|B) = \dfrac{4}{5}$, determine:

   a. $P(A)$ b. $P(B|A)$ c. $P(A \cap B)$ d. $P(A \cup B)$.

6. Identify which of the following are independent events.
   a. Event 1: Rolling a 5 on a die; Event 2: getting a head on a tossed coin
   b. Event 1: Removing a green ball from a bag containing green, yellow and red balls without replacement; Event 2: Removing a red ball from the same bag
   c. Event 1: A cyclone strikes Brisbane; Event 2: Brisbane airport is closed down
   d. Event 1: Drawing a black card from a deck of cards with replacement; Event 2: Drawing a picture card from the deck of cards

7. **WE13** Consider the experiment of tossing a coin twice.
   Let A be the event the first toss is a Tail, B the event of one Head and one Tail, and C the event of no more than one Tail.

   a. List the sample space and the set of outcomes in each of A, B and C.
   b. Determine whether A and B are independent.
   c. Determine whether B and C are independent.
   d. Use the addition formula to calculate $P(B \cup A)$.

8. Two unbiased six-sided dice are rolled. Let A be the event the same number is obtained on each die and B be the event the sum of the numbers on each die exceeds 8.

   a. Determine whether events A and B are mutually exclusive. Justify your answer.
   b. Determine whether events A and B are independent. Justify your answer.
   c. If C is the event the sum of the two numbers equals 8, determine whether B and C are:

      i. mutually exclusive
      ii. independent.

9. Events A and B are such that $P(B) = \dfrac{3}{5}$, $P(A|B) = \dfrac{1}{3}$ and $P(A \cup B) = \dfrac{23}{30}$.

   a. Calculate $P(A \cap B)$.
   b. Calculate $P(A)$.
   c. Determine whether events A and B are independent.

10. **WE14** Three underweight people, Ava, Bambi and Chi, commence a carbohydrate diet. The chances that each person sticks to the diet are 0.4, 0.9 and 0.6 respectively, independent of each other. Determine the probability that:

    a. all three people stick to the diet
    b. only Ava and Chi stick to the diet
    c. at least one of the three people does not stick to the diet.

11. A box of toy blocks contains 10 red blocks and 5 yellow blocks. A child draws out two blocks at random.

    a. Construct the tree diagram if the sampling is with replacement and calculate the probability that one block of each colour is obtained.
    b. Construct the tree diagram if the sampling is without replacement and calculate the probability that one block of each colour is obtained.
    c. If the child was to draw out three blocks, rather than two, calculate the probability of obtaining three blocks of the same colour if the sampling is with replacement.

12. A family owns two cars, A and B. Car A is used 65% of the time, car B is used 74% of the time and at least one of the cars is used 97% of the time. Determine whether the two cars are used independently.

Complex familiar

13. The probability that a male is colourblind is 0.05, and the probability that a female is colourblind is 0.0025. If there is an equal number of males and females in a population, determine the probability that a person selected at random from the population is:

    a. female and colourblind
    b. colourblind given that the person is female
    c. male given that the person is colourblind.

    If two people are chosen at random, determine the probability that:

    d. both are colourblind males
    e. one is colourblind given a male and a female are chosen.

14. Events A and B are independent such that $P(A \cup B) = 0.8$ and $P(A|\overline{B}) = 0.6$. Determine $P(B)$.

15. A survey of 200 people was carried out to determine the number of traffic violations committed by different age groups. The results are shown in the table below.

	Number of violations		
Age group	0	1	2
Under 25	8	30	7
25 − 45	47	15	2
45 − 65	45	18	3
65+	20	5	0

If one person is selected at random from the group, determine the probability that:

a. the person belongs to the under-25 age group
b. the person has had at least one traffic violation
c. given that the person has had at least 1 traffic violation, he or she has had only 1 violation
d. the person is 38 and has had no traffic violations
e. the person is under 25, given that he/she has had 2 traffic violations.

16. In an attempt to determine the efficacy of a test used to detect a particular disease, 100 subjects, of whom 27 had the disease, were tested. A positive result means the test detected the disease, and a negative result means the test did not detect the disease. Only 23 of the 30 people who tested positive actually had the disease. Construct a two-way table to show this information and hence determine the probability that a subject selected at random:

 a. does not have the disease
 b. tested positive but did not have the disease
 c. had the disease given that the subject tested positive
 d. did not have the disease, given that the subject tested negative

Complex unfamiliar

17. Six hundred people were surveyed about whether they watched movies on their TV or on their devices. They were classified according to age, with the following results.

	Age 15 to 30	Age 30 to 70	
TV	95	175	270
Device	195	135	330
	290	310	600

Based on these findings, determine whether the medium used to watch movies is independent of a person's age.

18. A popular fast-food restaurant has studied the customer service provided by a sample of 100 of its employees across Australia. They wanted to know if employees who were with the company longer received more positive feedback from customers than newer employees.
The results of their study are shown below.

	Positive customer feedback (P)	Negative customer feedback ($\bar{P}$)	
Employed 2 years or more (T)	34	16	50
Employed fewer than 2 years ($\bar{T}$)	24	26	50
	58	42	100

State whether the events P and T are independent. Justify your answer.

19. Roll two fair dice and record the number uppermost on each. Let A be the event of rolling a 6 on one die, B be the event of rolling a 3 on the other, and C be the event of the product of the numbers on the dice being at least 20.
Determine which (if any) pairs of events A, B and C are independent.

20. Explain whether it is possible for two events, A and B, to be mutually exclusive and independent.

Fully worked solutions for this chapter are available online.

LESSON
7.6 Review

doc-41802

7.6.1 Summary

Hey students! Now that it's time to revise this chapter, go online to:

📄 **Access the chapter summary**

✅ **Review your results**

A+ **Practise exam questions**

Find all this and MORE in jacPLUS

7.6 Exercise

learn on

7.6 Exercise	7.6 Exam practice on

Simple familiar	Complex familiar	Complex unfamiliar
1, 2, 3, 4, 5, 6, 7, 8, 9, 10, 11, 12	13, 14, 15, 16	17, 18, 19, 20

These questions are even better in jacPLUS!
- Receive immediate feedback
- Access sample responses
- Track results and progress

Find all this and MORE in jacPLUS

Simple familiar

1. The numbers 1 to 5 are written on the front of 5 cards that are turned face down. Michelle then chooses one card at random. She wants to choose a number greater than 2. List the sample space and all favourable outcomes.

2. From every 100 televisions on a production line, two are found to be defective. If a television is chosen at random, determine the relative frequency of defective televisions.

3. Given $P(A) = 0.4$, $P(A \cup B) = 0.58$ and $P(A \cap B) = 0.12$:
   a. calculate $P(B|A)$
   b. calculate $P(B)$
   c. determine whether A and B are independent.

4. **MC** Let the universal set S be the set of integers from 1 to 50 inclusive. Three subsets, A, B and C, are defined as:
   A is the set of positive multiples of 3 less than 50
   B is the set of positive prime numbers less than 50
   C is the set of positive multiples of 10 less than or equal to 50.
   Determine which of the following statements is correct.
   A. $B \subseteq C$
   B. $A \cap B = \emptyset$
   C. A and B are independent.
   D. B and C are mutually exclusive.

5. **MC** Hockey players Ash and Ben believe their independent chances of scoring a goal in a match are $\frac{2}{3}$ and $\frac{2}{5}$, respectively. Calculate the probability that in the next match they play, neither scores a goal.

A. $\frac{1}{5}$ B. $\frac{8}{15}$ C. $\frac{11}{15}$ D. $\frac{4}{5}$

6. From a bag that contains 6 red and 7 green balls, 2 balls are drawn without replacement. Calculate the probability that:
   a. one ball of each colour is chosen
   b. at least one green ball is chosen
   c. one ball of each colour is chosen given at least one green ball is chosen.

7. At a teacher's college, 70% of students are female. On average, 75% of female and 85% of male students graduate. A student who graduates is selected at random. Determine the probability that the student is male.

8. Consider the following sets:

$$S = \{1, 2, 3, 4, 5, 6, 7, 8, 9, 10, 11, 12, 13, 14, 15, 16, 17, 18, 19, 20\}$$

$$A = \{\text{evens in } S\}$$

$$B = \{\text{multiples of 3 in } S\}$$

   a. Explain whether the events A and B are independent.
   b. If S is changed to the integers from 1 to 10 only, explain whether the result from part a changes.

9. There are three coins in a box. One coin is a fair coin, one coin is biased with an 80% chance of landing Heads, and one is a biased coin with a 40% chance of landing Heads. A coin is selected at random and flipped.
   If the result is a Head, determine the probability that the fair coin was selected.

10. A card is drawn from a shuffled pack of 52 playing cards. Event A is drawing a club and event B is drawing an ace.
   a. Explain if events A and B are mutually exclusive.
   b. Calculate P(A), P(B) and P($A \cap B$).
   c. Calculate P($A \cup B$).

11. A tetrahedral die is numbered $0, 1, 2$ and 3. Two of these dice are rolled and the sum of the numbers (the number on the face that the die sits on) is taken.
   a. Show the possible outcomes in a two-way table.
   b. Determine if all the outcomes are equally likely.
   c. Determine which total has the least chance of being rolled.
   d. Determine which total has the best chance of being rolled.
   e. Determine which sums have the same chance of being rolled.

12. On grandparents day at a school, a group of grandparents was asked where they most like to take their grandchildren — the beach (*B*) or shopping (*Sh*). The results are illustrated in the Venn diagram. Use the Venn diagram to calculate the following probabilities relating to the place grandparents most like to take their grandchildren.

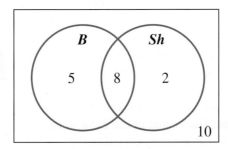

a. Determine the probability that a randomly selected grandparent preferred to take their grandchildren to the beach or shopping.

b. Determine the probability that a randomly selected grandparent liked to take their grandchildren to the beach, given that they liked to take their grandchildren shopping.

Complex familiar

13. One bag contains 2 green and 6 blue counters. A second identical bag contains 4 green counters. A bag is selected at random and a counter is drawn.
Determine the probability that the counter is green.

14. A class of 60 students is surveyed and it is found that 35 of the students study Japanese, 45 study General Mathematics and 10 study neither Japanese nor General Mathematics.
Calculate the probability that a student selected at random studies:

a. both Japanese and General Mathematics
b. Japanese given that they study General Mathematics.
c. General Mathematics.

15. A three-digit number is formed using the digits 2, 4 and 7.

a. Determine which is more likely to be formed: an odd number or an even number.
b. Determine which is more likely to be formed: a number less than 400 or a number greater than 400.

16. A new lie detector machine is being tested to determine its degree of accuracy. One hundred men, 20 of whom were known liars, were tested on the machine. Some of the results are shown below.

	Liars (L)	Honest people (H)	Totals
Correctly tested (C)	18		85
Incorrectly tested (I)			
Totals			100

Determine the probability that a person selected at random:

a. was incorrectly tested
b. was honest and correctly tested
c. was correctly tested, given that he was honest.

17. Matthew likes to collect differently shaped dice. Currently he has two tetrahedrons (4 sides), an icosahedron (20 sides), two dodecahedrons (12 sides) and an octahedron (8 sides), as well as two standard 6-sided cubes.

Matthew has decided to play a game of chance using the octahedral die (with sides numbered 1 to 8) and one dodecahedral die (with sides numbered 1 to 12). The dice are tossed simultaneously, and Matthew notes the number showing uppermost on both dice. This particular game of chance involves tossing the two dice simultaneously on three occasions. The winner of the game must obtain two primes with each of the three tosses. Calculate the probability of being a winner. Give your answer correct to 3 decimal places.

18. John lives in Mackay. Last year he flew to New York to visit his sister. He was about to collect his luggage at the JFK terminal when he bumped into an old friend from school. He thought, 'This is incredible. I probably only know about 100 people and I meet one on the other side of the world. The probability of this must be about 100 out of 7 000 000 000.' Assuming there are about 7 billion people in the world, comment on the likelihood of John's accidental meeting.

19. Jo, who owns a corner grocery, imports tins of chickpeas and lentils. When unpacking the tins, Jo finds that one box contains 10 tins that have lost their labels. The tins are identical but after looking through his invoices, Jo has calculated that 7 of the tins contain chickpeas and 3 contain lentils. He decides to take them home since he is unable to sell them without a label. He wants to use the chickpeas to make some hummus, so he opens the tins at random until he opens a tin of chickpeas.

a. Determine the minimum number of tins Jo must open to ensure he opens a tin of chickpeas.

b. In fact, Jo found the chickpeas in the second tin. After making the hummus, Jo decides he will open one tin each day and use whatever it contains.

Determine the probability that he opens at least one tin of each over the next 3 nights.

Justify your procedures and decisions.

20. Twenty-two items are to be sold at a small charity auction. The amount, in dollars, paid for each item is to be decided by the throw of a dart at a dartboard. The rules are as follows.

The price of any item is the number scored multiplied by $100.

- If more than one person wants to bid for an item, the person with the highest score wins.
- If a buyer misses the board, the buyer can throw the dart again until they hit the board, unless there are others who want to buy the same item, in which case the buyer does not get a second chance.
- Anyone hitting the inner bull will automatically win and will be sold the item for the modest price of $5000.
- Anyone hitting the outer bull will automatically win over any score other than an inner bull and will be sold the item for $2500.

Aino and Bryan attend the auction. The associated probabilities are as follows:

For Aino:

The probability of hitting any one of the numbers from 1 to 20 is 0.045; the probability of hitting the inner bull is 0.005; and the probability of hitting the outer bull is 0.006.

For Bryan:

The probability of hitting any one of the numbers from 1 to 20 is 0.046; the probability of hitting the inner bull is 0.004; and the probability of hitting the outer bull is 0.003.

a. Aino is the only buyer interested in a particular item. Determine the probability that she only needs the one throw but her item costs more than $1900.

Aino then decides to buy a second item in which Bryan is also interested. They each throw a dart.

b. Determine the probability that Bryan wins with his first dart.

c. Given that Bryan has already scored a 15 on his first throw, determine the probability that Aino will beat Bryan's score on her first throw.

Fully worked solutions for this chapter are available online.

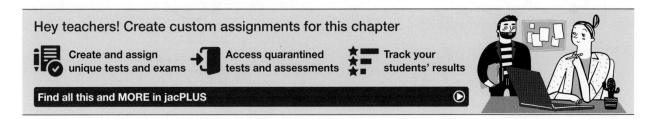

Answers

Chapter 7 Probability

7.2 Concepts of probability

7.2 Exercise

1. a. $A =$ {king of hearts, king of diamonds, king of spaceds, king of clubs}
   b. 13

2. a. $\dfrac{5}{23}$ b. $\dfrac{12}{23}$ c. $\dfrac{15}{23}$ d. 0

3. a. $\dfrac{1}{6}$ b. $\dfrac{5}{6}$ c. $\dfrac{1}{2}$
   d. $\dfrac{2}{3}$ e. $\dfrac{1}{3}$ f. 0

4. a. $\dfrac{1}{26}$ b. $\dfrac{5}{26}$ c. $\dfrac{3}{26}$
   d. $\dfrac{25}{26}$ e. 1 f. $\dfrac{9}{26}$

5. a. 1 b. $\dfrac{1}{2}$ c. $\dfrac{1}{4}$
   d. $\dfrac{7}{13}$ e. $\dfrac{12}{13}$

6. a. $\dfrac{1}{4}$ b. $\dfrac{1}{8}$ c. $\dfrac{9}{16}$

7. a.

   M (30) G (15)

   | 25 | 5 | 10 |

   2

   S (42)

   b. $\dfrac{25}{42}$

   c. $\dfrac{1}{21}$

   d. $\dfrac{5}{6}$

8. a.

	B	$\overline{B}$	
A	0.35	0.3	0.65
$\overline{A}$	0.15	0.2	0.35
	0.5	0.5	1

   b. 0.85

9. a. 0.8 b. 0.5

10. a.

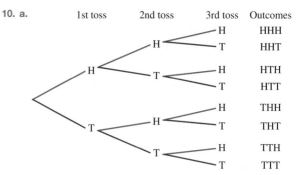

1st toss	2nd toss	3rd toss	Outcomes

H — H — H : HHH
H — H — T : HHT
H — T — H : HTH
H — T — T : HTT
T — H — H : THH
T — H — T : THT
T — T — H : TTH
T — T — T : TTT

   b. $\dfrac{7}{8}$

   c. $\dfrac{3}{4}$

11. a. $\dfrac{3}{8}$ b. $\dfrac{1}{4}$ c. $\dfrac{1}{2}$
    d. $\dfrac{7}{8}$ e. $\dfrac{1}{2}$

12. a. $\dfrac{89}{400}$ b. $\dfrac{53}{800}$ c. $\dfrac{21}{25}$ d. $\dfrac{439}{800}$

13. a. $\dfrac{23}{100}$ b. $\dfrac{81}{200}$

14. a. $\dfrac{4}{9}$ b. $\dfrac{5}{18}$ c. 1

15. a. See the worked solutions in the online resources for the proof.

   b.

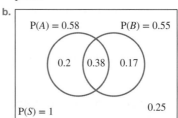

   $P(A) = 0.58$ $P(B) = 0.55$

   0.2   0.38   0.17

   $P(S) = 1$ 0.25

16. a.

   1st roll

	1	2	3	4	5	6
1	2	2	3	4	5	6
2	2	4	3	4	5	6
3	3	3	6	4	5	6
4	4	4	4	8	5	6
5	5	5	5	5	10	6
6	6	6	6	6	6	12

   (2nd roll labelled on left)

   b. i. $\dfrac{7}{18}$
      ii. $\dfrac{1}{12}$
      iii. 1

17. a.

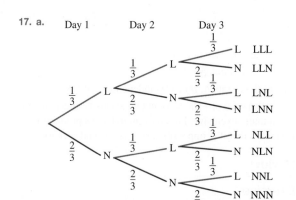

Day 1   Day 2   Day 3

L $\frac{1}{3}$ L LLL

$\frac{1}{3}$ L $\frac{2}{3}$ N LLN

$\frac{1}{3}$ L $\frac{2}{3}$ N $\frac{1}{3}$ L LNL

$\frac{2}{3}$ N LNN

$\frac{2}{3}$ N $\frac{1}{3}$ L NLL

$\frac{1}{3}$ L $\frac{2}{3}$ N NLN

$\frac{2}{3}$ N $\frac{1}{3}$ L NNL

$\frac{2}{3}$ N NNN

b. i. $\frac{7}{27}$

ii. $\frac{2}{3}$

iii. $\frac{8}{27}$

18. a. 288 students   **b.** 0.0784

19. $\frac{7}{13}$

20. 240 games

7.3 Relative frequency

7.3 Exercise

1. A

2. D

3. 0.74

4. 0.79

5. 0.375

6. a. 0.45   **b.** 0.55

7. 4%

8. a. 0.03   **b.** 0.97

9. 98.9%

10. a. 0.525   **b.** 0.4375   **c.** 0.0375

11. 80

12. $400

13. a. The assembly line will need upgrading.

  b. i. 2.5%;   **ii.** 51.5%;   **iii.** 17.5%

14. The maximum distance is 40 000 km.

15. Student investigation

16. a. $\frac{7}{20}$ **b.** $\frac{2}{5}$

17–20. Sample responses can be found in the worked solutions in the online resources.

7.4 Conditional probability

7.4 Exercise

1. a. $P\left(R_2|Y_1\right) = \frac{3}{8}$ **b.** $P\left(Y_2|Y_1\right) = \frac{1}{8}$

  c. $P\left(P_2|R_1\right) = \frac{1}{2}$ **d.** $P\left(R_2|\overline{R_1}\right) = \frac{3}{8}$

2. a. T   **b.** F   **c.** F
  d. T   **e.** T   **f.** F

3. $\frac{6}{25}$

4. a. $\frac{9}{14}$ **b.** $\frac{61}{81}$ **c.** $\frac{13}{25}$

5. a. 0.2   **b.** $\frac{2}{3}$ **c.** $\frac{2}{7}$ **d.** $\frac{5}{7}$

6. a. 0.3   **b.** $\frac{3}{5}$ **c.** $\frac{1}{2}$ **d.** $\frac{3}{5}$

7. a. 0.9   **b.** $\frac{4}{7}$ **c.** $\frac{3}{4}$ **d.** $\frac{1}{3}$

8. a. $\frac{1}{5}$ **b.** $\frac{5}{9}$ **c.** $\frac{7}{11}$ **d.** $\frac{11}{25}$

9. a. $\frac{5}{36}$ **b.** $\frac{2}{15}$ **c.** $\frac{1}{9}$ **d.** $\frac{4}{5}$

10. a. $\frac{9}{50}$ **b.** $\frac{21}{50}$ **c.** $\frac{47}{232}$ **d.** $\frac{143}{168}$

11. a.

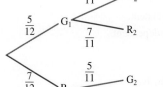

1st choice   2nd choice   Outcomes

$\frac{5}{12}$ G_1 $\frac{4}{11}$ G_2 $G_1 G_2$

$\frac{7}{11}$ R_2 $G_1 R_2$

$\frac{7}{12}$ R_1 $\frac{5}{11}$ G_2 $R_1 G_2$

$\frac{6}{11}$ R_2 $R_1 R_2$

  b. $\frac{35}{132}$

  c. $\frac{31}{66}$

12. a.

0.90 T $\frac{0.98}{}$ S TS = 0.882

0.02 $\overline{S}$ T$\overline{S}$ = 0.018

0.10 $\overline{T}$ 0.56 S $\overline{T}$S = 0.056

0.44 $\overline{S}$ $\overline{T}\overline{S}$ = 0.044

  b. 0.938

13. $\frac{2}{9}$

14. a. $\frac{1}{17}$ **b.** $\frac{15}{34}$ **c.** $\frac{2}{15}$ **d.** $\frac{4}{17}$

15. a. $\frac{7}{22}$ **b.** $\frac{28}{33}$

  c. $\frac{3}{8}$ **d.** $\frac{1}{2}$

16. a. 0.72   **b.** 0.98   **c.** 0.2

17. a. $\dfrac{427}{500}$ b. $\dfrac{32}{235}$ c. $\dfrac{32}{61}$

18. a. $\dfrac{267}{1000}$ b. $\dfrac{257}{1000}$ c. $\dfrac{82}{267}$ d. $\dfrac{185}{743}$

19. 0.61

20. 0.85

7.5 Independence

7.5 Exercise

1. $P(B|A) = P(B), P(A \cap B) = P(A) \times P(B)$

2. They are independent.

3. A and B are independent events.

4. a. $p = 0.4$ b. $p = 0.6$ c. $q = \dfrac{3}{4}$

5. a. $\dfrac{4}{5}$ b. $\dfrac{2}{3}$ c. $\dfrac{8}{15}$ d. $\dfrac{14}{15}$

6. a. Independent   b. Dependent
   c. Dependent   d. Independent

7. a. $S = \{HH, HT, TH, TT\}$
   $A = \{TH, TT\}$
   $B = \{HT, TH\}$
   $C = \{HH, HT, TH\}$
   b. A and B are independent.
   c. B and C are not independent.
   d. $\dfrac{3}{4}$

8. a. Not mutually exclusive
   b. Not independent
   c. i. B and C are mutually exclusive.
      ii. A and B are not independent.

9. a. $\dfrac{1}{5}$
   b. $\dfrac{11}{30}$
   c. The events are not independent.

10. a. 0.216   b. 0.024   c. 0.784

11. a. $\dfrac{4}{9}$ b. $\dfrac{10}{21}$ c. $\dfrac{1}{3}$

12. The cars are not used independently.

13. a. 0.001 25   b. 0.0025   c. 0.9524
    d. 0.000 625   e. 0.052 25

14. 0.5

15. a. $\dfrac{9}{40}$ b. $\dfrac{2}{5}$ c. $\dfrac{17}{20}$
    d. $\dfrac{47}{200}$ e. $\dfrac{7}{12}$

16. a. 0.73   b. $\dfrac{7}{100}$ c. $\dfrac{23}{30}$ d. $\dfrac{33}{35}$

17. Medium and age are not independent.

18. P and T are not independent.

19. None of the events A, B or C is independent of another.

20. Only if $P(A) = 0$ or $P(B) = 0$ can two events be independent and mutually exclusive. For an event to have a probability of 0 means that it is impossible, so it is a trivial scenario.

7.6 Review

7.6 Exercise

1. $S = \{1, 2, 3, 4, 5\}$; favourable outcomes $= \{3, 4, 5\}$

2. 0.02

3. a. 0.3
   b. 0.3
   c. A and B are independent.

4. D

5. A

6. a. $\dfrac{7}{13}$ b. $\dfrac{21}{26}$ c. $\dfrac{2}{3}$

7. $\dfrac{17}{52}$

8. a. A and B are independent.
   b. A and B are not independent in this situation.

9. $\dfrac{5}{17}$

10. a. No. It is possible to draw a card that is a club and an ace.
    b. $P(A) = \dfrac{1}{4}$ and $P(B) = \dfrac{1}{13}, P(A \cap B) = \dfrac{1}{52}$
    c. $\dfrac{4}{13}$

11. a. See the table at the bottom of the page*
    b. No
    c. 0 and 6
    d. 3
    e. 0 and 6, 1 and 5, 2 and 4

12. a. $\dfrac{15}{25} = \dfrac{3}{5}$ b. $\dfrac{8}{10} = \dfrac{4}{5}$

13. $\dfrac{5}{8}$

14. a. 0.5
    b. 0.67
    c. 0.86

*11

	0	1	2	3
0	$0 + 0 = 0$	$0 + 1 = 1$	$0 + 2 = 2$	$0 + 3 = 3$
1	$1 + 0 = 1$	$1 + 1 = 2$	$1 + 2 = 3$	$1 + 3 = 4$
2	$2 + 0 = 2$	$2 + 1 = 3$	$2 + 2 = 4$	$2 + 3 = 5$
3	$3 + 0 = 3$	$3 + 1 = 4$	$3 + 2 = 5$	$3 + 3 = 6$

15. a. Sample responses can be found in the worked solutions in the online resources.

b. Numbers greater than 400 are more likely to be formed.

16. a. $\dfrac{3}{20}$

b. $\dfrac{67}{100}$

c. $\dfrac{67}{80}$

17. P(win) = 0.009

18. Sample responses can be found in the worked solutions in the online resources.

19. a. 4   b. $\dfrac{9}{14}$

20. a. 0.056   b. 0.482 165   c. 0.236

UNIT

2 Calculus and further functions

Source: Mathematical Methods Senior Syllabus 2024 © State of Queensland (QCAA) 2024; licensed under CC BY 4.0.

8 Indices and index laws

LESSON SEQUENCE

Fully worked solutions for this chapter are available online.

EXAM PREPARATION

Access exam-style questions in every lesson, available online.

on Resources

Solutions	Solutions — Chapter 8 (sol-1275)
Exam questions	Exam question booklet — Chapter 8 (eqb-0296)
Digital documents	Learning matrix — Chapter 8 (doc-41790)
	Chapter summary — Chapter 8 (doc-41803)

LESSON
8.1 Overview

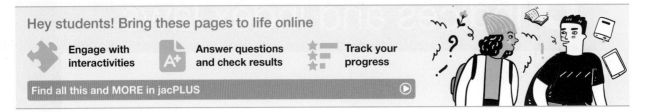
8.1.1 Introduction

Exponential (or indicial) notation provides a convenient means of writing multiples of the same variable in a succinct form. The French mathematician René Descartes — best known to us for his philosophy 'I think, therefore I am' — was the first to describe indices as we use them today. In his 1637 text *Discours de la méthode pour bien conduire sa raison et* chercher (roughly translated as 'Discourse on the methodology of logic and reasoning'), he described the notation that he used to write multiples of the same variable:

' … in order to multiply a by itself we write aa or a^2, and a^3 in order to multiply it once more by a, and thus to infinity …'

This notation is at its most useful when we are using numbers that involve very large or very small powers of ten. Hence, we write the speed of light as 3×10^8 metres per second rather than $300\,000\,000$ metres per second, and the mass of a carbon atom as 1.997×10^{-23} grams instead of $0.00\,000\,000\,000\,000\,000\,000\,001\,997$ grams. As these extraordinary numbers are most commonly used in fields such as chemistry and physics, this form of notation is referred to as scientific notation. (We shall examine this in more detail during the course of this chapter.)

Without exponential notation it would take us a very long time to write numbers such as the googol (equal to 10^{100}), let alone the googolplex, which is $10^{10^{100}}$. In fact, astrophysicist Carl Sagan once noted that to write the decimal representation of the googolplex would take a piece of paper bigger than the boundaries of the known universe!

8.1.2 Syllabus links

Lesson	Lesson title	Syllabus links
8.2	Index laws	◯ Use indices (including negative and fractional indices) and the index laws.
8.3	Negative and fractional indices	◯ Use indices (including negative and fractional indices) and the index laws. ◯ Convert radicals to and from fractional indices.
8.4	Scientific notation	◯ Understand and use scientific notation.

Source: Mathematical Methods Senior Syllabus 2024 © State of Queensland (QCAA) 2024; licensed under CC BY 4.0.

LESSON
8.2 Index laws

SYLLABUS LINKS

- Use indices (including negative and fractional indices) and the index laws.

Source: Mathematical Methods Senior Syllabus 2024 © State of Queensland (QCAA) 2024; licensed under CC BY 4.0.

8.2.1 Index or exponential form

Recall that if a number, a, is multiplied by itself x times, it can be represented in **index** or **exponential form**.

For any positive number n where $n = a^x$, the statement $n = a^x$ is called an index or exponential statement.

For our study:
- the base is a where $a \in R^+\{1\}$
- the exponent, or index, is x where $x \in R$
- the number n is positive, so $a^x \in R^+$.

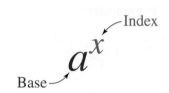

For example, when the number 8 is expressed as a power of 2, it is written as $8 = 2^3$.

In this form:
- the base is 2
- the power (also known as the index or exponent) is 3.

The words 'power', 'index' and 'exponent' are interchangeable.

Index laws control the simplification of expressions that have the same base.

8.2.2 Review of the index laws

Multiplication

When multiplying two numbers in index form with the same *base, add* the indices.

For example, $2^3 \times 2^4 = 2 \times 2 \times 2 \times 2 \times 2 \times 2 \times 2 = 2^7$.

Division

When dividing two numbers in index form with the same *base, subtract* the indices.

For example, $2^6 \div 2^2 = \dfrac{2 \times 2 \times 2 \times 2 \times 2 \times 2}{2 \times 2} = 2^4$.

Raising to a power

To raise an indicial expression to a power, *multiply* the indices.

For example, $\left(2^4\right)^3 = 2^4 \times 2^4 \times 2^4 = 2^{4+4+4} = 2^{(4 \times 3)} = 2^{12}$.

Raising to the power of zero

Any number raised to the power of zero is equal to *one*.

For example: $2^3 \div 2^3 = 2^{3-3} = 2^0$

or $\qquad 2^3 \div 2^3 = (2 \times 2 \times 2) \div (2 \times 2 \times 2) = 8 \div 8 = 1$

$\qquad\qquad \therefore 2^0 = 1$.

These index laws are summarised below.

> **Index laws**
>
> $$a^m \times a^n = a^{m+n}$$
>
> $$a^m \div a^n = a^{m-n}$$
>
> $$(a^m)^n = a^{mn}$$
>
> $$a^0 = 1, a \neq 0$$

8.2.3 Products and quotients and the index laws

Products

When the base is a product, raise every part of the product to the index outside the bracket.

For example:

$$
\begin{aligned}
(2 \times 3)^4 &= (2 \times 3) \times (2 \times 3) \times (2 \times 3) \times (2 \times 3) \\
&= 2 \times 3 \times 2 \times 3 \times 2 \times 3 \times 2 \times 3 \\
&= 2 \times 2 \times 2 \times 2 \times 3 \times 3 \times 3 \times 3 \\
&= 2^4 \times 3^4
\end{aligned}
$$

Quotients

When the base is a fraction, raise both the numerator and denominator to the index outside the bracket.

For example:

$$
\begin{aligned}
\left(\frac{2}{3}\right)^4 &= \left(\frac{2}{3}\right) \times \left(\frac{2}{3}\right) \times \left(\frac{2}{3}\right) \times \left(\frac{2}{3}\right) \\
&= \frac{2 \times 2 \times 2 \times 2}{3 \times 3 \times 3 \times 3} \\
&= \frac{2^4}{3^4}
\end{aligned}
$$

These index laws are summarised below.

> **Product and quotients and the index laws**
>
> $$(ab)^n = a^n b^n$$
>
> $$\left(\frac{a}{b}\right)^n = \frac{a^n}{b^n}$$

Simplify each of the following.

a. $2x^3y^2 \times 4x^2y$

b. $\left(2x^2y^3\right)^2 \times xy^4$

c. $(3a)^5 b^6 \div 9a^4b^3$

d. $\dfrac{8p^6m^2 \times (3p)^3 m^5}{6p^4m}$

THINK	WRITE
a. 1. Collect 'plain' numbers (2 and 4) and terms with the same base.	a. $2x^3y^2 \times 4x^2y$ $= 2 \times 4 \times x^3 \times x^2 \times y^2 \times y$
2. Simplify by multiplying plain numbers and adding powers with the same base.	$= 8x^5y^3$
b. 1. Remove the bracket by multiplying the powers.	b. $\left(2x^2y^3\right)^2 \times xy^4$ $= 2^2 \times x^4 \times y^6 \times xy^4$
2. Convert 2^2 to a plain number (4) first and collect terms with the same base.	$= 4 \times x^4 \times x \times y^6 \times y^4$
3. Simplify by adding powers with the same base.	$= 4x^5y^{10}$
c. 1. Write the quotient as a fraction.	c. $(3a)^5 b^6 \div 9a^4b^3 = \dfrac{(3a)^5 b^6}{9a^4b^3}$
2. Remove the bracket by multiplying the powers.	$= \dfrac{243a^5b^6}{9a^4b^3}$
3. Simplify by first cancelling plain numbers.	$= \dfrac{27a^5b^6}{a^4b^3}$
4. Complete simplification by subtracting powers with the same base. (*Note:* $a^1 = a$)	$= 27ab^3$
d. 1. Expand the brackets by raising each term to the power of 3.	d. $\dfrac{8p^6m^2 \times (3p)^3 m^5}{6p^4m} = \dfrac{8p^6m^2 \times 3^3 p^3 m^5}{6p^4m}$
2. Convert 3^3 to 27 and collect 'like' pronumerals.	$= \dfrac{8 \times 27 \times p^6 \times p^3 \times m^2 \times m^5}{6p^4m}$
3. Simplify by first reducing the plain numbers, then simplify the pronumerals by adding the indices for multiplication and subtracting the indices for division.	$= 36p^{6+3-4}m^{2+5-1}$
4. Simplify the indices of each base.	$= 36p^5m^6$

Simplify $\dfrac{6a^4b^3}{16a^7b^6} \div \left(\dfrac{3a^2b}{2a^3b^2}\right)^3$.

THINK	WRITE
1. Write the expression.	$\dfrac{6a^4b^3}{16a^7b^6} \div \left(\dfrac{3a^2b}{2a^3b^2}\right)^3$

2. Change the division sign to multiplication and replace the second term with its reciprocal (turn the second term upside down).

$$= \frac{6a^4b^3}{16a^7b^6} \times \left(\frac{2a^3b^2}{3a^2b}\right)^3$$

3. Remove the brackets by multiplying the powers.

$$= \frac{6a^4b^3}{16a^7b^6} \times \frac{2^3a^9b^6}{3^3a^6b^3}$$

4. Collect plain numbers and terms with the same base.

$$= \frac{6 \times 8a^{4+9-7-6}b^{3+6-6-3}}{16 \times 27}$$

5. Cancel plain numbers and apply index laws.

$$= \frac{a^0b^0}{9}$$

6. Simplify.

$$= \frac{1}{9}$$

Expressions involving just numbers and numerical indices can be simplified using index laws and then evaluated.

WORKED EXAMPLE 3 Evaluating expressions with indices

Write the following expressions in simplest index notation and evaluate.

a. $2^3 \times 16^2$

b. $\dfrac{9^5 \times 3^4}{27^3}$

THINK

WRITE

a. 1. Rewrite the bases in terms of their prime factors.

a. $2^3 \times 16^2 = 2^3 \times (2 \times 2 \times 2 \times 2)^2$

2. Simplify the brackets using index notation.

$$= 2^3 \times \left(2^4\right)^2$$

3. Remove the brackets by multiplying the powers.

$$= 2^3 \times 2^8$$

4. Simplify by adding the powers.

$$= 2^{11}$$

5. Evaluate as a basic number. Use technology if necessary.

$$= 2048$$

b. 1. Rewrite the bases in terms of their prime factors.

b. $\dfrac{9^5 \times 3^4}{27^3} = \dfrac{(3 \times 3)^5 \times 3^4}{(3 \times 3 \times 3)^3}$

2. Simplify the brackets using index notation.

$$= \frac{\left(3^2\right)^5 \times 3^4}{\left(3^3\right)^3}$$

3. Remove the brackets by multiplying the powers.

$$= \frac{3^{10} \times 3^4}{3^9}$$

4. Write in simplest index form.

$$= 3^5$$

5. Evaluate as a basic number.

$$= 243$$

Complex expressions involving terms with different bases have to be simplified by replacing each base with its prime factors.

Simplify $\dfrac{3^{4n} \times 18^{n+1}}{6^{3n-2}}$.

THINK	WRITE
1. Rewrite the bases in terms of their prime factors.	$\dfrac{3^{4n} \times 18^{n+1}}{6^{3n-2}} = \dfrac{3^{4n} \times (3 \times 3 \times 2)^{n+1}}{(2 \times 3)^{3n-2}}$
2. Simplify the brackets using index notation.	$= \dfrac{3^{4n} \times (3^2 \times 2^1)^{n+1}}{(2 \times 3)^{3n-2}}$
3. Remove the brackets by multiplying powers.	$= \dfrac{3^{4n} \times 3^{2n+2} \times 2^{n+1}}{2^{3n-2} \times 3^{3n-2}}$
4. Collect terms with the same base by adding the powers in the products and subtracting the powers in the quotients.	$= 3^{4n+2n+2-(3n-2)} \times 2^{n+1-(3n-2)}$
5. Simplify.	$= 3^{6n+2-3n+2} \times 2^{n+1-3n+2}$
	$= 3^{3n+4} \times 2^{3-2n}$

Exercise 8.2 Index laws

8.2 Exercise	8.2 Exam practice on

Simple familiar	Complex familiar	Complex unfamiliar
1, 2, 3, 4, 5, 6, 7, 8, 9, 10, 11	12, 13, 14, 15, 16	17, 18, 19, 20

Simple familiar

1. **WE1a** Simplify each of the following.
   a. $x^2 \times x^5 \times x^3$
   b. $m^3 \times m^2 p \times p^4$
   c. $4y^3 \times 2y \times y^7$

2. **WE1b** Simplify each of the following.
   a. $5^2 \times 5^7 \times (5^3)^3$
   b. $(xy)^3 \times x^4 y^5$
   c. $(2x^4)^2 \times (4x^2)^5$
   d. $3m^2 p^5 \times (mp^2)^3 \times 2m^4 p^6$
   e. $5x^2 y^3 \times (5xy^2)^4 \times (5x^2 y)^2$

3. **WE1c** Simplify each of the following.
   a. $a^7 b^8 \div a^2 b^5$
   b. $2a^{12} b^9 \div (2a)^3 b^4$
   c. $(3x^5) y^{11} \div 6x^2 y^2$
   d. $p^{13} q^{10} \div (pq^4)^2$
   e. $(4mn^4)^2 \div 14n^3$

4. Simplify each of the following.

a. $\dfrac{a^3 b^4}{ab^2}$

b. $25r^{15}s^{10}t^4 \div r^5 \left(s^5\right)^2 (5t)^3$

c. $\dfrac{15a^6 b^7}{3a^3 b^4}$

d. $\dfrac{24x^4 y^7}{20x^2 y^3}$

5. **WE1d** Simplify each of the following.

a. $\dfrac{6p^8 m^4 \times 2p^7 m^6}{9p^5 m^2}$

b. $\dfrac{(3x)^2 y^2 \times 5x^6 y^3}{10x^7 y}$

c. $\dfrac{14u^{11} v^9 \times \left(3u^2\right)^3 v}{21u^6 v^5}$

d. $\dfrac{\left(5e^3\right)^2 f^4 \times 8e^4 f^3}{20ef^5}$

e. $\dfrac{6w^2 t^7 \times 9w^4 t^{12}}{(3w)^5 t^{13}}$

6. Simplify each of the following.

a. $\dfrac{(2x)^4 y \times \left(3x^7 y\right)^2}{18x^5 (2y)^3}$

b. $\dfrac{\left(-3x^3 y^2\right)^3}{2x^3 y^6} \times \dfrac{6x^7 y^5}{\left(x^2 y\right)^2}$

c. $\dfrac{(-3mp)^2 \times 4m^4 p}{12\,(mp)^2}$

d. $\dfrac{m^3 p^4 \times \left(mp^3\right)^2}{\left(-mp^2\right)^4}$

e. $\dfrac{4\left(u^7 v^6\right)^3}{\left(-2u^3 v^2\right)^2 \times u^4 \left(3v^5\right)^2}$

7. **WE2** Simplify each of the following.

a. $\dfrac{15a^8 b^3}{9a^4 b^5} \div \left(\dfrac{2a^3 b}{3ab^2}\right)^2$

b. $\dfrac{5k^{12}d}{\left(2k^3\right)^2} \div \dfrac{6kd^4}{25\left(k^2 d^3\right)^3}$

c. $\dfrac{4g^4 \left(2p^{11}\right)^2}{g^3 p^7} \div \dfrac{8g^4 p}{(2gp)^3}$

d. $\left(\dfrac{3jn^2}{n^5}\right)^3 \div \dfrac{\left(4j^2 n\right)^2}{n^{13}(2j)^4}$

e. $\dfrac{x^4 y^7}{x^3 y^2} \div \dfrac{x^3 y^2}{x^5 y}$

f. $\dfrac{6x^3 y^8}{\left(x^2 y^3\right)^3} \div \dfrac{\left(2xy^3\right)^2}{8x^5 y^7}$

8. Simplify each of the following.

a. $\dfrac{3p^3 m^4}{p^1 m^2}$

b. $\dfrac{6x^6 y^5}{x^5 y^3} \times \dfrac{x^4}{(2y)^2}$

c. $\dfrac{3ab^3}{-ab} \div \left(\dfrac{a^2 b}{a^5}\right)^2$

9. **WE3** Write the following expressions in simplest index notation.

a. $2^4 \times 4^2 \times 8$

b. $3^7 \times 9^2 \times 27^3 \times 81$

c. $5^3 \times 15^2 \times 3^2$

10. Write the following expressions in simplest index notation.

a. $20^5 \times 8^4 \times 125$

b. $\dfrac{3^4 \times 27^2}{6^4 \times 3^5}$

c. $\dfrac{8 \times 5^2}{2^3 \times 10}$

11. Write the following in simplest index notation and evaluate. Use technology to verify your solutions.

a. $\dfrac{(625)^4}{\left(5^3\right)^5}$

b. $\dfrac{(25)^4}{(125)^3}$

c. $\dfrac{4^{11} \div 8^2}{16^3}$

d. $\dfrac{27^2 \times 81}{9^3 \times 3^5}$

Complex familiar

12. **WE4** Simplify each of the following.

a. $\dfrac{2^n \times 9^{2n+1}}{6^{n-2}}$

b. $\dfrac{25^{3n} \times 5^{n-3}}{5^{4n+3}}$

c. $\dfrac{12^{x-2} \times 4^x}{6^{x-2}}$

d. $\dfrac{12^{n-3} \times 27^{1-n}}{9^{2n} \times 8^{n-1} \times 16^n}$

e. $\dfrac{4^n \times 7^{n-3} \times 49^{3n+1}}{14^{n+2}}$

13. Simplify each of the following.

a. $\dfrac{35^2 \times 5^5 \times 7^6}{25^4 \times 49^3}$

b. $\dfrac{3^{5n-4} \times 16^n \times 9^3}{4^{n+1} \times 18^{1-n} \times 6^{3-2n}}$

14. Simplify each of the following.

a. $\dfrac{x^{n+1} \times y^5 \times z^{4-n}}{x^{n-2} \times y^{4-n} \times z^{3-n}}$

b. $\dfrac{\left(x^n y^{m+3}\right)^2}{x^{n+2} y^{3-m}} \times \dfrac{x^2 y}{x^{n-5} \times y^{5-3m}}$

15. Simplify the following.

$\dfrac{36^{2n} \times 6^{n+3}}{216^{n-2}}$

16. Write the following in simplest index notation and evaluate. Use technology to verify your solutions.

a. $\dfrac{4^5}{2^7}$

b. $9^4 \times 3^5 \div 27$

c. $\dfrac{\left(16^2\right)^3}{\left(2^5\right)^4}$

d. $\dfrac{27^2}{\left(3^2\right)^3}$

Complex unfamiliar

17. a. Explain if $2x$ is ever the same as x^2. Justify your reasoning.
    b. Explain, with reasoning, the difference between $3x^0$ and $(3x)^0$.

18. Using factorisation, simplify $\dfrac{3^n + 3^{n+1}}{3^n + 3^{n-1}}$.

19. Simplify $\dfrac{5^n - 5^{n+1}}{5^{n+1} + 5^n}$.

20. Identify the error in the solution below. Justify your decision and correct the solution.

$$\left(\frac{a^2 b^3 c}{a^2 b^2}\right)^3 \times \left(\frac{a^3 b^2 c^2}{a^2 b^3}\right)^2 = \left(\frac{b^3 c}{b^2}\right)^3 \times \left(\frac{ab^2 c^2}{b^3}\right)^2$$

$$= \left(\frac{bc}{1}\right)^3 \times \left(\frac{ac^2}{b}\right)^2$$

$$= \left(\frac{abc^3}{b}\right)^6$$

$$= \left(\frac{ac^3}{1}\right)^6$$

$$= a^6 c^{18}$$

Fully worked solutions for this chapter are available online.

LESSON
8.3 Negative and fractional indices

SYLLABUS LINKS

- Use indices (including negative and fractional indices) and the index laws.
- Convert radicals to and from fractional indices.

Source: Mathematical Methods Senior Syllabus 2024 © State of Queensland (QCAA) 2024; licensed under CC BY 4.0.

8.3.1 Negative indices

Consider the expression $\dfrac{a^3}{a^5}$. This expression can be simplified in two different ways.

1. Written in expanded form:
$$\dfrac{a^3}{a^5} = \dfrac{a \times a \times a}{a \times a \times a \times a \times a}$$
$$= \dfrac{1}{a \times a}$$
$$= \dfrac{1}{a^2}$$

2. Using division with indices:
$$\dfrac{a^3}{a^5} = a^{3-5}$$
$$= a^{-2}$$

Equating the results of both of these simplifications, we get $a^{-2} = \dfrac{1}{a^2}$.

In general,
$$\dfrac{1}{a^n} = \dfrac{a^0}{a^n} \quad (1 = a^0)$$
$$= a^{0-n} \quad \text{(using the division with indices)}$$
$$= a^{-n}$$

The converse of this law can be used to rewrite terms with positive indices only.

$$\dfrac{1}{a^{-n}} = a^n$$

It is also worth noting that applying a negative index to a fraction has the effect of swapping the numerator and denominator.

$$\left(\dfrac{a}{b}\right)^{-n} = \dfrac{b^n}{a^n}$$

Negative indices

$$a^{-n} = \dfrac{1}{a^n}, \; a \neq 0$$

$$\left(\dfrac{a}{b}\right)^{-n} = \dfrac{a^{-n}}{b^{-n}} = \dfrac{b^n}{a^n}$$

WORKED EXAMPLE 5 Simplifying expressions with negative indices

Express each of the following with positive index numbers.

a. $\left(\dfrac{5}{8}\right)^{-4}$

b. $\dfrac{x^4 y^{-2} \times \left(x^2 y\right)^{-5}}{x^{-3} y^3}$

THINK   **WRITE**

a. 1. Remove the brackets by raising the denominator and numerator to the power of -4.

a. $\left(\dfrac{5}{8}\right)^{-4} = \dfrac{5^{-4}}{8^{-4}}$

   2. Interchange the numerator and denominator, changing the signs of the powers.

   $= \dfrac{8^4}{5^4}$

   3. Simplify by expressing as a fraction to the power of 4.

   $= \left(\dfrac{8}{5}\right)^4$

b. 1. Remove the brackets by multiplying powers.

b. $\dfrac{x^4 y^{-2} \times \left(x^2 y\right)^{-5}}{x^{-3} y^3} = \dfrac{x^4 y^{-2} \times x^{-10} y^{-5}}{x^{-3} y^3}$

   2. Collect terms with the same base by adding the powers on the numerator and subtracting the powers on the denominator.

   $= \dfrac{x^{-6} y^{-7}}{x^{-3} y^3}$

   $= x^{-6-(-3)} y^{-7-3}$

   $= x^{-3} y^{-10}$

   3. Rewrite the answer with positive powers.

   $= \dfrac{1}{x^3 y^{10}}$

8.3.2 Fractional and radical indices

Up to now, we have only looked at cases in which the indices have all been integers. However, an index can be any number. Indices can be represented as fractions (such as $\dfrac{1}{3}$, 5 and $\dfrac{-3}{5}$) or as radicals (such as $a^{\frac{1}{2}}$ and $\sqrt{a}$).

Consider the expression $a^{\frac{1}{2}}$. Now consider what happens if we square that expression.

$$\left(a^{\frac{1}{2}}\right)^2 = a \ \left(\text{using } (a^m)^n = a^{m \times n}\right)$$

From our work on surds, we know that $\left(\sqrt{a}\right)^2 = a$.

Equating the two facts above, $\left(a^{\frac{1}{2}}\right)^2 = \left(\sqrt{a}\right)^2$. Therefore, $a^{\frac{1}{2}} = \sqrt{a}$.

Similarly, $b^{\frac{1}{3}} \times b^{\frac{1}{3}} \times b^{\frac{1}{3}} = \left(b^{\frac{1}{3}}\right)^3 = b$, implying that $b^{\frac{1}{3}} = \sqrt[3]{b}$.

This pattern can be continued and generalised to produce $a^{\frac{1}{n}} = \sqrt[n]{a}$.

Now consider: $a^{\frac{m}{n}} = a^{m \times \frac{1}{n}}$ or $a^{\frac{m}{n}} = a^{\frac{1}{n} \times m}$

$$= (a^m)^{\frac{1}{n}} \qquad = \left(a^{\frac{1}{n}}\right)^m$$

$$= \sqrt[n]{a^m} \qquad = \left(\sqrt[n]{a}\right)^m$$

Fractional and radical indices

$$a^{\frac{1}{n}} = \sqrt[n]{a}$$

$$a^{\frac{m}{n}} = \sqrt[n]{a^m} \quad \text{or} \quad a^{\frac{m}{n}} = \left(\sqrt[n]{a}\right)^m$$

WORKED EXAMPLE 6 Evaluating fractional indices

Evaluate each of the following without technology.

a. $16^{\frac{3}{2}}$

b. $\left(\dfrac{9}{25}\right)^{-\frac{3}{2}}$

THINK	WRITE
a. 1. Rewrite the base number in terms of its prime factors.	a. $16^{\frac{3}{2}} = \left(2^4\right)^{\frac{3}{2}}$
2. Remove the brackets by multiplying the powers.	$= 2^6$
3. Evaluate as a basic number.	$= 64$
b. 1. Rewrite the base numbers of the fraction in terms of their prime factors.	b. $\left(\dfrac{9}{25}\right)^{-\frac{3}{2}} = \left(\dfrac{3^2}{5^2}\right)^{-\frac{3}{2}}$
2. Remove the brackets by multiplying the powers.	$= \dfrac{3^{-3}}{5^{-3}}$
3. Rewrite with positive powers by interchanging the numerator and denominator.	$= \dfrac{5^3}{3^3}$
4. Evaluate the numerator and denominator as basic numbers.	$= \dfrac{125}{27}$

WORKED EXAMPLE 7 Converting and simplifying radicals to and from fractional indices

Simplify the following, expressing your answers with positive indices.

a. $\sqrt[3]{64} \times \sqrt[4]{512}$

b. $\sqrt[3]{x^2y^6} \div \sqrt{x^3y^5}$

c. $5^{\frac{2}{3}}$

d. $x^{\frac{3}{2}} \div y^{\frac{1}{2}}$

THINK	WRITE
a. 1. Write the expression.	**a.** $\sqrt[3]{64} \times \sqrt[4]{512}$
2. Write using fractional indices.	$= 64^{\frac{1}{3}} \times 512^{\frac{1}{4}}$
3. Write 64 and 512 in index form.	
4. Multiply the powers.	$= \left(2^6\right)^{\frac{1}{3}} \times \left(2^9\right)^{\frac{1}{4}}$
5. Simplify the powers.	$= 2^2 \times 2^{\frac{9}{4}}$
	$= 2^{\frac{17}{4}}$
b. 1. Write the expression.	**b.** $\sqrt[3]{x^2y^6} \div \sqrt{x^3y^5}$
2. Express the roots in index notation.	$= \left(x^2y^6\right)^{\frac{1}{3}} \div \left(x^3y^5\right)^{\frac{1}{2}}$
3. Remove the brackets by multiplying the powers.	$= x^{\frac{2}{3}}y^2 \div x^{\frac{3}{2}}y^{\frac{5}{2}}$
4. Collect terms with the same base by subtracting the powers.	$= x^{\frac{2}{3}-\frac{3}{2}}y^{2-\frac{5}{2}}$
5. Simplify the powers.	$= x^{-\frac{5}{6}}y^{-\frac{1}{2}}$
6. Rewrite with positive powers.	$= \dfrac{1}{x^{\frac{5}{6}}y^{\frac{1}{2}}}$
c. 1. Write the term using radical signs.	**c.** $5^{\frac{2}{3}} = \sqrt[3]{5^2}$
2. Simplify the powers.	$= \sqrt[3]{25}$
3. Express the term as a radical.	$5^{\frac{2}{3}} = \sqrt[3]{25}$
d. 1. Write the expression using radical signs.	**d.** $x^{\frac{3}{2}} \div y^{\frac{1}{2}} = \sqrt{x^3} \div \sqrt{y}$
2. Simplify as a single surd since both are square roots.	$= \sqrt{\dfrac{x^3}{y}}$
3. Express the expression as a radical.	$x^{\frac{3}{2}} \div y^{\frac{1}{2}} = \sqrt{\dfrac{x^3}{y}}$

All the index laws from Lessons 8.2 and 8.3 are summarised below.

Summary of index laws

$$a^m \times a^n = a^{m+n}$$

$$a^m \div a^n = a^{m-n}$$

$$a^0 = 1, \; a \neq 0$$

$$(a^m)^n = a^{mn}$$

$$(ab)^m = a^m b^m$$

$$\left(\frac{a}{b}\right)^m = \frac{a^m}{b^m}$$

$$a^{-n} = \frac{1}{a^n}$$

$$a^{\frac{m}{n}} = \sqrt[n]{a^m} = \left(\sqrt[n]{a}\right)^m$$

Exercise 8.3 Negative and fractional indices

learnon

8.3 Exercise	8.3 Exam practice on

Simple familiar	Complex familiar	Complex unfamiliar
1, 2, 3, 4, 5, 6, 7, 8, 9, 10, 11, 12	13, 14, 15, 16, 17, 18, 19	20

These questions are even better in jacPLUS!
- Receive immediate feedback
- Access sample responses
- Track results and progress

Find all this and MORE in jacPLUS

Simple familiar

1. **MC** The exact value of 6^{-2} is:

   A. -12 **B.** -36 **C.** $-\dfrac{1}{6}$ **D.** $\dfrac{1}{36}$

2. **MC** The exact value of $\left(\dfrac{27}{8}\right)^{-\frac{1}{3}}$ is:

   A. $-\dfrac{2}{3}$ **B.** 6 **C.** $\dfrac{2}{3}$ **D.** $-\dfrac{3}{2}$

3. **MC** $\sqrt[3]{25} \times \sqrt{125}$ simplifies to:

   A. $25^{\frac{5}{6}}$ **B.** $5^{\frac{7}{6}}$ **C.** $5^{\frac{3}{2}}$ **D.** $5^{\frac{13}{6}}$

4. **WE5a** Express each of the following with positive index numbers.

   a. 6^{-3} **b.** 5^{-4} **c.** $\left(\dfrac{3}{5}\right)^{-2}$ **d.** $\left(\dfrac{7}{4}\right)^{-5}$

5. Express each of the following with positive index numbers.

a. $\left(\dfrac{1}{9}\right)^{-2}$
b. $\left(64^{-2}\right)^{3}$
c. $(-3)^{-1}$
d. $\left(\dfrac{3^4}{2^3}\right)^{-4}$

6. **WE6** Evaluate the following without using technology.

a. $9^{\frac{1}{2}}$
b. $27^{\frac{1}{3}}$
c. $625^{\frac{1}{4}}$

d. $256^{\frac{1}{8}}$
e. $8^{\frac{2}{3}}$
f. $81^{\frac{3}{4}}$

7. Evaluate the following without using technology. Check your answers using technology.

a. $\left(\dfrac{16}{81}\right)^{\frac{1}{4}}$
b. $\left(\dfrac{25}{16}\right)^{\frac{3}{2}}$
c. $\left(\dfrac{27}{64}\right)^{\frac{2}{3}}$

d. $32^{-\frac{2}{5}}$
e. $81^{-\frac{3}{4}}$
f. $\left(\dfrac{8}{27}\right)^{-\frac{2}{3}}$

8. **WE7a** Simplify each of the following, expressing your answers with positive indices.

a. $\sqrt{9} \times \sqrt[3]{81}$
b. $x^{\frac{2}{3}} \times x^{\frac{1}{6}}$
c. $x^{-\frac{3}{4}} \times x^{\frac{9}{8}}$

d. $x^{\frac{5}{2}} \div \left(x^{\frac{1}{3}}\right)^{4}$
e. $\sqrt[3]{(xy^3)} \div \sqrt{(x^2y)}$
f. $\sqrt[5]{32} \times \sqrt[4]{8}$

9. Simplify each of the following, expressing your answers with positive indices.

a. $2^{\frac{5}{4}} \times 4^{-\frac{1}{2}} \times 8^{-\frac{2}{3}}$
b. $27^{-\frac{1}{4}} \times 9^{\frac{2}{3}} \times 3^{-\frac{5}{4}}$

c. $\dfrac{18^{\frac{1}{2}}}{9^{\frac{4}{3}} \times 4^{\frac{3}{4}}}$
d. $\left(\sqrt[4]{x^3}\right)^{\frac{2}{3}} \times \left(\sqrt[3]{x^4}\right)^{\frac{3}{8}}$

e. $\dfrac{\left(64m^6\right)^{\frac{4}{3}}}{4m^{-2}}$
f. $\dfrac{\sqrt{x^3}}{\sqrt{x}}$

10. Express the following in index (exponent) form.

a. $\sqrt{a^3b^4}$
b. $\sqrt{\dfrac{a^5}{b^{-4}}}$
c. $\sqrt[3]{a^2b}$

11. **WE7b** Express the following in radical form.

a. $a^{\frac{1}{2}} \div b^{\frac{3}{2}}$
b. $2^{\frac{5}{2}}$
c. $3^{-\frac{2}{5}}$

12. **WE5b** Simplify each of the following, expressing your answers with positive index numbers.

a. $\dfrac{(-2)^3 \times 2^{-4}}{2^{-3}}$
b. $\dfrac{\left(x^{-2}\right)^3 \times \left(y^4\right)^{-2}}{x^{-5} \times \left(y^{-2}\right)^3}$
c. $\dfrac{(-m)^2 \times m^{-3}}{\left(p^{-2}\right)^{-1} \times p^{-4}}$

Complex familiar

13. Simplify each of the following, expressing your answers with positive index numbers.

a. $\dfrac{x^5}{x^{-3}} \div \dfrac{\left(x^4\right)^{-2}}{\left(x^2\right)^{-3}}$
b. $\dfrac{\left(3^{-2}\right)^2 \times \left(2^{-5}\right)^{-1}}{\left(2^4\right)^{-2} \times \left(3^4\right)^{-3}}$
c. $\dfrac{x^3y^{-2} \times \left(xy^2\right)^{-3}}{\left(2x^3\right)^2 \times \left(y^{-3}\right)^2}$

14. Evaluate the following without using technology. Check your answers with technology.

a. $4^{\frac{3}{2}}$

b. $3^{-1} + 5^0 - 2^2 \times 9^{-\frac{1}{2}}$

c. $2^3 \times \left(\dfrac{4}{9}\right)^{-\frac{1}{2}} \div \left(6 \times \left(3^{-2}\right)^2\right)$

d. $\dfrac{15 \times 5^{\frac{3}{2}}}{125^{\frac{1}{2}} - 20^{\frac{1}{2}}}$

15. Simplify the following, expressing your answers with positive indices.

a. $\dfrac{3\left(x^2 y^{-2}\right)^3}{\left(3x^4 y^2\right)^{-1}}$

b. $\dfrac{2a^{\frac{2}{3}} b^{-3}}{3a^{\frac{1}{3}} b^{-1}} \times \dfrac{3^2 \times 2 \times (ab)^2}{\left(-8a^2\right)^2 b^2}$

c. $\dfrac{\left(2mn^{-2}\right)^{-2}}{m^{-1} n} \div \dfrac{10n^4 m^{-1}}{3\left(m^2 n\right)^{\frac{3}{2}}}$

d. $\dfrac{4m^2 n^{-2} \times -2\left(m^2 n^{\frac{3}{2}}\right)^2}{\left(-3m^3 n^{-2}\right)^2}$

16. Express $\dfrac{2^{1-n} \times 8^{1+2n}}{16^{1-n}}$ as a power of 2.

17. Simplify each of the following, expressing your answers with positive indices.

a. $\dfrac{1}{\sqrt{x^{-4}}}$

b. $\dfrac{(x+1)^2}{\sqrt{x+1}}$

c. $(y-4)\sqrt{y-4}$

d. $(p+3)(p+3)^{-\frac{2}{5}}$

18. Simplify each of the following, expressing your answers with positive indices.

a. $\sqrt{x} - \dfrac{1}{\sqrt{x}}$

b. $\sqrt{x+2} + \dfrac{x}{\sqrt{x+2}}$

19. Simplify the following, expressing the answers with positive indices.

a. $\dfrac{m^{-1} - n^{-1}}{m^2 - n^2}$

b. $\sqrt{4x-1} - 2x(4x-1)^{-\frac{1}{2}}$

Complex unfamiliar

20. A scientist has discovered a piece of paper with a complex formula written on it. She thinks that someone has tried to disguise a simpler formula. The formula is:

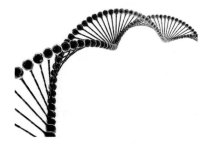

$$\dfrac{\sqrt[4]{a^{13}} a^2 \sqrt{b^3}}{\sqrt{a^1 b}} \times b^3 \times \left(\dfrac{\sqrt{a^3 b}}{ab^2}\right)^2 \times \left(\dfrac{b^2}{a^2 \sqrt{b}}\right)^3$$

Considering only integers, evaluate the smallest value for a for which the expression will give a rational answer.

Fully worked solutions for this chapter are available online.

LESSON
8.4 Scientific notation

SYLLABUS LINKS

- Understand and use scientific notation.

Source: Mathematical Methods Senior Syllabus 2024 © State of Queensland (QCAA) 2024; licensed under CC BY 4.0.

8.4.1 Scientific notation

Index notation provides a convenient way to express numbers that are either very large or very small. Writing a number as $a \times 10^b$ (the product of a number a where $1 \leq a < 10$ and a power of 10) is known as writing the number in **scientific notation** (or **standard form**). The age of the Earth since the Big Bang is estimated to be 4.54×10^9 years, and the mass of a carbon atom is approximately 1.994×10^{-23} grams. These numbers are written in scientific notation.

To convert scientific notation back to a basic numeral, apply the following:
- Move the decimal point b places to the **right** if the power of 10 has a positive index to obtain the large number $a \times 10^b$ represents
- Move the decimal point b places to the **left** if the power of 10 has a negative index to obtain the small number $a \times 10^{-b}$ represents. Remember multiplying by 10^{-b} is equivalent to dividing by 10^b.

WORKED EXAMPLE 8 Expressing numbers in scientific notation

Write each of these numbers in scientific notation.
a. 827.2

b. 51 920 000 000

c. 0.0051

d. 0.000 000 007 648

THINK

a. 1. The numerical part must be written so it is a number between 1 and 10. This means we must move the decimal point 2 steps to the left.

  2. Moving the decimal point 2 steps left corresponds to the power 10^2.
  Note that the number of steps moved is equal to the exponent.

b. 1. Even though the decimal point is not written, we know that it lies after the final zero.
  2. To get a number between 1 and 10, we move the decimal 10 steps to the left. The corresponding power will be 10^{10}.

c. To get a number between 1 and 10, we must move the decimal point 3 steps to the right. This corresponds to the power 10^{-3}. Note that moving to the right gives a negative index.

d. To get a number between 1 and 10, we must move the decimal point 9 steps to the right. This corresponds to the power 10^{-9}.

WRITE

a.
$$8.2\,7\,2$$
Decimal point moves 2 steps left

8.272×10^2

b. 5.192×10^{10}

c. 5.1×10^{-3}

d. 7.648×10^{-9}

8.4.2 Significant figures

When a number is expressed in scientific notation as either $a \times 10^b$ or $a \times 10^{-b}$, the number of digits in a determines the number of **significant figures** in the basic numeral. The age of the Earth is 4.54×10^9 years in scientific notation or 4 540 000 000 years to three significant figures. To one significant figure, the age would be 5 000 000 000 or 5×10^9 years.

WORKED EXAMPLE 9 Expressing decimal numbers in scientific notation

a. Express each of the following numerals in scientific notation and state the number of significant figures each numeral contains.
   i. **3 266 400**
   ii. **0.009 876 03**
b. Express the following as basic numerals.
   i. **4.54×10^9**
   ii. **1.037×10^{-5}**

THINK	WRITE
a. i. 1. Write the given number as a value between 1 and 10 multiplied by a power of 10, discarding any trailing zeroes. *Note:* The number is large, so the power of 10 should be positive.	a. i. In scientific notation, $3\,266\,400 = 3.2664 \times 10^6$.
2. Count the number of digits in the number a in the scientific notation form $a \times 10^b$ and state the number of significant figures.	There are 5 significant figures in the number 3 266 400.
ii. 1. Write the given number as a value between 1 and 10 multiplied by a power of 10. *Note:* The number is small, so the power of 10 should be negative.	ii. $0.009\,876\,03 = 9.876\,03 \times 10^{-3}$
2. Count the number of digits in the number a in the scientific notation form and state the number of significant figures.	0.009 876 03 has 6 significant figures.
b. i. 1. Perform the multiplication. *Note:* The power of 10 is positive, so a large number should be obtained.	b. i. 4.54×10^9 Move the decimal point 9 places to the right. $\therefore 4.54 \times 10^9 = 4\,540\,000\,000$
ii. 2. Perform the multiplication. *Note:* The power of 10 is negative, so a small number should be obtained.	ii. 1.037×10^{-5} Move the decimal point 5 places to the left. $\therefore 1.037 \times 10^{-5} = 0.000\,010\,37$

TI	THINK	WRITE	CASIO	THINK	WRITE

TI | THINK

a. i. 1. On the Home screen, select:
5: Settings
2: Document settings
Then change the Exponential Format to Scientific and select OK.

2. On a Calculator page, complete the entry line as:
3 266 400E0
then press ENTER.
Note: The EE button is located above the π button.

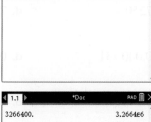

3. The answer appears on the screen.

$3\,266\,400 = 3.2664 \times 10^6$

b. i. 1. Change the Exponential Format back to Normal.

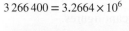

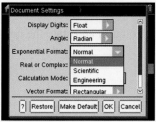

2. On a Calculator page, complete the entry line as:
4.54×10^9
then press ENTER.

3. The answer appears on the screen.

$4.54 \times 10^9 = 4\,540\,000\,000$

CASIO | THINK

a. i. 1. On a Run-Matrix screen, press SHIFT then MENU. Change the Display mode to Scientific, then type '0' and press ENTER. Press EXIT.

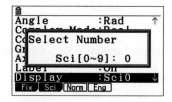

2. Complete the entry line as:
3 266 400
then press EXE.

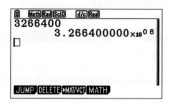

3. The answer appears on the screen.

$3\,266\,400 = 3.2664 \times 10^6$

b. i. 1. On the Run-Matrix screen, change the Display mode back to Normal.

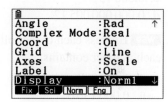

2. Complete the entry line as:
4.54×10^9
then press EXE.

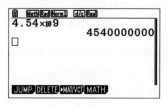

3. The answer appears on the screen.

$4.54 \times 10^9 = 4\,540\,000\,000$

Exercise 8.4 Scientific notation

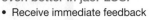
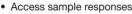

8.4 Exercise	8.4 Exam practice on

Simple familiar	Complex familiar	Complex unfamiliar
1, 2, 3, 4, 5, 6, 7, 8, 9, 10, 11	12	N/A

These questions are even better in jacPLUS!
- Receive immediate feedback
- Access sample responses
- Track results and progress

Find all this and MORE in jacPLUS ▶

Simple familiar

1. **WE8** Write these numbers in scientific notation.
   - a. 72.5
   - b. 725
   - c. 7250
   - d. 725 000 000

2. Write these numbers in scientific notation.
   - a. 0.231
   - b. 0.003 62
   - c. 0.000 731
   - d. 0.063

3. Write these numbers in decimal notation.
   - a. 7.34×10^5
   - b. 7.1414×10^6
   - c. 3.51×10
   - d. 8.05×10^4

4. Write these numbers in decimal notation.
   - a. 8.273×10^{-2}
   - b. 7.295×10^{-2}
   - c. 2.9142×10^{-3}
   - d. 3.753×10^{-5}

5. Write these numbers in scientific notation, correct to 4 significant figures.
   - a. 43.792
   - b. 5317
   - c. 258.95
   - d. 110.11
   - e. 1 632 000
   - f. 1 million

6. Write these numbers in scientific notation, correct to 3 significant figures.
   - a. 0.006 731
   - b. 0.142 57
   - c. 0.000 068 3
   - d. 0.000 000 005 12
   - e. 0.0509
   - f. 0.012 46

7. **WE9** Express each of the following numbers in scientific notation and state the number of significant figures each number contains.
   - a. i. 1 409 000
      ii. 0.000 130 6
   - b. Express the following as basic numerals.
      i. 3.04×10^5
      ii. 5.803×10^{-2}

8. Express in scientific notation:
   - a. $-0.000 000 050 6$
   - b. the diameter of the Earth, given its radius is 6370 km
   - c. $3.2 \times 10^4 \times 5 \times 10^{-2}$
   - d. the distance between Roland Garros and Kooyong tennis stadiums, 16 878.7 km.

9. Express each of the following as a basic numeral.
   - a. $6.3 \times 10^{-4} + 6.3 \times 10^4$
   - b. $\left(1.44 \times 10^6\right)^{\frac{1}{2}}$

10. Write each of the following sets of numbers in ascending order.
    a. 8.31×10^2, 3.27×10^3, 9.718×10^2, 5.27×10^2
    b. 7.95×10^2, 4.09×10^2, 7.943×10^2, 4.37×10^2
    c. 5.31×10^{-2}, 9.29×10^{-3}, 5.251×10^{-2}, 2.7×10^{-3}
    d. 8.31×10^2, 3.27×10^3, 7.13×10^{-2}, 2.7×10^{-3}

11. Express the following to 2 significant figures.

    a. 60 589 people attended a football match.
    b. The probability of winning a competition is 1.994×10^{-2}.
    c. The solution to an equation is $x = -0.006\,34$.
    d. The distance flown per year by the Royal Flying Doctor Service is 26 597 696 km.

Complex familiar

12. If Earth is approximately 1.496×10^8 km from the Sun and Mercury is about 5.8×10^7 km from the Sun, evaluate the distance from Mercury to Earth.

Fully worked solutions for this chapter are available online.

LESSON
8.5 Review

8.5 Exercise

learn on

8.5 Exercise	**8.5 Exam practice** on

Simple familiar	Complex familiar	Complex unfamiliar
1, 2, 3, 4, 5, 6, 7, 8, 9, 10, 11, 12	13, 14, 15, 16, 17	18, 19, 20

Simple familiar

1. **MC** When simplified, $\dfrac{\left(2xy^3\right)^2}{7x^3} \times \dfrac{3x^5y^2}{4y}$ is equal to:

   A. $\dfrac{x^4y^7}{7}$
   B. $\dfrac{3x^4y^7}{7}$
   C. $\dfrac{3y^7}{x^2}$
   D. $\dfrac{3x^4}{y^6}$

2. **MC** $\dfrac{5m^4p^2}{2m^3p} \div \dfrac{\left(5m^2p^6\right)^3}{3m^7p}$ may be simplified to:

   A. $\dfrac{m^2}{47p^{16}}$
   B. $\dfrac{3m^{10}}{2p^{32}}$
   C. $\dfrac{3m^2}{50p^{16}}$
   D. $\dfrac{m^{15}}{p^{29}}$

3. **MC** The value of $5^{-2}\left(\dfrac{64}{125}\right)^{-\frac{1}{3}}$ is:

   A. $\dfrac{1}{20}$
   B. 5
   C. $\dfrac{4}{5}$
   D. $\dfrac{5}{4}$

4. **MC** Select the expression that is *not* the same as $(4xy)^{\frac{3}{2}}$.

   A. $8x^{\frac{3}{2}}y^{\frac{3}{2}}$
   B. $\left(\sqrt{4xy}\right)^3$
   C. $\sqrt{64x^3y^3}$
   D. $4xy^{\frac{1}{2}} \times \left(2xy^2\right)^{\frac{1}{2}}$

5. **MC** The expression $\dfrac{x^2y}{\left(2xy^2\right)^3} \div \dfrac{xy}{16x^0}$ is equal to:

   A. $\dfrac{2}{x^2y^6}$
   B. $\dfrac{2x^2}{b^6}$
   C. $2x^2y^6$
   D. $\dfrac{2}{xy^6}$

6. **MC** In scientific notation, $(3.2 \times 10^{-2}) \times (5 \times 10^5)$ would equal:

    A. 2.56×10^8 **B.** 16×10^{-10} **C.** 1.6×10^{-9} **D.** 1.6×10^4

7. Express the following in surd form.

    a. $a^{\frac{1}{2}} \div b^{\frac{3}{2}}$ **b.** $2^{\frac{5}{2}}$ **c.** $3^{-\frac{2}{5}}$

8. Calculate $(4 \times 10^6)^2 \times (5 \times 10^{-3})$.

9. Simplify the following.

    a. $\dfrac{5a^2b^3}{(2a^3b)^3}$ **b.** $\dfrac{4x^5y^6}{(2xy^3)^4}$ **c.** $\dfrac{(3m^2n^3)^3}{(2m^5n^5)^7}$

10. Simplify the following.

    a. $\left(\dfrac{4x^3y^{10}}{2x^7y^4}\right)^6$ **b.** $\dfrac{3a^3b^{-5}}{(2a^7b^4)^{-3}}$ **c.** $\left(\dfrac{3g^2h^5}{2g^4h}\right)^3$

11. Simplify the following.

    a. $\dfrac{6x^3y^2 \times 4x^6y}{9xy^5 \times 2x^3y^6}$ **b.** $\dfrac{(6x^3y^2)^4}{9x^5y^2 \times 4xy^7}$ **c.** $\dfrac{5x^2y^3 \times 2xy^5}{10x^3y^4 \times x^4y^2}$

12. Simplify the following.

    a. $\left(\dfrac{4a^9}{b^6}\right)^3 \div \left(\dfrac{3a^7}{2b^5}\right)^4$

    b. $\dfrac{5x^2y^6}{(2x^4y^5)^2} \div \dfrac{(4x^6y)^3}{10xy^3}$

    c. $\left(\dfrac{x^5y^{-3}}{2xy^5}\right)^{-4} \div \dfrac{4x^6y^{-10}}{(3x^{-2}y^2)^{-3}}$

Complex familiar

13. Simplify $(9a^3b^{-4})^{\frac{1}{2}} \times 2\left(a^{\frac{1}{2}}b^{-2}\right)^{-2}$.

14. Evaluate $27^{-\frac{2}{3}} + \left(\dfrac{49}{81}\right)^{\frac{1}{2}}$.

15. Simplify the following expression with positive indices.

$$\left(16x^{-6}y^{10}\right)^{\frac{1}{2}} \div \sqrt[3]{(27x^3y^9)}$$

16. Evaluate the following.

    a. $4^{\frac{3}{2}}$

    b. $3^{-1} + 5^0 - 2^2 \times 9^{-\frac{1}{2}}$

17. Simplify $\dfrac{20p^5}{m^3q^{-2}} \div \dfrac{5(p^2q^{-3})^2}{-4m^{-1}}$.

18. Simplify $\dfrac{a - a^{-1}}{a + 1}$, expressing your answer with positive indices.

19. If $x = 3^{\frac{1}{3}} + 3^{-\frac{1}{3}}$, show that $x^3 - 3x = \dfrac{10}{3}$.

20. If $\left(\dfrac{2x^2}{3a}\right)^{n-1} \div \left(\dfrac{3x}{a}\right)^{n+1} = \left(\dfrac{x}{4}\right)^3$, determine the values of the constants a and n.

Fully worked solutions for this chapter are available online.

Hey teachers! Create custom assignments for this chapter

Create and assign unique tests and exams

Access quarantined tests and assessments

Track your students' results

Find all this and MORE in jacPLUS

Answers

Chapter 8 Indices and index laws

8.2 Index laws

8.2 Exercise

1. a. x^{10} b. $m^5 p^5$ c. $8y^{11}$
2. a. 5^{18} b. $x^7 y^8$ c. $2^{12}x^{18}$
   d. $6m^9 p^{17}$ e. $5^7 x^{10} y^{13}$
3. a. $a^5 b^3$ b. $\dfrac{a^9 b^5}{4}$ c. $\dfrac{x^3 y^9}{2}$
   d. $p^{11} q^2$ e. $\dfrac{8m^2 n^5}{7}$
4. a. $a^2 b^2$ b. $\dfrac{r^{10} t}{5}$ c. $5a^3 b^3$ d. $\dfrac{6x^2 y^4}{5}$
5. a. $\dfrac{4p^{10}m^8}{3}$ b. $\dfrac{9xy^4}{2}$ c. $18u^{11}v^5$
   d. $10e^9 f^2$ e. $\dfrac{2wt^6}{9}$
6. a. x^{13} b. $-81x^9 y^3$ c. $3m^4 p$
   d. mp^2 e. $\dfrac{u^{11}v^4}{9}$
7. a. $\dfrac{15}{4}$ b. $\dfrac{125k^{11}d^6}{24}$ c. $16p^{17}$
   d. $27j^3 n^2$ e. $x^3 y^4$ f. 12
8. a. $3p^2 m^2$ b. $\dfrac{3x^5}{2}$ c. $-3a^6$
9. a. 2^{11} b. 3^{24} c. $5^5 \times 3^4$
10. a. $2^{22} \times 5^8$ b. $\dfrac{3}{2^4}$ c. $\dfrac{5}{2}$
11. a. 5 b. $\dfrac{1}{5}$ c. 16 d. $\dfrac{1}{3}$
12. a. $2^2 \times 3^{3n+4}$ b. 5^{3n-6}
    c. 2^{3x-2} d. $2^{-5n-3} \times 3^{-6n}$
    e. $2^{n-2} \times 7^{6n-3}$
13. a. $\dfrac{49}{5}$ b. $3^{9n-3} \times 2^{5n-6}$
14. a. $x^3 y^{1+n} z$ b. $x^5 y^{6m-1}$
15. 6^{2n+9}
16. a. 8 b. 59 049 c. 16 d. 1
17. a. Only if $x = 2$
    b. Sample responses are available in the worked solutions in your online resources.
18. 3
19. $-\dfrac{2}{3}$
20. The student made a mistake when multiplying the two brackets in line 3. Individual brackets should be expanded first.

8.3 Negative and fractional indices

8.3 Exercise

1. D
2. C
3. D
4. a. $\dfrac{1}{6^3}$ b. $\dfrac{1}{5^4}$ c. $\dfrac{5^2}{3^2}$ d. $\dfrac{4^5}{7^5}$
5. a. 9^2 b. $\dfrac{1}{64^6}$ c. $\dfrac{1}{-3}$ d. $\dfrac{2^{12}}{3^{16}}$
6. a. 3 b. 3 c. 5 d. 2 e. 4 f. 27
7. a. $\dfrac{2}{3}$ b. $\dfrac{125}{64}$ c. $\dfrac{9}{16}$ d. $\dfrac{1}{4}$ e. $\dfrac{1}{27}$ f. $\dfrac{9}{4}$
8. a. $3^{\frac{7}{3}}$ b. $x^{\frac{5}{6}}$ c. $x^{\frac{3}{8}}$ d. $x^{\frac{7}{6}}$ e. $\dfrac{y^{\frac{1}{2}}}{x^{\frac{2}{3}}}$ f. $2^{\frac{7}{4}}$
9. a. $\dfrac{1}{2^{\frac{7}{4}}}$ b. $\dfrac{1}{3^{\frac{2}{3}}}$ c. $\dfrac{1}{2\times 3^{\frac{5}{3}}}$ d. x e. $64\,m^{10}$ f. x
10. a. $a^{\frac{3}{2}}b^2$ b. $a^{\frac{5}{2}}b^2$ c. $a^{\frac{2}{3}}b^{\frac{1}{3}}$
11. a. $\sqrt{\dfrac{a}{b^3}}$ b. $\sqrt{32}$ c. $\sqrt[5]{\dfrac{1}{9}}$
12. a. -2^2 b. $\dfrac{1}{xy^2}$ c. $\dfrac{p^2}{m}$
13. a. x^{10} b. $3^8 \times 2^{13}$ c. $\dfrac{1}{4x^6 y^2}$
14. a. 8 b. 0 c. 162 d. 25
15. a. $\dfrac{9x^{10}}{y^4}$ b. $\dfrac{3}{16a^{\frac{5}{3}}b^2}$ c. $\dfrac{3m^3 n^{\frac{1}{2}}}{40}$ d. $\dfrac{-8n^5}{9}$
16. 2^{9n}
17. a. x^2 b. $(x+1)^{1\frac{1}{2}}$ c. $(y-4)^{\frac{3}{2}}$ d. $(p+3)^{\frac{3}{5}}$
18. a. $\dfrac{x-1}{x^{\frac{1}{2}}}$ b. $\dfrac{2x+2}{(x+2)^{\frac{1}{2}}}$
19. a. $\dfrac{-1}{mn(m+n)}$ b. $\dfrac{2x-1}{(4x-1)^{\frac{1}{2}}}$
20. $a^{-\frac{1}{4}} \times b^{\frac{13}{2}}$

No, because you can't take the fourth root of a negative number.

$a = 1$

8.4 Scientific notation

8.4 Exercise

1. a. 7.25×10^1 b. 7.25×10^2
   c. 7.25×10^3 d. 7.25×10^8

2. a. 2.31×10^{-1} b. 3.62×10^{-3}
   c. 7.31×10^{-4} d. 6.3×10^{-2}

3. a. $734\,000$ b. $7\,141\,400$
   c. 35.1 d. $80\,500$

4. a. $0.082\,73$ b. $0.072\,95$
   c. $0.002\,914\,2$ d. $0.000\,037\,53$

5. a. 4.379×10^1 b. 5.317×10^3
   c. 2.590×10^2 d. 1.101×10^2
   e. 1.632×10^6 f. 1.000×10^6

6. a. 6.73×10^{-3} b. 1.43×10^{-1}
   c. 6.83×10^{-5} d. 5.12×10^{-9}
   e. 5.09×10^{-2} f. 1.25×10^{-2}

7. a. i. 1.409×10^6; 4 significant figures
      ii. 1.306×10^{-4}; 4 significant figures
   b. i. $304\,000$
      ii. $0.058\,03$

8. a. -5.06×10^{-8} b. 1.274×10^4 km
   c. 1.6×10^3 d. $1.687\,87 \times 10^4$ km

9. a. $63\,000.000\,63$ b. 1200

10. a. 5.27×10^2, 8.31×10^2, 9.718×10^2, 3.27×10^3
    b. 4.09×10^2, 4.37×10^2, 7.943×10^2, 7.95×10^2
    c. 2.7×10^{-3}, 9.29×10^{-3}, 5.251×10^{-2}, 5.31×10^{-2}
    d. 2.7×10^{-3}, 7.13×10^{-2}, 8.31×10^2, 3.27×10^3

11. a. $61\,000$ b. 0.020
    c. -0.0063 d. $27\,000\,000$ km

12. 9.16×10^7 km

12. a. $\dfrac{1024b^2}{81a}$ b. $\dfrac{25}{128x^{23}y^4}$ c. $\dfrac{4y^{36}}{27x^{16}}$

13. $6a^{\frac{1}{2}}b^2$

14. $\dfrac{8}{9}$

15. $\dfrac{4y^2}{3x^4}$

16. a. 8
    b. 0

17. $\dfrac{-16pq^8}{m^4}$

18. $\dfrac{a-1}{a}$

19. Sample responses are available in the worked solutions in your online resources.

20. $a = \pm\left(3^6 \times 2^{-\frac{11}{2}}\right), n = 6$

8.5 Review

8.5 Exercise

1. B
2. C
3. A
4. D
5. A
6. D

7. a. $\sqrt{\dfrac{a}{b^3}}$ b. $\sqrt{32}$ c. $\sqrt[5]{\dfrac{1}{9}}$

8. 8×10^{10}

9. a. $\dfrac{5}{8a^7}$ b. $\dfrac{x}{4y^6}$ c. $\dfrac{27}{128m^{29}n^{26}}$

10. a. $\dfrac{64y^{36}}{x^{24}}$ b. $24a^{24}b^7$ c. $\dfrac{27h^{12}}{8g^6}$

11. a. $\dfrac{4x^5}{3y^8}$ b. $\dfrac{36x^6}{y}$ c. $\dfrac{y^2}{x^4}$

9 Introduction to exponential functions

LESSON SEQUENCE

Fully worked solutions for this chapter are available online.

EXAM PREPARATION

Access exam-style questions in every lesson, available online.

on Resources

Solutions	Solutions — Chapter 9 (sol-1276)
Exam questions	Exam question booklet — Chapter 9 (eqb-0297)
Digital documents	Learning matrix — Chapter 9 (doc-41791)
	Chapter summary — Chapter 9 (doc-41804)

LESSON
9.1 Overview

9.1.1 Introduction

There is a very old story, which has been told in one form or another in many different cultures, of an emperor who wished to grant a reward to a humble peasant for a good deed. The peasant looked at the emperor's chess board and asked that he be given coins such that he received one coin for the first square of the chessboard, two coins for the second, four coins for the third, eight coins for the fourth and so on. The emperor quickly agreed, thinking that he would end up giving the peasant far less than he was originally planning to give him, not realising that the modest amounts would rapidly increase — so much so, that there were not enough coins in the empire to fulfil the total amount that the peasant was really asking for (the final square on the chessboard would correspond to nearly 2×10^{19} coins!). The growth of the coins in this story is just one example of exponential growth.

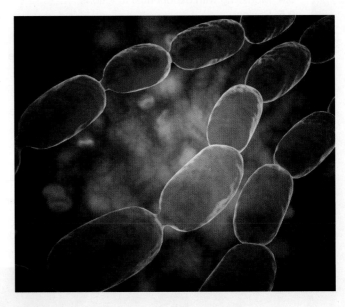

Many processes in nature can be modelled by exponential growth. Bacteria, for example, reproduce by binary fission whereby one organism splits to become two. These two organisms then also split to form four and so on, reproducing rapidly to form colonies of millions under the right conditions. Conversely, there are many examples of elements decreasing rapidly.

9.1.2 Syllabus links

Lesson	Lesson title	Syllabus links
9.2	**Solving exponential equations**	○ Solve equations involving exponential functions, with and without technology.
9.3	**Graphs of exponential functions**	○ Recognise and determine the qualitative features of the graph of $y = r^x$ (where $r > 0$), including asymptote and intercept.
		○ Recognise and determine the effect of the parameters h, k and r on the graph of $y = r^{(x-h)} + k$ (where $r > 0$), with and without technology.
		○ Sketch the graphs of exponential functions, with and without technology.
9.4	**Modelling with exponential functions**	○ Model and solve problems that involve exponential functions, with and without technology.

Source: Mathematical Methods Senior Syllabus 2024 © State of Queensland (QCAA) 2024; licensed under CC BY 4.0.

LESSON
9.2 Solving exponential equations

SYLLABUS LINKS

- Solve equations involving exponential functions, with and without technology.

Source: Mathematical Methods Senior Syllabus 2024 © State of Queensland (QCAA) 2024; licensed under CC BY 4.0.

An exponential equation (or indicial equation) has the unknown variable as an exponent or index. In this section we shall consider exponential equations that have rational solutions or require technology to determine the solutions.

9.2.1 Method of equating indices

If index laws can be used to express both sides of an equation as single powers of the same base, then this allows indices to be equated. For example, if an equation can be simplified to the form $2^{3x} = 2^4$, then for the equality to hold, $3x = 4$. Solving this linear equation gives the solution to the indicial equation as $x = \dfrac{4}{3}$.

Solving exponential equations

To solve for the exponent x in equations of the form $a^x = n$:
1. express both sides as powers of the same base
2. equate the indices and solve the equation formed to obtain the solution to the exponential equation.

WORKED EXAMPLE 1 Solving exponential equations

Solve for x in each of the following equations.

a. $3^x = 81$

b. $4^{x-1} = 256$

THINK	WRITE
a. 1. Write the equation.	a. $\qquad 3^x = 81$
2. Express both sides to the same base.	$\qquad 3^x = 3^4$
3. Equate the powers.	Therefore, $x = 4$.
b. 1. Write the equation.	b. $\qquad 4^{x-1} = 256$
2. Express both sides to the same base.	$\qquad 4^{x-1} = 4^4$
3. Equate the powers.	Therefore, $x - 1 = 4$.
4. Solve the linear equation for x by adding 1 to both sides.	$\qquad x = 5$

The index laws, studied in Chapter 8, may need to be applied to the question to enable both sides to be expressed as powers of the same base.

WORKED EXAMPLE 2 Solving exponential equations using index laws

Solve $5^{3x} \times 25^{4-2x} = \dfrac{1}{125}$ for x.

THINK	WRITE
1. Use the index laws to express the left-hand side of the equation as a power of a single base.	$5^{3x} \times 25^{4-2x} = \dfrac{1}{125}$ $5^{3x} \times 5^{2(4-2x)} = \dfrac{1}{125}$ $5^{3x+8-4x} = \dfrac{1}{125}$ $5^{8-x} = \dfrac{1}{125}$
2. Express the right-hand side as a power of the same base.	$5^{8-x} = \dfrac{1}{5^3}$ $5^{8-x} = 5^{-3}$
3. Equate indices and calculate the required value of x.	Equating indices, $8 - x = -3$ $x = 11$

TI | THINK	WRITE	CASIO | THINK	WRITE
1. On a Calculator page, press MENU, then select: 3: Algebra 1: Numerical Solve Complete the entry line as: nSolve $\left(5^{3x} \times 25^{4-2x} = \dfrac{1}{125}, x\right)$ then press ENTER.	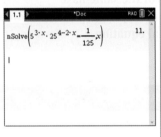	1. On the Run-Matrix screen, press OPTN, then select CALC by pressing F4. Select SolveN by pressing F5, then complete the entry line as: SolveN $\left(5^{3x} \times 25^{4-2x} = \dfrac{1}{125}, x\right)$ and press EXE.	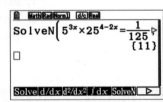
2. The answer appears on the screen.	$x = 11$	2. The answer appears on the screen.	$x = 11$

9.2.2 Exponential equations that reduce to quadratic form

The technique of substitution to form a quadratic equation may be applicable to indicial equations.

To solve equations of the form $p \times a^{2x} + q \times a^x + r = 0$, use the following steps.
1. Note that $a^{2x} = (a^x)^2$.
2. Reduce the indicial equation to quadratic form by using a substitution for a^x.
3. Solve the quadratic and then substitute back for a^x.
4. Since $a > 0$, a^x must always be positive, so solutions for x can only be obtained for $a^x > 0$. Reject any negative or zero values for a^x.

WORKED EXAMPLE 3 Solving exponential equations in quadratic form

Solve $3^{2x} - 6 \times 3^x - 27 = 0$ for x.

THINK	WRITE
1. Use a substitution technique to reduce the indicial equation to quadratic form. *Note:* The subtraction signs prevent the use of index laws to express the left-hand side as a power of a single base.	$3^{2x} - 6 \times 3^x - 27 = 0$ Let $a = 3^x$. $\therefore a^2 - 6a - 27 = 0$
2. Solve the quadratic equation.	$(a - 9)(a + 3) = 0$ $a = 9, a = -3$
3. Substitute back and solve for x.	Replace a with 3^x. $\therefore 3^x = 9$ or $3^x = -3$ (reject negative value) $3^x = 9$ $3^x = 3^2$ $x = 2$

9.2.4 Solving exponential equations with technology

Not all solutions to exponential equations are rational. In order to obtain the solution to an equation such as $2^x = 5$, we need to use technology. Later, in Chapter 10, you will be able to find the exact solutions.

WORKED EXAMPLE 4 Solving exponential equations using technology

Determine the value of x in each of the following equations, correct to 2 decimal places.
a. $2^x = 5$
b. $3 \times 2^{(5-x)} = 7$

THINK	WRITE
a. 1. Write the equation	$2^x = 5$
2. Recognise that the equation cannot be solved using the index laws; hence, use technology as shown below.	$\therefore x = 2.321\,928\,1$
3. Write the answer correct to 2 decimal places.	$x = 2.32$
b. 1. Write the equation.	$3 \times 2^{(5-x)} = 7$
2. Recognise that the equation cannot be solved using the index laws; hence, use technology.	$\therefore x = 3.777\,607\,6$
3. Write the answer correct to 2 decimal places.	$x = 3.78$

TI	THINK	WRITE		CASIO	THINK	WRITE
a. 1. On a Calculator page, press MENU, then select: 3: Algebra 1: Numerical Solve Complete the entry line as: nSolve $(2^x = 5, x)$ and press ENTER.	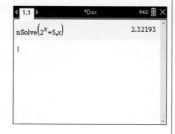		a. 1. On the Run-Matrix screen, press OPTN, then select CALC by pressing F4. Select SolveN by pressing F5, then complete the entry line as: SolveN $(2^x = 5, x)$ and press EXE.	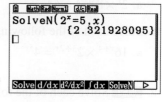
2. The answer appears on the screen.	$x = 2.32$ (correct to 2 decimal places)		2. The answer appears on the screen.	$x = 2.32$ (correct to 2 decimal places)

Exercise 9.2 Solving exponential equations

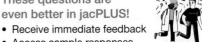

9.2 Exercise	9.2 Exam practice on

Simple familiar	Complex familiar	Complex unfamiliar
1, 2, 3, 4, 5, 6, 7, 8	9, 10, 11, 12, 13, 14, 15, 16, 17, 18	19, 20

These questions are even better in jacPLUS!
- Receive immediate feedback
- Access sample responses
- Track results and progress

Find all this and MORE in jacPLUS ▶

Simple familiar

1. **WE1** Solve each of the following equations for x.
   a. $2^x = 32$
   b. $5^x = 625$
   c. $3^x = 243$
   d. $10^{-x} = \dfrac{1}{100}$

2. Solve each of the following equations for x.
   a. $4^{-x} = 16$
   b. $6^x = \dfrac{1}{216}$
   c. $3^{-x} = \dfrac{1}{81}$
   d. $2^{-x} = 1$

3. Solve each of the following equations for n.
   a. $2^{3n+1} = 64$
   b. $5^{2n+3} = 25$
   c. $3^{2-n} = 27$

4. Solve each of the following equations for n.
   a. $16^{n+3} = 2^3$
   b. $49^{5-3n} = \dfrac{1}{7}$
   c. $36^{4n-3} = 216$

5. **WE2** Solve each of the following equations for x.
   a. $4^{2x} = 8^{x-1}$
   b. $27^{4-x} = 9^{2x+1}$
   c. $16^{3x+1} = 128^{x-2}$
   d. $25^{2x-3} = \dfrac{1}{125}$

6. Solve each of the following equations for x.
   a. $32^{5-x} = 4^{3x+2}$
   b. $64^{2-3x} = 16^{x+1}$
   c. $9^{3x+5} = \dfrac{1}{243}$
   d. $16^{4-3x} = \dfrac{1}{8^{x+3}}$

7. **WE4a** Determine the value of x in each of the following equations, correct to 3 decimal places.
   a. $3^x = 8$
   b. $4^x = 12$
   c. $5^x = 10$
   d. $5^{2x} = 15$

8. **WE4b** Determine the value of x in each of the following equations, correct to 2 decimal places.
   a. $4 \times 3^{(x+2)} = 11$
   b. $5 \times 4^{(x-2)} = 28$
   c. $3^{(x+2)} + 8 = 37$
   d. $2 \times 3^{(4x-5)} = 25$

Complex familiar

9. Solve $\dfrac{2^{5x-3} \times 8^{9-2x}}{4^x} = 1$ for x.

10. Solve each of the following equations for x.
    a. $2^x \times 8^{3x-1} = 64$
    b. $5^{2x} \times 125^{3-x} = 25$
    c. $3^{4x} \times 27^{x+3} = 81$

11. Solve each of the following equations for x.
    a. $16^{x+4} \times 2^{3+2x} = 4^{5x}$
    b. $3125 \times 25^{2x+1} = 5^{3x+4}$
    c. $\dfrac{81^{2-x}}{27^{x+3}} = 9^{2x}$

12. **WE3** Solve each of the following equations for x.
    a. $3^{2x} - 4(3^x) + 3 = 0$
    b. $2^{2x} - 6(2^x) + 8 = 0$
    c. $3(2^{2x}) - 36(2^x) + 96 = 0$

13. Solve each of the following equations for x.
    a. $2(5^{2x}) - 12(5^x) + 10 = 0$
    b. $3(4^{2x}) = 15(4^x) - 12$
    c. $25^x - 30(5^x) + 125 = 0$

14. Use a suitable substitution to solve the following equations.

    a. $3^{2x} - 10 \times 3^x + 9 = 0$

    b. $24 \times 2^{2x} + 61 \times 2^x = 2^3$

15. Solve each of the following for x.

    a. $2^{2x} \times 8^{2-x} \times 16^{-\frac{3x}{2}} = \dfrac{2}{4^x}$

    b. $9^x \div 27^{1-x} = \sqrt{3}$

16. Solve each of the following equations for x.

    a. $25^x + 5^{2+x} - 150 = 0$

    b. $(2^x + 2^{-x})^2 = 4$

17. Solve each of the following equations for x.

    a. $10^x - 10^{2-x} = 99$

    b. $2^{3x} + 3 \times 2^{2x-1} - 2^x = 0$

18. Consider the equation $2 \times 25^x + 9 \times 5^x = 35$. Determine the value of x, correct to 3 decimal places.

Complex unfamiliar

19. Solve the pair of simultaneous equations for x and y.

$$5^{2x-y} = \frac{1}{125}$$
$$10^{2y-6x} = 0.01$$

20. Solve the pair of simultaneous equations for a and k.

$$a \times 2^{k-1} = 40$$
$$a \times 2^{2k-2} = 10$$

Fully worked solutions for this chapter are available online.

LESSON
9.3 Graphs of exponential functions

SYLLABUS LINKS

- Recognise and determine the qualitative features of the graph of $y = r^x$ (where $r > 0$), including asymptote and intercept.
- Recognise and determine the effect of the parameters h, k and r on the graph of $y = r^{(x-h)} + k$ (where $r > 0$), with and without technology.
- Sketch the graphs of exponential functions, with and without technology.

Source: Mathematical Methods Senior Syllabus 2024 © State of Queensland (QCAA) 2024; licensed under CC BY 4.0.

Exponential functions are functions of the form $f : R \to R$, $f(x) = r^x$, $r \in R^+ \backslash \{1\}$. They provide mathematical models of exponential growth and exponential decay situations, such as population increase and radioactive decay, respectively.

9.3.1 The graph of $y = r^x$ where $r > 1$

Before sketching such a graph, consider the table of values for the function with rule $y = 2^x$.

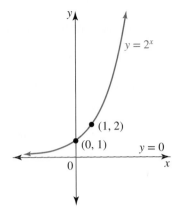

x	-3	-2	-1	0	1	2	3
$y = 2^x$	$\dfrac{1}{8}$	$\dfrac{1}{4}$	$\dfrac{1}{2}$	1	2	4	8

From the table it is evident that $2^x > 0$ for all values of x, and that as $x \to -\infty$, $2^x \to 0$. This means that the graph will have a horizontal asymptote with equation $y = 0$.

It is also evident that as $x \to \infty$, $2^x \to \infty$, with the values increasing rapidly.

Since these observations are true for any function $y = r^x$ where $r > 1$, the graph of $y = 2^x$ will be typical of the basic graph of any exponential with base larger than 1.

The graph of $y = r^x$ where $r > 1$

Key features of the graph of $y = r^x$ where $r > 1$:
- **There is a horizontal asymptote with equation $y = 0$.**
- **The y-intercept is $(0, 1)$.**
- **The shape is of 'exponential growth'.**
- **The domain is R.**
- **The range is R^+.**

For $y = 2^x$, the graph contains the point $(1, 2)$; for the graph of $y = r^x$, $r > 1$, the graph contains the point $(1, r)$, showing that as the base increases, the graph becomes steeper more quickly for values $x > 0$. This is illustrated by the graphs of $y = 2^x$ and $y = 10^x$, with the larger base giving the steeper graph.

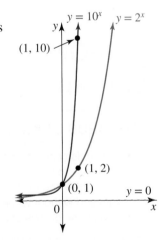

9.3.2 The graph of $y = r^x$ where $0 < r < 1$

An example of a function whose rule is in the form $y = r^x$ where $0 < r < 1$ is $y = \left(\dfrac{1}{2}\right)^x$. Since $\left(\dfrac{1}{2}\right)^x = 2^{-x}$, the rule for the graph of this exponential function $y = \left(\dfrac{1}{2}\right)^x$, where the base lies between 0 and 1, is identical to the rule $y = 2^{-x}$, where the base is greater than 1.

The graph of $y = 2^{-x}$ shown is typical of the graph of $y = r^{-x}$ where $r > 1$ and of the graph of $y = r^x$ where $0 < r < 1$.

Note: The substitution of $-x$ for x always indicates a reflection in the y-axis.

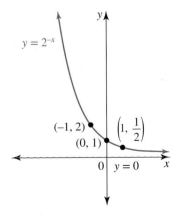

> **The graph of $y = r^x$, where $0 < r < 1$ or $y = r^{-x}$, where $r > 1$**
>
> Key features of the graph with the rule expressed as either $y = r^x$ where $0 < r < 1$ or as $y = r^{-x}$ where $r > 1$:
>
> - There is a horizontal asymptote with equation $y = 0$.
> - The y-intercept is $(0, 1)$.
> - The shape is of 'exponential decay'.
> - The domain is R.
> - The range is R^+.

The basic shape of an exponential function is either one of 'growth' or 'decay'.

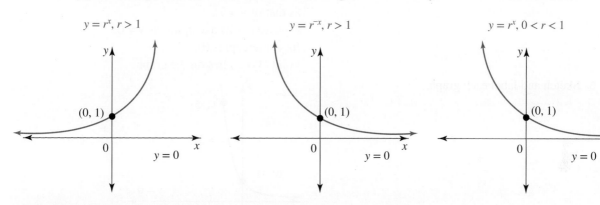

As with other functions, the graph of $y = -r^x$ will be inverted (reflected in the x-axis).

$$y = -r^x \Leftrightarrow y = -(r^x)$$

on Resources

⬦ **Interactivity** Exponential functions (int-5959)

WORKED EXAMPLE 5 Sketching exponential graphs

a. On the same set of axes, sketch the graphs of $y = 5^x$ and $y = -5^x$, identifying their ranges.

b. Give a possible equation for the graph shown.

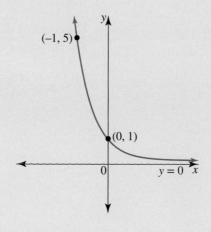

THINK

a. 1. Identify the asymptote of the first function.

2. Find the y-intercept.

3. Calculate the coordinates of a second point.

4. Use the relationship between the two functions to deduce the key features of the second function.

5. Sketch and label each graph.

6. State the range of each graph.

b. 1. Use the shape of the graph to suggest a possible form for the rule.

2. Use a given point on the graph to calculate r.

3. State the equation of the graph.

WRITE

a. $y = 5^x$

The asymptote is the line with equation $y = 0$.

y-intercept: when $x = 0$, $y = 1 \Rightarrow (0, 1)$

Let $x = 1$.
$y = 5^1$
$\quad = 5$
$\Rightarrow (1, 5)$

$y = -5^x$
This is the reflection of $y = 5^x$ in the x-axis.
The graph of $y = -5^x$ has the same asymptote as that of $y = 5^x$.
Equation of its asymptote is $y = 0$.
Its y-intercept is $(0, -1)$.
Point $(1, -5)$ lies on the graph.

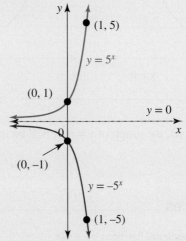

The range of $y = 5^x$ is R^+ and the range of $y = -5^x$ is R^-.

b. The graph has a 'decay' shape. Let the equation be $y = r^{-x}$.

The point $(-1, 5) \Rightarrow 5 = r^1$
$\qquad\qquad \therefore r = 5$

The equation of the graph could be $y = 5^{-x}$.
The equation could also be expressed as
$$y = \left(\frac{1}{5}\right)^x \text{ or } y = 0.2^x.$$

9.3.3 Translations of exponential graphs

Once the basic exponential growth or exponential decay shapes are known, the graphs of exponential functions can be transformed in similar ways to graphs of any other functions previously studied.

Vertical translations

The graph of $y = r^x + k$ where $r > 1$

The key features of the graph of $y = r^x + k$ where $r > 1$ are:
- vertical translation of k units
  - if $k > 0$, translated vertically upwards
  - if $k < 0$, translated vertically downwards
- a horizontal asymptote with equation $y = k$
- y-intercept $(0, 1 + k)$
- x-intercept, if $k < 0$, found by solving the exponential equation $r^x + k = 0$
- domain R
- range (k, ∞)
- 'exponential growth' shape.

Horizontal translations

The graph of $y = r^{(x-h)}$ where $r > 1$

The key features of the graph of $y = r^{(x-h)}$ where $r > 1$ are:
- horizontal translation of h units
  - if $h > 0$, translated horizontally to the right
  - if $h < 0$, translated horizontally to the left
- a horizontal asymptote, unchanged, with equation $y = 0$
- an additional helpful point $(h, 1)$
- domain R
- range $(0, \infty)$ *or* R^+
- 'exponential growth' shape.

Consider the graph $y = 3^x$, shown in the diagram.

A vertical translation of the graph of $y = 3^x$ is illustrated by the graph of $y = 3^x - 1$; under the vertical translation of 1 unit down, the point $(1, 3) \rightarrow (1, 2)$.

A horizontal translation of the graph of $y = 3^x$ is illustrated by the graph of $y = 3^{(x+1)}$; under the horizontal translation of 1 unit to the left, the point $(0, 1) \rightarrow (-1, 1)$.

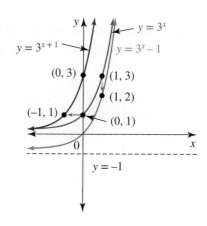

Sketch the graphs of each of the following and identify the range of each.

a. $y = 2^x - 4$

b. $y = 10^{-(x+1)}$

THINK

WRITE

a. 1. State the equation of the asymptote.

a. $y = 2^x - 4$

The vertical translation 4 units down affects the asymptote.

The asymptote has the equation $y = -4$.

2. Calculate the y-intercept.

y-intercept: let $x = 0$.

$y = 1 - 4$

$\quad = -3$

The y-intercept is $(0, -3)$.

3. Calculate the x-intercept.

x-intercept: let $y = 0$.

$2^x - 4 = 0$

$\quad 2^x = 4$

$\quad 2^x = 2^2$

$\quad\quad x = 2$

The x-intercept is $(2, 0)$.

4. Sketch the graph and state the range.

A 'growth' shape is expected since the coefficient of x is positive.

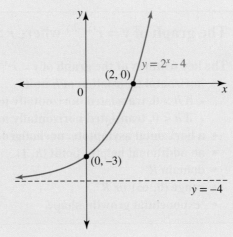

The range is $(-4, \infty)$.

b. 1. Identify the key features from the given equation.

b. $y = 10^{-(x+1)}$

Reflection in the y-axis, horizontal translation 1 unit to the left

The asymptote will not be affected.

Asymptote: $y = 0$

There is no x-intercept.

y-intercept: let $x = 0$.

$y = 10^{-1}$

$\quad = \dfrac{1}{10}$

The y-intercept is $(0, 0.1)$.

2. Calculate the coordinates of a second point on the graph.

Let $x = -1$.
$y = 10^0$
$ = 1$
The point $(-1,\ 1)$ lies on the graph.

3. Sketch the graph and state the range.

A 'decay' shape is expected since the coefficient of x is negative.

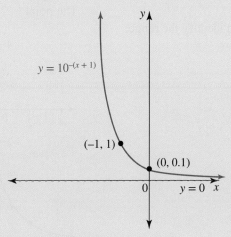

The range is R^+.

Combinations of transformations

Exponential functions with equations of the form $y = r^{(x-h)} + k$ (where $r > 0$) are derived from the basic graph of $y = r^x$ (where $r > 0$) by applying a combination of transformations. The key features to identify in order to sketch the graphs of such exponential functions are:

- the asymptote
- the y-intercept
- the x-intercept, if there is one
- the expected shape, either 'growth' or 'decay'.

Another point that can be obtained simply could provide assurance about the shape. **Always** aim to show at least two points on the graph.

WORKED EXAMPLE 7 Sketching combined transformations of exponential graphs

Sketch the graph of $y = 2^{(x+2)} - 8$ and identify the range.

THINK	WRITE

1. Identify the key features using the given equation:
   asymptote
 	
   y-intercept
 	
   x-intercept
	$y = 2^{(x+2)} - 8$
	Asymptote: $y = -8$ as vertical translation downwards.
	For y-intercept: let $x = 0$.
	$y = 2^2 - 8 = -4$
	y-intercept: $(0, -4)$
	For x-intercept: let $y = 0$.
	$\quad 0 = 2^{(x+2)} - 8$
	$2^{(x+2)} = 8$
	$2^{(x+2)} = 2^3$
	$\quad x + 2 = 3$
	$\quad\quad\quad x = 1$

▶

basic shape.

2. Calculate the coordinates of another point.

3. Sketch the graph and identify the range.

x-intercept: $(1, 0)$

'Growth' shape, translated vertically downwards by 8 units and translated horizontally to the left by 2 units. Since the graph is translated horizontally to the left by 2 units, let $x = -2$:

$y = 2^{(0)} - 8 = 1 - 8 = -7$

The point $(-2, -7)$ lies on the graph.

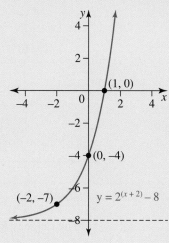

Range: $y \in (-8, \infty)$

TI | THINK	WRITE	CASIO | THINK	WRITE
1. On a Graphs screen, enter the equation as: $f1\,(x) = 2^{(x+2)} - 8$ then press ENTER.		1. On a Graph screen, enter the equation as: Y1: $2^{(x+2)} - 8$ Press EXE then press F6 to Draw. *Note:* You can use the V-Window function by pressing F3 to zoom out.	
2. To find the y-intercept, press MENU, then select: 5: Trace 1: Graph Trace Move to the y-axis and press ENTER.		2. To find the y-intercept, press G-Solve F5 then F4 for Y-ICEPT, then press EXE.	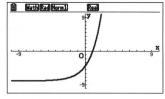
3. To find the y-intercept, press MENU, then select: 6: Analyse Graph 1: Zero Select the lower bound and higher bound on both sides of the x-intercept, and press ENTER.		3. To find the x-intercept, press G-Solve F5 then F1 for ROOT, then press EXE.	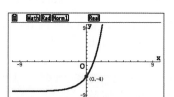
4. The answer appears on the screen.	The exponential graph approaches an asymptote of -8. Therefore, the range is $y \in (-8, \infty)$.	4. The answer appears on the screen.	The exponential graph approaches an asymptote of -8. Therefore, the range is $y \in (-8, \infty)$.

9.3 Exercise	9.3 Exam practice **on**	These questions are even better in jacPLUS!

Simple familiar	**Complex familiar**	**Complex unfamiliar**
1, 2, 3, 4, 5, 6, 7, 8, 9, 10, 11, 12	13, 14, 15, 16	17, 18, 19, 20

These questions are even better in jacPLUS!
- Receive immediate feedback
- Access sample responses
- Track results and progress

Find all this and MORE in jacPLUS ⊙

Simple familiar

1. **WE5** a. On the same set of axes, sketch the graphs of $y = 3^x$ and $y = -3^x$, identifying their ranges.

   b. Determine a possible equation for the graph shown.

2. a. Sketch, on the same set of axes, the graphs of $y = 4^x$, $y = 6^x$ and $y = 8^x$.

   b. Describe the effect produced by increasing the base.

3. a. Sketch, on the same set of axes, the graphs of $y = \left(\dfrac{1}{4}\right)^x$,

   $y = \left(\dfrac{1}{6}\right)^x$ and $y = \left(\dfrac{1}{8}\right)^x$.

   b. Express each rule in a form that uses negative indices.

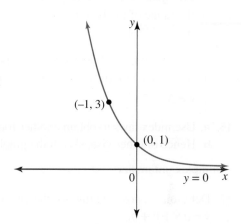

(−1, 3)

(0, 1)

0   y = 0   x

4. a. Sketch, on the same set of axes, the graphs of $y = 5^{-x}$, $y = 7^{-x}$ and $y = 9^{-x}$.

   b. Describe the effect produced by increasing the base.

5. a. Sketch, on the same set of axes, the graphs of $y = (0.8)^x$, $y = (1.25)^x$ and $y = (0.8)^{-x}$.

   b. Describe the relationships between the three graphs.

6. **MC** The rule for the graph shown is:

   A. $y = 3^{x-2}$ **B.** $y = 3^x$

   C. $y = 2^{x-3}$ **D.** $y = 3^{x+2}$

7. **MC** State which one of the following is *not* a key feature of the graph $y = 2^x$ or any such function $y = r^x$ where $r > 1$.

   A. Horizontal asymptote with equation $y = 0$

   B. y-intercept $(0, 1)$

   C. Domain R

   D. Range R

8. **MC** A possible equation for the graph shown is:

   A. $y = 10^x$ **B.** $y = 10^{-x}$

   C. $y = 2^x$ **D.** $y = -2^x$

9. **WE6** Sketch the graphs of each of the following and state the range of each.

   a. $y = 4^x - 2$ b. $y = 3^{-(x+2)}$

10. Sketch each of the following graphs, showing the asymptote and labelling any intersections with the coordinate axes with their exact coordinates.

   a. $y = 5^{-x} + 1$ b. $y = 1 - 4^x$

   c. $y = 3^x - 27$ d. $y = 6.25 - (2.5)^{-x}$

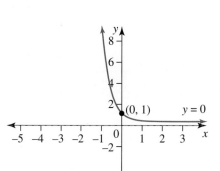

(0, 1)   y = 0

11. Sketch the graphs of each of the following.

    a. $y = 2^{x-2}$ **b.** $y = -3^{x+2}$ **c.** $y = 4^{x-0.5}$ **d.** $y = 7^{1-x}$

12. Sketch the graphs of $y = (1.5)^x$ and $y = \left(\dfrac{2}{3}\right)^x$ on the same set of axes.

Complex familiar

13. **WE7** Sketch the graph of $y = 4^{x-2} + 1$ and state its range.

14. For each of the following exponential functions, state:

    i. the equation of its asymptote   **ii.** the coordinates of its y-intercept
    iii. the range of the function.

      a. $y = 2^{-x} + 3$ **b.** $y = 2 - 5^{1-x}$

15. Sketch the graphs of the following exponential functions and identify their ranges. Where appropriate, any intersections with the coordinate axes should be given to 1 decimal place.

    a. $y = 3^x - 2$ **b.** $y = 5 - 2^{3-x}$

16. **a.** Use index laws to obtain another form of the rule for $y = 2 \times 4^{0.5x}$.
    b. Hence or otherwise, sketch the graph of $y = 2 \times 4^{0.5x}$.

Complex unfamiliar

17. Determine a possible rule for the given graph in the form $y = a \times 10^x + b$.

18. The graph of an exponential function of the form $y = a \times 3^{kx}$ contains the points $(1, 36)$ and $(0, 4)$. Determine its rule and state the equation of its asymptote.

19. Use a graphical means to determine the number of intersections between:

    a. $y = 2^x$ and $y = -x$, specifying an interval in which the x-coordinate of any point of intersection lies
    b. $y = 2^x$ and $y = x^2$
    c. $y = 2^{2x-1}$ and $y = \dfrac{1}{2} \times 16^{\frac{x}{2}}$, giving the coordinates of any points of intersection.

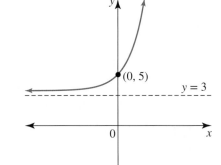

20. The graph shown has the equation $y = a \times 3^x + b$.

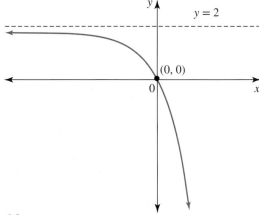

    Determine the values of a and b.

Fully worked solutions for this chapter are available online.

LESSON
9.4 Modelling with exponential functions

SYLLABUS LINKS

- Model and solve problems that involve exponential functions, with and without technology.

Source: Mathematical Methods Senior Syllabus 2024 © State of Queensland (QCAA) 2024; licensed under CC BY 4.0.

The importance of exponential functions lies in the frequency with which they occur in models of phenomena involving growth and decay situations, in chemical and physical laws of nature, and in higher-level mathematical analysis.

9.4.1 Exponential growth and decay models

For time t, the exponential function defined by $y = b \times a^{nt}$, where $a > 1$, represents exponential growth over time if $n > 0$ and exponential decay over time if $n < 0$. The domain of this function would be restricted according to the way the independent time variable t is defined. The rule $y = b \times a^{nt}$ may also be written as $y = b \cdot a^{nt}$.

In some mathematical models such as population growth, the initial population may be represented by a symbol such as N_0. For an exponential decay model, the time it takes for 50% of the initial amount of the substance to decay is called its half-life.

WORKED EXAMPLE 8 Evaluating exponential decay models

The decay of a radioactive substance is modelled by $Q(t) = Q_0 \times 2.7^{-kt}$, where Q kg is the amount of the substance present at time t years and Q_0 and k are positive constants.
a. Demonstrate that the constant Q_0 represents the initial amount of the substance.
b. If the half-life of the radioactive substance is 100 years, calculate k to 1 significant figure.
c. If initially there was 25 kilograms of the radioactive substance, determine how many kilograms would decay in 10 years. Give your answer to 2 decimal places. Use the value of k from part b in your calculations.

THINK	WRITE
a. 1. Calculate the initial amount.	a. $Q(t) = Q_0 \times 2.7^{-kt}$ The initial amount is the value of Q when $t = 0$. Let $t = 0$: $Q(0) = Q_0 \times 2.7^0$ $\qquad\qquad = Q_0$ Therefore, Q_0 represents the initial amount of the substance.
b. 1. Form an equation in k from the given information. *Note:* It does not matter that the value of Q_0 is unknown, since the Q_0 terms cancel.	b. The half-life is the time it takes for 50% of the initial amount of the substance to decay. Since the half-life is 100 years, when $t = 100$, $Q(100) = 50\%$ of Q_0. $Q(100) = 0.50 Q_0$ ..... (1) From the equation, $Q(t) = Q_0 \times 2.7^{-kt}$. When $t = 100$, $Q(100) = Q_0 \times 2.7^{-k(100)}$. $Q(100) = Q_0 \times 2.7^{-100k}$ .... (2) Equate equations (1) and (2): $0.50 Q_0 = Q_0 \times 2.7^{-100k}$ Cancel Q_0 from each side: $\quad 0.50 = 2.7^{-100k}$

2. Solve the exponential equation to obtain k to the required accuracy using technology.

$k \approx 0.007$

c. 1. Use the values of the constants to state the actual rule for the exponential decay model.

c. $Q_0 = 25$, $k = 0.007$
$\therefore\ Q(t) = 25 \times 2.7^{-0.007t}$

2. Calculate the amount of the substance present at the time given.

When $t = 10$,
$Q(10) = 25 \times 2.7^{-0.07}$
≈ 23.32

3. Calculate the amount that has decayed.
Note: Using a greater accuracy for the value of k would give a slightly different answer for the amount decayed.

Since $25 - 23.32 = 1.68$, in 10 years approximately 1.68 kg will have decayed.

TI	THINK	WRITE	CASIO	THINK	WRITE

b. 1. Set up an equation to be solved.

When $t = 100$, $Q = 0.5Q_0$. Substituting these into the equation $Q(t) = Q_0 \times 2.7^{-kt}$ gives $0.5Q_0 = Q_0 \times 2.7^{-100k}$. Dividing both sides by Q_0 gives $0.5 = 2.7^{-100k}$.

b. 1. Set up an equation to be solved.

When $t = 100$, $Q = 0.5Q_0$. Substituting these into the equation $Q(t) = Q_0 \times 2.7^{-kt}$ gives $0.5Q_0 = Q_0 \times 2.7^{-100k}$. Dividing both sides by Q_0 gives $0.5 = 2.7^{-100k}$.

2. On a Calculator page, press MENU, then select:
3: Algebra
1: Numerical Solve
Complete the entry line as:
nSolve($0.5 = 2.7^{-100k}, k$)
then press ENTER.

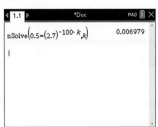

On an Equation screen, select Solver by pressing F3. Complete the entry line for the equation as:
$0.5 = 2.7^{-100k}$
then press EXE. Select SOLVE by pressing F6.

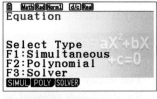

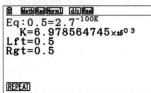

3. The answer appears on the screen.

$k = 0.007$
(1 significant figure)

3. The answer appears on the screen.

$k = 0.007$ (1 significant figure)

c. 1. On a Calculator page, press MENU, then select
1: Actions
1: Define
then complete the entry line as:
Define $q(t) = 25 \times 2.7^{-0.007t}$
then press ENTER.

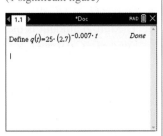

c. 1. On a Run-Matrix screen, complete the next entry line as:
$25 \times 2.7^{-0.007 \times 10}$
then press EXE. Interpret the result.

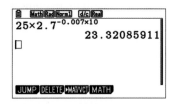

When $t = 10$, $Q \approx 23.32$.

2. Complete the next entry line as:
$q(10)$
then press ENTER. Interpret the result.

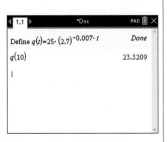

When $t = 10$, $Q \approx 23.32$.

2. Complete the next entry line as:
25−Ans
then press EXE.

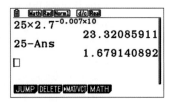

3. Complete the next entry line as: 25-Ans then press ENTER.	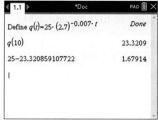	3. The answer appears on the screen.	*Approximately 1.68 kg will have decayed in 10 years.*

4. The answer appears on the screen. *Approximately 1.68 kg will have decayed in 10 years.*

Exercise 9.4 Modelling with exponential functions

learn on

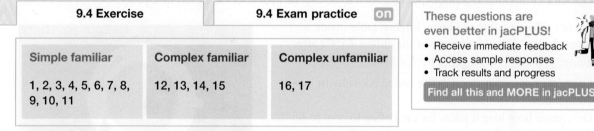

9.4 Exercise	9.4 Exam practice on

Simple familiar	Complex familiar	Complex unfamiliar
1, 2, 3, 4, 5, 6, 7, 8, 9, 10, 11	12, 13, 14, 15	16, 17

These questions are even better in jacPLUS!
- Receive immediate feedback
- Access sample responses
- Track results and progress

Find all this and MORE in jacPLUS ▶

Simple familiar

1. **WE8** The decay of a radioactive substance is modelled by $Q(t) = Q_0 \times 1.7^{-kt}$, where Q is the amount of the substance present at time t years and Q_0 and k are positive constants.

    a. Show that the constant Q_0 represents the initial amount of the substance.
    b. If the half-life of the radioactive substance is 300 years, calculate k to 1 significant figure.
    c. If initially there was 250 kilograms of the radioactive substance, determine how many kilograms would decay in 10 years. Use the value of k from part **b** in your calculations.

2. The manager of a small business is concerned about the amount of time she spends dealing with the growing number of emails she receives. The manager starts keeping records and finds the average number of emails received per day can be modelled by $D = 42 \times 2^{\frac{t}{16}}$, where D is the average number of emails received per day t weeks from the start of the records.

    a. Determine how many emails per day, on average, the manager was receiving when she commenced her records.
    b. Determine how many weeks the model predicts it will take for the average number of emails received per day to double.

3. The value V of a new car depreciates so that its value after t years is given by $V = V_0 \times 2^{-kt}$.

    a. If 50% of the purchase value is lost in 5 years, calculate k.
    b. Determine how long it will take for the car to lose 75% of its purchase value.

4. The number of *Drosophila* (fruit flies), N, in a colony after t days of observation is modelled by $N = 30 \times 2^{0.072t}$. Give whole number answers to the following.

    a. Determine how many *Drosophila* were present when the colony was initially observed.
    b. Determine how many of the insects were present after 5 days.
    c. Determine how many days it takes the population number to double from its initial value. Round your answer to the nearest full day.

d. Sketch a graph of N versus t to show how the population changes.

e. Determine how many days it will take for the population to first exceed 100.

5. The value of an investment that earns compound interest can be calculated from the formula

$A = P \left(1 + \dfrac{r}{n}\right)^{nt}$, where P is the initial investment, r the interest rate per annum (yearly), n the number of times per year the interest is compounded and t the number of years of the investment.

An investor deposits \$2000 in an account where interest is compounded monthly.

a. If the interest rate is 3% per annum:

   i. demonstrate that the formula giving the value of the investment is $A = 2000\,(1.0025)^{12t}$
   ii. calculate how much the investment is worth after a 6-month period
   iii. determine what time period would be needed for the value of the investment to reach \$2500.

b. The investor would like the \$2000 to grow to \$2500 in a shorter time period. Determine what the interest rate, still compounded monthly, would need to be for the goal to be achieved in 4 years.

6. A cup of coffee is left to cool on a kitchen table inside a Brisbane home. The temperature of the coffee T (°C) after t minutes is thought to be given by $T = 85 \times 3^{-0.008t}$.

a. Calculate how many degrees the coffee cools by in 10 minutes.

b. Determine how long it takes for the coffee to cool to 65 °C.

c. Sketch a graph of the temperature of the coffee for $t \in [0, 40]$.

d. By considering the temperature the model predicts the coffee will eventually cool to, explain why the model is not realistic in the long term.

7. The contents of a meat pie immediately after being heated in a microwave have a temperature of 95 °C. The pie is removed from the microwave and left to cool. A model for the temperature of the pie as it cools is given by $T = a \times 3^{-0.13t} + 25$, where T is the temperature after t minutes of cooling.

a. Calculate the value of a.

b. Calculate the temperature of the contents of the pie after being left to cool for 2 minutes.

c. Determine how long, to the nearest minute, it will take for the contents of the meat pie to cool to 65 °C.

d. Sketch the graph showing the temperature over time and state the temperature to which this model predicts the contents of the pie will eventually cool if left unattended.

8. The barometric pressure P, measured in kilopascals, at height h above sea level, measured in kilometres, is given by $P = P_0 \times 10^{-kh}$, where P_0 and k are positive constants. The pressure at the top of Mount Everest is approximately one-third that of the pressure at sea level.

a. Given the height of Mount Everest is approximately 8848 metres, calculate the value of k to 2 significant figures.
   Use the value obtained for k for the remainder of this question.

b. Mount Kilimanjaro has a height of approximately 5895 metres. If the atmospheric pressure at its summit is approximately 48.68 kilopascals, calculate the value of P_0 to 3 decimal places.

c. Use the model to determine the atmospheric pressure to 2 decimal places at the summit of Mont Blanc, 4810 metres in height, and of Mount Kosciuszko, 2228 metres in height.

d. Construct a graph of atmospheric pressure against height, showing the readings for the four mountains from the above information. Give your answer to 2 decimal places.

9. The common Indian mynah bird was introduced into Australia in order to control insects affecting market gardens in Melbourne. It is now considered to be Australia's most important pest problem. In 1976, the species was introduced to an urban area in New South Wales. By 1991 the area averaged 15 birds per square kilometer, and by 1994 the density reached an average of 75 birds per square kilometre.

A model for the increasing density of the mynah bird population is thought to be $D = D_0 \times 10^{kt}$, where D is the average density of the bird per square kilometre t years after 1976 and D_0 and k are constants.

a. Use the given information to construct a pair of simultaneous equations in D and t.

b. Determine that $k \approx 0.233$ and $D_0 \approx 0.005$.

c. A project was introduced in 1996 to curb the growth in numbers of these birds. Determine what the model predicts was the average density of the mynah bird population at the time the project was introduced in the year 1996. Use $k \approx 0.233$ and $D_0 \approx 0.005$ and round the answer to the nearest whole number.

d. Sometime after the project was successfully implemented, a different model for the average density of the bird population became applicable. This model is given by $D = 30 \times 10^{-\frac{t}{3}} + b$. Four years later, the average density was reduced to 40 birds per square kilometre. Determine by how much we can expect the average density to be reduced,

10. Carbon dating enables estimates of the age of fossils of once-living organisms to be ascertained by comparing the amount of the radioactive isotope carbon-14 remaining in the fossil with the normal amount present in the living entity, which can be assumed to remain constant during the organism's life. It is known that carbon-14 decays with a half-life of approximately 5730 years according to an exponential model of the form $C = C_0 \times \left(\frac{1}{2}\right)^{kt}$, where C is the amount of the isotope remaining in the fossil t years after death and C_0 is the normal amount of the isotope that would have been present when the organism was alive.

a. Calculate the exact value of the positive constant k.

b. The bones of an animal are unearthed during digging explorations by a mining company. The bones are found to contain 83% of the normal amount of the isotope carbon-14. Determine how old the bones are.

11. The data shown in the table gives the population of Australia, in millions, in years since 1960.

	1975	1990
t (years since 1960)	15	30
p (population in millions)	13.9	17.1

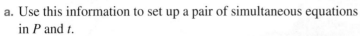

A model for increasing population is thought to be $P = P_0 \times 10^{kt}$, where P is the population (in millions) t years after 1960 and P_0 and k are constant.

a. Use this information to set up a pair of simultaneous equations in P and t.
b. Show that $k \approx 0.006$ and $P_0 \approx 10.3$.
c. Calculate how many years it took for the population to reach double the 1960 population.
d. It is said that the population of Australia is likely to exceed 28 million by the year 2030. Explain whether this model supports this claim.

Complex familiar

12. Experimental data yielded the following table of values.

x	1	1.5	2	2.5	3	3.5	4
y	5.519	6.483	7.615	8.994	10.506	12.341	14.496

Use technology to determine the exponential rule that best models the data.

13. The weight, W grams, of bacteria in a culture t hours after it was first observed is given by the formula $W = W_0\, 3^{\frac{t}{10}}$, where W_0 is a positive constant.

Determine the time taken for the weight of the bacteria to double. Give your answer correct to the nearest minute.

14. The number of bacteria present in a culture at any time, t hours, can be modelled by the equation $N = N_0 \times 2.7^{kt}$, where N_0 and k are positive constants. Since the original number doubled in 3 hours, determine the original number if there were 2500 bacteria after 4 hours. Express your answer to the nearest 10.

15. Following a fall from his bike, Stephan is feeling some shock but not, initially, a great deal of pain. However, his doctor gives him an injection for relief from the pain that he will start to feel once the shock of the accident wears off. The amount of pain Stephan feels over the next 10 minutes is modelled by the function $P(t) = (200t + 16) \times 2.7^{-t}$, where P is the measure of pain on a scale from 0 to 100 that Stephan feels t minutes after receiving the injection.

a. Determine the measure of pain Stephan is feeling:
   i. at the time the injection is administered
   ii. 15 seconds later when his shock is wearing off but the injection has not reached its full effect.

b. Determine, correct to 2 decimal places:
   i. the maximum measure of pain he feels
   ii. the number of seconds it takes for the injection to start lowering his pain level
   iii. his pain levels after 5 minutes and after 10 minutes have elapsed.

c. Over the 10-minute interval, determine when the effectiveness of the injection was greatest.
d. At the end of the 10 minutes, Stephan receives a second injection modelled by $P(t) = (100(t - 10) + a) \times 2.7^{-(t-10)}$, $10 \leq t \leq 20$.
   i. Determine the value of a.
   ii. Sketch the pain measure over the time interval $t \in [0, 20]$ and label the maximum points with their coordinates.

16. The diameter of a tree for a period of its growth can be modelled by the equation $D = D_0 \times 2.5^{kt}$, where D_0 and k are positive constants.

The diameter of the tree grew from 50 cm to 60 cm in the first 2 years that measurements were taken. How long will it take for the tree's diameter to double. Give your answer correct to the nearest year.

17. The decay of a radioactive substance can be modelled by the equation $M = M_0 \times 2^{-kt}$, where M_0 and k are positive constants. After 10 years the mass of the substance is 98 grams and after 20 years the mass is 96 grams.

Determine the mass of the substance after 50 years.
Give your answer correct to the nearest gram.

Fully worked solutions for this chapter are available online.

LESSON
9.5 Review

📄 9.5.1 Summary

doc-
41804

9.5 Exercise

 learn on

9.5 Exercise	9.5 Exam practice on

Simple familiar	Complex familiar	Complex unfamiliar
1, 2, 3, 4, 5, 6, 7, 8, 9, 10, 11, 12	13, 14, 15, 16	17, 18, 19, 20

Simple familiar

1. **MC** The solution set of the equation $(2^x - 1)(2^{2x} - 4) = 0$ is:

   A. $\{0, 1\}$.   **B.** $\{0, 2\}$.   **C.** $\{1, 2\}$.   **D.** $\{1, 4\}$.

2. **MC** The transformations required to change $y = 3^x$ into $y = 3^{x+4} + 5$ are:

   A. horizontally right by 4, vertically upwards by 5
   B. horizontally right by 4, vertically downwards by 5
   C. horizontally left by 4, vertically upwards by 5
   D. horizontally left by 4, vertically downwards by 5

3. **MC** Identify which of the following graphs best represents the function $y = 2^{x+3} - 1$.

   A.
   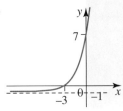

   B.
   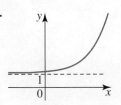

   C.
   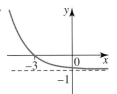

   D.
   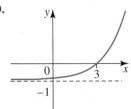

4. **MC** If $27^{3-2x} = 729^x$, then x is equal to:

   A. 6   **B.** $\dfrac{3}{4}$ **C.** $\dfrac{1}{3}$ **D.** 1

5. **MC** The decay of a radioactive substance is modelled by the equation $Q = 15 \times 5^{-0.0025t}$, where Q kg is the amount of the substance present at time t years. The initial amount of the substance was:

   A. 5   B. 15   C. 75   D. 0.0025

6. **MC** If $25^{2-x} = 125$, then x is equal to:

   A. 1   B. $\dfrac{1}{2}$ C. -1 D. 2

7. Solve the following equations.

   a. $2x^5 = 486$

   b. $8^{x+1} \times 2^{2x} = 4^{3x-1}$

8. Sketch the graph of $y = 3^{x-4} + 1$, identifying the equation of the asymptote, the coordinates of any points of intersection with the axes, and the domain and range.

9. Sketch the graph of $y = 2^{x+4} - 3$, identifying the equation of the asymptote, the coordinates of any points of intersection with the axes, and the domain and range.
   If necessary, express points to 1 decimal place.

10. Sue calculates that a cup of tea cools at a rate of $T = 90 \times 3^{-0.01t}$, where T (°C) is the temperature of the tea after t minutes. Determine the temperature of the tea, correct to the nearest degree, if it is left for 30 minutes.

11. A number of deer, N, are introduced to a reserve and its population can be predicted by the model

   $$N = 120\,(1.1^t),$$

   where t is the number of years since introduction.

   a. Determine the initial number of deer in the reserve.
   b. Determine the number of deer after:

       i. 2 years   ii. 4 years   iii. 6 years.

12. Before a mouse plague that lasts 6 months, the population of mice in a country region was estimated to be 10 000. The mice population doubles every month during the plague. If P represents the mice population and t is the number of months after the plague starts:

   a. express P as a function of t
   b. determine the population after:

       i. 3 months   ii. 6 months
   c. calculate how long it takes the population to reach 100 000 during the plague.

Complex familiar

13. Solve the following equations for x.

   a. $3^{1-7x} = 81^{x-2} \times 9^{2x}$

   b. $2^{2x} - 6 \times 2^x - 16 = 0$

14. The number of bacteria (N) in a culture is given by the exponential function $N = 12\,000\,(2^{0.125t})$, where t is the number of days.

   a. Determine the initial number of bacteria in the culture.
   b. Determine the time taken for the bacteria to reach 32 000. Give your answer correct to 2 decimal places.

When the bacteria reach a certain number, they are treated with an anti-bacterial serum. The serum destroys bacteria according to the exponential function

$$D = N_0 \times 3^{-0.789t},$$

where D is the number of bacteria remaining after time t and N_0 is the number of bacteria present at the time the serum is added. The culture is considered cured when the number of bacteria drops below 1000.

c. If the bacteria are treated with the serum when their numbers reach 32 000, determine the number of days it takes for the culture to be classed as cured.

d. Determine how much longer it would take for the culture to be cured if the serum is applied after 6 weeks.

15. The number of lions, L, in a wildlife park is given by

$$L = 20 \left(10^{0.1t} \right),$$

where t is the number of years since counting started. At the same time the number of cheetahs, C, is given by

$$C = 25 \left(10^{0.05t} \right).$$

Determine after how many months the populations will be equal, and calculate this population value.

16. It is thought that a new car depreciates in value by approximately 15% per year, giving the formula

$$V = V_0 (0.85)^t,$$

where V is the value of the car after time t and V_0 is a positive constant. A new car was bought for $45 000 in 2020 and will be traded for another new car when its value falls below $20 000. Determine, using two different methods, when the car will be traded for another new car.

Complex unfamiliar

17. A cup of soup cools to the temperature of the surrounding air. Newton's Law of Cooling can be written as

$$T - T_s = (T_0 - T_s) \times 3^{-kt},$$

where T is the temperature of the object after t minutes and T_s is the temperature of the surrounding air. If the soup cooled from 90 °C to 70 °C after 6 minutes in a room with an air temperature of 15 °C, determine how long it will take for the soup to be 40 °C. Give your answer to the nearest minute.

18. The graph of the function defined by $f : R \to R$, $f(x) = 2^{x+h} + k$ is shown.

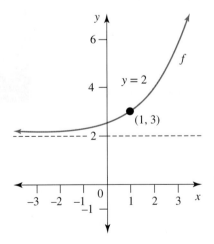

The graph has an asymptote at $y = 2$ and contains the point (1, 3).
The point A lies on the curve where $f(x) = 6$.
Determine the distance from the point A to the y-intercept of $y = f(x)$.

19. A school in the suburb of Bienvenue opened with 30 students in February 2005. It has been found for the first years after opening that the number of students, B, enrolled in the school t years after opening can be modelled by the equation

$$B = B_0 \times 3^{kt},$$

where B_0 and k are positive constants.

Another school in the suburb of En Baisse has a declining student population, with 1000 enrolled in February 2005. The number of students, E, enrolled at any time, t years, can be modelled by the equation

$$E = E_0 \times 3^{-pt},$$

where E_0 and p are positive constants.

In 2006, Bienvenue had 45 students enrolled and En Baisse had 900 students enrolled. Determine how many years it will take for the two schools to have approximately the same number of students.

20. Polly fills a kettle with water, planning to make a pot of tea. She switches the kettle on and the water heats to its boiling point of 100 °C at 10 am, when the kettle automatically switches off. However, Polly is distracted by reading her email and forgets she has put the kettle on. The water in the kettle begins to cool in such a way that the temperature, T °C, can be modelled by

$$T = a \times \left(\frac{16}{5} \right)^{-kt},$$

where t is the number of minutes since 10 am and a and k are constants.

a. The temperature of the water in the kettle was 75 °C at 10:12 am. Determine the temperature, to the nearest degree, of the water in the kettle at 10:30 am.

b. At 10.30 am Polly switches the kettle back on and the water is reheated. If the temperature of the water t minutes after 10:30 am is described by $T = 50 \times 2^{\frac{t}{9}}$, determine the exact time at which the water will re-reach its boiling point.

Fully worked solutions for this chapter are available online.

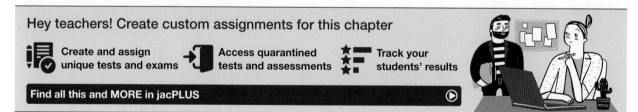

Hey teachers! Create custom assignments for this chapter

Create and assign unique tests and exams

Access quarantined tests and assessments

Track your students' results

Find all this and MORE in jacPLUS

Answers

Chapter 9 Introduction to exponential functions

9.2 Solving exponential equations

9.2 Exercise

1. a. 5   b. 4   c. 5   d. 2

2. a. -2 b. -3 c. 4   d. 0

3. a. $\dfrac{5}{3}$ b. $-\dfrac{1}{2}$ c. -1

4. a. $-\dfrac{9}{4}$ b. $\dfrac{11}{6}$ c. $\dfrac{9}{8}$

5. a. -3 b. $\dfrac{10}{7}$ c. $-\dfrac{18}{5}$ d. $\dfrac{3}{4}$

6. a. $\dfrac{21}{11}$ b. $\dfrac{4}{11}$ c. $-\dfrac{5}{2}$ d. $\dfrac{25}{9}$

7. a. 1.893   b. 1.792   c. 1.431   d. 0.841

8. a. -1.08 b. 3.24   c. 1.07   d. 1.82

9. 8

10. a. $\dfrac{9}{10}$ b. 7   c. $\dfrac{-5}{7}$

11. a. $\dfrac{19}{4}$ b. -3 c. $\dfrac{-1}{11}$

12. a. $x = 0$ or $x = 1$ b. $x = 2$ or $x = 1$ c. $x = 3$ or $x = 2$

13. a. $x = 0$ or $x = 1$ b. $x = 0$ or $x = 1$ c. $x = 2$ or $x = 1$

14. a. $x = 0$ or $x = 2$ b. $x = -3$

15. a. $x = 1$ b. $\dfrac{7}{10}$

16. a. $x = 1$ b. $x = 0$

17. a. $x = 2$ b. $x = -1$

18. $x = 0.569$

19. $x = 4, y = 11$

20. $a = 160, k = -1$

9.3 Graphs of exponential functions

9.3 Exercise

1. a.

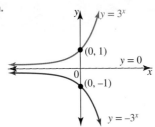

For $y = 3^x$, the range is R^+.

b. $y = 3^{-x}$

For $y = -3^x$, the range is R^-. The asymptote is $y = 0$ for both graphs.

2. a.

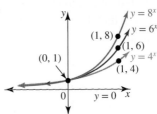

b. For $x > 0$, as the base increases, the steepness of the graph increases.

3. a.

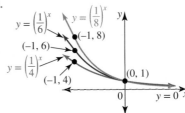

b. The rules for the graphs can be expressed as $y = 4^{-x}$, $y = 6^{-x}$ and $y = 8^{-x}$.

4. a.

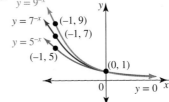

b. As the base increases, the decrease of the graph is steeper for $x < 0$.

5. a.

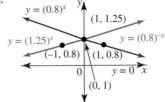

b. The graphs of $y = (0.8)^{-x}$ and $y = (1.25)^x$ are the same. The graph of $y = (0.8)^{-x}$ is the reflection in the y-axis of the graph $y = (0.8)^x$.

6. A

7. D

8. B

9. a.

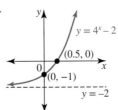

The range is $(-2, \infty)$.

b.

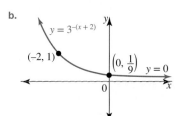

$y = 3^{-(x+2)}$
$(-2, 1)$
$\left(0, \dfrac{1}{9}\right)$
$y = 0$

The range is R^+.

10. a.

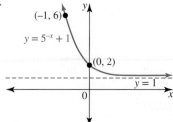

$(-1, 6)$
$y = 5^{-x} + 1$
$(0, 2)$
$y = 1$

b.

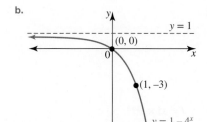

$y = 1$
$(0, 0)$
$(1, -3)$
$y = 1 - 4^x$

c.

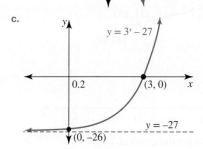

$y = 3^x - 27$
0.2
$(3, 0)$
$y = -27$
$(0, -26)$

d.

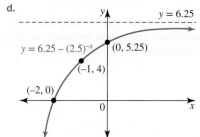

$y = 6.25$
$y = 6.25 - (2.5)^{-x}$
$(0, 5.25)$
$(-1, 4)$
$(-2, 0)$

11. a.

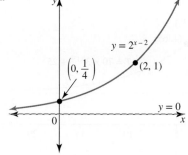

$y = 2^{x-2}$
$\left(0, \dfrac{1}{4}\right)$
$(2, 1)$
$y = 0$

b.

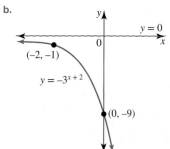

$y = 0$
$(-2, -1)$
$y = -3^{x+2}$
$(0, -9)$

c.

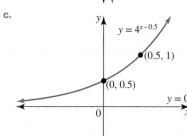

$y = 4^{x-0.5}$
$(0.5, 1)$
$(0, 0.5)$
$y = 0$

d.

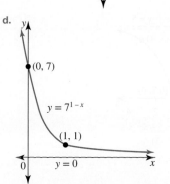

$(0, 7)$
$y = 7^{1-x}$
$(1, 1)$
$y = 0$

12.

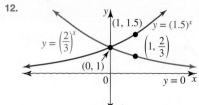

$(1, 1.5)$ $y = (1.5)^x$
$y = \left(\dfrac{2}{3}\right)^x$ $\left(1, \dfrac{2}{3}\right)$
$(0, 1)$
$y = 0$

13.

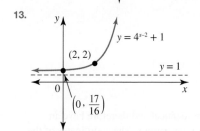

$y = 4^{x-2} + 1$
$(2, 2)$
$y = 1$
$\left(0, \dfrac{17}{16}\right)$

The range is $(1, \infty)$.

14. a. i. $y = 3$ **ii.** $(0, 4)$ **iii.** $(3, \infty)$
  b. i. $y = 2$ **ii.** $(0, -3)$ **iii.** $(-\infty, 2)$

15. a.

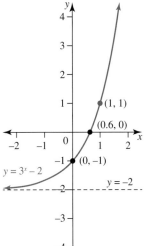

$y = 3^x - 2$

Points: (1, 1), (0.6, 0), (0, −1); asymptote $y = -2$

b.

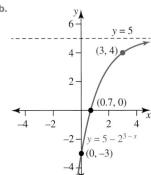

$y = 5 - 2^{3-x}$

Points: (3, 4), (0.7, 0), (0, −3); asymptote $y = 5$

16. a. $y = 2^{x+1}$

b.

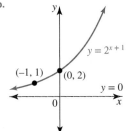

$y = 2^{x+1}$

Points: (−1, 1), (0, 2); asymptote $y = 0$

17. $y = 2 \times 10^x + 3$

18. $y = 4 \times 3^{2x}$; asymptote at $y = 0$

19. a. One

   b. Three

   c. The two curves are identical and therefore have an infinite number of intersections. The coordinates of the points of intersection are of the form $(t, 2^{2t-1})$, $t \in R$.

20. $y = -2 \times 3^x + 2$ and $a = -2$, $b = 2$

9.4 Modelling with exponential functions

9.4 Exercise

1. a. $Q(0) = Q_0$ **b.** $k = 0.004$
   c. 5.3 kg

2. a. 42 emails per day   **b.** 16 weeks

3. a. $k = 0.2$ **b.** 10 years

4. a. 30

   b. Approximately 39

   c. Approximately 14 days

   d.

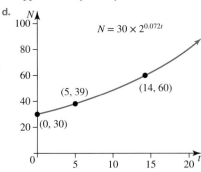

$N = 30 \times 2^{0.072t}$

Points: (0, 30), (5, 39), (14, 60)

   e. 25 days

5. a. i. $A = 2000 (1.0025)^{12t}$

     ii. \$2030.19

    iii. 7.45 years

   b. 5.6 %

6. a. 7 degrees

   b. 30.5 minutes

   c.

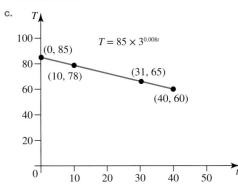

$T = 85 \times 3^{0.008t}$

Points: (0, 85), (10, 78), (31, 65), (40, 60)

   d. The asymptote for the graph of $T = 8.5 \times 3^{-0.008t}$ is $T = 0$.
   This model therefore predicts that the temperature will approach zero degrees. This makes the model unrealistic, particularly in Brisbane!

7. a. $a = 70$

   b. 77.6 degrees

   c. 4 minutes

   d. 25 degrees

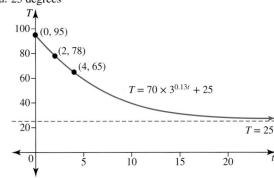

$T = 70 \times 3^{0.13t} + 25$

Points: (0, 95), (2, 78), (4, 65); asymptote $T = 25$

8. a. $k = 0.054$

   b. 101.317 kPa

c. 55.71 kPa; 76.80 kPa

d.

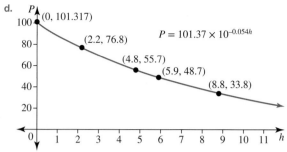

9. a. $15 = D_0 \times 10^{15t}$ and $75 = D_0 \times 10^{18t}$

b. Correct to 3 decimal places, $k = \dfrac{1}{3} \log(5) = 0.233$ and $D_0 = 3 \times 5^{-4} = 0.005$.

c. 229 birds per square kilometre

d. Unlikely to be reduced below 39 birds per square kilometre

10. a. $k = \dfrac{1}{5730}$

b. 1540 years old

11. a. Sample responses can be found in the worked solutions in the online resources.

b. Sample responses can be found in the worked solutions in the online resources.

c. Approximately 50 years

d. The model supports the claim.

12. $y = 4.0033e^{0.3219x}$ or $y = 4 \times 1.38^x$

13. 6 hours 19 minutes

14. 990

15. a. i. 16   ii. 51.5

b. i. 80.2   ii. 55.6 seconds

iii. 7.08 and 0.10

c. Greatest after 10 minutes

d. i. $a = 0.10$

ii.

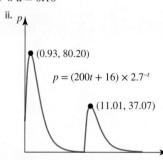

16. 8 years

17. 90 grams

9.5 Review

9.5 Exercise

1. A

2. C

3. A

4. B

5. B

6. B

7. a. 3   b. 5

8. Asymptote: $y = 1$
Domain: $x \in R$
Range: $y \in (1, \infty)$

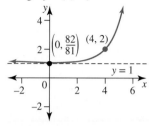

9. Asymptote: $y = -3$
Domain: $x \in R$
Range: $y \in (-3, \infty)$

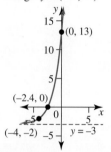

10. 65 °C

11. a. 120

b. i. 145   ii. 176   iii. 213

12. a. $P(t) = 10\,000 \left(2^t\right)$

b. i. 80 000   ii. 640 000

c. 3.32 months

13. a. $\dfrac{3}{5}$ b. 3

14. a. 12 000   b. 11.32 days

c. 4 days   d. 3 more days

15. 23 months, $L = C = 31$

16. 5 years

17. 21 minutes

18. $\dfrac{\sqrt{85}}{2}$ units

19. 7 years

20. a. 50 degrees   b. 10:39 am

10 Logarithms and logarithmic functions

LESSON SEQUENCE

Fully worked solutions for this chapter are available online.

EXAM PREPARATION

Access exam-style questions in every lesson, available online.

on Resources

Solutions	Solutions — Chapter 10 (sol-1277)
Exam questions	Exam question booklet — Chapter 10 (eqb-0298)
Digital documents	Learning matrix — Chapter 10 (doc-41792)
	Chapter summary — Chapter 10 (doc-41805)

LESSON
10.1 Overview

Hey students! Bring these pages to life online

Engage with interactivities

Answer questions and check results

Track your progress

Find all this and MORE in jacPLUS ▶

10.1.1 Introduction

Understanding logarithms and logarithmic functions is crucial for handling big numerical ranges. They're the reverse of the exponentials we covered in the previous chapter, simplifying complex calculations in growth and decay situations.

Logarithms come into play when we're dealing with things such as sound intensity and earthquake magnitudes. Logarithmic scales make sense here because our ears naturally perceive changes in sound intensity on a logarithmic scale — a small increase in the scale corresponds to a big change in what we hear. The Richter scale also relies on logarithms to measure the energy released during an earthquake. Each whole number increase on the Richter scale means the earthquake is ten times more powerful and releases around 31.6 times more energy.

10.1.2 Syllabus links

Lesson	Lesson title	Syllabus links
10.2	**Logarithms and logarithmic laws**	○ Define logarithms as indices, where $a^x = b$ is equivalent to $x = \log_a(b)$, and convert between both forms. ○ Use logarithmic laws and definitions: • $\log_a(x) + \log_a(y) = \log_a(xy)$ • $\log_a(x) - \log_a(y) = \log_a\left(\dfrac{x}{y}\right)$ • $\log_a(x^n) = n\log_a(x)$ • $\log_a(x) = \dfrac{\log_b(x)}{\log_b(a)}$ • $\log_a(a) = 1$ • $\log_a(1) = 0$
10.3	**Indicial equations and logarithms**	○ Solve equations involving indices using logarithms, with and without technology. ○ Solve equations involving logarithmic functions with and without technology.
10.4	**Graphs of logarithmic functions**	○ Recognise and determine the qualitative features of the graph of $y = \log_a(x)$ (where $a > 1$), including asymptote and intercept. ○ Recognise and determine the effect of the parameters a, h and k on the graph of $y = \log_a(x - h) + k$ (where $a > 1$), with and without technology. ○ Sketch graphs of logarithmic functions, with and without technology.
10.5	**Modelling with logarithmic functions**	○ Model and solve problems that involve logarithmic functions, e.g. decibels in acoustics and the Richter scale for earthquake magnitude, with and without technology.

Source: Mathematical Methods Senior Syllabus 2024 © State of Queensland (QCAA) 2024; licensed under CC BY 4.0.

LESSON
10.2 Logarithms and logarithmic laws

SYLLABUS LINKS

- Define logarithms as indices, where $a^x = b$ is equivalent to $x = \log_a(b)$, and convert between both forms.
- Use logarithmic laws and definitions:
  - $\log_a(x) + \log_a(y) = \log_a(xy)$
  - $\log_a(x) - \log_a(y) = \log_a\left(\dfrac{x}{y}\right)$
  - $\log_a(x^n) = n\log_a(x)$
  - $\log_a(x) = \dfrac{\log_b(x)}{\log_b(a)}$
  - $\log_a(a) = 1$
  - $\log_a(1) = 0$

Source: Mathematical Methods Senior Syllabus 2024 © State of Queensland (QCAA) 2024; licensed under CC BY 4.0.

10.2.1 Defining logarithms

The index, power or exponent (x), in the indicial equation $a^x = b$ is also known as a **logarithm**.

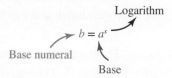

The index statement, $a^x = b$, with base a and index x, can be expressed with the index as the subject. This is called the logarithm statement and is written as $x = \log_a(b)$, which is read as 'the logarithm of b to the base a is equal to x'.

For example: $3^2 = 9$ can be written as $\log_3(9) = 2$.

$\qquad 10^5 = 100\,000$ can be written as $\log_{10}(100\,000) = 5$.

Logarithms

The statements $a^x = b$ and $x = \log_a(b)$ are equivalent.

$$a^x = b \quad \Leftrightarrow \quad x = \log_a(b)$$

where:

the base $a \in R^+\{1\}$ or $a > 0$ and $a \neq 1$

the number $b \in R^+$ or $b > 0$

the logarithm, or index, $x \in R$

Using the indicial equivalent, it is possible to find the exact value of some logarithms.

For example: $\log_2(8) = 3$, since $2^3 = 8$.

Not all expressions containing logarithms are rational.

For example: $\log_2(5)$ is an irrational number, since the powers of 2 are $1, 2, 4, 8, \cdots$

$\qquad$ Using a scientific calculator, $\log_2(5) \approx 2.3219$ to 5 significant figures.

Use of a calculator

Scientific calculators have two inbuilt logarithmic functions.

Base 10 logarithms are obtained from the log key. Thus, $\log_{10}(2)$ is evaluated as log (2), giving the value of 0.3010 to 4 decimal places. Base 10 logarithms are called common logarithms.

Base e logarithms are obtained from the ln key. Thus, $\log_e(2)$ is evaluated as ln (2), giving the value 0.6931 to 4 decimal places. Base e logarithms are called natural or Naperian logarithms. These logarithms occur extensively in calculus. The number e itself is known as Euler's number and, like π, it is a transcendental irrational number that has great importance in higher mathematical studies. Further exploration of e is formally included in Unit 3.

WORKED EXAMPLE 1 Converting between logarithmic or exponential equivalents

a. Express $3^4 = 81$ as a logarithm statement.

b. Express $\log_7\left(\dfrac{1}{49}\right) = -2$ as an index statement.

c. Solve the equation $10^x = 12.8$, expressing the exponent x to 1 decimal place.

d. Solve the equation $\log_5(x) = 2$ for x.

THINK	WRITE
a. 1. Identify the given form.	a. $3^4 = 81$ The index form is given with base 3, index or logarithm 4, and number 81.
2. Convert to the equivalent form.	Since $b = a^x \Leftrightarrow x = \log_a(b)$, $81 = 3^4 \Rightarrow 4 = \log_3(81)$. The logarithm statement is $4 = \log_3(81)$ or $\log_3(81) = 4$.
b. 1. Identify the given form.	b. $\log_7\left(\dfrac{1}{49}\right) = -2$ The logarithm form is given with base 7, number $\dfrac{1}{49}$, and logarithm or index of -2.
2. Convert to the equivalent form.	Since $x = \log_a(b) \Leftrightarrow b = a^x$, $-2 = \log_7\left(\dfrac{1}{49}\right) \Rightarrow \dfrac{1}{49} = 7^{-2}$ The index statement is $\dfrac{1}{49} = 7^{-2}$ or $7^{-2} = \dfrac{1}{49}$.
c. 1. Convert to the equivalent form.	c. $10^x = 12.8$ $\therefore x = \log_{10}(12.8)$
2. Evaluate using a calculator and state the answer to the required accuracy. *Note:* The base is 10, so use the LOG key on the calculator.	≈ 1.1 to 1 decimal place
d. 1. Convert to the equivalent form.	d. $\log_5(x) = 2$
2. Calculate the answer.	$\therefore x = 5^2$ $\therefore x = 25$

Converting between indicial and logarithmic equations allows you to evaluate the logarithm of a number to a certain base

WORKED EXAMPLE 2 Evaluating a logarithm

Evaluate the following without technology.

a. $\log_6(216)$

b. $\log_2\left(\dfrac{1}{8}\right)$

THINK	WRITE
a. 1. Let x equal the quantity we wish to find.	a. Let $x = \log_6(216)$.
2. Express the logarithmic equation as an indicial equation.	$6^x = 216$
3. Express both sides of the equation to the same base.	$6^x = 6^3$
4. Equate the powers.	$x = 3$
b. 1. Write the logarithm as a logarithmic equation.	b. Let $x = \log_2\left(\dfrac{1}{8}\right)$.
2. Express the logarithmic equation as an indicial equation.	$2^x = \dfrac{1}{8}$ $= \left(\dfrac{1}{2}\right)^3$ $= \left(2^{-1}\right)^3$
3. Express both sides of the equation to the same base.	$2^x = 2^{-3}$
4. Equate the powers.	$x = -3$

10.2.2 Logarithm laws

Since logarithms are indices, unsurprisingly there is a set of laws that control simplification of logarithmic expressions with the same base.

Logarithmic laws

For any $a, x, y > 0, a \neq 1$, the laws are:

- $\log_a(x) + \log_a(y) = \log_a(xy)$
- $\log_a(x) - \log_a(y) = \log_a\left(\dfrac{x}{y}\right)$
- $\log_a(x^n) = n\log_a(x)$
- $\log_a(a) = 1$
- $\log_a(1) = 0$

Note that there is no logarithm law for either the product or quotient of logarithms, or for expressions such as $\log_a(x \pm y)$.

To change bases, the following law can be applied.

Change of base law

$$\log_a(x) = \frac{\log_b(x)}{\log_b(a)}$$

where a, b, $x > 0$ and $a \neq 1$, $b \neq 1$.

This form enables decimal approximations of logarithms to be easily calculated on scientific calculators.

For example: $\log_2(8) = 3$

However, $\log_2(8) = \dfrac{\log_{10}(8)}{\log_{10}(2)} = \dfrac{0.903\,089\,99\ldots}{0.301\,029\,99\ldots} = 3$, the same result.

Similarly: $\log_2(10) = \dfrac{\log_{10}(10)}{\log_{10}(2)} = \dfrac{1}{0.301\,03} \approx 3.3219$ (to 5 significant figures)

Convention

There is a convention that if the base of a logarithm is not stated, this implies it is base 10. As it is on a calculator, $\log(x)$ represents $\log_{10}(x)$. When working with base 10 logarithms, it can be convenient to adopt this convention.

WORKED EXAMPLE 3 Simplifying using logarithmic laws

Simplify the following using the logarithm laws.

a. $\log_{10}(5) + \log_{10}(2)$

b. $\log_2(80) - \log_2(5)$

c. $\log_3(2a^4) + 2\log_3\left(\sqrt{\dfrac{a}{2}}\right)$

d. $\dfrac{\log_a\left(\frac{1}{4}\right)}{\log_a(2)}$

THINK	WRITE
a. 1. Apply the appropriate logarithm law.	a. $\log_{10}(5) + \log_{10}(2) = \log_{10}(5 \times 2)$
	$= \log_{10}(10)$
2. Simplify and write the answer.	$= 1 \left(\text{since } \log_a(a) = 1\right)$
b. 1. Apply the appropriate logarithm law.	b. $\log_2(80) - \log_2(5) = \log_2\left(\dfrac{80}{5}\right)$
	$= \log_2(16)$
2. Further simplify the logarithmic expression.	$= \log_2\left(2^4\right)$
	$= 4\log_2(2)$
3. Calculate the answer.	$= 4 \times 1$
	$= 4$

c. 1. Apply a logarithm law to the second term.
Note: As with indices, there is often more than one way to approach the simplification of logarithms.

c. $\log_3\left(2a^4\right) + 2\log_3\left(\sqrt{\dfrac{a}{2}}\right)$

$= \log_3\left(2a^4\right) + 2\log_3\left(\dfrac{a}{2}\right)^{\frac{1}{2}}$

$= \log_3\left(2a^4\right) + 2\times\dfrac{1}{2}\log_3\left(\dfrac{a}{2}\right)$

$= \log_3\left(2a^4\right) + \log_3\left(\dfrac{a}{2}\right)$

2. Apply an appropriate logarithm law to combine the two terms as one logarithmic expression.

$= \log_3\left(2a^4\times\dfrac{a}{2}\right)$

$= \log_3\left(a^5\right)$

3. State the answer.

$= 5\log_3\left(a\right)$

d. 1. Express the numbers in the logarithm terms in index form.
Note: There is no law for division of logarithms.

d. $\dfrac{\log_a\left(\frac{1}{4}\right)}{\log_a\left(2\right)} = \dfrac{\log_a\left(2^{-2}\right)}{\log_a\left(2\right)}$

2. Simplify the numerator and denominator separately.

$= \dfrac{-2\log_a\left(2\right)}{\log_a\left(2\right)}$

3. Cancel the common factor in the numerator and denominator.

$= \dfrac{-2\cancel{\log_a\left(2\right)}}{\cancel{\log_a\left(2\right)}}$

$= -2$

Exercise 10.2 Logarithms and logarithmic laws

10.2 Exercise	**10.2 Exam practice** on

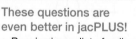

Simple familiar	Complex familiar	Complex unfamiliar
1, 2, 3, 4, 5, 6, 7, 8, 9, 10, 11, 12, 13, 14	15, 16, 17, 18	19, 20

Simple familiar

1. **WE1** **a.** Express $5^4 = 625$ as a logarithm statement.

   b. Express $\log_{36}\left(6\right) = \dfrac{1}{2}$ as an index statement.

   c. Solve the equation $10^x = 8.52$, expressing the exponent x to 2 significant figures.

   d. Solve the equation $\log_3\left(x\right) = -1$ for x.

2. **a.** Evaluate $\log_e(5)$ to 4 significant figures and write the equivalent index statement.

   b. Evaluate $10^{3.5}$ to 4 significant figures and write the equivalent logarithm statement.

3. Express the following indicial equations in logarithmic form.

   a. $2^3 = 8$

   b. $3^5 = 243$

   c. $5^0 = 1$

   d. $0.01 = 10^{-2}$

   e. $b^n = a$

   f. $2^{-4} = \dfrac{1}{16}$

4. Express the following logarithmic equations in indicial form.

   a. $\log_4(16) = 2$

   b. $\log_{10}(1\,000\,000) = 6$

   c. $\log_2\left(\dfrac{1}{2}\right) = -1$

   d. $\log_3(27) = 3$

   e. $\log_5(625) = 4$

   f. $\log_2(128) = 7$

5. **WE2** Evaluate the following without using technology.

   a. $\log_2(16)$

   b. $\log_3(81)$

   c. $\log_5(125)$

   d. $\log_2\left(\dfrac{1}{4}\right)$

   e. $\log_{10}(1000)$

   f. $\log_{10}(0.000\,01)$

6. Evaluate the following without using technology.

   a. $\log_2(0.25)$

   b. $\log_3\left(\dfrac{1}{243}\right)$

   c. $\log_2(32)$

   d. $\log_2\left(\dfrac{1}{64}\right)$

   e. $\log_3(-3)$

   f. $\log_n\left(n^5\right)$

7. Simplify, and evaluate where possible, each of the following without technology.

   a. $\log_2(8) + \log_2(10)$

   b. $\log_3(7) + \log_3(15)$

   c. $\log_{10}(20) + \log_{10}(5)$

   d. $\log_6(8) + \log_6(7)$

   e. $\log_2(20) - \log_2(5)$

   f. $\log_3(36) - \log_3(12)$

8. Simplify, and evaluate where possible, each of the following without technology.

   a. $\log_4(25) + \log_4\left(\dfrac{1}{5}\right)$

   b. $\log_{10}(5) - \log_{10}(20)$

   c. $\log_3\left(\dfrac{4}{5}\right) - \log_3\left(\dfrac{1}{5}\right)$

   d. $\log_2(9) + \log_2(4) - \log_2(12)$

   e. $\log_3(8) - \log_3(2) + \log_3(5)$

   f. $\log_4(24) - \log_4(2) - \log_4(6)$

9. **WE3** Simplify the following using the logarithm laws.

   a. $\log_{12}(3) + \log_{12}(4)$

   b. $\log_2(192) - \log_2(12)$

   c. $\log_3\left(3a^3\right) - 2\log_3\left(a^{\frac{3}{2}}\right)$

   d. $\dfrac{\log_a(8)}{\log_a(4)}$

10. Simplify the following.

    a. $3\log_{10}(5) + \log_{10}(2)$

    b. $2\log_2(8) + 3\log_2(3)$

    c. $2\log_3(2) + 3\log_3(1)$

    d. $\log_5(12) - 2\log_5(2)$

    e. $4\log_{10}(2) - 2\log_{10}(8)$

    f. $\log_3(4^2) - 3\log_3(2)$

11. Simplify the following.

a. $\dfrac{\log_3(25)}{\log_3(125)}$

b. $\dfrac{\log_2(81)}{\log_2(9)}$

c. $\dfrac{\log_4(36)}{\log_4(6)}$

d. $\dfrac{2\log_{10}(8)}{\log_{10}(16)}$

e. $\dfrac{3\log_5(27)}{2\log_5(9)}$

f. $\dfrac{4\log_3(32)}{5\log_3(4)}$

12. Use the logarithm laws to evaluate the following.

a. $\log_5(5 \div 5)$

b. $\log_{10}(5) + \log_{10}(2)$

c. $\log_3\left(\dfrac{1}{3}\right)$

d. $\log_2(32)$

e. $\log_4\left(\dfrac{7}{32}\right) - \log_4(14)$

f. $\log_6(9) + \log_6(8) - \log_6(2)$

13. a. Express the following as logarithm statements with the index as the subject.

    i. $2^5 = 32$

    ii. $4^{\frac{3}{2}} = 8$

    iii. $10^{-3} = 0.001$

  b. Express the following as index statements.

    i. $\log_2(16) = 4$

    ii. $\log_9(3) = \dfrac{1}{2}$

    iii. $\log_{10}(0.1) = -1$

14. Simplify the following.

a. $\log_7(7)$

b. $\log_6(3 \div 3)$

c. $\log_2(8)$

d. $\log_{10}\left(1 + 3^2\right)$

e. $\log_9(3)$

f. $2\log_2\left(\dfrac{1}{4}\right)$

Complex familiar

15. Simplify the following.

a. $\log_2(x - 4) + 3\log_2(x)$

b. $2\log_{10}(x + 3) - \log_{10}(x - 2)$

c. $\dfrac{\log_3\left(x^6\right)}{\log_3\left(x^2\right)}$

d. $\dfrac{\log_{10}\left(x^3\right)}{\log_{10}\left(\sqrt{x}\right)}$

e. $\dfrac{\log_5\left(x^{\frac{3}{2}}\right)}{\log_5\left(\sqrt{x}\right)}$

f. $\dfrac{2\log_2(x + 1)^3}{\log_2(x + 1)}$

16. Rewrite each of the following in the equivalent index or logarithm form and hence calculate the value of x.

a. $x = \log_2\left(\dfrac{1}{8}\right)$

b. $\log_{25}(x) = -0.5$

c. $10^{(2x)} = 4$

d. $3 = e^{-x}$

e. $\log_x(125) = 3$

f. $\log_x(25) = -2$

17. Use the logarithm laws to simplify the following.

a. $\log_3\left(x^3\right) - \log_3\left(x^2\right)$

b. $\log_a\left(2x^5\right) + \log_a\left(\dfrac{x}{2}\right)$

c. $-\dfrac{1}{2}\log_{10}(a) + \log_{10}\left(\sqrt{a}\right)$

d. $\dfrac{1}{2}\log_b\left(16a^4\right) - \dfrac{1}{3}\log_b\left(8a^3\right)$

e. $3\log_{10}(x) - 2\log_{10}\left(x^3\right) + \dfrac{1}{2}\log_{10}\left(x^5\right)$

f. $\dfrac{\log_3\left(x^6\right)}{\log_3\left(x^2\right)}$

18. Use the logarithm laws to evaluate the following.

   a. $\log_9(3) + \log_9(27)$

   b. $\log_9(3) - \log_9(27)$

   c. $2\log_2(4) + \log_2(6) - \log_2(12)$

   d. $\log_5\left(\log_3(3)\right)$

   e. $\log_{11}\left(\dfrac{7}{3}\right) + 2\log_{11}\left(\dfrac{1}{3}\right) - \log_{11}\left(\dfrac{11}{3}\right) - \log_{11}\left(\dfrac{7}{9}\right)$

   f. $\log_2\left(\dfrac{\sqrt{x} \times x^{\frac{3}{2}}}{x^2}\right)$

Complex unfamiliar

19. Given $\log_a(2) = 0.3$ and $\log_a(5) = 0.7$, evaluate the following.

   a. $\log_a(0.5)$

   b. $\log_a(2.5)$

   c. $\log_a(20)$

20. Given $\log_a(3) = p$ and $\log_a(5) = q$, express the following in terms of p and q.

   a. $\log_a(15)$

   b. $\log_a(125)$

   c. $\log_a(45)$

   d. $\log_a(0.6)$

   e. $\log_a\left(\dfrac{25}{81}\right)$

   f. $\log_a\left(\sqrt{5}\right) \times \log_a\left(\sqrt{27}\right)$

Fully worked solutions for this chapter are available online.

LESSON
10.3 Indicial equations and logarithms

SYLLABUS LINKS

- Solve equations involving indices using logarithms, with and without technology.
- Solve equations involving logarithmic functions with and without technology.

Source: Mathematical Methods Senior Syllabus 2024 © State of Queensland (QCAA) 2024; licensed under CC BY 4.0.

10.3.1 Logarithms as operators

Just as both sides of an equation may be raised to a power and the equality will still hold, taking logarithms of both sides of an equation maintains the equality.

If $m = n$, then it is true that $\log_a(m) = \log_a(n)$ and vice versa, provided the same base is used for the logarithms of each side.

This application of logarithms can provide an important tool when solving indicial equations.

Consider the equation $2^x = 5$, which converts to $x = \log_2(5)$ by definition of logarithms.

Take base 10 logarithms of both sides of this equation:

$$2^x = 5$$

$$\log_{10}(2^x) = \log_{10}(5)$$

Using logarithm laws: $\quad x\log_{10}(2) = \log_{10}(5)$

$$x = \frac{\log_{10}(5)}{\log_{10}(2)}$$

Using the change of base law: $\quad x = \log_2(5)$

Hence, for the indicial equation $2^x = 5$:

- the solution $x = \log_2(5)$ is exact
- the solution $x = \dfrac{\log_{10}(5)}{\log_{10}(2)}$ is used to evaluate the solution on a scientific calculator.

Change of base law for calculator use

$$\log_a(x) = \frac{\log_{10}(x)}{\log_{10}(a)}$$

WORKED EXAMPLE 4 Solving indicial equations using logarithms

a. State the exact solution to $5^x = 8$ and calculate its value to 3 decimal places.
b. Calculate the exact value and the value to 3 decimal places of the solution to the equation $2^{1-x} = 6^x$.

THINK	WRITE
a. 1. Convert to the equivalent form and state the exact solution.	a. $5^x = 8$ $\therefore x = \log_5(8)$ The exact solution is $x = \log_5(8)$.
2. Use the change of base law to express the answer in terms of base 10 logarithms. Alternatively, $\log_5(8)$ can be evaluated directly on some calculators without changing the base.	Since $\log_a(x) = \dfrac{\log_{10}(x)}{\log_{10}(a)}$ then $\log_5(8) = \dfrac{\log_{10}(8)}{\log_{10}(5)}$ $x = \dfrac{\log_{10}(8)}{\log_{10}(5)}$
3. Calculate the approximate value.	$\therefore x \approx 1.292$ to 3 decimal places.
b. 1. Take base 10 logarithms of both sides. *Note*: The convention is not to write the base 10.	b. $2^{1-x} = 6^x$ Take logarithms to base 10 of both sides: $\log(2^{1-x}) = \log(6^x)$
2. Apply the logarithm law so that x terms are no longer exponents.	$(1-x)\log(2) = x\log(6)$

3. Solve the linear equation in x.
Note: This is no different to solving any other linear equation of the form $a - bx = cx$ except the constants a, b, c are expressed as logarithms.

Expand:
$$\log(2) - x\log(2) = x\log(6)$$
Collect x terms together:
$$\log(2) = x\log(6) + x\log(2)$$
$$= x(\log(6) + \log(2))$$
$$x = \frac{\log(2)}{\log(6) + \log(2)}$$

This is the exact solution.

4. Calculate the approximate value.
Note: Remember to place brackets around the denominator for the division.

$x \approx 0.279$ to 3 decimal places.

TI | THINK	WRITE	CASIO | THINK	WRITE
a. 1. On a Calculator page, press MENU, then select: 3: Algebra 1: Numerical Solve Complete the entry line as: nSolve($5^x = 8, x$) then press ENTER.		**a. 1.** On a Run-Matrix screen, press OPTN, F4 (CALC), F5 (SolveN) and complete the entry line before pressing EXE. Note that the message 'More solutions may exist' might be displayed to indicate that other solutions than the ones displayed by SolveN might exist.	 
2. The answer appears on the screen.	$x = 1.292$ (3 decimal places)	**2.** The answer appears on the screen.	$x = 1.292$ (3 decimal places)
b. 1. On a Calculator page, press MENU, then select: 3: Algebra 1: Numerical Solve Complete the entry line as: nSolve($2^{1-x} = 6^x, x$) then press ENTER.		**b. 1.** On a Run-Matrix screen, press OPTN, F4 (CALC), F5 (SolveN) and complete the entry line before pressing EXE.	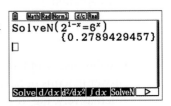 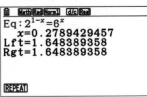
2. The answer appears on the screen.	$x = 0.279$ (3 decimal places)	**2.** The answer appears on the screen.	$x = 0.279$ (3 decimal places)

10.3.2 Equations containing logarithms

Indicial equations with rational solutions were studied in the previous chapter. Worked example 4 demonstrated how logarithms enable equations with irrational solutions, both exact and approximate, to be solved.

Solving logarithmic equations involves the use of the logarithm laws as well as converting to index form. As $\log_a(x)$ is only defined for $x > 0$ and $a \in R^+ \{1\}$, always check the validity of your solution, as you cannot take the log of a negative value. This means any value of x that, when substituted back into the original equation, creates a '$\log_a$ (negative number)' term must be rejected as a solution. Otherwise, normal algebraic approaches together with the index and logarithm laws are the techniques for solving such equations.

WORKED EXAMPLE 5 Solving equations containing logarithms

Find x if $\log_3 9 = x - 2$.

THINK	WRITE
1. Write the equation.	$\log_3 9 = x - 2$
2. Simplify the logarithm using the 'logarithm of a power' law and the fact that $\log_3 3 = 1$.	$\log_3 3^2 = x - 2$ $2 \log_3 3 = x - 2$ $2 = x - 2$,
3. Solve for x by adding 2 to both sides.	$x = 4$

WORKED EXAMPLE 6 Solving equations with logarithms

Solve the equation $\log_6(x) + \log_6(x - 1) = 1$ for x.

THINK	WRITE
1. Apply the logarithm law that reduces the equation to one logarithm term.	$\log_6(x) + \log_6(x - 1) = 1$ $\therefore \log_6(x(x-1)) = 1$ $\therefore \log_6(x^2 - x) = 1$
2. Convert the logarithm form to its equivalent form. *Note:* An alternative method is to write $\log_6(x^2 - x) = \log_6(6)$, from which $x^2 - x = 6$ is obtained.	Converting from logarithm form to index form gives: $x^2 - x = 6^1$ $x^2 - x = 6$
3. Solve the quadratic equation.	$x^2 - x - 6 = 0$ $(x - 3)(x + 2) = 0$ $\therefore x = 3, \ x = -2$
4. Check the validity of both solutions in the original equation.	Check in $\log_6(x) + \log_6(x - 1) = 1$. If $x = 3$, $\text{LHS} = \log_6(3) + \log_6(2)$ $\qquad = \log_6(6)$ $\qquad = 1$ $\qquad = \text{RHS}$ If $x = -2$, $\text{LHS} = \log_6(-2) + \log_6(-3)$, which is not admissible. Therefore, reject $x = -2$.
5. Write the answer.	The solution is $x = 3$.

TI	THINK	WRITE	CASIO	THINK	WRITE

TI | THINK

1. On a Calculator page, press MENU, then select:
   3: Algebra
   1: Numerical Solve
   Complete the entry line as:
   $\text{nSolve}(\log_6(x) + \log_6(x-1) = 1, x)$
   then press ENTER.

2. The answer appears on the screen.

$x = 3$

CASIO | THINK

1. On a Run-Matrix screen, press OPTN, F4 (CALC), F5 (SolveN) and complete the entry line before pressing EXE.
   Note: The $\log_a b$ template can be found by pressing OPTN, selecting CALC by pressing F2, then selecting $\log_a b$ by pressing F4.

2. The answer appears on the screen.

$x = 3$

The following example demonstrates using the logarithmic law $\log_a(a) = 1$ and the method of solving quadratic equations by substitution.

WORKED EXAMPLE 7 Using substitution to solve logarithmic equations

Solve the following equations for x.

a. $\log_2(3x) + 3 = \log_2(x-2)$

b. $(\log_2(x))^2 = 3 - 2\log_2(x)$

THINK

WRITE

a. 1. Rewrite 3 in log form, given $\log_2(2) = 1$.

a.
$$\log_2(3x) + 3 = \log_2(x-2)$$
$$\log_2(3x) + 3\log_2(2) = \log_2(x-2)$$

2. Apply the law $\log_a(m^n) = n\log_a(m)$.

$$\log_2(3x) + \log_2(2^3) = \log_2(x-2)$$

3. Simplify the left-hand side by applying $\log_a(mn) = \log_a(m) + \log_a(n)$.

$$\log_2(3x \times 8) = \log_2(x-2)$$

4. Equate the logs and simplify.

$$24x = x - 2$$
$$23x = -2$$
$$x = -\frac{2}{23}$$

b. 1. Identify the quadratic form of the log equation. Let $u = \log_2(x)$ and rewrite the equation in terms of a.

b.
$$(\log_2(x))^2 = 3 - 2\log_2(x)$$
Let $u = \log_2(x)$.
$$u^2 = 3 - 2u$$

2. Solve the quadratic.

$$u^2 + 2u - 3 = 0$$
$$(u-1)(u+3) = 0$$
$$\therefore u = 1, -3$$

3. Substitute $u = \log_2(x)$ and solve for x.

$$\log_2(x) = 1 \quad \log_2(x) = -3$$
$$x = 2^1 \quad\quad x = 2^{-3}$$

$$\therefore x = 2, \frac{1}{8}$$

Exercise 10.3 Indicial equations and logarithms

10.3 Exercise	10.3 Exam practice on

Simple familiar	Complex familiar	Complex unfamiliar
1, 2, 3, 4, 5, 6, 7, 8, 9, 10, 11, 12	13, 14, 15, 16, 17, 18	19, 20

These questions are even better in jacPLUS!
- Receive immediate feedback
- Access sample responses
- Track results and progress

Find all this and MORE in jacPLUS ⊙

Simple familiar

1. **WE4** a. Determine the exact solution to $7^x = 15$ and calculate its value to 3 decimal places.

   b. Calculate the exact value and the value to 3 decimal places of the solution to the equation $3^{2x+5} = 4^x$.

2. Solve the following equations correct to 3 decimal places.

   a. $2^x = 11$ b. $2^x = 0.6$ c. $3^x = 20$

   d. $3^x = 1.7$ e. $5^x = 8$ f. $0.7^x = 3$

3. Solve the following equations correct to 3 decimal places.

   a. $10^{x-1} = 18$ b. $3^{x+2} = 12$ c. $2^{2x+1} = 5$

   d. $4^{3x+1} = 24$ e. $10^{-2x} = 7$ f. $8^{2-x} = 0.75$

4. **WE5** Determine x in each of the following.

   a. $\log_2 4 = x$ b. $\log_9 1 = x$ c. $\log_3 27 = x$

   d. $\log_4 256 = x$ e. $\log_{10} \dfrac{1}{10} = x$ f. $\log_3 \dfrac{1}{9} = x$

5. Solve the following for x.

   a. $\log_5 (x-1) = 2$ b. $\log_2 (2x+1) = -1$ c. $\log_x \left(\dfrac{1}{49}\right) = -2$ d. $\log_x (36) - \log_x (4) = 2$

6. If $\log_2 (3) - \log_2 (2) = \log_2 (x) + \log_2 (5)$, solve for x.

7. **WE6** Solve the equation $\log_3 (x) + \log_3 (2x+1) = 1$ for x.

8. Solve the equation $\log_6 (x) - \log_6 (x-1) = 2$ for x.

9. Solve the following for x.

   a. $\log_{10} (x+5) = \log_{10} (2) + 3\log_{10} (3)$ b. $\log_4 (2x) + \log_4 (5) = 3$

   c. $\log_{10} (x+1) = \log_{10} (x) + 1$ d. $2\log_6 (3x) + 3\log_6 (4) - 2\log_6 (12) = 2$

10. Solve the following for x.

    a. $\log_2 (2x+1) + \log_2 (2x-1) = 3\log_2 (3)$ b. $\log_3 (2x) + \log_3 (4) = \log_3 (x+12) - \log_3 (2)$

    c. $\log_2 (2x+12) - \log_2 (3x) = 4$ d. $\log_2 (x) + \log_2 (2-2x) = -1$

    e. $\left(\log_{10} (x) + 3\right)\left(2\log_4 (x) - 3\right) = 0$ f. $2\log_3 (x) - 1 = \log_3 (2x-3)$

11. Express each of the following in the equivalent index or logarithm form, and hence calculate the value of x.

    a. $x = \log_2 \left(\dfrac{1}{8}\right)$

    b. $\log_{25} (x) = -0.5$

    c. $10^{(2x)} = 4$. Express the answer to 2 decimal places.

    d. $3 = e^{-x}$. Express the answer to 2 decimal places.

    e. $\log_x (125) = 3$

    f. $\log_x (25) = -2$

12. Express y in terms of x.

a. $\log_{10}(y) = \log_{10}(x) + 2$

b. $\log_2\left(x^2\sqrt{y}\right) = x$

c. $2\log_2\left(\dfrac{y}{2}\right) = 6x - 2$

d. $x = 10^{y-2}$

e. $\log_{10}\left(10^{3xy}\right) = 3$

f. $10^{3\log_{10}(y)} = xy$

Complex familiar

13. **WE7** Solve the following for x.

a. $\log_5(125) = x$

b. $\log_4(x-1) + 2 = \log_4(x+4)$

c. $3\left(\log_2(x)\right)^2 - 2 = 5\log_2(x)$

d. $\log_5(4x) + \log_5(x-3) = \log_5(7)$

14. Solve the following equations, giving exact solutions.

a. $2^{2x} - 14 \times 2^x + 45 = 0$

b. $5^{-x} - 5^x = 4$

c. $9^{2x} - 3^{1+2x} + 2 = 0$

d. $\log_a\left(x^3\right) + \log_a\left(x^2\right) - 4\log_a(2) = \log_a(x)$

e. $\left(\log_2(x)\right)^2 - \log_2\left(x^2\right) = 8$

f. $\dfrac{\log_{10}\left(x^3\right)}{\log_{10}(x+1)} = \log_{10}(x)$

15. State the exact solution and then give the approximate solution to 4 significant figures for each of the following indicial equations.

a. $11^x = 18$

b. $5^{-x} = 8$

c. $7^{2x} = 3$

16. Solve the following equations.

a. $2^{\log_5(x)} = 8$

b. $2^{\log_2(x)} = 7$

17. Solve the indicial equations to obtain the value of x to 2 decimal places.

a. $7^{1-2x} = 4$

b. $10^{-x} = 5^{x-1}$

c. $5^{2x-9} = 3^{7-x}$

d. $10^{3x+5} = 6^{2-3x}$

e. $0.25^{4x} = 0.8^{2-0.5x}$

f. $4^{x+1} \times 3^{1-x} = 5^x$

18. Express $\log_2(10)$ in terms of $\log_{10}(2)$.

Complex unfamiliar

19. Evaluate $\log_y\left(x^2\right) \times \log_x\left(y^2\right)$ and explain how the result is obtained.

20. Solve the equation $8\log_x(4) = \log_2(x)$ for x.

Fully worked solutions for this chapter are available online.

LESSON
10.4 Graphs of logarithmic functions

SYLLABUS LINKS

- Recognise and determine the qualitative features of the graph of $y = \log_a(x)$ (where $a > 1$), including asymptote and intercept.
- Recognise and determine the effect of the parameters a, h and k on the graph of $y = \log_a(x - h) + k$ (where $a > 1$), with and without technology.
- Sketch graphs of logarithmic functions, with and without technology.

Source: Mathematical Methods Senior Syllabus 2024 © State of Queensland (QCAA) 2024; licensed under CC BY 4.0.

10.4.1 The graph of $y = \log_a(x)$ where $a > 1$

The graph of the logarithmic function $f : R^+ \to R, f(x) = \log_a(x), a > 1$ has the following characteristics.

The graph of $y = \log_a(x)$

For $f(x) = \log_a(x), a > 1$:
- the domain is $(0, \infty)$
- the range is R
- the graph is an increasing function
- the graph cuts the x-axis at $(1, 0)$
- as $x \to 0, y \to -\infty$, so the line $x = 0$ is an asymptote
- as a increases, the graph rises more steeply for $x \in (0, 1)$ and is flatter for $x \in (1, \infty)$.

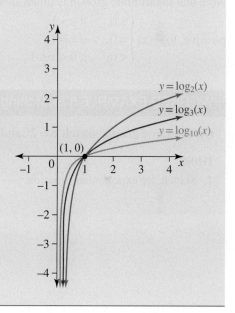

The parameter a

The graphs shown above illustrate the effect of the parameter a on the graph of $y = \log_a(x)$

10.4.2 The graphs of $y = \log_a(x)$ and $y = a^x$, where $a > 1$

Exponential and logarithmic functions are connected since, by definition, $a^c = b$ is equivalent to $c = \log_a(b)$. The relationships between the two functions $y = \log_a(x)$ and $y = a^x$, where $a > 1$, are shown.

The shape of the basic logarithmic graph with rule $y = \log_a(x), a > 1$ is shown as the reflection in the line $y = x$ of the exponential graph with rule $y = a^x, a > 1$.

The key features of the graph of $y = \log_a(x)$ can be deduced from those of the exponential graph.

The key features are summarised in the table.

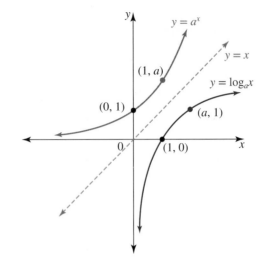

$y = a^x$	$y = \log_a(x)$
Horizontal asymptote with equation $y = 0$	Vertical asymptote with equation $x = 0$
y-intercept $(0, 1)$	x-intercept $(1, 0)$
The point $(1, a)$ lies on the graph.	The point $(a, 1)$ lies on the graph.
Range R^+	Domain R^+
Domain R	Range R

Note that logarithmic growth is much slower than exponential growth. Also note that unlike a^x, which is always positive, $\log_a(x) \begin{cases} > 0, & \text{if } x > 1 \\ = 0, & \text{if } x = 1 \\ < 0, & \text{if } 0 < x < 1 \end{cases}$.

WORKED EXAMPLE 8 Sketching exponential and logarithmic graphs

On the same graph, sketch $y = 2^x$ and $y = \log_2(x)$. Identify the domain of each graph.

THINK

1. Sketch the exponential function.

WRITE

$y = 2^x$
Asymptote: $y = 0$
y-intercept: $(0, 1)$
Second point: let $x = 1$.
$\therefore y = 2$
The point $(1, 2)$ is on the graph.

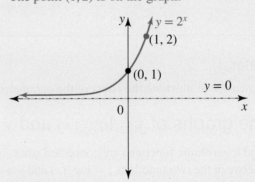

2. Sketch the logarithmic function.
   (Remember the axis of symmetry about the line $y = x$ makes it easier to sketch.)

$y = \log_2(x)$
Asymptote: $x = 0$
x-intercept: $(1, 0)$
Second point: $(2, 1)$

3. Identify the domain of each graph.

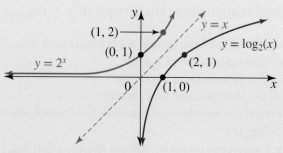

For $y = 2^x$: domain $x \in R$

For $y = \log_2(x)$: domain $x \in R^+$

10.4.3 The graph of $y = \log_a(x - h) + k$ where $a > 1$

Once the basic logarithmic graph is known, the graphs of logarithmic functions can be translated in similar ways to graphs of any other function previously studied. Knowledge of the effects of the parameters a, h and k on the graph of $y = \log_a(x - h) + k$ enables the graph of any logarithmic function to be obtained from the basic graph of $y = \log_a(x)$.

The parameter h and the logarithmic graph

As with other functions, the transformation of $y = \log_a(x)$ to $y = \log_a(x - h)$ is a horizontal translation of h units.

Horizontal translation: the graph of $y = \log_a(x - h)$

The key features of the graph of $y = \log_a(x - h)$ are:
- **horizontal translation of h units**
  - **if $h > 0$, translated to the right**
  - **if $h < 0$, translated to the left**
- **a vertical asymptote with equation $x = h$**
- **x-intercept $(1 + h, 0)$**
- **domain $x \in (h, \infty)$.**

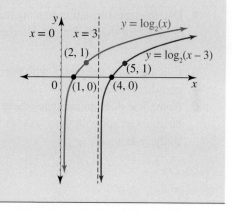

These changes are illustrated in the diagram by the graph of $y = \log_2(x)$ and its image, $y = \log_2(x - 3)$, after a horizontal translation of 3 units to the right.

The diagram shows that the domain of $y = \log_2(x - 3)$ is $(3, \infty)$. Its range is unaffected by the horizontal translation and remains R.

It is important to realise that the domain and the asymptote position can be calculated algebraically, since we only take logarithms of positive numbers. For example, the domain of $y = \log_2(x - 3)$ can be calculated by solving the inequation $x - 3 > 0 \Rightarrow x > 3$. This means that the domain is $(3, \infty)$ as the diagram shows. The equation of the asymptote of $y = \log_2(x - 3)$ can be calculated from the equation $x - 3 = 0 \Rightarrow x = 3$.

The parameter k and the logarithmic graph

As with other functions, the transformation of $y = \log_a(x)$ to $y = \log_a(x) + k$ is a vertical translation of k units.

Vertical translation: the graph of $y = \log_a(x) + k$

The key features of the graph of $y = \log_a(x) + k$ are:
- **vertical translation of k units**
  - **if $k > 0$, translated upwards**
  - **if $k < 0$, translated downwards**
- **vertical asymptote with equation $x = 0$, unchanged from $y = \log_a(x)$**
- **an x-intercept found by solving the equation $\log_a(x) + k = 0$**
- **an additional helpful point $(1, k)$**
- **domain $x \in R^+$, unchanged from $y = \log_a(x)$**
- **range R.**

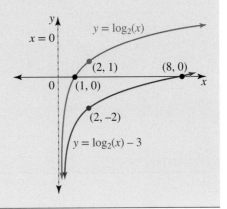

The graph of $y = \log_2(x) - 3$ is a vertical translation down by 3 units of the graph of $y = \log_2(x)$. Solving $\log_2(x) - 3 = 0$ gives $x = 2^3$, so the graph cuts the x-axis at $x = 8$, as illustrated.

WORKED EXAMPLE 9 Sketching logarithmic graphs

a. Sketch the graph of $y = \log_2(x + 2)$ and identify its domain.

b. Sketch the graph of $y = \log_{10}(x) + 1$ and identify its domain.

THINK

a. 1. Identify the transformation involved.

2. Use the transformation to state the equation of the asymptote and the domain.

3. Calculate any intercepts with the coordinate axes.
Note: The domain indicates there will be an intercept with the y-axis as well as the x-axis.

4. Sketch the graph.

WRITE

a. $y = \log_2(x + 2)$
Horizontal translation 2 units to the left
The vertical line $x = 0 \rightarrow$ the vertical line $x = -2$ under the horizontal translation.
The domain is $\{x : x > -2\}$.

y-intercept: when $x = 0$,
$y = \log_2(2)$
$\quad = 1$
y-intercept $(0, 1)$
x-intercept: when $y = 0$,
$\log_2(x + 2) = 0$
$$x + 2 = 2^0$$
$$x + 2 = 1$$
$$x = -1$$
x-intercept $(-1, 0)$
Check: the point $(1, 0) \rightarrow (-1, 0)$ under the horizontal translation.

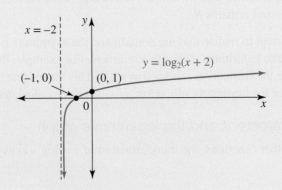

b. 1. Identify the transformation involved.

b. $y = \log_{10}(x) + 1$

Vertical translation of 1 unit upwards

2. Identify the equation of the asymptote and the domain.

The vertical transformation does not affect either the position of the asymptote or the domain.

Hence, the equation of the asymptote is $x = 0$. The domain is R^+.

3. Obtain any intercept with the coordinate axes.

Since the domain is R^+, there is no y-intercept.

x-intercept: when $y = 0$,

$\log_{10}(x) + 1 = 0$

$\log_{10}(x) = -1$

$x = 10^{-1}$

$\qquad = \dfrac{1}{10}$ or 0.1

The x-intercept is $(0.1, 0)$.

4. Calculate the coordinates of a second point on the graph.

Point: let $x = 1$.

$y = \log_{10}(1) + 1$

$\quad = 0 + 1$

$\quad = 1$

The point $(1, 1)$ lies on the graph.

Check: the point $(1, 0) \rightarrow (1, 1)$ under the vertical translation.

5. Sketch the graph.

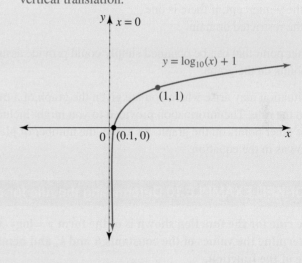

TI	THINK	WRITE	CASIO	THINK	WRITE

b. 1. On a Graphs screen, complete the function as:
$f1(x) = \log(x) + 1$
and press ENTER.

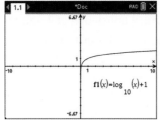

2. The graph and the domain appear on the screen.

There is an asymptote at $x = 0$; therefore, the domain is R^+.

b. 1. On a Graph screen, complete the equation as:
$Y1 = \log(x) + 1$
and press EXE.

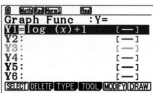

2. To graph the equation, press F6 for DRAW.

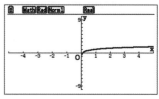

3. The graph and the domain appear on the screen.

There is an asymptote at $x = 0$; therefore, the domain is R^+.

Combinations of transformations

Logarithmic functions with equations of the form $y = \log_a(x - h) + k$ (where $a > 1$) are derived from the basic graph of $y = \log_a(x)$ (where $a > 1$) by applying a combination of transformations. The key features to identify in order to sketch the graphs of such logarithmic functions are:

- the asymptote
- the x-intercept
- the y-intercept, if there is one
- the restricted domain

Another point that can be obtained simply could provide assurance about the shape. Always aim to show at least two points on the graph.

The situation may arise where you are given the graph of a translated logarithmic function and you are required to find the rule. The information provided to you might include the equation of the asymptote, the intercepts and/or other points on the graph. As a rule, the number of pieces of information is equivalent to the number of unknowns in the equation.

WORKED EXAMPLE 10 Determining the rule for a graph

The rule for the function shown is of the form $y = \log_2(x - h) + k$.
Determine the values of the constants h and k, and hence identify the rule of the function.

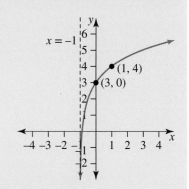

THINK

1. The vertical asymptote corresponds to the value of h.

WRITE

The vertical asymptote is $x = -1$.
Therefore, $h = -1$, giving the rule
$y = \log_2(x + 1) + k$.

2.	Substitute the given point to determine the value of k.	$4 = \log_2(1+1) + k$ $4 = \log_2 2 + k$ $4 = 1 + k$ $k = 3$
3.	Write the answer for the values of h and k.	$h = -1, k = 3$
4.	Write the rule of the function.	$\therefore y = \log_2(x+1) + 3$

Exercise 10.4 Graphs of logarithmic functions

learn

10.4 Exercise	10.4 Exam practice on

Simple familiar	Complex familiar	Complex unfamiliar
1, 2, 3, 4, 5, 6, 7, 8, 9, 10	11, 12, 13, 14, 15, 16	17, 18, 19, 20

These questions are even better in jacPLUS!
- Receive immediate feedback
- Access sample responses
- Track results and progress

Find all this and MORE in jacPLUS ▶

Simple familiar

1. **WE9** a. Sketch the graph of $y = \log_{10}(x-1)$ and state its domain.
   b. Sketch the graph of $y = \log_5(x) - 1$ and state its domain.

2. Consider the function with rule $y = \log_3(x+9)$.
   a. State the domain.
   b. State the equation of the vertical asymptote.
   c. Calculate the coordinates of the y-intercept.
   d. Calculate the coordinates of the x-intercept.
   e. Sketch the graph of $y = \log_3(x+9)$.
   f. State the range of the graph.

3. Consider the function with rule $y = \log_{10}(x-5)$.
   a. State the domain.
   b. State the equation of the vertical asymptote.
   c. Calculate the coordinates of any intercepts with the coordinate axes.
   d. Sketch the graph, stating its range.
   e. On the same diagram, sketch the graph of $y = -\log_{10}(x-5)$.

4. State the maximal domain and the equation of the vertical asymptote of each of the following logarithmic functions.
   a. $y = \log_{10}(x+6)$
   b. $y = \log_{10}(x-6)$
   c. $y = \log_{10}(6x)$
   d. $y = -\log_4(2x-3) + 1$

5. Sketch the graphs of the following transformations of the graph of $y = \log_a(x)$, stating the domain and range, the equation of the asymptote, and any points of intersection with the coordinate axes.
   a. $y = \log_5(x) - 2$
   b. $y = \log_5(x-2)$

6. Sketch the graphs of the following transformations of the graph of $y = \log_a(x)$, stating the domain and range, the equation of the asymptote, and any points of intersection with the coordinate axes.
   a. $y = \log_{10}(x) + 1$
   b. $y = \log_3(x+1)$

7. Sketch the graphs of the following, demonstrating clearly all important features.

   a. $y = \log_2 (x - 4)$

   b. $y = \log_2 (x) - 4$

8. **WE10** The rule for the function shown is of the form
   $y = \log_2 (x - h) + k$.

   Determine the values of the constants h and k, and identify the rule of the function.

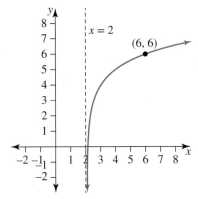

9. The graph of the function $f(x) = -\log_2 (x + b)$ is shown. Determine the value of b.

10. **WE8** On the same graph, sketch $y = 3^x$ and $y = \log_3 (x)$. Identify the domain of each graph.

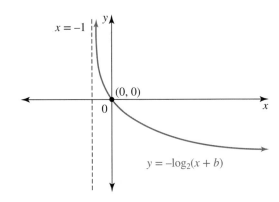

Complex familiar

11. On the same graph, sketch $y = 2^x - 1$ and $y = \log_2 (x + 1)$. Identify any points of intersection.

12. The graph of the function with equation $y = a \log_7 (bx)$ contains the points $(2, 0)$ and $(14, 14)$. Determine its equation.

13. The graph of the function with equation $y = a \log_3 (x) + b$ contains the points $\left(\dfrac{1}{3}, 8\right)$ and $(1, 4)$.

    Determine its equation.

14. For the graph illustrated in the diagram, determine a possible equation in the form $y = a \log_2 (x - b) + c$.

15. Sketch the following graphs, clearly showing any axis intercepts and asymptotes.
    Give intercepts to 2 decimal places when necessary.

    a. $y = \log_e (x) + 3$
    b. $y = \log_e (x) - 5$
    c. $y = \log_e (x) + 0.5$

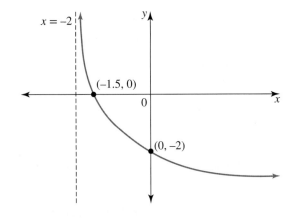

16. Sketch the following graphs, clearly showing any axis intercepts and asymptotes. Give intercepts to 2 decimal places when necessary.

  a. $y = \log_e (x - 4)$
  b. $y = \log_e (x + 2)$
  c. $y = \log_e (x + 0.5)$

Complex unfamiliar

17. Determine the value of x if $4 - x = \log_2 (x - 1)$.

18. Determine the value of x, correct to 3 decimal places, if $5 - 2x = \log_3 (x + 2)$.

19. The graph of $y = m \log_2 (nx)$ passes through the points $(-2, 3)$ and $\left(-\dfrac{1}{2}, \dfrac{1}{2}\right)$. Show that the values of m and n are 1.25 and $-2^{\frac{7}{5}}$ respectively.

20. The equation $y = a \log_e (bx)$ relates x to y. The table shows values for x and y.

x	1	2	3
y	$\log_e (2)$	0	w

Determine the value of w, expressing your answer in exact form and correct to 4 decimal places.

Fully worked solutions for this chapter are available online.

LESSON
10.5 Modelling with logarithmic functions

SYLLABUS LINKS

• Model and solve problems that involve logarithmic functions, e.g. decibels in acoustics and the Richter scale for earthquake magnitude, with and without technology.

Source: Mathematical Methods Senior Syllabus 2024 © State of Queensland (QCAA) 2024; licensed under CC BY 4.0.

10.5.1 Logarithmic scales

Many scientific quantities are measured in terms of scales that are logarithmic rather than linear.

Examples of logarithmic scales include:
  • decibels in acoustics as a function of a sound's intensity
  • the Richter scale, which describes the amount of energy released by the seismic waves of earthquakes in terms of magnitude
  • the frequencies of musical notes
  • the pH scale in chemistry
  • the intensity of the brightness of stars.

Loudness, L, in decibels (dB), is related to the intensity, I, of a sound by the equation

$$L = 10 \log_{10} \left(\frac{I}{I_0} \right)$$

where I_0 is equal to 10^{-12} watts per square metre $\left(W/m^2 \right)$. (This value is the lowest intensity of sound that can be heard by human ears.)

a. **An ordinary conversation has a loudness of 60 dB. Calculate the intensity in W/m^2.**
b. **If the intensity is doubled, determine the change in the loudness, correct to 2 decimal places.**

THINK	WRITE
a. 1. Substitute $L = 60$ and simplify.	a. $L = 10 \log_{10} \left(\dfrac{I}{I_0} \right)$
	$60 = 10 \log_{10} \left(\dfrac{I}{10^{-12}} \right)$
	$60 = 10 \log_{10}(10^{12}I)$
	$6 = \log_{10}(10^{12}I)$
2. Convert the logarithm to index form and solve for I.	$10^6 = 10^{12}I$
	$I = 10^{-6} \, W/m^2$
b. 1. Determine an equation for L_1.	b. $L_1 = 10 \log_{10} \left(\dfrac{I_1}{10^{-12}} \right)$
	$= 10 \log_{10} \left(10^{12}I_1 \right)$
	$= 10 \log_{10} \left(10^{12} \right) + 10 \log_{10} (I_1)$
	$= 120 \log_{10} (10) + 10 \log_{10} (I_1)$
	$= 120 + 10 \log_{10} (I_1)$
2. The intensity has doubled; therefore, $I_2 = 2I_1$. Determine an equation for L_2.	$L_2 = 10 \log_{10} \left(\dfrac{2I_1}{10^{-12}} \right)$
	$= 10 \log_{10}(2 \times 10^{12}I_1)$
	$= 10 \log_{10}(2) + 10 \log_{10}(10^{12}) + 10 \log_{10}(I_1)$
	$= 3.010 + 120 \log_{10}(10) + 10 \log_{10}(I_1)$
	$= 3.01 + 120 + 10 \log_{10}(I_1)$
3. Replace $120 + 10 \log_{10} (I_1)$ with L_1.	$= 3.01 + L_1$
Write the answer.	Doubling the intensity increases the loudness by 3.01 dB.

10.5.2 Application to exponential models

Exponential functions can be used to model many real-life situations. Exponential growth and decay were studied in Chapter 9, where technology was required to solve the equations formed.

Logarithmic functions can be used to solve exponential functions of the form $M = M_0 a^{kt}$, where M_0 represents the initial value, t represents the time taken and k is a constant. In exponential models, the functions are often expressed with base e (approximately 2.718 28); that is, $M = M_0 e^{kt}$.

For some exponential models that are functions of time, the behaviour (or limiting value) as $t \to \infty$ may be of interest.

WORKED EXAMPLE 12 Evaluating exponential decay models

A tennis ball is dropped from a height of 100 cm, bounces and rebounds to 80% of its previous height. The height, h cm, of the ball after n bounces is given by the formula

$$h = A \times a^n$$

where A cm is the height from which the ball is dropped and a is the percentage of the height reached by the ball on the previous bounce.

a. Determine the values of A and a, and hence write the formula for h in terms of n.

b. Determine the height that the ball will reach after 5 bounces. Give your answer to 1 decimal place.

c. Determine how many bounces it will take before the ball reaches less than 1 cm.

THINK	WRITE
a. 1. The ball is dropped from a height of A cm.	**a.** $A = 100$
2. The percentage of the height reached by the ball on the previous bounce is a.	$a = 80\% = \dfrac{80}{100} = 0.8$
3. Substitute the values for A and a into the formula $h = A \times a^n$.	$h = A \times a^n \quad h = 100 \times (0.8)^n$
b. 1. Substitute 5 for n.	**b.** $h = 100 \times (0.8)^5$
2. Evaluate using a calculator.	$= 32.768$
3. Write your answer in a sentence.	The ball bounces to 32.8 cm after 5 bounces.
c. 1. Substitute 1 for h.	**c.** $\qquad 1 = 100 \times (0.8)^n$
2. Divide both sides by 100. Note that for a tech-active question, you could simply use solveN here.	$(0.8)^n = 0.01$
3. Take the log of both sides to base 10.	$\log_{10}(0.8)^n = \log_{10}(0.01)$
4. Use $\log_a(m^p) = p\log_a(m)$ to simplify.	$n \log_{10}(0.8) = \log_{10}(0.01)$
5. Divide both sides by $\log_{10}(0.8)$.	$n = \dfrac{\log_{10}(0.01)}{\log_{10}(0.8)}$
6. Evaluate.	$= 20.64$
7. Bounces must be in whole numbers.	≈ 21
8. After 20 bounces the ball reaches more than 1 cm, but after 21 bounces the ball reaches less than 1 cm, because it bounces to a smaller height each time. Write the answer in a sentence.	The ball reaches less than 1 cm after 21 bounces.

WORKED EXAMPLE 13 Applying exponential growth

If P dollars is invested into an account that earns interest at a rate of r for t years and the interest is compounded continuously, then $A = Pe^{rt}$, where A is the accumulated dollars.

A deposit of $6000 is invested at the Western Bank, and $9000 is invested at the Common Bank at the same time. Western offers compound interest continuously at a nominal rate of 6%, whereas the Common Bank offers compound interest continuously at a nominal rate of 5%.

Determine how many years, correct to 1 decimal place, it will take for the two investments to be the same.

THINK	WRITE
1. Write the compound interest equation for each of the two investments.	$A = Pe^{rt}$ Western Bank: $A = 6000e^{0.06t}$ Common Bank: $A = 9000e^{0.05t}$
2. Equate the two equations and solve for t. Your calculator could also be used to determine the answer.	$6000e^{0.06t} = 9000e^{0.05t}$ $\dfrac{e^{0.06t}}{e^{0.05t}} = \dfrac{9000}{6000}$ $e^{0.01t} = \dfrac{3}{2}$ $0.01t = \log_e\left(\dfrac{3}{2}\right)$ $0.01t = 0.4055$ $t = \dfrac{0.4055}{0.01}$ $t = 40.5$ years

WORKED EXAMPLE 14 Using technology with data to test a logarithmic model

Newton's law of cooling states that the rate of heat loss of a body, $\dfrac{dT}{dt}$, where T is the temperature of the body, is directly proportional to the difference between the temperatures of the body and its environment.

A Physics student constructs the following model for their data:

$$\log\left(T - T_r\right) = kt + \log_{10}\left(T_0 - T_r\right)$$

where k is a constant.

The student decides to test this model by measuring the temperature over time of a cooling cup of tea in a room at a constant temperature, $T_r = 18\,°C$.

The initial temperature of the cup is $T_0 = 100\,°C$, and the experimental data collected by the student is as follows.

Time (min)	0	1	3	5	10	20	30	60	90	120
Temperature (°C)	100	98	95	92	85	73	62	42	31	25

Use technology to test the student's model.

TI | THINK

WRITE

CASIO | THINK

WRITE

1. From Graph entry/Edit, select scatterplot and enter the provided data. In Window/Zoom, select Zoom – Fit to display the student's data, T as a function of t, on a scatter diagram, and check that there is no linear relationship between T and t.
Note that you can also create 2 predefined variables containing the time and temperature values to plot, for example using the Lists & Spreadsheet.

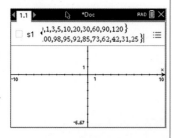

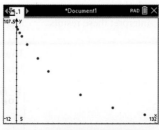

1. From the main menu, enter **Statistics** mode and input the data into a list: enter t values into LIST 1 and T values into LIST 2. Plot the provided data, T as a function of t, on a scatter diagram using GRAPH to check that there is no linear relationship between T and t.

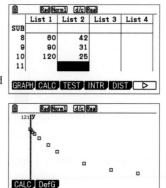

2. Enter the time and temperature values in Lists & Spreadsheet and in a third column enter the formula to calculate $\log(T - T_r)$ (where $T_r = 18$). Give a name to your variables in the columns heading and then from the Graph Entry/Edit menu, select Scatter Plot. You can now, using the var button, select the variables for the scatterplot. In Window/Zoom, select Zoom – Fit. This will plot $\log(T - T_r)$ as a function of t and you can observe a linear relationship, as expected from the student's model.

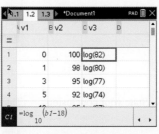

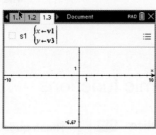

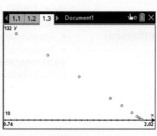

2. Enter log (LIST 2-18) into LIST 3 to calculate $\log(T - T_r)$ (where $T_r = 18$) and plot $\log(T - T_r)$ as a function of t to verify that there is a linear relationship between $\log(T - T_r)$ and t, as this is what is expected from the student's model.

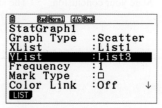

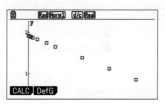

3. From your spreadsheet, go into the Statistics menu. Select Stat Calculation and LinearRegression (mx+b) to choose the regression model. Select your column for the time for the X-list, your column for $\log(T - T_r)$ for the Y-List and an empty column for the linear regression.

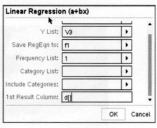

3. Select the linear regression in CALC. The answers appear on the screen.

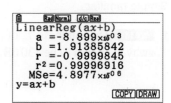

$a = -8.899 \times 10^{-3}$ and $b = 1.914$
(to 3 decimal places).

The answers appear on the screen for $y = a + bx$. Write them for $y = ax + b$.

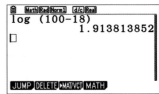

$a = -8.899 \times 10^{-3}$ and $b = 1.914$ (to 3 decimal places).

4. Interpret these results.

$\log(T - 18) = -8.899 \times 10^{-3}t$
$\qquad + 1.914$
Thus, $k = 8.899 \times 10^{-3}$.
Moreover, $\log(T_0 - T_r) \approx 1.914$.

5. State your conclusion.

The student's model,
$\log(T - T_r) = kt + \log(T_0 - T_r)$, with $k = 8.899 \times 10^{-3}$, correctly models their data.

4. Interpret these results.

$\log(T - 18) = -8.899 \times 10^{-3}t$
$\qquad + 1.914$
Thus, $k = 8.899 \times 10^{-3}$.
Moreover, $\log(T_0 - T_r) \approx 1.914$.

5. State your conclusion.

The student's model,
$\log(T - T_r) = kt + \log(T_0 - T_r)$, with $k = 8.899 \times 10^{-3}$, correctly models their data.

Exercise 10.5 Modelling with logarithmic functions

learn on

10.5 Exercise	10.5 Exam practice on

These questions are even better in jacPLUS!
- Receive immediate feedback
- Access sample responses
- Track results and progress

Find all this and MORE in jacPLUS ▶

Simple familiar	Complex familiar	Complex unfamiliar
1, 2, 3, 4, 5, 6, 7, 8, 9, 10	11, 12, 13, 14, 15, 16, 17	18, 19, 20

Simple familiar

1. **WE11a** The loudness, L, of a jet taking off about 30 m away is known to be 130 dB. Using the formula

$$L = 10 \log_{10}\left(\frac{I}{I_0}\right)$$

where I is the intensity measured in W/m^2 and I_0 is equal to 10^{-12} W/m^2, calculate the intensity in W/m^2 for this situation.

2. The moment magnitude scale measures the magnitude, M, of an earthquake in terms of energy released, E, in joules, according to the formula

$$M = 0.67 \, \log_{10} \left(\frac{E}{K} \right)$$

where K is the minimum amount of energy used as a basis of comparison.

An earthquake that measures 5.5 on the moment magnitude scale releases 10^{13} joules of energy. Determine the value of K, correct to the nearest integer.

3. Two earthquakes, about 10 kilometres apart, occurred in Iran on 11 August 2012. One measured 6.3 on the moment magnitude scale, and the other was 6.4 on the same scale. Use the formula from question **2** to compare the energy released, in joules, by the two earthquakes.

4. An earthquake of magnitude 9.0 occurred in Japan in 2011, releasing about 10^{17} joules of energy. Use the formula from question **2** to determine the value of K correct to 2 decimal places.

5. To the human ear, determine how many decibels louder a $500 \, \text{W/m}^2$ amplifier is compared to a $20 \, \text{W/m}^2$ model. Use the formula

$$L = 10 \, \log_{10} \left(\frac{I}{I_0} \right)$$

where L is measured in dB, I is measured in W/m^2 and $I_0 = 10^{-12} \, \text{W/m}^2$. Give your answer correct to 2 decimal places.

6. Your eardrum can be ruptured if it is exposed to a noise that has an intensity of $10^4 \, \text{W/m}^2$. Using the formula

$$L = 10 \log_{10} \left(\frac{I}{I_0} \right)$$

where I is the intensity measured in W/m^2 and I_0 is equal to $10^{-12} \, \text{W/m}^2$, calculate the loudness, L, in decibels that would cause your eardrum to be ruptured.

Questions 7–9 relate to the following information.

Chemists define the acidity or alkalinity of a substance according to the formula

$$\text{pH} = - \log_{10} \, [\text{H}^+]$$

where $[\text{H}^+]$ is the hydrogen ion concentration measured in moles/litre.

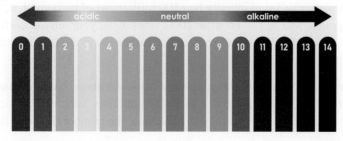

Solutions with a pH less than 7 are acidic, whereas solutions with a pH greater than 7 are alkaline. Solutions with a pH of 7, such as pure water, are neutral.

7. Lemon juice has a hydrogen ion concentration of 0.001 moles/litre. Determine the pH and determine whether lemon juice is acidic or alkaline.

8. Calculate the hydrogen ion concentration for each of the following.
   a. Battery acid has a pH of 0.
   b. Tomato juice has a pH of 4.
   c. Sea water has a pH of 8.
   d. Soap has a pH of 12.

9. Shampoos and hair conditioners can have different hydrogen ion concentrations.

   a. A brand of hair conditioner has a hydrogen ion concentration of 0.000 015 8 moles/litre. Calculate the pH of the hair conditioner.

   b. A brand of shampoo has a hydrogen ion concentration of 0.000 002 75 moles/litre. Calculate the pH of the shampoo.

10. **WE12** The value of a machine is depreciated by 15% per year. The machine was initially valued at $180 000 and will be replaced when its value is less than $60 000.

    a. Determine its value after 3 years.
    b. Determine when the machine will be replaced. Round your answer to the nearest whole year.

Complex familiar

11. **WE13** A deposit of $4200 is invested at the Western Bank, and $5500 is invested at the Common Bank at the same time. Western offers compound interest continuously at a nominal rate of 5%, whereas the Common bank offers compound interest continuously at a nominal rate of 4.5%.
    Determine how many years it will take for the two investments to be the same. Give your answer to the nearest year.

12. a. An investment triples in value over 15 years. Determine the interest rate that this investment earns if it is compounded continuously. Give your answer correct to 2 decimal places.
    b. An investment of $2000 earns 4.5% interest compounded continuously. Determine how long it will take for the investment to grow to $9000. Give your answer to the nearest month.

13. If $1000 is invested for 10 years at 5% interest compounded continuously, determine how much money will have accumulated after the 10 years.

14. Carbon-14 dating works by measuring the amount of carbon-14, a radioactive element, that is present in a fossil. All living things have a constant level of carbon-14 in them. Once an organism dies, the carbon-14 in its body starts to decay according to the rule

$$Q = Q_0 e^{-0.000124t}$$

where t is the time in years since death, Q_0 is the amount of carbon-14 in milligrams present at death, and Q is the quantity of carbon-14 in milligrams present after t years.

    a. If it is known that a particular fossil initially had 100 milligrams of carbon-14, determine how much carbon-14, in milligrams, will be present after 1000 years. Give your answer correct to 1 decimal place.
    b. Determine how long it will take before the amount of carbon-14 in the fossil is halved. Give your answer correct to the nearest year.

15. Glottochronology is a method of dating a language at a particular stage, based on the theory that over a long period of time linguistic changes take place at a fairly constant rate. Suppose a particular language originally has W_0 basic words and that at time t, measured in millennia, the number, $W(t)$, of basic words in use is given by

$$W(t) = W_0 (0.805)^t$$

    a. Calculate the percentage of basic words lost after ten millennia.
    b. Calculate the length of time it would take for the number of basic words lost to be one-third of the original number of basic words. Give your answer correct to 2 decimal places.

16. **WE14** Due to the relatively short half-life of carbon-14 (5730 years), the application of radiocarbon-14 dating is limited to organic matter (containing carbon) formed less than 60 000 years ago.

To estimate the age of geological materials that might have been formed as far as 4.5 billion years ago, other radioactive isotopes with longer half-lives can be used, such as potassium-40, uranium-235 or uranium-238.

For example, uranium-235 (U-235) decays into lead-207 (Pb-207). By measuring the uranium-to-lead ratio in a rock sample, its age can be determined.

The table below displays the ratio of Pb-207 to U-235 and the age of a few rock samples.

Ratio of Pb-207 to U-235	$\frac{1}{3}$	$\frac{2}{5}$	1	9	15	31	100	200	500
Age of sample (in 10^6 years)	3.5	5.4	7.0	23.4	28.2	35.2	46.8	53.8	63.1

Using technology, show that the age of a rock sample t, in millions of years, can be modelled by

$$t = k \ln(x) + C$$

where k and c are constants and x is the ratio of Pb-207 to U-235.

17. Octaves in music can be measured in cents, n. The frequencies of two notes, f_1 and f_2, are related by the equation

$$n = 1200 \log_{10}\left(\frac{f_2}{f_1}\right)$$

Middle C on the piano has a frequency of 256 hertz; the C an octave higher has a frequency of 512 hertz. Calculate the number of cents between these two Cs.

C   D   E   F   G   A   B

Complex unfamiliar

18. Prolonged exposure to sounds above 85 decibels can cause hearing damage or loss. A gunshot from a .22 rifle has an intensity of about $\left(2.5 \times 10^{13}\right) I_0$.
Calculate the loudness, in decibels, of the gunshot sound and state if ear protection should be worn when a person goes to a rifle range for practice shooting.
Use the formula

$$L = 10 \log_{10}\left(\frac{I}{I_0}\right)$$

where I_0 is equal to 10^{-12} W/m^2, and give your answer correct to 2 decimal places.

19. Early in the 20th century, San Francisco had an earthquake that measured 8.3 on the magnitude scale. In the same year, another earthquake was recorded in South America that was four times stronger than the one in San Francisco. Using the equation

$$M = 0.67 \log_{10}\left(\frac{E}{K}\right)$$

calculate the magnitude of the earthquake in South America, correct to 1 decimal place.

20. In 1947 a cave with beautiful prehistoric paintings was discovered in Lascaux, France.

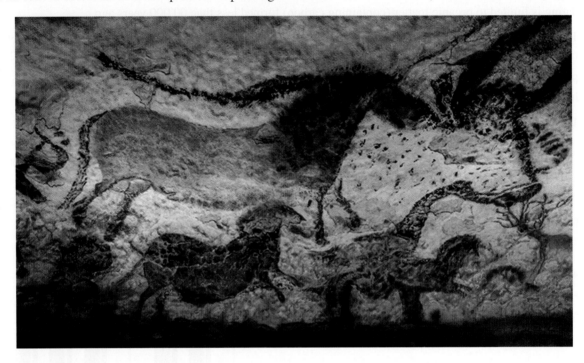

Some charcoal found in the cave contained 20% of the carbon-14 that would be expected in living trees. Determine the age of the paintings to the nearest whole number if

$$Q = Q_0 e^{-0.000\,124t}$$

where Q_0 is the amount of carbon-14 originally present and t is the time in years since the death of the prehistoric material. Give your answer correct to the nearest year.

Fully worked solutions for this chapter are available online.

LESSON
10.6 Review

doc-41805

10.6.1 Summary

Hey students! Now that it's time to revise this chapter, go online to:

 Access the chapter summary

Review your results

A+ Practise exam questions

Find all this and MORE in jacPLUS

10.6 Exercise

learn **on**

10.6 Exercise	10.6 Exam practice **on**

Simple familiar	Complex familiar	Complex unfamiliar
1, 2, 3, 4, 5, 6, 7, 8, 9, 10, 11, 12	13, 14, 15, 16	17, 18, 19, 20

These questions are even better in jacPLUS!
- Receive immediate feedback
- Access sample responses
- Track results and progress

Find all this and MORE in jacPLUS

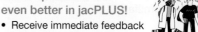

Simple familiar

1. **MC** The statement $3^5 = 243$ expressed in logarithm form would be:

   A. $\log_3(5) = 243$ **B.** $\log_5(3) = 243$ **C.** $\log_5(243) = 3$ **D.** $\log_3(243) = 5$

2. **MC** When expressed in log form, $5^x = 250$ becomes:

   A. $\log_x(5) = 250$ **B.** $\log_5(x) = 250$ **C.** $\log_5(250) = x$ **D.** $\log_x(250) = 5$

3. **MC** The value of $\log_7(49) + 3\log_2(8) - 4$ is:

   A. 3   **B.** 7   **C.** 0   **D.** 69

4. **MC** Choose the correct value for x in the equation $3 + \log_2(3) = \log_2(x)$.

   A. $x = 0$ **B.** $x = 3$ **C.** $x = 9$ **D.** $x = 24$

5. **MC** The equation $y = 5^{3x}$ is equivalent to:

   A. $x = \dfrac{1}{3}\log_5(y)$ **B.** $x = \log_5\left(y^{\frac{1}{3}}\right)$ **C.** $x = \log_5(y) - 3$ **D.** $x = 5^{(y-3)}$

6. Evaluate the following.

   a. $\log_6(9) - \log_6\left(\dfrac{1}{4}\right)$

   b. $2\log_a(4) + 0.5\log_a(16) - 6\log_a(2)$

   c. $\dfrac{\log_a(27)}{\log_a(3)}$

7. Sketch the graph of $y = \log_5(x - 2) + 1$ and identify the domain and range.

8. Solve the following for x.

   a. $\log_5(x-1)=2$

   b. $\log_2(2x+1)=-1$

   c. $\log_x\left(\dfrac{1}{49}\right)=-2$

   d. $\log_x(36)-\log_x(4)=2$

9. Solve the following for x.

   a. $\log_{10}(x+5)=\log_{10}(2)+3\log_{10}(3)$

   b. $\log_4(2x)+\log_4(5)=3$

   c. $\log_{10}(x+1)=\log_{10}(x)+1$

   d. $2\log_6(3x)+3\log_6(4)-2\log_6(12)=2$

10. Solve the equation $\log_3(x)+\log_3(2x+1)=1$ for x.

11. Solve the equation $\log 6(x)-\log_6(x-1)=2$ for x.

12. Determine the maximal domain and the equation of the vertical asymptote for each of the following logarithmic functions.

   a. $y=\log_8(x-6)$

   b. $y=\log_8(x+6)$

   c. $y=\log_8(x)-6$

   d. $y=\log_8(x)+6$

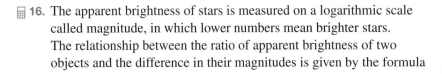

Complex familiar

13. Solve the following equations for x.

   a. $\log_5(x+2)+\log_5(x-2)=1$

   b. $2\log_{10}(x)-\log_{10}(101x-10)=-1$

14. Jan's new neighbours are very noisy. It seems to Jan that the neighbours practise playing their two electric guitars most evenings until quite late at night. The measure of loudness in decibels (dB) is given by

$$L=10\log_{10}\left(I\times10^{12}\right)$$

where L is the number of decibels and I is the intensity of the sound measured in watts per square metre. Given the sound level produced by a guitar is 70 decibels, answer the following questions.

   a. Calculate the intensity of the sound each guitar produces.

   b. Determine the decibel reading when both guitars are played together.

15. If $\log_a(2)=0.43$ and $\log_a(3)=0.68$, determine, correct to 2 decimal places:

   a. $\log_a(6)$

   b. $\log_a(4.5)$

16. The apparent brightness of stars is measured on a logarithmic scale called magnitude, in which lower numbers mean brighter stars. The relationship between the ratio of apparent brightness of two objects and the difference in their magnitudes is given by the formula

$$m_2-m_1=-2.5\log_{10}\left(\dfrac{b_2}{b_1}\right)$$

where m is the magnitude and b is the apparent brightness. Determine how many times brighter a magnitude 2.0 star is than a magnitude 3.0 star.

17. The Richter scale is used to describe the energy of earthquakes.

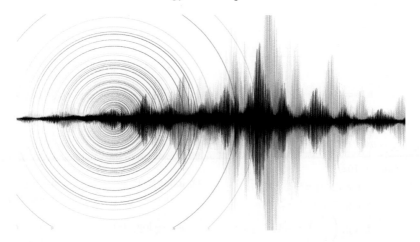

A formula for the Richter scale is

$$R = \frac{2}{3} \log_{10}(K) - 0.9$$

where R is the Richter scale value for an earthquake that releases K kilojoules (kJ) of energy.
Explain why an earthquake measuring 8 on the Richter scale is so much more devastating than one that measures 5.

18. If $\log_a(2) = \log_b(16)$, show that $b = a^4$.

19. $n! = n \times (n-1) \times (n-2) \times \ldots 3 \times 2 \times 1$
Show that $\log(n!) = \log(2) + \log(3) + \ldots + \log(n)$ for $n \in N$ and hence, evaluate $\log(10!) - \log(9!)$.

20. For any integer $x > 1$, it was established in the late nineteenth century that the number of prime numbers less than or equal to x approaches the ratio $\dfrac{x}{\log_e(x)}$ as x becomes large.

Let the function $p(x) = \dfrac{x}{\log_e(x)}$ be an estimate of the number of prime numbers less than or equal to x.

Obtain $p(10)$ and $p(30)$, and compare the estimates with the actual number of primes in each case.

Fully worked solutions for this chapter are available online.

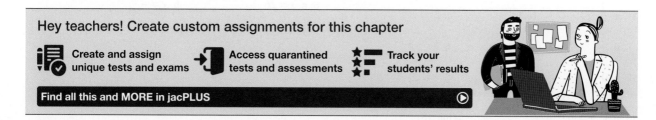

Hey teachers! Create custom assignments for this chapter

Create and assign unique tests and exams

Access quarantined tests and assessments

Track your students' results

Find all this and MORE in jacPLUS

Answers

Chapter 10 Logarithms and logarithmic functions

10.2 Logarithms and logarithmic laws

10.2 Exercise

1. a. $4 = \log_5(625)$ b. $6 = 36^{\frac{1}{2}}$

   c. $x \approx 0.93$ d. $x = \dfrac{1}{3}$

2. a. $\log_e(5) \approx 1.609$; $5 = e^{1.609}$
   b. $10^{3.5} \approx 3162$; $3.5 = \log_{10}(3162)$

3. a. $\log_2(8) = 3$ b. $\log_3(243) = 5$
   c. $\log_5(1) = 0$ d. $\log_{10}(0.01) = -2$

   e. $\log_b(a) = n$ f. $\log_2\left(\dfrac{1}{16}\right) = -4$

4. a. $16 = 4^2$ b. $1\,000\,000 = 10^6$

   c. $\dfrac{1}{2} = 2^{-1}$ d. $27 = 3^3$

   e. $625 = 5^4$ f. $128 = 2^7$

5. a. 4 b. 4 c. 3
   d. -2 e. 3 f. -5

6. a. -2 b. -5 c. 5
   d. -6 e. Undefined f. 5

7. a. $\log_2(80)$ b. $\log_3(105)$ c. 2
   d. $\log_6(56)$ e. 2 f. 1

8. a. $\log_4(5)$ b. $-2\log_{10}(2)$
   c. $2\log_3(2)$ d. $\log_2(3)$
   e. $\log_3(20)$ f. $\log_4(2)$

9. a. 1 b. 4 c. 1 d. $\dfrac{3}{2}$

10. a. $\log_{10}(250)$ b. $\log_2(1728)$
    c. $\log_3(4)$ d. $\log_5(3)$

    e. $\log_{10}\left(\dfrac{1}{4}\right)$ f. $\log_3(2)$

11. a. $\dfrac{2}{3}$ b. 2 c. 2

    d. $\dfrac{3}{2}$ e. $\dfrac{9}{4}$ f. 2

12. a. 0 b. 1 c. -1
    d. 5 e. -3 f. 2

13. a. i. $5 = \log_2(32)$ ii. $\dfrac{3}{2} = \log_4(8)$

    iii. $-3 = \log_{10}(0.001)$

    b. i. $2^4 = 16$ ii. $9^{\frac{1}{2}} = 3$

    iii. $10^{-1} = 0.1$

14. a. 1 b. 0 c. 3
    d. 1 e. $\dfrac{1}{2}$ f. -4

15. a. $\log_2\left(x^4 - 4x^3\right)$ b. $\log_{10}\dfrac{(x+3)^2}{x-2}$

    c. 3 d. 6
    e. 3 f. 6

16. a. $2^x = \dfrac{1}{8}$; $x = -3$

    b. $25^{-0.5} = x$; $x = \dfrac{1}{5}$

    c. $2x = \log_{10}(4)$; $x = \log_{10}(2) = 0.30$ (to 2 d. p.)
    d. $-x = \log_e(3)$; $x = -\log_e(3) = -1.10$ (to 2 d. p.)
    e. $x^3 = 125$; $x = 5$

    f. $x^{-2} = 25$; $x = \dfrac{1}{5}$

17. a. $\log_3(x)$ b. $6\log_a(x)$
    c. 0 d. $\log_b(2a)$

    e. $-\dfrac{1}{2}\log_{10}(x)$ f. 3

18. a. 2 b. -1 c. 3
    d. 0 e. -1 f. 0

19. a. -0.3 b. 0.4 c. 1.3

20. a. $p + q$ b. $3q$
    c. $2p + q$ d. $p - q$
    e. $2q - 4p$ f. $\dfrac{3}{4}pq$

10.3 Indicial equations and logarithms

10.3 Exercise

1. a. $\log_7(15) = 1.392$

   b. $\dfrac{5\log(3)}{\log(4) - 2\log(3)} = -6.774$

2. a. 3.459 b. -0.737 c. 2.727
   d. 0.483 e. 1.292 f. -3.080

3. a. 2.255 b. 0.262 c. 0.661
   d. 0.431 e. -0.423 f. 2.138

4. a. 2 b. 0 c. 3
   d. 4 e. -1 f. -2

5. a. $x = 26$ b. $x = -\dfrac{1}{4}$

   c. $x = 7$ d. $x = 3$

6. $x = \dfrac{3}{10}$

7. $x = 1$

8. $x = \dfrac{36}{35}$

9. a. $x = 49$ b. $x = 6.4$

   c. $x = \dfrac{1}{9}$ d. $x = 3$

10. a. $\sqrt{7}$ b. $\dfrac{4}{5}$ c. $\dfrac{6}{23}$

    d. $\dfrac{1}{2}$ e. $0.001, 8$ f. 3

11. a. -3 **b.** $\dfrac{1}{5}$ **c.** 0.30

d. -1.10 **e.** 5 **f.** $\dfrac{1}{5}$

12. a. $y = 100x$ **b.** $y = 2^{2x} \times x^{-4}$
c. $y = 2^{3x}$ **d.** $y = \log_{10} x + 2$

e. $y = \dfrac{1}{x}$ **f.** $y = \sqrt{x},\ x > 0$

13. a. 3 **b.** $\dfrac{4}{3}$

c. $2^{-\frac{1}{3}},\ 4$ **d.** $\dfrac{7}{2}$

14. a. $\log_2(5)$ or $\log_2(9)$ **b.** $\log_5\left(\sqrt{5} - 2\right)$
c. 0 or $\log_9(2)$ **d.** 2

e. $\dfrac{1}{4}$ or 16 **f.** 1 or 999

15. a. $\log_{11}(18) \approx 1.205$ **b.** $-\log_5(8) \approx -1.292$

c. $\dfrac{1}{2}\log_7(3) \approx 0.2823$

16. a. 125 **b.** 7

17. a. 0.14 **b.** 0.41 **c.** 5.14
d. -0.65 **e.** 0.08 **f.** 1.88

18. $\dfrac{1}{\log_{10}(2)}$

19. 4

20. $16,\ \dfrac{1}{16}$

10.4 Graphs of logarithmic functions

10.4 Exercise

1. a. Asymptote $x = 1$; domain $(1, \infty)$; x-intercept $(2, 0)$

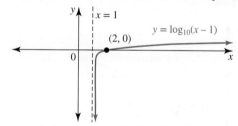

b. Asymptote $x = 0$; domain R^+; x-intercept $(5, 0)$

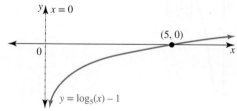

2. a. $(-9, \infty)$

b. $x = -9$

c. $(0, 2)$

d. $(-8, 0)$

e.

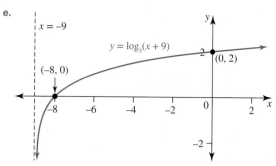

f. R

3. a. $(5, \infty)$

b. $x = 5$

c. $(6, 0)$, no y-intercept

d. and e.

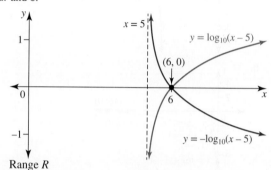

Range R

4. a. $(-6, \infty)$, $x = -6$ **b.** $(6, \infty)$, $x = 6$

c. $(0, \infty)$, $x = 0$ **d.** $\left(-\infty, \dfrac{3}{2}\right)$, $x = \dfrac{3}{2}$

5. a.

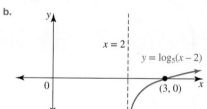

b.

6. a.

b.

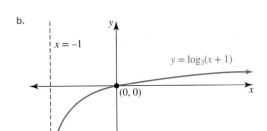

7. a.

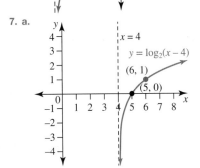

b.

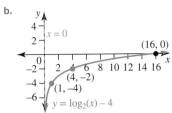

8. $h = 2$, $k = 4$, $y = \log_2(x - 2) + 4$

9. $b = 1$

10.

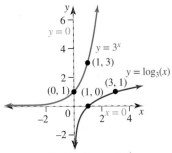

For $y = 3^x$: domain $x \in R$
For $y = \log_3(x)$: domain $x \in R^+$

11.

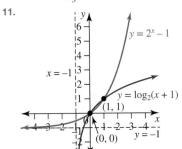

Points of intersection: $(0, 0)$ and $(1, 1)$

12. $y = 14 \log_7\left(\dfrac{x}{2}\right)$

13. $y = -4 \log_3(x) + 4$

14. $y = -\log_2(x + 2) - 1$

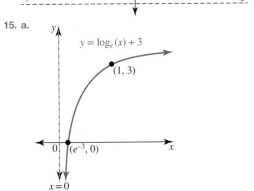

15. a.

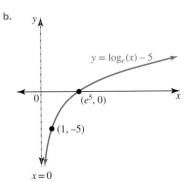

b.

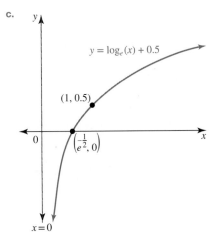

c.

16. a.

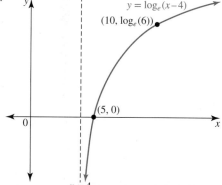

b.

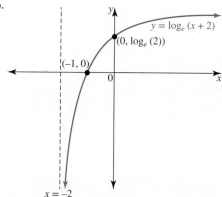

c.

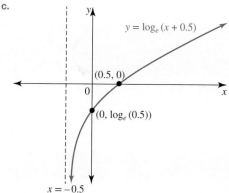

17. $x = 3$

18. $x = 1.883$

19. Sample responses can be found in the worked solutions in the online resources.

20. $w = -\log_e\left(\dfrac{3}{2}\right) \approx -0.4055$

10.5 Modelling with logarithmic functions

10.5 Exercise

1. 10 W/m^2

2. $K = 61\,808$ joules

3. The 6.4 earthquake was 1.41 times bigger than the 6.3 earthquake.

4. $K = 3691.17$ joules

5. A 500-watt amplifier is 13.98 dB louder than a 20-watt amplifier.

6. 160 dB

7. pH $= 3$ (acidic)

8. a. $\left[H^+\right] = 1$ mole/litre

b. $\left[H^+\right] = 0.0001$ moles/litre

c. $\left[H^+\right] = 10^{-8}$ moles/litre

d. $\left[H^+\right] = 10^{-12}$ moles/litre

9. a. 4.8 (acidic) **b.** 5.56 (acidic)

10. a. \$110543 **b.** 7 years

11. 54 years

12. a. 7.32% **b.** 33 years 5 months

13. \$1648.72

14. a. 88.3 mg **b.** 5590 years

15. a. 88.57% lost **b.** 1.87 millennia

16. Sample responses can be found in the worked solutions in the online resources.

17. $n = 361$ cents

18. Ear protection should be worn as $L = 133.98$ dB.

19. 8.7

20. 12 979 years

10.6 Review

10.6 Exercise

1. D

2. C

3. B

4. D

5. A

6. a. 2 **b.** 0 **c.** 3

7.

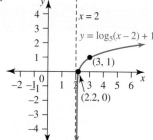

Domain: $x > 2$ or $x \in (2, \infty)$
Range: $y \in R$

8. a. $x = 26$ **b.** $x = -\dfrac{1}{4}$

c. $x = 7$ **d.** $x = 3$

9. a. $x = 49$ **b.** $x = 6.4$

c. $x = \dfrac{1}{9}$ **d.** $x = 3$

10. $x = 1$

11. $x = \dfrac{36}{35}$

12. a. Domain: $x > 6$
       Asymptote: $x = 6$
    b. Domain: $x > -6$
       Asymptote: $x = -6$
    c. Domain: $x > 0$
       Asymptote: $x = 0$
    d. Domain: $x > 0$
       Asymptote: $x = 0$

13. a. 3
    b. $\dfrac{1}{10}$ or 10

14. a. 1×10^{-5} watts/m^2
    b. 73 dB

15. a. 1.11
    b. 0.93

16. Approximately 2.5 times brighter.

17. It releases 31.62^3 times more energy.

18. Sample responses can be found in the worked solutions in the online resources.

19. 1

20. $p(10) \cong 4$; $p(30) \cong 9$
    Although the equation correctly predicted the number of primes equal to 10 or less, it underestimated the number of primes equal to 30 or less.

11 Rates of change and the concept of derivatives

LESSON SEQUENCE

Fully worked solutions for this chapter are available online.

EXAM PREPARATION

Access exam-style questions in every lesson, available online.

on Resources

Solutions	Solutions – Chapter 11 (sol-1278)
Exam questions	Exam question booklet – Chapter 11 (eqb-0299)
Digital documents	Learning matrix – Chapter 11 (doc-41793)
	Chapter summary – Chapter 11 (doc-41806)

LESSON
11.1 Overview

11.1.1 Introduction

Understanding differential calculus is not just about solving mathematical problems; it's about interpreting the world around us. This branch of mathematics helps us analyse rates of change — how fast or slow things happen. For example, it allows us to calculate the speed of a car at a specific moment or the rate at which a plant grows. By learning calculus, you're not just learning a subject, you're gaining a tool to understand and describe natural phenomena, economics, engineering and more. It's a gateway to many fields and can empower you to make informed decisions in real-world situations.

Moreover, calculus is the language of engineers, scientists and economists. It is used to create models of real-life situations and to solve problems that are not just theoretical but have practical applications. Whether it's optimising the design of a new product, predicting weather patterns or understanding the dynamics of the stock market, calculus plays a crucial role.

11.1.2 Syllabus links

Lesson	Lesson title	Syllabus links
11.2	Average rates of change	○ Determine average rate of change in a variety of practical contexts.
11.3	Differentiation from first principles	○ Use the rule $f'(x) = \lim\limits_{h \to 0} \dfrac{f(x+h) - f(x)}{h}$ to determine the derivative of simple power functions and polynomial functions from first principles.
11.4	Differentiation by rule	○ Use the rule $\dfrac{d}{dx}(x^n) = nx^{n-1}$ for positive integers. ○ Recognise and use properties of the derivative $\dfrac{d}{dx}(f(x) + g(x)) = \dfrac{d}{dx}f(x) + \dfrac{d}{dx}g(x).$
11.5	Interpreting the derivatives	○ Interpret the derivative as the instantaneous rate of change. ○ Interpret the derivative as the gradient of a tangent line of the graph of $y = f(x).$ ○ Understand the concept of the derivative as a function.
11.6	Differentiation of power functions	○ Use the rule $\dfrac{d}{dx}(x^n) = nx^{n-1}$ for positive integers. ○ Calculate derivatives of power and polynomial functions.

Source: Mathematical Methods Senior Syllabus 2024 © State of Queensland (QCAA) 2024; licensed under CC BY 4.0.

LESSON
11.2 Average rates of change

SYLLABUS LINKS

• Determine average rate of change in a variety of practical contexts.

Source: Mathematical Methods Senior Syllabus 2024 © State of Queensland (QCAA) 2024; licensed under CC BY 4.0.

11.2.1 Constant and variable rates of change

It is useful for us to consider change as either variable or constant. A situation with a constant rate of change could be a person walking at a constant speed. They start walking at 5 km/h, end walking at 5 km/h and stay at 5 km/h the whole time in-between. In contrast, a sprinter's speed is a variable rate of change. They initially rush up to a high speed and then gradually lose speed as they head to the finish. Graphs of the distance versus time for these two situations could look like the graphs below.

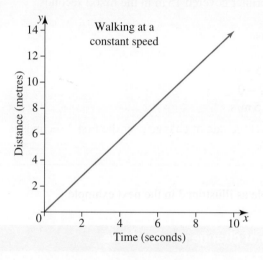
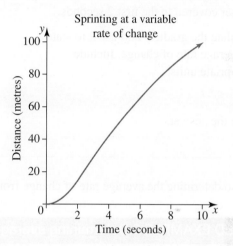

As can be seen, constant rates have a linear function and variable rates have a non-linear function. The rate of change is given by the gradient of the function. For linear functions, the gradient is always the same and calculated by $m = \dfrac{\text{rise}}{\text{run}}$. For variable rates of change, we need to work out what the gradient is at any particular point on the graph. This will be considered later in the chapter.

11.2.2 Average rates of change

One way is to work out what the average gradient is, or the average rate of change, between two points. Basically, we work out the equivalent constant rate of change that would give the same result as the variable rate of change. Consider the graph of our sprinter shown above. Over the course of the race they covered 100 metres in 10 seconds. Average rate of change:

$$
\begin{aligned}
m &= \frac{\text{rise}}{\text{run}} \\
&= \frac{100 - 0}{10 - 0} \\
&= 10 \, \text{m/s}
\end{aligned}
$$

We can therefore conclude that on average they ran at 10 m/s. In other words, if they had run at a constant speed of 10 m/s, they would have ended at the same distance in the same time.

WORKED EXAMPLE 1 Calculating average rate of change from a graph

Calculate the average rate of change for the sprinter over the first 2 seconds.

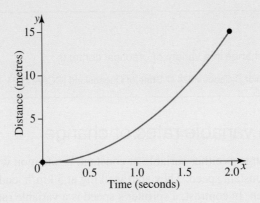

THINK	WRITE
1. From the graph identify the distance the sprinter covered in the first 2 seconds.	The sprinter covered 15 m in the first 2 seconds.
2. Calculate the gradient and use it to state the average rate of change. Include appropriate units.	$m = \dfrac{\text{rise}}{\text{run}}$ $= \dfrac{15-0}{2-0}$ $= 7.5 \text{ m/s}$
3. Write the answer.	The average rate of change over the first 2 seconds is 7.5 m/s.

We can also determine the average rate of change from a table as illustrated in the next example.

WORKED EXAMPLE 2 Determining average rate of change given a table

Determine the average rate of change of height between $t = 1$ and $t = 3$ from the table below.

t (min)	0	1	2	3	4	5
d (m)	20	60	90	130	140	145

THINK	WRITE
Calculate the average rate of change of height with respect to time by considering the change in each quantity. When the time changes from $t = 1$ min to $t = 3$ min, the height changes from 60 m to 130 m.	Average rate of change of height $= \dfrac{\text{change in height}}{\text{change in time}}$ $= \dfrac{(130-60) \text{ m}}{(3-1) \text{ min}}$ $= \dfrac{70 \text{ m}}{2 \text{ min}}$ $= 35 \text{ m/min}$

For any function $y = f(x)$, the **average rate of change** of the function over the interval $x \in [a, b]$ is calculated as $\dfrac{\text{rise}}{\text{run}} = \dfrac{f(b) - f(a)}{b - a}$. This is the gradient of the line, or chord, joining the end points of the interval.

Average rate of change

For the function $y = f(x)$,

$$\text{average rate of change} = \frac{\text{rise}}{\text{run}} = \frac{f(b) - f(a)}{b - a} \text{ for } x \in [a, b].$$

WORKED EXAMPLE 3 Determining the average rate of change of a function

Over a period of 6 hours, the temperature of a room is described by the function $T(h) = h^2 - 4h + 22$ where T is the temperature in degrees Celsius after h hours.
a. Calculate the initial temperature of the room.
b. Sketch the graph of the function over the given time interval.
c. Construct a chord between the points where $h = 1$ and $h = 5$.
d. Determine the gradient of this chord.
e. Determine the average rate of change of temperature between $h = 1$ and $h = 5$.

THINK	WRITE
a. The initial temperature is the temperature at the start of the time period. Substitute $h = 0$ into the function to calculate $T(0)$.	a. When $h = 0$, $T(0) = 0 - 0 + 22$ $= 22$ The initial temperature is 22 °C.
b. Use a graphics calculator or rewrite the function in turning point form.	b. $T(h) = h^2 - 4h + 22$ $= (h^2 - 4h + 4) - 4 + 22$ $= (h - 2)^2 + 18$ The turning point of the parabola is $(2, 18)$.
c. 1. Determine the required points: $T(1) = 1^2 - 4(1) + 22 = 19$ $T(5) = 5^2 - 4(5) + 22 = 27$ 2. Indicate the points $(1, 19)$ and $(5, 27)$ on the graph and join with a straight line.	c.
d. Use gradient $= \dfrac{\text{rise}}{\text{run}} = \dfrac{T(5) - T(1)}{5 - 1}$ and the points $(1, 19)$ and $(5, 27)$.	d. Gradient $= \dfrac{27 - 19}{5 - 1}$ $= \dfrac{8}{4}$ $= 2$
e. Use the gradient to state the average rate of change. Include appropriate units.	e. The average rate of change is 2 °C/h.

11.2 Exercise	**11.2 Exam practice** on

Simple familiar	Complex familiar	Complex unfamiliar
1, 2, 3, 4, 5, 6, 7, 8, 9, 10, 11, 12, 13, 14	15, 16, 17, 18	19, 20

These questions are even better in jacPLUS!
- Receive immediate feedback
- Access sample responses
- Track results and progress

Find all this and MORE in jacPLUS ▶

Simple familiar

1. Identify the following rates as constant or variable.

   a. A person's pulse rate when running 3 km
   b. The rate of growth of Australia's population
   c. A person's pulse rate when lying down
   d. The daily hire rate of a certain car
   e. The rate of growth of a baby
   f. The rate of temperature change during the day

2. Identify the following rates as constant or variable.

   a. The commission rate of pay of a salesperson
   b. The rate at which the Earth spins on its axis
   c. The rate at which students arrive at school in the morning
   d. The rate at which water runs into a bath when the tap is left on
   e. The number of hours of daylight per day over the course of the year
   f. Walking speed up a steep hill

3. Identify whether the following graphs show constant or variable rates of change.

   a. 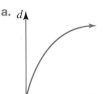 b. 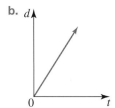 c. 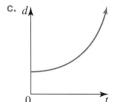 d. 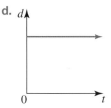 e.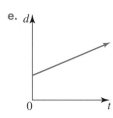

4. Identify whether the following graphs show constant or variable rates of change.

   a. 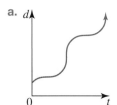 b. 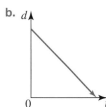 c. 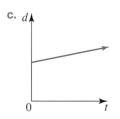 d. 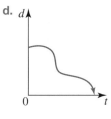 e.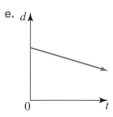

5. **WE1** For each of the following distance–time graphs:

   i. draw a chord to the graph for the interval $t = 1$ to $t = 3$
   ii. calculate the gradient of this chord

iii. hence, determine the average speed from $t = 1$ to $t = 3$.

a.
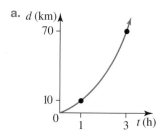

b.
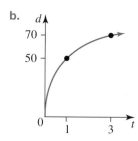

6. The total number of people at the zoo at various times of the day is shown in the table.

	am		pm					
T (time of day)	10:00	11:00	12:00	1:00	2:00	3:00	4:00	5:00
N (number of people at the zoo)	0	200	360	510	540	550	550	550

a. Construct the graph of N versus T.
b. Construct chords to the graph for the interval:
    i. 10:00 am to 1:00 pm
    ii. 1:00 pm to 3:00 pm
    iii. 3:00 pm to 5:00 pm.
c. Calculate the gradient of each of these chords.
d. Determine the average rate of change from:
    i. 10:00 am to 1:00 pm
    ii. 1:00 pm to 3:00 pm
    iii. 3:00 pm to 5:00 pm.
e. Briefly describe what these rates suggest about the number of people attending the zoo during the course of the day.

7. **WE2** The height, h metres, reached by a balloon released from ground level after t minutes, is shown in the table.

t (min)	0	2	4	6	8	10
h (m)	0	220	360	450	480	490

a. Without drawing the graph, calculate the average rate of change of height with respect to time between:
    i. $t = 0$ and $t = 2$
    ii. $t = 2$ and $t = 4$
    iii. $t = 4$ and $t = 6$
    iv. $t = 6$ and $t = 8$
    v. $t = 8$ and $t = 10$.
b. Determine whether the average rate of change for each 2-minute interval is increasing or decreasing.

8. The following tables of values show distance travelled, d km, at various times, t hours. Determine whether the rate of change of distance with respect to time appears 'constant' or 'variable'. Justify your answer.

a.
t	0	1	2	3	4
d	0	5	10	15	20

b.
t	0	1	2	3	4
d	0	15	20	45	80

c.
t	0	1	2	3	4
d	0	6	12	18	24

d.
t	0	1	2	3	4
d	0	4	12	24	40

e.
t	0	1	2	3	4
d	0	1.5	4	8.5	11

f.
t	1	2	3	4	5
d	6	9	13	16	20

9. Murray calculates that the rate of change at $t = 2$ for the graph shown below is 4 °C/hour.
   a. Determine what error he has made.
   b. Calculate a more reasonable value.

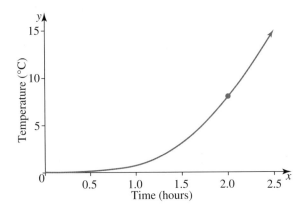

10. Calculate the average rates of change between points A and B, B and C, and A and C for the following graphs. Deduce a general relationship that exists between the three gradients.

a.

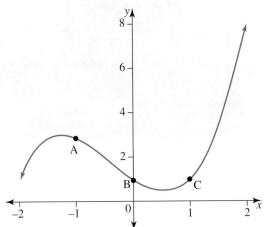

b.

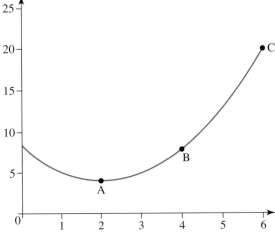

11. The number of crates of fruit picked by a fruit picker over the course of a day is shown in the graph. If the fruit picker is paid $12 per crate, answer the following.

   a. Determine the rate of pay per hour in the first 3 hours.
   b. Explain what probably happened between 12:00 pm and 1:00 pm.
   c. Determine the rate of pay per hour in the last 4 hours.
   d. Propose two possible reasons why the line is not as steep in the afternoon.
   e. Determine how much the fruit picker earned for the day

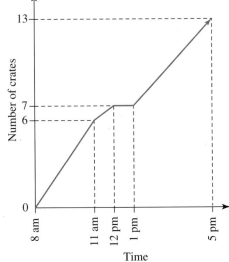

12. The weight of a person over a 40-week period is illustrated in the graph.

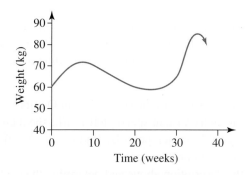

a. Determine the maximum weight and the time at which it occurs.
b. Estimate the average rate of change of weight between week 10 and week 20.
c. Estimate when the person was experiencing their greatest increase in weight.
d. Discuss whether in this situation average or instantaneous rates of change of weight would be of more significance to the person.

13. **WE3** The temperature of an iron rod placed in a furnace is described by the function $T(t) = t^2 + 20$ between $t = 0$ and $t = 10$.
    T represents the temperature of the rod in degrees Celsius and t is the time in seconds.

a. Calculate the initial temperature of the iron rod.
b. Sketch the graph of the function over the given time interval.
c. Construct a chord between the points where $t = 2$ and $t = 8$.
d. Determine the gradient of this chord.
e. Determine the average rate of change of temperature between $t = 2$ and $t = 8$.

14. Calculate the average rate of change of the function $f(x) = x^2 + 3$ over the interval for which $x \in [1, 4]$.

Complex familiar

15. A Petri dish contains 40 bacterial cells that reproduce steadily so that after half an hour there are 100 cells in the dish. Show this information on a graph and calculate the rate at which the cells are reproducing with respect to time.

16. After a review of its operations, part of a company's strategic plan is to improve efficiency in one of its administrative departments. The operating costs for this section of the company are modelled by the function $c(t) = \dfrac{4000}{t+2}$, where t months after the strategic plan came into effect, the operating cost is c. Sketch the graph of the operating costs over time and calculate the average rate of change of the operating costs over the first 2 months.

17. One Australian dollar was worth 0.67 euros in November; by August of the following year one Australian dollar was worth 0.83 euros. Calculate the average rate of change per month in the exchange value of the Australian dollar with respect to the euro over this time period.

18. A small boat sets out to sea away from a jetty. Its distance in metres from the jetty after t hours is given by $d = \dfrac{200t}{t+1}$, $t \geq 0$.
    Calculate the average speed of the boat over the first 4 hours.

19. Jasmine and Jesse drive from Abingdon to Boulia at a speed of 90 km/h, and from Boulia to Clarenvale at an average speed of 100 km/h. If the total time taken for the journey from Abingdon to Clarenvale was 2 hours, and the average speed for the entire journey was 96 km/h, determine the distance between Abingdon and Boulia.

20. A cyclist riding along a straight road returns to the starting point O after 9 hours.
    She travels from O to A in 2 hours at a constant speed of 10 km/h; she then has a rest break for 0.5 hours, after which she rides at constant speed to reach point B, 45 km from O. After a lunch break of 1 hour, the cyclist takes 2.5 hours to return to O at constant speed.
    Construct a graph to represent the description above and determine the cyclist's average speed for the ride.

Fully worked solutions for this chapter are available online.

LESSON
11.3 Differentiation from first principles

SYLLABUS LINKS

- Use the rule $f'(x) = \lim_{h \to 0} \dfrac{f(x+h) - f(x)}{h}$ to determine the derivative of simple power functions and polynomial functions from first principles.

Source: Mathematical Methods Senior Syllabus 2024 © State of Queensland (QCAA) 2024; licensed under CC BY 4.0.

11.3.1 The limit

In mathematics it is important to understand the concept of a limit. This concept is especially important in the study of calculus. In everyday life we use the term *limit* to describe a restriction put on a quantity. For example, the legal blood alcohol concentration limit for a driver is normally 0.05 g/100 mL. As the number of standard alcoholic drinks consumed in 1 hour approaches 2, the average adult male's blood alcohol concentration approaches 0.05. Likewise, some time after a celebration, a person who has been drinking heavily at an earlier time may have a blood alcohol concentration that is approaching the legal limit of 0.05 from a higher level as the number of drinks not yet metabolised by their body approaches

2. We could say that as the number of standard drinks remaining in the body approaches 2, the blood alcohol concentration approaches 0.05. In essence the blood alcohol concentration is a function, say $f(x)$, of the number of drinks, x, remaining in the body.

Add the following series of numbers and identify what value it is approaching.

$$\frac{1}{2} + \frac{1}{4} + \frac{1}{8} + \frac{1}{16} + \frac{1}{32} + \ldots$$

THINK

1. Add the first 2 terms.

2. Add the first 3 terms.

3. Add the first 4 terms.

4. Add the first 5 terms.

5. Add the first 6 terms.

6. Give the upper limit.

WRITE

The sum of the first 2 terms is $\frac{3}{4}$ ($=0.750$).

The sum of the first 3 terms is $\frac{7}{8}$ ($=0.875$).

The sum of the first 4 terms is $\frac{15}{16}$ (≈ 0.938).

The sum of the first 5 terms is $\frac{31}{32}$ (≈ 0.969).

The sum of the first 6 terms is $\frac{63}{64}$ (≈ 0.984).

The sum is approaching 1.

Expressing limits in mathematical language, we say a limit can be used to describe the behaviour of a function, $f(x)$, as the independent variable, x, approaches a certain value, say a. As an example, the notation:

$$\lim_{x \to a} f(x) = b$$

is read as 'the limit of $f(x)$ as x approaches a is equal to b'. This notation is illustrated in the following worked example.

By investigating the behaviour of the function $f(x) = x + 3$ in the vicinity of $x = 2$, show that $\lim_{x \to 2} f(x) = 5$.

THINK

1. Create a table of values for x and $f(x)$ in the vicinity of $x = 2$.

2. Consider the values taken by $f(x)$ as x approaches 2.

WRITE

x	1.95	1.99	1.995	2	2.005	2.01	2.05
$f(x)$	4.95	4.99	4.995	5	5.005	5.01	5.05

As x approaches 2 from the left and the right, $f(x)$ approaches a value of 5. So $\lim_{x \to 2} f(x) = 5$.

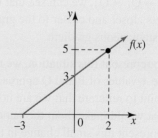

TI	THINK	WRITE	CASIO	THINK	WRITE

TI | THINK

1. On a Graphs page, complete the entry line for function 1 as:
$f1(x) = x + 3$
then press ENTER.

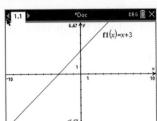

WRITE

CASIO | THINK

1. On a Graph screen, complete the entry line for Y1 as:
$Y1 = x + 3$
then press EXE.
Select DRAW by pressing F6.

WRITE

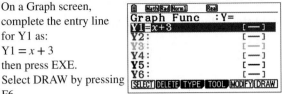

2. Press MENU, then select:
5: Trace
3: Trace Step…
Set the Trace Step to 0.005, then select OK.
Press MENU, then select:
5: Trace
1: Graph Trace
Type '1.95' and press ENTER, then use the right arrow to investigate the behavior of the function as x approaches 2 from the left.
Type '2.05' and press ENTER, then use the left arrow to investigate the behavior of the function as x approaches 2 from the right.

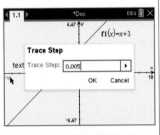

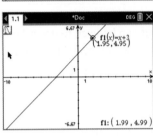

2. Select Trace by pressing SHIFT then F1. Use the right arrow to investigate the behavior of the function as x approaches 2 from the left.
Use the left arrow to investigate the behavior of the function as x approaches 2 from the right.

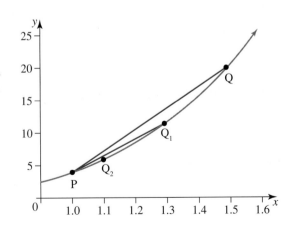

3. The answer appears on the screen.

As x approaches 2 from the left and the right, $f(x)$ approaches a value of 5. So $\lim_{x \to 2} f(x) = 5$.

3. The answer appears on the screen.

As x approaches 2 from the left and the right, $f(x)$ approaches a value of 5. So $\lim_{x \to 2} f(x) = 5$.

11.3.2 The gradient as a limit

We can use this same approach to get a better idea of the value of our gradient, or **instantaneous rate of change**, using the average rate of change. Consider the graph shown. As the distance between two points P and Q gets smaller and smaller, $Q \to Q_1 \to Q_2$, we can see that the gradient between the points gets closer and closer to the gradient of the tangent at P, the instantaneous gradient.

To represent this situation, we make the difference between the x-values of P and Q approach 0. It is important at this point to reiterate that we are not making the difference 0, as that would make them the same point. We use the letter h to represent this small change in the x-value.

Therefore, for any function $f(x)$, P is the point $(x, f(x))$ and Q is the point $(x + h, f(x + h))$. So, if we use our gradient formula to calculate the gradient between our close points, we get the following.

$$
\begin{aligned}
PQ &= \frac{\text{rise}}{\text{run}} \\
&= \frac{f(x + h) - f(x)}{x + h - x} \\
&= \frac{f(x + h) - f(x)}{h}
\end{aligned}
$$

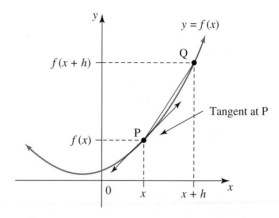

The closer h gets to 0, the closer the two points are to each other, and the more accurate our value for the instantaneous rate of change is.

WORKED EXAMPLE 6 Predicting the gradient of the tangent using limits

Consider the function $f(x) = x^2 + 2x + 1$.
a. Complete the following table of values.

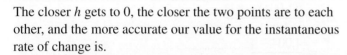

x	1.1	1.01	1.001	1.0001	1
y					

b. Using $x = 1$ as the initial point, state the value of h for each other point.
c. Calculate the average rate of change from $x = 1$ to:
   i. $x = 1.1$ ii. $x = 1.01$ iii. $x = 1.001$ iv. $x = 1.0001$.
d. Predict the gradient of the tangent to the curve at $x = 1$.

THINK

a. Substitute $x = 1.1, 1.01, 1.001, 1.0001, 1$ into $f(x) = x^2 + 2x + 1$.

b. Subtract $x = 1$ from each x-value.

WRITE

a.
$$
\begin{aligned}
f(1.1) &= (1.1)^2 + 2(1.1) + 1 \\
&= 4.41
\end{aligned}
$$

$$
\begin{aligned}
f(1.01) &= (1.01)^2 + 2(1.01) + 1 \\
&= 4.0401
\end{aligned}
$$

$$
\begin{aligned}
f(1.001) &= (1.001)^2 + 2(1.001) + 1 \\
&= 4.004
\end{aligned}
$$

$$
\begin{aligned}
f(1.0001) &= (1.0001)^2 + 2(1.0001) + 1 \\
&= 4.0004
\end{aligned}
$$

$$
\begin{aligned}
f(1) &= (1)^2 + 2(1) + 1 \\
&= 4
\end{aligned}
$$

b.
$$
\begin{aligned}
h_1 &= 1.1 - 1 \\
&= 0.1
\end{aligned}
$$

$$
\begin{aligned}
h_2 &= 1.01 - 1 \\
&= 0.01
\end{aligned}
$$

$$
\begin{aligned}
h_3 &= 1.001 - 1 \\
&= 0.001
\end{aligned}
$$

$$
\begin{aligned}
h_4 &= 1.0001 - 1 \\
&= 0.0001
\end{aligned}
$$

c. Substitute the values into the gradient formula.

$$m_1 = \frac{f(1.1) - f(1)}{1.1 - 1}$$
$$= \frac{4.41 - 4}{0.1}$$
$$= 4.1$$

$$m_2 = \frac{f(1.01) - f(1)}{1.01 - 1}$$
$$= \frac{4.0401 - 4}{0.01}$$
$$= 4.01$$

$$m_3 = \frac{f(1.001) - f(1)}{1.001 - 1}$$
$$= \frac{4.004 - 4}{0.001}$$
$$= 4.001$$

$$m_4 = \frac{f(1.0001) - f(1)}{1.0001 - 1}$$
$$= \frac{4.0004 - 4}{0.0001}$$
$$= 4.0001$$

d. Write the answer.

The gradient at $x = 1$ appears to be approaching 4.

11.3.3 The derivative function

As we considered previously, for any function $y = f(x)$, where P is the point $(x, f(x))$ and Q is the point $(x + h, f(x + h))$, the gradient joining P and Q is given by:

$$PQ = \frac{\text{rise}}{\text{run}}$$
$$= \frac{f(x + h) - f(x)}{x + h - x}$$
$$= \frac{f(x + h) - f(x)}{h}$$

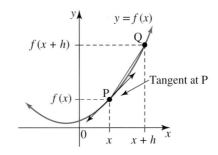

As h becomes smaller, the point Q moves closer and closer to the point P; the line joining the points becomes closer and closer to the tangent at P.

Therefore, as h approaches 0, the gradient of PQ approaches the value of the gradient of the tangent at P. Mathematically, we write this as $\lim_{h \to 0}$, which reads as 'the limit as h approaches 0'.

Hence, the gradient of the tangent at P is $\lim_{h \to 0} \frac{f(x + h) - f(x)}{h}$.

For the function $y = f(x)$, its gradient or **derivative function**, $y = f'(x)$, is defined as the limiting value of the average rate of change as $h \to 0$.

Gradient or derivative function

For the function $y = f(x)$,

$$f'(x) = \lim_{h \to 0} \frac{f(x+h) - f(x)}{h}$$

That is, $f'(x)$ is the gradient function of $f(x)$.

The process of obtaining the gradient or derivative function is called **differentiation by first principles** or determining the derivative from first principles.

The derivative is calculated from forms such as $f'(x) = \lim_{h \to 0} \dfrac{f(x+h) - f(x)}{h}$.

The derivative with respect to x of the function $y = f(x)$ is $\dfrac{dy}{dx} = f'(x)$.

Once the derivative is known, the gradient of the tangent to the curve $y = f(x)$ at a point where $x = a$ is calculated as $\left.\dfrac{dy}{dx}\right|_{x=a}$ or $f'(a)$. Alternatively, this gradient value could be calculated directly from the limit definition using $f'(a) = \lim_{h \to 0} \dfrac{f(a+h) - f(a)}{h}$.

WORKED EXAMPLE 7 Calculating the derivative from first principles

Calculate the derivative of $x^2 - 2x$ using first principles.

THINK	WRITE
1. Define $f(x)$.	$f(x) = x^2 - 2x$
2. The derivative is equal to $f'(x) = \lim_{h \to 0} \dfrac{f(x+h) - f(x)}{h}$.	$f'(x) = \lim_{h \to 0} \dfrac{f(x+h) - f(x)}{h}$
3. Simplify the numerator $f(x+h) - f(x)$.	$f(x+h) - f(x)$ $= (x+h)^2 - 2(x+h) - (x^2 - 2x)$ $= x^2 + 2xh + h^2 - 2x - 2h - x^2 + 2x$ $= 2xh + h^2 - 2h$
4. Factorise the numerator $f(x+h) - f(x)$.	$= h(2x + h - 2)$
5. Simplify $\lim_{h \to 0} \dfrac{f(x+h) - f(x)}{h}$ by cancelling the common factor of h.	$f'(x) = \lim_{h \to 0} \dfrac{f(x+h) - f(x)}{h}$ $= \lim_{h \to 0} \dfrac{h(2x + h - 2)}{h}$ $= \lim_{h \to 0}(2x + h - 2), \ h \neq 0$
6. Evaluate the limit by substituting $h = 0$.	$= 2x - 2$.

The following example illustrates where the derived function can be applied.

WORKED EXAMPLE 8 Determining and applying the derived function

If $g(x) = 2x^2 + 5x - 2$, calculate:

a. the gradient function using first principles

b. the value(s) of x where the gradient equals 0.

THINK	WRITE
a. 1. Let $g(x) = 2x^2 + 5x - 2$.	a. $g(x) = 2x^2 + 5x - 2$
2. The derivative is equal to $g'(x) = \lim\limits_{h \to 0} \dfrac{g(x+h) - g(x)}{h}$.	$g'(x) = \lim\limits_{h \to 0} \dfrac{g(x+h) - g(x)}{h}$
3. Simplify the numerator $g(x+h) - g(x)$.	$g(x+h) - g(x)$ $= 2(x+h)^2 + 5(x+h) - 2 - (2x^2 + 5x - 2)$ $= 2(x^2 + 2xh + h^2) + 5x + 5h - 2 - 2x^2 - 5x + 2$ $= 2x^2 + 4xh + 2h^2 + 5x + 5h - 2 - 2x^2 - 5x + 2$ $= 4xh + 2h^2 + 5h$
4. Factorise the numerator $g(x+h) - g(x)$.	$= h(4x + 2h + 5)$
5. Simplify $\lim\limits_{h \to 0} \dfrac{g(x+h) - g(x)}{h}$ by cancelling the common factor of h.	$g'(x) = \lim\limits_{h \to 0} \dfrac{g(x+h) - g(x)}{h}$ $= \lim\limits_{h \to 0} \dfrac{h(4x + 2h + 5)}{h}$ $= \lim\limits_{h \to 0}(4x + 2h + 5), \; h \neq 0$ $= 4x + 5$
6. Evaluate the limit by substituting $h = 0$.	So $g'(x) = 4x + 5$.
b. Solve $g'(x) = 0$.	b. $g'(x) = 0$ $4x + 5 = 0$ $4x = -5$ $x = -\dfrac{5}{4}$ So the gradient equals 0 when $x = -\dfrac{5}{4}$.

Simple familiar

1. **WE4** Add the following series of numbers and identify what value it is approaching.

$$1 + \frac{1}{3} + \frac{1}{9} + \frac{1}{27} + \frac{1}{81} + \ldots$$

2. **a.** Determine the value of $\frac{1}{n}$ as n gets infinitely large.

   b. Write this using limit notation.

3. **WE5** By investigating the behaviour of the function $f(x) = x + 5$ in the vicinity of $x = 3$, demonstrate that $\lim\limits_{x \to 3} f(x) = 8$.

4. **a.** If $S = 1 + \frac{1}{2} + \frac{1}{4} + \frac{1}{8} + \frac{1}{16} + \ldots$ and n represents the number of terms to be summed in the series, copy and complete the following table.

n	1	2	3	4	5	6	10
S	1	$1\frac{1}{2}$					

   b. **MC** Identify which of the following is equal to $\lim\limits_{n \to \infty} S$.

   A. 1.75 **B.** 1.95 **C.** 2 **D.** 1

5. For $f(x) = x^2 + 3x$:

   a. calculate the gradient $m = \dfrac{f(x+h) - f(x)}{h}$ for $x = 1$ and $h = 0.1, 0.01, 0.001$

   b. use these results to predict the gradient at $x = 1$.

6. The temperature of a meteor, $T\,°C$, t minutes after reaching the Earth's atmosphere is $T = 2t^3 + 5t^2 + 200$ where $t \in [0, 10]$.
   Using $h = 0.001$, the rate of change of the temperature of the meteor after:

   a. 1 minute

   b. 10 minutes.

7. **WE6** Consider the function $f(x) = x^3 - 3x^2 + 2$.

   a. Complete the following table of values.

x	2.6	2.51	2.501	2.5001	2.5
y					

   b. Using $x = 2.5$ as the initial point, state the value of h for each other point.

c. Calculate the average rate of change from $x = 2.5$ to:

    i. $x = 2.6$ ii. $x = 2.51$ iii. $x = 2.501$ iv. $x = 2.5001$.

d. Predict the gradient of the tangent to the curve at $x = 2.5$.

8. Consider the function $f(x) = 0.5x^2 - 3x + 1$.

a. Complete the following table of values.

x	2.1	2.01	2.001	2.0001	2
y					

b. Using $x = 2$ as the initial point, state the value of h for each other point.

c. Calculate the average rate of change from $x = 2$ to:

    i. $x = 2.1$ ii. $x = 2.01$ iii. $x = 2.001$ iv. $x = 2.0001$.

d. Predict the gradient of the tangent to the curve at $x = 2$.

9. **WE7** Calculate the derivatives of the following using first principles.

    a. $5x - 7$ b. $x^2 + 10x$ c. $x^2 - 8x$ d. $x^3 + 2x$

10. **WE8** a. If $f(x) = x^3 - 8$, calculate the gradient function using first principles.

b. Hence, determine the value(s) of x where the gradient function is equal to 12.

11. If $g(x) = x^2 - 6x$, calculate:

a. the gradient function using first principles
b. the value(s) of x where the gradient equals 0.

12. Derive the following functions and calculate where their gradient is equal to -1.

    a. $7x + 5$ b. $x^2 + 4x$ c. $x^2 - 3x + 2$ d. $x^3 - 5$.

13. For the function $f(x) = 4x - x^2$:

a. sketch the function
b. identify the value of x where the gradient is equal to 0
c. determine the derivative of y using the method of first principles
d. calculate the derivative at the value identified in **b**.

14. Identify which of the following denote the gradient at any point on a graph. (One or more answers may be correct.)

    a. $f'(x)$ b. $\lim_{h \to 0} \dfrac{f(x+h) - f(x)}{h}$ c. $\lim_{h \to \infty} \dfrac{f(x+h) - f(x)}{h}$

    d. $\lim_{h \to 0} \dfrac{f(x+h) - f(x)}{(x+h) - x}$ e. $\dfrac{f(x+h) - f(x)}{h}$

15. **MC** The most accurate method for calculating the gradient when $x = 3$ for the function $f(x) = x^2 + 2x$ is by:

    A. sketching the graph and drawing a tangent at $x = 3$ to calculate the gradient
    B. calculating the gradient of the secant to the curve joining the points where $x = 3$ and $x = 3.1$
    C. calculating the derivative using first principles and evaluating at $x = 3$
    D. guessing.

Complex familiar

16. **WE4** Consider the function $f(x) = 2x^2 + x + 1$.

a. Determine the rule of the gradient function.
b. Hence, obtain the gradient of the function at the point $(-1, 2)$.
c. Evaluate the instantaneous rate of change of the function when $x = 0$.

17. Given $f(x) = (x+5)^2$:
    a. determine $f'(x)$ using first principles
    b. calculate $f'(-5)$ and explain its geometric meaning
    c. calculate the gradient of the tangent to the curve $y = f(x)$ at its y-intercept
    d. calculate the instantaneous rate of change of the function $y = f(x)$ at $(-2, 9)$.

18. Given $f(x) = 2(x-4)(x+2)$:
    a. sketch the graph of $y = f(x)$ showing its key points
    b. determine $f'(x)$ using first principles
    c. calculate $f'(1)$ and hence draw the tangent at $x = 1$ on the graph
    d. calculate the gradient of the tangent to the curve at each of its x-intercepts and draw these tangents on the graph.

19. Examine the functions and their derivatives shown in the table below.

Function	Derivative function
$f(x) = 9$	$f'(x) = 0$
$f(x) = 4x$	$f'(x) = 4$
$f(x) = x^2$	$f'(x) = 2x$
$f(x) = 5x^3$	$f'(x) = 15x^2$
$f(x) = 4x^6$	$f'(x) = 24x^5$

    a. Describe, in everyday language or mathematically, a rule for calculating the derivative without using the method of first principles.
    b. Test your rule by applying it to the following functions. Confirm by differentiating using first principles.
        i. $f(x) = 9x - 3$
        ii. $f(x) = x^2 - 8x + 2$

20. Determine, from first principles, the derivatives of:
    a. $\dfrac{1}{x}, \; x \neq 0$
    b. $\dfrac{1}{x^2}, \; x \neq 0$
    c. $\sqrt{x}, \; x > 0$

Fully worked solutions for this chapter are available online.

LESSON
11.4 Differentiation by rule

SYLLABUS LINKS

- Use the rule $\dfrac{d}{dx}(x^n) = nx^{n-1}$ for positive integers.

- Recognise and use properties of the derivative $\dfrac{d}{dx}\big(f(x) + g(x)\big) = \dfrac{d}{dx}f(x) + \dfrac{d}{dx}g(x)$.

Source: Mathematical Methods Senior Syllabus 2024 © State of Queensland (QCAA) 2024; licensed under CC BY 4.0.

11.4.1 Differentiation by rule

Differentiation using the limit definition is an important procedure, yet it can be quite lengthy and for non-polynomial functions it can be quite difficult.

The results obtained from some questions in the previous exercise are compiled into the following table.

Function, $f(x)$	Derivative function, $f'(x)$
$7x + 5$	7
$x^2 + 4x$	$2x + 4$
$x^2 - 3x + 2$	$2x - 3$
$x^3 - 5$	$3x^2$
$2x^2 + x + 1$	$4x + 1$
8	0

Derivative of x^n, $n \in N$

Close examination of these results suggests that the derivative of the polynomial function, $f(x) = x^n$, $n \in N$ is $f'(x) = nx^{n-1}$, a polynomial of one degree less. This observation can be proven to be correct by differentiating x^n using the limit definition.

> **Derivative of $f(x) = x^n$**
>
> If $f(x) = x^n$, $n \in N$,
>
> then $f'(x) = nx^{n-1}$.

It also follows that any coefficient of x^n will be unaffected by the differentiation operation.

> **Derivative of $f(x) = ax^n$**
>
> If $f(x) = ax^n$, $n \in N, a \in R$,
>
> then $f'(x) = anx^{n-1}$.

However, if a function is itself a constant, then its graph is a horizontal line with zero gradient. The derivative with respect to x of such a function must be zero.

> **Derivative of a constant**
>
> If $f(x) = c$ where c is a constant,
> then $f'(x) = 0$.

Where the polynomial function consists of the sum or difference of a number of terms of different powers of x, the derivative is calculated as the sum or difference of the derivative of each of these terms.

> **Derivative of polynomial functions**
>
> If $f(x) = a_n x^n + a_{n-1} x^{n-1} + \ldots + a_1 x + a_0$,
>
> then $f'(x) = a_n n x^{n-1} + a_{n-1}(n-1)x^{n-2} + \ldots + a_1$.

The notation for the derivative has several forms including $f'(x)$, $\dfrac{dy}{dx}$, $\dfrac{d}{dx}f(x)$.

For functions of variables other than x, such as $p = f(t)$, the derivative would be $\dfrac{dp}{dt} = f'(t)$.

Differentiate each of the following.

a. $y = x^8$ b. $y = 3x^2$ c. $y = 7$ d. $y = 5x^3 - 6x^2$

THINK	WRITE

a. 1. Write the expression.

$$y = x^8$$

2. Use the rules of differentiation.

$$\frac{dy}{dx} = 8 \times x^{8-1}$$
$$= 8 \times x^7$$
$$= 8x^7$$

b. 1. Write the expression.

$$y = 3x^2$$

2. Use the rules of differentiation.

$$\frac{dy}{dx} = 3 \times 2x^{2-1}$$
$$= 6x$$

c. 1. Write the expression.

$$y = 7$$

2. Use the rule of differentiation of a constant.

$$\frac{dy}{dx} = 0$$

d. 1. Write the expression.

$$y = 5x^3 - 6x^2$$

2. Use the rules of differentiation.

$$\frac{dy}{dx} = 5 \times 3x^{3-1} - 6 \times 2x^{2-1}$$
$$= 15x^2 - 12x$$

Sometimes it will be necessary to change the way a function is written so that it is in an appropriate form to apply the rules of differentiation. This is illlustrated in the following worked example.

a. **Calculate** $\dfrac{d}{dx}\left(5x^3 - 8x^2\right)$.

b. **Differentiate** $\dfrac{1}{15}x^5 - 7x + 9$ **with respect to** x.

c. **If** $p(t) = (2t + 1)^2$, **calculate** $p'(t)$.

d. **If** $y = \dfrac{x - 4x^5}{x}, x \neq 0$, **calculate** $\dfrac{dy}{dx}$.

THINK	WRITE

a. Apply the rules for differentiation of polynomials.

a. This is a difference of two functions.

$$\frac{d}{dx}\left(5x^3 - 8x^2\right) = \frac{d}{dx}\left(5x^3\right) - \frac{d}{dx}\left(8x^2\right)$$
$$= 5 \times 3x^{3-1} - 8 \times 2x^{2-1}$$
$$= 15x^2 - 16x$$

▶

b. Choose which notation to use and then differentiate.

b. Let $f(x) = \dfrac{1}{15}x^5 - 7x + 9$.

$$f'(x) = \dfrac{1}{15} \times 5x^{5-1} - 7 \times 1x^{1-1} + 0$$

$$= \dfrac{1}{3}x^4 - 7$$

c. 1. Express the function in expanded polynomial form.

c. $p(t) = (2t+1)^2$

$$= 4t^2 + 4t + 1$$

2. Differentiate the function.

$$\therefore p'(t) = 8t + 4$$

d. 1. Express the function in partial fraction form and simplify.
Note: An alternative method is to factorise the numerator as $y = \dfrac{x(1-4x^4)}{x}$ and cancel to obtain $y = 1 - 4x^4$.
Another alternative is to use an index law to write $y = x^{-1}(x - 4x^5)$ and expand using index laws to obtain $y = 1 - 4x^4$.

d. $y = \dfrac{x - 4x^5}{x}, \; x \neq 0$

$$= \dfrac{x}{x} - \dfrac{4x^5}{x}$$

$$= 1 - 4x^4$$

2. Calculate the derivative.

$$y = 1 - 4x^4$$
$$\dfrac{dy}{dx} = -16x^3$$

11.4.2 Properties of the derivative

The rules for the derivative of x^n, $n \in N$ are known as the linearity properties of the derivative, giving the important property for the derivative of the sum of two functions.

Derivative of sum of polynomial functions

If $y = f(x) + g(x)$,

then $\dfrac{dy}{dx} = \dfrac{d}{dx}(f(x) + g(x))$

$$= \dfrac{d}{dx}f(x) + \dfrac{d}{dx}g(x)$$

This property is illustrated in the following example.

WORKED EXAMPLE 11 Determining the derivative of the sum of two functions

Consider the functions $f(x) = 4x^2 + 3x - 6$ and $g(x) = x^3 - 5x^2 + 3x - 9$.

If $y = f(x) + g(x)$, demonstrate that $\dfrac{d}{dx}(f(x) + g(x)) = \dfrac{d}{dx}f(x) + \dfrac{d}{dx}g(x)$.

THINK

1. Determine $f(x) + g(x)$.

2. Collect like terms.

WRITE

$y = f(x) + g(x)$

$$= (4x^2 + 3x - 6) + (x^3 - 5x^2 + 3x - 9)$$

$$= x^3 - x^2 + 6x - 15$$

3. Differentiate the function.

$$\frac{dy}{dx} = 3x^2 - 2x + 6$$

4. Determine $\frac{d}{dx}f(x)$.

$$f(x) = 4x^2 + 3x - 6$$
$$f'(x) = 8x + 3$$

5. Determine $\frac{d}{dx}g(x)$.

$$g(x) = x^3 - 5x^2 + 3x - 9$$
$$g'(x) = 3x^2 - 10x + 3$$

6. Determine the sum and simplify.

$$\frac{d}{dx}f(x) + \frac{d}{dx}g(x) = (8x + 3) + \left(3x^2 - 10x + 3\right)$$
$$= 3x^2 - 2x + 6$$

7. Equate both expressions.

Hence,

$$\frac{d}{dx}(f(x) + g(x)) = \frac{d}{dx}f(x) + \frac{d}{dx}g(x)$$

as both equal $3x^2 - 2x + 6$.

11.4.3 Applying the rule to rates of change

Differential calculus is concerned with the analysis of rates of change. At the start of this topic, we explored rates of change and differentiation from first principles as a method to estimate the instantaneous rate of change of a function. Now, at least for polynomial functions, we can calculate these rates of change using calculus. The instantaneous rate of change is obtained by evaluating the derivative function at a particular point. An average rate of change, however, is the gradient of the line joining the end points of an interval, and its calculation does not involve the use of calculus.

WORKED EXAMPLE 12 Calculating rates of change

Consider the polynomial function with equation $y = \dfrac{x^3}{3} - \dfrac{2x^2}{3} + 7$.

a. Calculate the rate of change of the function at the point $(3, 10)$.

b. Calculate the average rate of change of the function over the interval $x \in [0, 3]$.

c. Obtain the coordinates of the point(s) on the graph of the function where the gradient is $-\dfrac{4}{9}$.

THINK

a. 1. Differentiate the function.

WRITE

a. $y = \dfrac{x^3}{3} - \dfrac{2x^2}{3} + 7$

$$\frac{dy}{dx} = \frac{3x^2}{3} - \frac{4x}{3}$$

$$= x^2 - \frac{4x}{3}$$

2. Calculate the rate of change at the given point.

At the point $(3, 10)$, $x = 3$.

The rate of change is the value of $\dfrac{dy}{dx}$ when $x = 3$.

$$\frac{dy}{dx} = (3)^2 - \frac{4(3)}{3}$$
$$= 5$$

The rate of change of the function at $(3, 10)$ is 5.

b. 1. Calculate the end points of the interval.

b. $y = \dfrac{x^3}{3} - \dfrac{2x^2}{3} + 7$ over $[0, 3]$

When $x = 0$, $y = 7 \Rightarrow (0, 7)$
When $x = 3$, $y = 10 \Rightarrow (3, 10)$ (given)

2. Calculate the average rate of change over the given interval.

The gradient of the line joining the end points of the interval is:

$$m = \frac{10 - 7}{3 - 0}$$
$$= 1$$

Therefore, the average rate of change is 1.

c. 1. Form an equation from the given information.

c. When the gradient is $-\dfrac{4}{9}$, $\dfrac{dy}{dx} = -\dfrac{4}{9}$.

$$\therefore x^2 - \frac{4x}{3} = -\frac{4}{9}$$

2. Solve the equation.

Multiply both sides by 9.
$$9x^2 - 12x = -4$$
$$9x^2 - 12x + 4 = 0$$
$$(3x - 2)^2 = 0$$
$$3x - 2 = 0$$
$$x = \frac{2}{3}$$

3. Calculate the y-coordinate of the point.

$$y = \frac{1}{3}x^3 - \frac{2}{3}x^2 + 7$$

When $x = \dfrac{2}{3}$,

$$y = \frac{1}{3} \times \left(\frac{2}{3}\right)^3 - \frac{2}{3} \times \left(\frac{2}{3}\right)^2 + 7$$
$$= \frac{8}{81} - \frac{8}{27} + 7$$
$$= -\frac{16}{81} + 7$$
$$= 6\frac{65}{81}$$

At the point $\left(\dfrac{2}{3}, 6\dfrac{65}{81}\right)$, the gradient is $-\dfrac{4}{9}$.

TI | THINK	WRITE	CASIO | THINK	WRITE
a. 1. On a Calculator page, press MENU, then select: 4: Calculus 1: Numerical Derivative at a Point … Complete the fields as: Variable: x Value: 3 Derivative: 1st Derivative then select OK. Complete the entry line as: $\dfrac{d}{dx}\left(\dfrac{x^3}{3}-\dfrac{2x^2}{3}+7\right)\Big\|x=3$ then press ENTER.	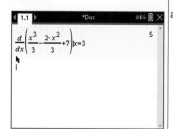	**a. 1.** On a Run-Matrix screen, select MATH by pressing F4, then select $\dfrac{d}{dx}$ by pressing F4. Complete the entry line as: $\dfrac{d}{dx}\left(\dfrac{x^3}{3}-\dfrac{2x^2}{3}+7\right)$ $\|x=3$ then press EXE.	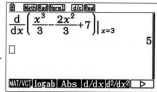
2. The answer appears on the screen.	The rate of change at the point $(3, 10)$ is 5.	**2.** The answer appears on the screen.	The rate of change at the point $(3, 10)$ is 5.

Exercise 11.4 Differentiation by rule

11.4 Exercise	11.4 Exam practice on

Simple familiar	Complex familiar	Complex unfamiliar
1, 2, 3, 4, 5, 6, 7, 8, 9, 10, 11, 12, 13, 14	15, 16, 17	18, 19

These questions are even better in jacPLUS!
- Receive immediate feedback
- Access sample responses
- Track results and progress

Find all this and MORE in jacPLUS ⊙

Simple familiar

1. **WE9** Differentiate each of the following.

   a. $y = x^6$ b. $y = 7x^2$ c. $y = -5x$ d. $y = \dfrac{2}{3}x^2$

2. Match the following functions to their derivatives.

$p = -6w$	$\dfrac{dp}{dw} = -6w^2$
$p = -6$	$\dfrac{dp}{dw} = -6$
$p = -2w^3$	$\dfrac{dp}{dw} = 0$
$p = -3w^2$	$\dfrac{dp}{dw} = -6w$

3. Calculate the value of the derivative at $x = -3$ for each of the following functions.

   a. $y = 0.5x^3$ b. $r = \dfrac{x^2}{3}$ c. $w = 4x$ d. $q = \dfrac{2x^2}{\sqrt{2}}$

4. Calculate $f'(x)$ for each of the following polynomial functions.

    a. $f(x) = 16$ **b.** $f(x) = 21x + 9$ **c.** $f(x) = 3x^2 - 12x + 2$

    d. $f(x) = 0.3x^4$ **e.** $f(x) = -\dfrac{2}{3}x^6$ **f.** $f(x) = 10x^4 - 2x^3 + 8x^2 + 7x + 1$

5. Express each function in expanded polynomial form and then calculate $\dfrac{dy}{dx}$.

    a. $y = -2x(5x - 4)$ **b.** $y = (6x + 5)(4x + 1)$ **c.** $y = \left(\dfrac{x^2}{2} + 1\right)\left(\dfrac{x^2}{2} - 1\right)$

    d. $y = (2x - 3)\left(4x^2 + x + 9\right)$ **e.** $y = (5x + 4)^2$ **f.** $y = 0.5x^3(2 - x)^2$

6. **WE10** **a.** Calculate $\dfrac{d}{dx}\left(5x^8 + \dfrac{1}{2}x^{12}\right)$.

    b. Differentiate $2t^3 + 4t^2 - 7t + 12$ with respect to t.
    c. If $f(x) = (2x + 1)(3x - 4)$, calculate $f'(x)$.

    d. If $y = \dfrac{4x^3 - x^5}{2x^2}$, calculate $\dfrac{dy}{dx}$.

7. **a.** If $f(x) = x^7$, determine $f'(x)$.   **b.** If $y = 3 - 2x^3$, determine $\dfrac{dy}{dx}$.

    c. Calculate $\dfrac{d}{dx}\left(8x^2 + 6x - 4\right)$.   **d.** If $f(x) = \dfrac{1}{6}x^3 - \dfrac{1}{2}x^2 + x - \dfrac{3}{4}$, determine $f'(x)$.

    e. Calculate $\dfrac{d}{du}\left(u^3 - 1.5u^2\right)$.   **f.** If $z = 4\left(1 + t - 3t^4\right)$, determine $\dfrac{dz}{dt}$.

8. Differentiate the following with respect to x.

    a. $(2x + 7)(8 - x)$ **b.** $5x(3x + 4)^2$ **c.** $(x - 2)^4$

    d. $250\left(3x + 5x^2 - 17x^3\right)$ **e.** $\dfrac{3x^2 + 10x}{x}, \; x \neq 0$ **f.** $\dfrac{4x^3 - 3x^2 + 10x}{2x}, \; x \neq 0$

9. **WE11** Confirm for $f(x) = x^2 + 3x - 1$ and $g(x) = 5x^3 + 2x^2 - 9x$ that

$$\dfrac{d}{dx}(f(x) + g(x)) = \dfrac{d}{dx}f(x) + \dfrac{d}{dx}g(x).$$

10. Calculate each of the following.

    a. $f'(-1)$ given $f(x) = \left(3x^4\right)^2$

    b. $g'(2)$ given $g(x) = \dfrac{2x^3 + x^2}{x}, \; x \neq 0$

    c. The gradient of the tangent drawn to the curve $y = -\dfrac{1}{3}x^3 + 4x^2 + 8$ at the point $(6, 80)$

    d. The coordinates of the point on the curve $h = 10 + 20t - 5t^2$ where $\dfrac{dh}{dt} = 0$

11. **WE12** Consider the polynomial function with equation $y = \dfrac{2x^3}{3} - x^2 + 3x - 1$.

    a. Calculate the rate of change of the function at the point $(6, 125)$.
    b. Calculate the average rate of change of the function over the interval $x \in [0, 6]$.
    c. Obtain the coordinates of the point(s) on the graph of the function where the gradient is 3.

12. Consider the function $f(x) = 3x^2 - 2x$.

    a. Calculate the instantaneous rate of change of the function at the point where $x = -1$.
    b. Calculate the average rate of change of the function over the interval $x \in [-1, 0]$.

13. Consider the function $g(x) = x^2 \left(2 + x - x^2\right)$.

  a. Calculate the rate of change of the function at the point $(1, 2)$.

  b. Calculate the average rate of change of the function over the interval $x \in [0, 2]$.

14. Identify and correct any errors made in the following derivative calculation.

$$f(x) = 4.1 \times 10^6 x$$
$$= 6 \times 4.1 \times 10^5 x$$
$$f'(x) = 2.46 \times 10^5 x$$

Complex familiar

15. A rectangular fish tank has a square base with its height being equal to half the length of its base.

  a. Express the length and width of the base in terms of its height, h.

  b. Hence, express the volume, $V\,\text{m}^3$, in terms of the height, h, only.

  c. Determine the rate of change of V when:

  i. $h = 1$ m   **ii.** $h = 2$ m   **iii.** $h = 3$ m.

16. For the triangular package shown, determine:

  a. x in terms of h

  b. the volume, V, as a function of h only

  c. the rate of change of V when:

  i. $h = 0.5$ m

  ii. $h = 1$ m.

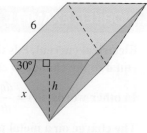

17. Water falling from a leaking roof spreads out on the ground in a circular manner.

  a. Express the relationship between the area of ground, $A\,\text{cm}^2$, over which the water has spread and its radius, r cm.

  b. Determine the rate of change of A with respect to r.

  c. Calculate the rate of change of A when the radius is 5 cm.

  d. Calculate the rate of change of A when the area covered is $30\,\text{cm}^2$.

Complex unfamiliar

18. The instantaneous rate of change of a function is given by $f'(x) = 3x^2 + 2x - 4$. Determine $f(x)$ given that the point $(3, 20)$ lies on the graph line that it describes.

19. A spherical balloon is being inflated. Calculate (to 3 decimal places) the balloon's volume in cubic metres when the rate of change of its volume is equal to $0.36\pi\,\text{m}^3/\text{m}$.

20. A frustum is a truncated cone as shown here. The volume of a frustum can be found from the values R, r and h using the equation $V = \dfrac{1}{3}\pi h \left(R^2 + Rr + r^2\right)$.

A 40-cm-tall glass vase in the shape of a frustum is being filled with water. The opening of the vase has a diameter of 32 cm and the base is 16 cm wide. Determine the rate at which the volume, V, is changing with respect to the depth of water, d, when 10 litres of water have been added to the vase. Give your answer correct to 1 decimal place.

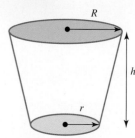

Fully worked solutions for this chapter are available online.

LESSON
11.5 Interpreting the derivative

11.5.1 Interpreting the derivative as the instantaneous rate of change

It is important to remember what it is we are determining when we calculate the derivative. The derivative is the function that expresses the instantaneous rate of change at every point on the original function. Many quantities you may be familiar with are defined as a rate of change.

WORKED EXAMPLE 13 Determining instantaneous rate of change

Electrical current, i, is the flow of electrical charge in a circuit. The current, measured in amperes, is equal to the change in charge, q measured in coulombs, over the change in time, t, in seconds.

In other words, $i(t) = \dfrac{dq}{dt}$.

The charge on a metal plate is changing according to the function $q(t) = 9 - t^2, 0 \le t \le 3$.
Calculate the current flowing across the plate at $t = 2$ seconds.

THINK	WRITE
1. Recognise that current is the derivative of charge with respect to time.	$i(t) = \dfrac{dq}{dt}$
2. Calculate the derivative function $\dfrac{dq}{dt}$.	$q(t) = 9 - t^2$
	$\dfrac{dq}{dt} = -2t$
3. Write $i(t)$.	$\therefore i(t) = -2t$
4. Calculate the current at $t = 2$ seconds.	$i(2) = -2 \times 2$
	$= -4$
5. Write the answer. Include appropriate units.	The current at $t = 2$ s is -4 amperes.

TI | THINK	**WRITE**	**CASIO | THINK**	**WRITE**
1. On a Calculator page, press MENU, then select: 4: Calculus 1: Numerical Derivative at a Point … Complete the fields as: Variable: t Value: 2 Derivative: 1^{st}Derivative then select OK. Complete the entry line as: $\dfrac{d}{dt}(9-t^2)\,\lvert t=2$ then press ENTER.	 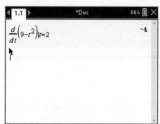	1. On a Run-Matrix screen, select MATH by pressing F4, then select d/dx by pressing F4. Complete the entry line as: $\dfrac{d}{dx}(9-x^2)\,\lvert x=2$ then press EXE.	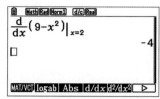
2. The answer appears on the screen.	The current at $t=2$ s is -4 A.	2. The answer appears on the screen.	The current at $t=2$ s is -4 A.

11.5.2 Interpreting the derivative as the gradient of a tangent line

Tangents and normals

As mentioned earlier in this chapter, the derivative $f'(x)$ is actually the gradient function. This means that the value of the gradient at any particular point on a curve is equal to the numerical value of the derivative at that point.

Recall that if the gradient of a tangent to a curve at point P is m_T, then the gradient of a normal (a straight line perpendicular to the tangent)

is $m_N = -\dfrac{1}{m_T}$.

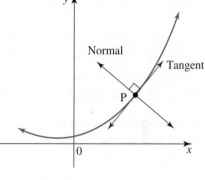

Also recall that the equation of a straight line given a point (x_1, y_1) and gradient m is:

$$y - y_1 = m(x - x_1)$$

We can use the instantaneous rate of change to determine the gradient of the tangent at any point and then the equation of the tangent line at any point on the curve.

WORKED EXAMPLE 14 Determining the equation of the tangent to a curve at a point

a. Determine the equation of the tangent to the curve $f(x) = x^2 + 6x - 8$ at the point where the gradient has a value of 8.
b. Hence, determine the equation of the normal at this point.

THINK	**WRITE**
a. 1. Determine the gradient function of the curve, $f'(x)$.	a. $f(x) = x^2 + 6x - 8$ $f'(x) = 2x + 6$
2. Determine the value of x where $f'(x) = 8$.	For gradient $= 8$ $\qquad f'(x) = 8$ $\qquad 2x + 6 = 8$ $\qquad\quad 2x = 2$ $\qquad\quad\ x = 1$

3. Determine the coordinates of point.

When $x = 1$:
$$f(1) = (1)^2 + 6(1) - 8$$
$$= -1$$
Point on curve: $(1, -1)$

4. Determine the equation of the tangent at $(1, -1)$ with $m = 8$ using the general equation of a straight line, $y - y_1 = m(x - x_1)$.

Equation of the tangent at $(1, -1)$, $m = 8$
$$y - (-1) = 8(x - 1)$$
$$y + 1 = 8x - 8$$
$$y = 8x - 9$$

b. 1. Identify the gradient of the normal using $m_N = -\dfrac{1}{m_T}$.

b. $m_N = -\dfrac{1}{8}$

2. Simplify the equation $y - y_1 = m(x - x_1)$ to determine the equation of the normal.

The equation of the normal at the point $(1, -1)$ is $m = -\dfrac{1}{8}$.

$$y - (-1) = -\frac{1}{8}(x - 1)$$
$$y = -\frac{1}{8}x - \frac{7}{8}$$
$$y + 1 = -\frac{1}{8}x + \frac{1}{8}$$
$$8y = -x - 7$$
$$x + 8y + 7 = 0$$

11.5.3 The derivative as a function

As already observed, the derivative of a polynomial function in x is also a polynomial function, but with one degree less than the original polynomial. Hence, standard function notation can be used for the derivative function. In function notation, the derivative of $f(x)$ is denoted by $f'(x)$.

In function notation, to identify the value of the derivative at a specific value for x, we write f'(value). So, the value of the derivative at $x = -3$ is written as $f'(-3)$.

WORKED EXAMPLE 15 Applying the derivative and function notation

For $f(x) = 4x^3$, determine:
a. the gradient function
b. the instantaneous gradient at $x = 3$.

THINK

a. 1. Use the rule $f'(x) = nax^{n-1}$.

2. Solve for the derivative function.

b. 1. Substitute into the derivative function for $x = 3$.

2. Calculate the derivative and use it to state the instantaneous gradient.

WRITE

a. $f'(x) = 3 \times 4x^{3-1}$

$= 12x^2$

b. $f'(3) = 12(3)^2$

$= 108$

The instantaneous gradient at $x = 3$ is 108.

11.5 Exercise	11.5 Exam practice on

Simple familiar	Complex familiar	Complex unfamiliar
1, 2, 3, 4, 5, 6, 7, 8, 9, 10, 11, 12, 13	14, 15, 16, 17	18, 19, 20

Simple familiar

1. **WE13** The charge across a metal wire varies according to the function $q(t) = 0.1t + 0.6$. Calculate the current at $t = 3$ seconds if current is equal to $\dfrac{dq}{dt}$.

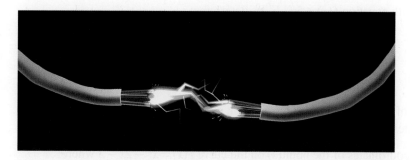

2. Voltage, v, is the difference in electrical potential energy between two points per unit of charge. It is measured in volts, V, and is equal to the change in energy (work done) over the change in charge, $v(q) = \dfrac{dw}{dq}$.

   a. Given $w = 0.2q^2 - 0.1q + 0.5$, derive the voltage function.
   b. Calculate the voltage at $q = 2.5$.

3. **WE14** a. Determine the equation of the tangent to the curve $f(x) = x^2 - 6$ at the point where the gradient has a value of 6.

   b. Hence, determine the equation of the normal at this point.

4. Determine the equation of the tangent line to the function $f(x) = x^2 - 3x$ at $x = 4$.

5. The graph of the function $y = 2x - 0.5x^2$ is shown.

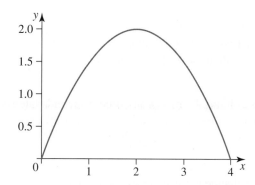

   a. Calculate the gradient of the tangent at $x = 1$.
   b. Sketch the tangent line on the graph and determine the point where the tangent crosses the y-axis.

6. Determine the equation of the normal to the curve $y = 2x^2 - 2x + 5$ at the point where the curve crosses the y-axis.

7. Electrical power, P, of a circuit is the rate at which electrical energy is transferred. It is measured in watts, W, and calculated by the formula $P = \dfrac{dw}{dt}$, where w is the energy, in joules, and t is time, in seconds.

a. Given $w = 0.1t^2 + 0.3t + 0.2$, derive the power function.
b. Calculate the power output at $t = 6$.
c. Sketch the energy and power functions on the same axes.
d. Comment on the relationship between the two functions.

8. **WE15** For each function below, calculate:

   i. the gradient function for $f(x)$
   ii. the instantaneous gradient at $x = 2$.

   a. $f(x) = x^2$ b. $f(x) = \dfrac{x}{6}$ c. $f(x) = 3x^5$ d. $f(x) = 9$

9. For the curve with equation $f(x) = 5x - \dfrac{3}{4}x^2$, determine:

a. $f'(-6)$
b. $f'(0)$
c. $\{x : f'(x) = 0\}$
d. $\{x : f'(x) > 0\}$
e. the gradient of the curve at the points where it cuts the x-axis
f. the coordinates of the point where the tangent has a gradient of 11.

10. Determine the coordinates of the point(s) on the curve $f(x) = x^2 - 2x - 3$ where:

a. the gradient is zero
b. the tangent is parallel to the line $y = 5 - 4x$
c. the tangent is perpendicular to the line $x + y = 7$
d. the slope of the tangent is half that of the slope of the tangent at point $(5, 12)$.

11. Assume an oil spill from an oil tanker is circular and remains that way.

a. Write down a relationship between the area of the spill, A m^2, and the radius, r metres.
b. Determine the rate of change of A with respect to the radius, r.
c. Calculate the rate of change of A when the radius is:

   i. 10 m   ii. 50 m   iii. 100 m.

d. Explain whether the area is increasing more rapidly as the radius increases, and if so, why.

12. The height of a ball thrown in the air is given by the function $h(t) = 8t - t^2$. The rate of change of height with respect to time is the velocity.

a. Derive the function.
b. Calculate the rate of change at $t = 0$, 2, 4, 6 and 8.
c. Determine the velocity of the ball at $t = 0$, 2, 4, 6 and 8.
d. Describe the flight of the ball.

13. A balloon is inflated so that its volume, V cm^3, at any time, t seconds later is:

$$V = -\frac{8}{5}t^2 + 24t$$

a. Calculate the average rate of change between $t = 0$ and $t = 10$.
b. Calculate the rate of change of volume when:

   i. $t = 0$ ii. $t = 5$ iii. $t = 10$.

c. Compare your results from a and b.

14. A bushfire burns out A hectares of land t hours after it started according to the rule:

$$A = 90t^2 - 3t^3.$$

    a. Determine the rate, in hectares per hour, at which the fire is spreading at any time, t.
    b. Calculate the rate when t equals:
      i. 0   **ii.** 4   **iii.** 8   **iv.** 10   **v.** 12   **vi.** 16
    c. Briefly explain how the rate of burning changes during the first 20 hours.
    d. Explain why there isn't a negative rate of change in the first 20 hours.
    e. Describe what happens after 20 hours.
    f. Determine after how long the rate of change is equal to 756 hectares per hour.

15. a. For the graph of $y = f(x)$ shown, state:
      i. $\{x : f'(x) = 0\}$
      ii. $\{x : f'(x) < 0\}$
      iii. $\{x : f'(x) > 0\}$.

    b. Sketch a possible graph for $y = f'(x)$.

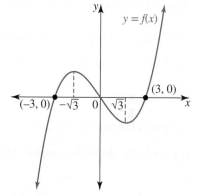

16. For the curve defined by $f(x) = (x - 1)(x + 2)$, determine the coordinates of the point where the tangent is parallel to the line with equation $3x + 3y = 4$.

17. The size of a population of ants in a kitchen cupboard at time t hours after a pot of honey is spilled is given by $N = 0.5(t^2 + 1)^2 + 2.5$.
    a. Calculate the rate at which the population is growing at times $t = 1$ and $t = 2$.
    b. Calculate the average rate of growth over the first 2 hours.
    c. Calculate the time at which the population of ants is increasing at 60 ants/hour.

18. Show that the gradient of the curve $f(x) = x^3 + 9x^2 + 30x + c, c \in R$ is always positive.

19. The radius, r metres, of the circular area covered by an oil spill after t days is $r = at^2 + bt$. After the first day the radius is growing at 6 metres/day, and after 3 days the rate is 14 metres/day. Determine the rate the radius is growing after another 3 days.

20. A section of roller-coaster track is defined by the function $y = 4x - x^2 + 8, 0 \le x \le 4$. The section that joins at $x = 4$ is a straight piece that has to have the same gradient for the pieces to join smoothly. Determine the equation of the straight piece that joins at $x = 4$.

Fully worked solutions for this chapter are available online.

LESSON
11.6 Differentiation of power functions

SYLLABUS LINKS

- Use the rule $\dfrac{d}{dx}(x^n) = nx^{n-1}$ for positive integers.
- Calculate derivatives of power and polynomial functions.

Source: Mathematical Methods Senior Syllabus 2024 © State of Queensland (QCAA) 2024; licensed under CC BY 4.0.

11.6.1 The derivative of $y = x^{-n}$, $n \in N$

A **power function** is of the form $y = x^n$, where n is a rational number. The hyperbola $y = \dfrac{1}{x} = x^{-1}$ and the square

root function $y = \sqrt{x} = x^{\frac{1}{2}}$ are examples of power functions that we have already studied.

Let $f(x) = x^{-n}$, $n \in N$.

Then $f(x) = \dfrac{1}{x^n}$.

Since this function is undefined and therefore not continuous at $x = 0$, it cannot be differentiated at $x = 0$.

For $x \neq 0$, the derivative can be calculated using the limit definition of differentiation.

The derivative of $y = x^{-n}$

For $x \neq 0$, if $f(x) = x^{-n}$ where $n \in N$, then

$$f'(x) = -nx^{-n-1}.$$

This rule means the derivative of the hyperbola $y = \dfrac{1}{x}$
can be obtained in a very similar way to differentiating a polynomial function.

$$y = \frac{1}{x} = x^{-1}$$
$$\frac{dy}{dx} = (-1)x^{-1-1}$$
$$= -x^{-2}$$
$$= -\frac{1}{x^2}$$

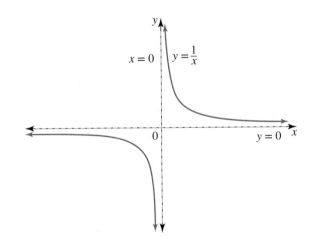

Since this derivative is always negative, it tells us that the hyperbola $y = \dfrac{1}{x}$ is always decreasing, as the gradient of its graph is always negative. This is consistent with what we already know about this hyperbola.

11.6.2 The derivative of $y = x^{\frac{1}{n}}$, $n \in N$

Derivative of $y = x^{\frac{1}{n}}$

If $y = x^{\frac{1}{n}}$, $n \in N$, then

$$\frac{dy}{dx} = \frac{1}{n}x^{\frac{1}{n}-1}.$$

This rule means the derivative of the square root function $y = \sqrt{x}$ can also be obtained in a very similar way to differentiating a polynomial function.

In index form, $y = \sqrt{x}$ is written as $y = x^{\frac{1}{2}}$.

Its derivative is calculated as:

$$\frac{dy}{dx} = \frac{1}{2}x^{\frac{1}{2}-1}$$
$$= \frac{1}{2}x^{-\frac{1}{2}}$$
$$= \frac{1}{2\sqrt{x}}$$

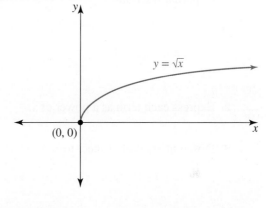

This derivative is always positive, which tells us that the square root function $y = \sqrt{x}$ is an increasing function, as its gradient is always positive. This is consistent with our knowledge of the graph of $y = \sqrt{x}$.

Although the domain of the function $y = \sqrt{x}$ is $R^+ \cup \{0\}$, the domain of its derivative is R^+ as $\dfrac{1}{2\sqrt{x}}$ is undefined when $x = 0$.

11.6.3 The derivative of any power function

The two previous types of power functions considered lead us to generalise that any power function can be differentiated using the same basic rule that was applied to polynomial functions. In fact, $y = x^n$ for any real number n is differentiated using that same basic rule.

Differentiation of any power function

For $n \in R$, if $f(x) = x^n$, then

$$f'(x) = nx^{n-1}.$$

The derivative function will be undefined at any value of x for which the tangent to the graph of the function is undefined.

Before the derivative can be calculated, index laws may be required to express power functions as a sum of terms where each x term is placed in the numerator with a rational index.

For example, $y = \dfrac{1}{x^2} + \sqrt[3]{x}$ would be expressed as $y = x^{-2} + x^{\frac{1}{3}}$, and then its derivative can be calculated using the rule of differentiation.

a. If $y = \dfrac{2 + x^2 - 8x^5}{2x^3}$, calculate $\dfrac{dy}{dx}$ and state its domain.

b. If $f(x) = 2\sqrt{3x}$:

   i. calculate $f'(x)$ and state its domain

   ii. calculate the gradient of the tangent at the point where $x = 3$.

THINK

a. 1. Express the rational function as the sum of its partial fractions.

   2. Express each term as a power of x.

   3. Differentiate with respect to x.

   4. Express each term in x with a positive index.

   5. State the domain of the derivative.

b. i. 1. Express the term in x with a rational index.

   2. Calculate the derivative.

   3. Express x with a positive index.

   4. State the domain of the derivative.

WRITE

a. $y = \dfrac{2 + x^2 - 8x^5}{2x^3}$

$= \dfrac{2}{2x^3} + \dfrac{x^2}{2x^3} - \dfrac{8x^5}{2x^3}$

$= \dfrac{1}{x^3} + \dfrac{1}{2x} - 4x^2$

$\therefore y = x^{-3} + \dfrac{1}{2}x^{-1} - 4x^2$

$\dfrac{dy}{dx} = -3x^{-3-1} - \dfrac{1}{2}x^{-1-1} - 8x^{2-1}$

$= -3x^{-4} - \dfrac{1}{2}x^{-2} - 8x$

$\therefore \dfrac{dy}{dx} = \dfrac{-3}{x^4} - \dfrac{1}{2x^2} - 8x$

There are terms in x in the denominator, so $x \neq 0$.

The domain of the derivative is $R \setminus \{0\}$.

b. i. $f(x) = 2\sqrt{3x}$

$f(x) = 2\sqrt{3}\sqrt{x}$

$= 2\sqrt{3}x^{\frac{1}{2}}$

$f'(x) = 2\sqrt{3} \times \dfrac{1}{2}x^{\frac{1}{2}-1}$

$= \sqrt{3}x^{-\frac{1}{2}}$

$f'(x) = \dfrac{\sqrt{3}}{\sqrt{x}}$

$= \sqrt{\dfrac{3}{x}}$

Due to the square root term on the denominator, $x > 0$.

The domain of the derivative is R^+.

ii. Use the derivative to calculate the gradient at the given point.

ii. The gradient at $x = 3$ is $f'(3)$.

$$f'(3) = \sqrt{\frac{3}{3}}$$
$$= \sqrt{1}$$
$$= 1$$

Therefore, the gradient is 1.

Exercise 11.6 Differentiation of power functions

learn on

11.6 Exercise	11.6 Exam practice on	These questions are even better in jacPLUS!

Simple familiar	Complex familiar	Complex unfamiliar
1, 2, 3, 4, 5, 6, 7, 8, 9, 10, 11, 12, 13, 14	15, 16, 17, 18	19, 20

These questions are even better in jacPLUS!
- Receive immediate feedback
- Access sample responses
- Track results and progress

Find all this and MORE in jacPLUS ▶

Simple familiar

1. Calculate $f'(x)$ for each of the following rational functions.

   a. $f(x) = x^{-3}$

   b. $f(x) = x^{-4}$

   c. $f(x) = x^{\frac{5}{2}}$

   d. $f(x) = x^{\frac{2}{3}}$

   e. $f(x) = 4x^{-2}$

   f. $f(x) = \frac{1}{2}x^{-\frac{6}{5}}$

2. Differentiate the following with respect to x.

   a. $y = 4x^{-1} + 5x^{-2}$

   b. $y = 4x^{\frac{1}{2}} - 3x^{\frac{2}{3}}$

   c. $y = 2 + 8x^{-\frac{1}{2}}$

   d. $y = 0.5x^{1.8} - 6x^{3.1}$

3. Calculate $\dfrac{dy}{dx}$ for each of the following.

   a. $y = 1 + \dfrac{1}{x} + \dfrac{1}{x^2}$

   b. $y = 5\sqrt{x}$

   c. $y = 3x - \sqrt[3]{x}$

   d. $y = x\sqrt{x}$

   e. $y = \dfrac{3}{x^8}$

   f. $y = \dfrac{4}{x} - 5x^3$

4. **WE16** a. If $y = \dfrac{4 - 3x + 7x^4}{x^4}$, calculate $\dfrac{dy}{dx}$ and state its domain.

   b. If $f(x) = 4\sqrt{x} + \sqrt{2x}$:

   i. calculate $f'(x)$ and state its domain

   ii. calculate the gradient of the tangent at the point where $x = 1$.

5. Calculate $f'(x)$, expressing the answer with positive indices, if:

   a. $f(x) = \dfrac{3x^2 + 5x - 9}{3x^2}$

   b. $f(x) = \left(\dfrac{x}{5} + \dfrac{5}{x}\right)^2$

   c. $f(x) = \sqrt[5]{x^2} + \sqrt{5x} + \dfrac{1}{\sqrt{x}}$

   d. $f(x) = 2x^{\frac{3}{4}}(4 + x - 3x^2)$.

6. Calculate $f'(6)$ for the following functions, accurate to 2 decimal places.

   a. $f(x) = 4\sqrt{x} - 13x$

   b. $f(x) = 2x^{-\frac{1}{2}} - 3x^{\frac{2}{3}} - x$

7. a. Calculate the gradient of the tangent to the graph of $y = 3\sqrt{x}$ at the point where $x = 9$.

   b. Calculate the gradient of the tangent to the graph of $y = \sqrt[3]{x} + 10$ at the point $(-8, 8)$.

8. a. Determine the gradient of the tangent to the curve $y = \dfrac{5}{x} - 1$ at its x-intercept.

   b. Obtain the coordinates of the point on the curve $y = 6 - \dfrac{2}{x^2}$ where the gradient of the tangent is $\dfrac{1}{2}$.

9. a. Determine the points on the curve $y = \dfrac{4}{3x}$ at which the tangent to the curve has a gradient of -12.

   b. Calculate the coordinates of the point on $y = 3 - 2\sqrt{x}$ where the tangent is parallel to the line $y = -\dfrac{2}{3}x$.

10. Consider the function $f : R\setminus\{0\} \to R, f(x) = x^2 + \dfrac{2}{x}$.

    a. Evaluate $f'(2)$.
    b. Determine the coordinates of the point for which $f'(x) = 0$.
    c. Determine the exact values of x for which $f'(x) = -4$.

11. A function f is defined as $f : [0, \infty) \to R, \ f(x) = 4 - \sqrt{x}$.

    a. Define the derivative function $f'(x)$.
    b. Obtain the gradient of the graph of $y = f(x)$ as the graph cuts the x-axis.
    c. Calculate the gradient of the graph of $y = f(x)$ when $x = 0.0001$ and when $x = 10^{-10}$.
    d. Describe what happens to the gradient as $x \to 0$.

12. Consider the hyperbola defined by $y = 1 - \dfrac{3}{x}$.

    a. At its x-intercept, the gradient of the tangent is g. Calculate the value of g.
    b. Calculate the coordinates of the other point where the gradient of the tangent is g.
    c. Sketch the hyperbola showing both tangents.
    d. Express the gradients of the hyperbola at the points where $x = 10$ and $x = 10^3$ in scientific notation. Describe what is happening to the tangent to the curve as $x \to \infty$.

13. Danielle calculates a derivative as follows.

$$y = \frac{1}{2x^2}$$
$$= 2x^{-2}$$
$$\frac{dy}{dx} = -2 \times 2x^{-2-1}$$
$$= -4x^{-3}$$
$$= -\frac{4}{x^3}$$

Identify the error(s) she has made and correctly calculate the derivative.

14. The weekly profit, P (hundreds of dollars), of a factory is given by $P = 4.5n - n^{\frac{3}{2}}$, where n is the number of employees.

    a. Determine $\dfrac{dP}{dn}$.
    b. Hence, calculate the rate of change of profit, in dollars per employee, if the number of employees is:

       i. 4   ii. 16   iii. 25.

    c. Determine n when the rate of change is zero.

15. The height of a magnolia tree, in metres, is modelled by $h = 0.5 + \sqrt{t}$, where h is the height t years after the tree was planted.

   a. Calculate how tall the tree was when it was planted.
   b. Determine the rate at which the tree is growing 4 years after it was planted.
   c. Calculate when the tree will be 3 metres tall.
   d. Determine the average rate of growth of the tree over the time period from planting to a height of 3 metres.

16. On a warm day in a garden, water in a bird bath evaporates in such a way that the volume, V mL, at time t hours is given by

$$V = \frac{60t + 2}{3t}, \, t > 0$$

   a. Show that $\dfrac{dV}{dt} < 0$.
   b. Calculate the rate at which the water is evaporating after 2 hours.
   c. Sketch the graph of $V = \dfrac{60t + 2}{3t}$ for $t \in \left[\dfrac{1}{3}, 2\right]$.
   d. Calculate the gradient of the chord joining the end points of the graph for $t \in \left[\dfrac{1}{3}, 2\right]$ and explain what the value of this gradient measures.

17. a. Graph the function $y = \sqrt{x} + \dfrac{1}{x}$ using technology.

   b. Compare the behaviour of $y = \sqrt{x} + \dfrac{1}{x}$ to the functions $y = \sqrt{x}$ and $y = \dfrac{1}{x}$ as $x \to 0$ and $x \to \infty$.
   c. Derive the gradient function of y.
   d. Calculate the domain and range of y.

18. a. Sketch the graph of $y = x^{\frac{2}{3}}$ and $y = x^{\frac{5}{3}}$ using technology and determine the coordinates of the points of intersection.

   b. Compare the gradients of the tangents to each curve at the points of intersection.

19. At a chocolate factory, sugar is added to the chocolate mix according to the function $m(t) = 800\sqrt{t} + t^3$, $0 \le t \le 10$, where m is the mass of chocolate (in grams) and t is the time (in seconds). Calculate the time at which the flow rate is equal to the average flow rate over the whole 10 seconds.

20. The volume of a sphere is given by the formula $V = \dfrac{4}{3}\pi r^3$. Air is being pumped into a bubble at a constant rate.

   a. Define the equation that expresses how the radius of the bubble changes with respect to the volume.
   b. Determine the radius of the bubble when the bubble's radius is changing by less than 0.025 cm/cm^3.

Fully worked solutions for this chapter are available online.

LESSON
11.7 Review

11.7.1 Summary

Hey students! Now that it's time to revise this chapter, go online to:

Access the chapter summary

Review your results

Practise exam questions

Find all this and MORE in jacPLUS

11.7 Exercise

learn on

11.7 Exercise	**11.7 Exam practice** on

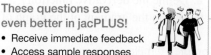

Simple familiar	Complex familiar	Complex unfamiliar
1, 2, 3, 4, 5, 6, 7, 8, 9, 10, 11, 12	13, 14, 15, 16, 17	18, 19, 20

These questions are even better in jacPLUS!
- Receive immediate feedback
- Access sample responses
- Track results and progress

Find all this and MORE in jacPLUS ▶

Simple familiar

1. **MC** Identify which one of the following is a constant rate.

   A. The number of people entering a zoo per hour
   B. The number of days it rains in Brisbane per year
   C. The hourly rate of pay of a tutor
   D. The number of crates of fruit picked per hour by a fruit picker

2. For the table below, calculate the average rate of change of H between $t = 2$ and $t = 5$.

t (h)	0	1	2	3	4	5	6
H (m)	0	20	40	70	120	190	280

3. The amount of substance, A kg, in a container at any time, t hours, is $A = t^2 - 3t + 4$, $t \in [0, 5]$.

   a. Calculate the average rate of change during the first 4 hours.
   b. Determine the rate at which the amount is changing after 4 hours.

4. **MC** If $f(x) = x^2 - 2x$, then $\lim\limits_{h \to 0} \dfrac{f(x + h) - f(x)}{h}$ equals:

   A. $2x - 2$ **B.** $2x + h$ **C.** $2x$ **D.** -2

5. **MC** If $f(x) = 10 - 7x^2 - 3x^3$, then $f'(-1)$ equals:

   A. 6   **B.** 5   **C.** 2   **D.** 0

6. **MC** If $y = (2x^2 - 5)(2x^2 + 5)$, then $\dfrac{dy}{dx}$ equals:

   A. $8x$ **B.** $6x$ **C.** $4x^2$ **D.** $16x^3$

7. **MC** $\dfrac{d}{dx}\left(2\left(x^3 - 5x^2 + 1\right)\right)$ equals:

    A. $2x^3 - 10x^2$ **B.** $6x^3 - 20x^2$ **C.** $2\left(3x - 5\right)$ **D.** $2x\left(3x - 10\right)$

8. **MC** If $y = 4x\sqrt{x}$, $\dfrac{dy}{dx}$ is:

    A. $6\sqrt{x}$ **B.** $4\sqrt{x}$ **C.** $\dfrac{6}{\sqrt{x}}$ **D.** $\dfrac{4}{\sqrt{x}}$

9. For the function defined by $f(x) = 3x^3 - 4x^2 + 2x + 5$, calculate:

    a. $f'(2)$ **b.** the values of x for which $f'(x) = 3$.

10. Determine the equation of the line tangent to the function $f(x) = 2x^2 + x$ at $x = 3$.

11. Obtain the equation of the normal to the curve $y = 6 - x^2$ at the point where the curve crosses the x-axis.

12. A function f is defined as $f : [0, \infty) \to R, f(x) = 3 - 2\sqrt{x}$.

    a. Define the derivative function f'.

    b. Obtain the gradient of the graph of $y = f(x)$ as the graph cuts the x-axis.

Complex familiar

13. The height, h metres, of a bird in flight is shown in the table below.

t (s)	0	2	4	6	8	10	12
h (m)	20	18	8	2	4	5	5

    a. Calculate the average rate of change of height with respect to time over each 2-second interval.

    b. Identify whether the rate of change is increasing or decreasing for each interval.

14. The first part of a children's water slide is defined by the function $y = 0.5x^2 - 2x + 2, 0 \le x \le 1.5$, where y is the height above ground level, in metres, and x is the distance from the start of the slide, in metres. The second section that joins at $x = 1.5$ is a straight piece that has the same gradient, so the sections join smoothly.

    a. Determine the equation of the straight piece that joins at $x = 1.5$.

    b. Determine the horizontal length of the slide.

15. A new estate is to be established on the side of a hill.

Regulations will not allow houses to be built on slopes where the gradient is greater than 0.45. The equation of the cross-section of the hill is

$$y = -0.000\,02x^3 + 0.006x^2$$

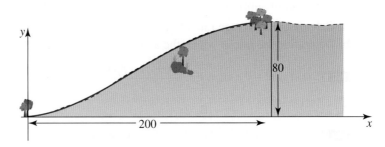

Determine:

    a. the gradient of the slope $\dfrac{dy}{dx}$

    b. the gradient of the slope when x equals:

        i. 160   **ii.** 100   **iii.** 40   **iv.** 20

    c. the values of x where the gradient is 0.45

    d. the range of heights for which houses cannot be built on the hill.

16. A mountain trail can be modelled by the curve with equation $y = 1.8 + 0.16x - 0.005x^4$, where x and y are, respectively, the horizontal and vertical distances measured in kilometres, $0 < x < 3$.

   a. Determine the gradient at the beginning and end of the trail.
   b. Calculate the point where the gradient is 0.
   c. Hence, state the maximum height of the path.

17. Consider the function defined by $y = x^4 + 2x^2$.

   a. Sketch the graphs of $y = x^4 + 2x^2$ and its derivative.
   b. Explain whether or not the derivative graph has a stationary point of inflection at the origin.
   c. Obtain, to 4 decimal places, any non-zero values of x where $\dfrac{dy}{dx} = y$.
   d. Form an equation where the solution(s) to the equation would give the x-values when the graphs of $y = x^4 + 2x^2$ and its derivative are parallel. Give the solution(s) correct to 2 decimal places where necessary.

Complex unfamiliar

18. The population of Town N is thought to be modelled by $P_N = 2\left(1 + 4n + \dfrac{n^2}{4}\right)$, where P_N is the population in thousands n years after 2010.

The model for the population of neighbouring Town W is $P_W = 2\left(1 + n - \left(\dfrac{n}{4}\right)^2\right)$, where P_W is the population in thousands n years after 2010.

Determine the year in which the population of Town N was growing at 12 times the rate of Town W.

19. Determine the equation of the tangent to the curve $f(x) = -\dfrac{2}{x^2} + 1$ that is perpendicular to the line $2y - 2 = -4x$.

20. A company's income each week is $\$\left(500n + 1800\sqrt{n} - 10n^2\right)$, where n is the number of employees. If each employee is paid $750 per week, determine the optimum number of employees for the company to make the most profit. Then determine the profit that will be generated with the optimum number of employees.

Fully worked solutions for this chapter are available online.

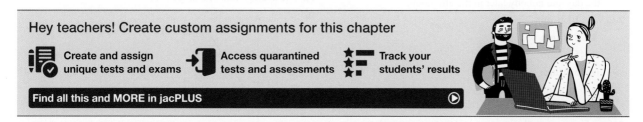

Hey teachers! Create custom assignments for this chapter

Create and assign unique tests and exams

Access quarantined tests and assessments

Track your students' results

Find all this and MORE in jacPLUS

Answers

Chapter 11 Rates of change and the concept of derivatives

11.2 Average rates of change

11.2 Exercise

1. a. Variable
   b. Variable
   c. Constant
   d. Constant
   e. Variable
   f. Variable

2. a. Constant
   b. Constant
   c. Variable
   d. Constant
   e. Variable
   f. Variable

3. a. Variable
   b. Constant
   c. Variable
   d. Constant
   e. Constant

4. a. Variable
   b. Constant
   c. Constant
   d. Variable
   e. Constant

5. a. ii. 30   iii. 30 km/h
   b. ii. 10   iii. 10 km/h

6. a. and b.

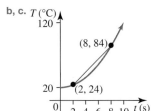

   c. i. 170
      ii. 20
      iii. 0
   d. i. 170 people/h
      ii. 20 people/h
      iii. 0 people/h
   e. Most people arrive in the morning, few in the middle of the day and nobody later in the afternoon.

7. a. i. 110 m/min
      ii. 70 m/min
      iii. 45 m/min
      iv. 15 m/min
      v. 5 m/min
   b. Decreasing

8. a, c: Constant — the increase in distance over each time interval is constant. b, d, e, f: Variable — the increase in distance over each interval is not constant.

9. a. He has calculated the average rate of change.
   b. 11.4 °C/hour (answers may vary slightly)

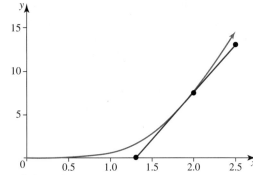

10. a. $-2, 0, -1$
    b. 2, 6, 4
       The average rate of change over AC is the average of the rates of change from AB and BC.

11. a. $24/h
    b. Took a lunch break
    c. $18/h
    d. The picker is getting tired or there is less fruit to find.
    e. $156

12. a. 85 kg at 35 weeks
    b. -1 kg/week
    c. Around 31 weeks
    d. More likely average change, as weight loss is usually a long-term process.

13. a. 20 °C
    b, c.

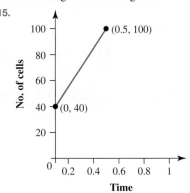

    d. 10
    e. 10 °C/s

14. The average rate of change is 5.

15.

Linear graph with positive gradient; the rate of growth is 120 bacterial cells per hour.

16.

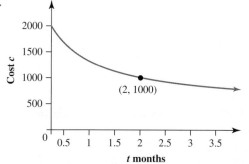

Hyperbola shape for $t \geq 0$; costs decreased at an average rate of $500 per month over the first 2 months.

17. Growing at approximately 0.018 euros per month

18. 40 m/h

19. 72 km

20. 10 km/h

11.3 Differentiation from first principles

11.3 Exercise

1. 3/2

2. a. Approaches 0   **b.** $\lim\limits_{n \to \infty} \dfrac{1}{n} = 0$

3. Solutions will vary but should include a table similar to the following.

x	2.95	2.99	2.995	3	3.005	3.01	3.05
$f(x)$	7.95	7.99	7.995	8	8.005	8.01	8.05

4. a.

n	1	2	3	4	5	6	10
s	1	$1\frac{1}{2}$	$1\frac{3}{4}$	$1\frac{7}{8}$	$1\frac{15}{16}$	$1\frac{31}{32}$	$1\frac{511}{512}$

     b. C

5. a. 5.1, 5.01, 5.001   **b.** 5

6. a. 16 °C/min   **b.** 700 °C/min

7. a.

x	2.6	2.51	2.501	2.5001	2.5
y	-0.704	$-1.087\,05$	$-1.121\,25$	$-1.124\,62$	-1.125

     b. 0.1, 0.01, 0.001, 0.0001, 0

     c. i. 4.21   **ii.** 3.7951

        iii. 3.7545   **iv.** 3.750 45

     d. 3.75

8. a.

x	2.1	2.01	2.001	2.0001	2
y	-3.095	$-3.009\,95$	-3.001	-3.0001	-3

     b. 0.1, 0.01, 0.001, 0.0001, 0

     c. i. -0.95,   **ii.** -0.995,

        iii. -0.9995,   **iv.** $-0.999\,95$

     d. -1

9. a. 5   **b.** $2x + 10$

     c. $2x - 8$ **d.** $3x^2 + 2$

10. a. $3x^2$ **b.** $-2, 2$

11. a. $2x - 6$ **b.** 3

12. a. Never   **b.** $-\dfrac{5}{2}$

     c. 1   **d.** Never

13. a.

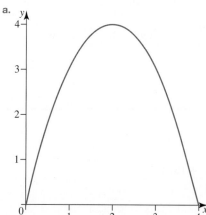

     b. 2

     c. $4 - 2x$

     d. 0

14. a, b, d

15. C

16. a. $f'(x) = 4x + 1$

     b. -3

     c. 1

17. a. $f'(x) = 2x + 10$

     b. 0; tangent at $x = -5$ has zero gradient

     c. 10

     d. 6

18. a.

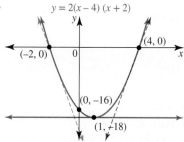

     x-intercepts at $x = -2$, $x = 4$; y-intercept at $y = -16$; minimum turning point $(1, -18)$

     b. $f'(x) = 4x - 4$

     c. 0; tangent at turning point is horizontal

     d. $-12; 12$

19. a. $f'(x) = nax^{n-1}$

     b. i. 9   **ii.** $2x - 8$

20. a. $-\dfrac{1}{x^2}$, $x \neq 0$ **b.** $-\dfrac{2}{x^3}$, $x \neq 0$

     c. $\dfrac{1}{2\sqrt{x}}$, $x > 0$

11.4 Differentiation by rule

11.4 Exercise

1. a. $6x^5$ b. $14x$ c. -5 d. $\dfrac{4}{3}x$

2. $p = -6w$, $\dfrac{dp}{dw} = -6$; $p = -6$, $\dfrac{dp}{dw} = 0$;

   $p = -2w^3$, $\dfrac{dp}{dw} = -6w^2$; $p = -3w^2$, $\dfrac{dp}{dw} = -6w$

3. a. 13.5 b. -2 c. 4 d. $-6\sqrt{2}$

4. a. 0
   b. 21
   c. $6x - 12$
   d. $1.2x^3$
   e. $-4x^5$
   f. $40x^3 - 6x^2 + 16x + 7$

5. a. $-20x + 8$
   b. $48x + 26$
   c. x^3
   d. $24x^2 - 20x + 15$
   e. $50x + 40$
   f. $6x^2 - 8x^3 + 2.5x^4$

6. a. $40x^7 + 6x^{11}$
   b. $6t^2 + 8t - 7$
   c. $12x - 5$
   d. $2 - \dfrac{3}{2}x^2$

7. a. $f'(x) = 7x^6$
   b. $\dfrac{dy}{dx} = -6x^2$
   c. $16x + 6$
   d. $D_x(f) = \dfrac{1}{2}x^2 - x + 1$
   e. $3u^2 - 3u$
   f. $\dfrac{dz}{dt} = 4\left(1 - 12t^3\right)$

8. a. $-4x + 9$
   b. $135x^2 + 240x + 80$
   c. $4x^3 - 24x^2 + 48x - 32$
   d. $250\left(3 + 10x - 51x^2\right)$
   e. 3
   f. $4x - \dfrac{3}{2}$

9. Sample responses can be found in the worked solutions in the online resources.

10. a. -72 b. 9 c. 12 d. $(2, 30)$

11. a. 63 b. 21 c. $(0, -1)$, $\left(1, \dfrac{5}{3}\right)$

12. a. -8 b. -5

13. a. 3 b. 0

14. $f'(x) = 4.1 \times 10^6 x^{1-1}$
    $= 4.1 \times 10^6$

15. a. Length $= 2h$, width $= 2h$
    b. $V = 4h^3$
    c. i. $12\,\text{m}^3/\text{m}$ ii. $48\,\text{m}^3/\text{m}$ iii. $108\,\text{m}^3/\text{m}$

16. a. $x = 2h$
    b. $V = 6\sqrt{3}h^2$
    c. i. $\dfrac{dV}{dh} = 6\sqrt{3}$ ii. $\dfrac{dV}{dh} = 12\sqrt{3}$

17. a. $A = \pi r^2$ b. $\dfrac{dA}{dr} = 2\pi r$
    c. $10\pi\,\text{cm}^2/\text{cm}$ d. 19.416 (to 3 d.p.)

18. $f(x) = x^3 + x^2 - 4x - 4$

19. $0.113\,\text{m}^3$

20. 389.6

11.5 Interpreting the derivative

11.5 Exercise

1. 0.1

2. a. $0.4q - 0.1$ b. 0.9

3. a. $y = 6x - 15$ b. $y = -\dfrac{1}{6}x + \dfrac{7}{2}$

4. $y = 5x - 16$

5. a. 1
   b. 0.5

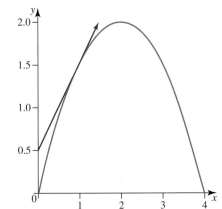

6. $y = \dfrac{1}{2}x + 5$

7. a. $0.2t + 0.3$
   b. 1.5 W

c.

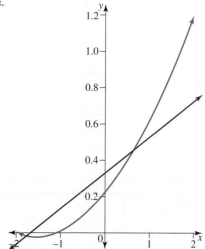

d. The derivative function equals 0 when the function's gradient is 0.

8. a. i. $f'(x) = 2x$ **ii.** 4

  b. i. $f'(x) = \dfrac{1}{6}$ **ii.** $\dfrac{1}{6}$

  c. i. $f'(x) = 15x^4$ **ii.** 240

  d. i. $f'(x) = 0$ **ii.** 0

9. a. 14   **b.** 5

  c. $\left\{\dfrac{10}{3}\right\}$ **d.** $\left\{x: x < \dfrac{10}{3}\right\}$

  e. 5 and -5 **f.** $(-4, -32)$

10. a. $(1, -4)$ **b.** $(-1, 0)$

  c. $\left(\dfrac{3}{2}, -\dfrac{15}{4}\right)$ **d.** $(3, 0)$

11. a. $A = \pi r^2$

  b. $\dfrac{dA}{dr} = 2\pi r$

  c. i. $20\pi \text{ m}^2/\text{m}$ **ii.** $100\pi \text{ m}^2/\text{m}$
    iii. $200\pi \text{ m}^2/\text{m}$

  d. Yes, because $\dfrac{dA}{dr}$ is increasing

12. a. $8 - 2t$

  b. $8, 4, 0, -4, -8$

  c. Same as **b**

  d. Slows travelling up, then stops, then drops and gets faster as it falls.

13. a. $8 \text{ cm}^3/\text{s}$

  b. i. 24   **ii.** 8   **iii.** -8

  c. Initially the rate is faster than the average. Towards the end is it negative, indicating the balloon was deflated. The average rate of change was equal to the instantaneous rate of change at $t = 5$.

14. a. $\dfrac{dA}{dt} = 180t - 9t^2$ hectares/hour

  b. i. 0   **ii.** 576
    iii. 864   **iv.** 900
    v. 864   **vi.** 576

c. The fire spreads at an increasing rate in the first 10 hours, then at a decreasing rate in the next 10 hours.

d. The fire is spreading; the area burnt out by a fire does not decrease.

e. The fire stops spreading; that is, the fire is put out or contained to the area already burnt.

f. $t = 6$ and $t = 14$ hours .

15. a. i. $\left\{\pm\sqrt{3}\right\}$

    ii. $\left\{x: -\sqrt{3} < x < \sqrt{3}\right\}$

    iii. $\left\{x: x < -\sqrt{3}\right\} \cup \left\{x : x > \sqrt{3}\right\}$

  b.

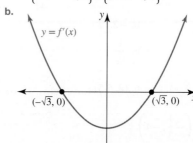

Concave up parabola with x-intercepts at $x = \pm\sqrt{3}$

16. $(-1, -2)$

17. a. 4 ants/hour when $t = 1$ and 20 ants/hour when $t = 2$

  b. 6 ants/hour

  c. 3 hours

18. Sample responses can be found in the worked solutions in the online resources.

19. 26 metres/day

20. $y = 16 - 4x$

11.6 Differentiation of power functions

11.6 Exercise

1. a. $f'(x) = -3x^{-4}$ **b.** $f'(x) = -4x^{-5}$

  c. $f'(x) = \dfrac{5}{2}x^{\frac{3}{2}}$ **d.** $f'(x) = \dfrac{2}{3}x^{-\frac{1}{3}}$

  e. $f'(x) = -8x^{-3}$ **f.** $f'(x) = -\dfrac{3}{5}x^{-\frac{11}{5}}$

2. a. $\dfrac{dy}{dx} = -4x^{-2} - 10x^{-3}$ **b.** $\dfrac{dy}{dx} = 2x^{-\frac{1}{2}} - 2x^{-\frac{1}{3}}$

  c. $\dfrac{dy}{dx} = -4x^{-\frac{3}{2}}$ **d.** $\dfrac{dy}{dx} = 0.9x^{0.8} - 18.6x^{2.1}$

3. a. $\dfrac{dy}{dx} = -\dfrac{1}{x^2} - \dfrac{2}{x^3}$ **b.** $\dfrac{dy}{dx} = \dfrac{5}{2\sqrt{x}}$

  c. $\dfrac{dy}{dx} = 3 - \dfrac{1}{3x^{\frac{2}{3}}}$ **d.** $\dfrac{dy}{dx} = \dfrac{3\sqrt{x}}{2}$

  e. $\dfrac{dy}{dx} = -\dfrac{24}{x^9}$ **f.** $\dfrac{dy}{dx} = -\dfrac{4}{x^2} - 15x^2$

4. a. $\dfrac{dy}{dx} = -\dfrac{16}{x^5} + \dfrac{9}{x^4}$; domain $R\backslash\{0\}$

b. i. $f'(x) = \dfrac{2}{\sqrt{x}} + \dfrac{\sqrt{2}}{2\sqrt{x}}$; domain R^+

ii. Gradient $= \dfrac{4 + \sqrt{2}}{2}$

5. a. $\dfrac{6}{x^3} - \dfrac{5}{3x^2}$ **b.** $\dfrac{2x}{25} - \dfrac{50}{x^3}$

c. $\dfrac{2}{5\sqrt[5]{x^3}} + \dfrac{\sqrt{5}}{2\sqrt{x}} - \dfrac{1}{2\sqrt{x^3}}$ **d.** $-\dfrac{33}{2}x^{\frac{7}{4}} + \dfrac{7}{2}x^{\frac{3}{4}} + \dfrac{6}{x^{\frac{1}{4}}}$

6. a. -12.18 **b.** -2.17

7. a. $\dfrac{1}{2}$ **b.** $\dfrac{1}{12}$

8. a. $-\dfrac{1}{5}$ **b.** $(2, 5.5)$

9. a. $\left(\dfrac{1}{3}, 4\right)$ and $\left(-\dfrac{1}{3}, -4\right)$.

b. $\left(\dfrac{9}{4}, 0\right)$

10. a. 3.5

b. $(1, 3)$

c. $x = -1, x = \dfrac{-1 \pm \sqrt{5}}{2}$

11. a. $f'(x) = \dfrac{-1}{2\sqrt{x}}$ **b.** $-\dfrac{1}{8}$

c. $-50,\ -50\,000$ **d.** Approaches $-\infty$

12. a. $g = \dfrac{1}{3}$

b. $(-3, 2)$

c.

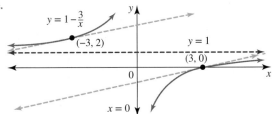

Asymptotes $x = 0,\ y = 1$

d. 3×10^{-2}, 3×10^{-6}, tangent approaches the horizontal asymptote.

13. $\dfrac{1}{2x^2} = \dfrac{1}{2}x^{-2}$; therefore, the solution is $-\dfrac{1}{x^3}$

14. a. $\dfrac{dp}{dn} = 4.5 - 1.5n^{\frac{1}{2}}$

b. i. \$37.50 **ii.** $-\$9.38$ **iii.** $-\$12.00$

c. $n = 9$

15. a. 0.5 metres **b.** 0.25 metres/year

c. $6\dfrac{1}{4}$ years **d.** 0.4 metres/year

16. a. $\dfrac{dV}{dt} = -\dfrac{2}{3t^2}$, which is < 0.

b. $\dfrac{1}{6}$ mL/h

c.

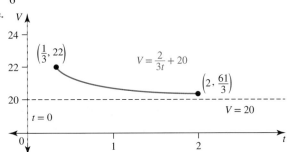

Part of hyperbola with asymptotes at $V = 20, t = 0$; end points $\left(\dfrac{1}{3}, 22\right), \left(2, \dfrac{61}{3}\right)$

d. -1, average rate of evaporation

17. a.

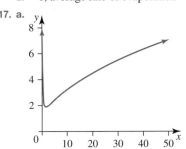

b. As $x \to 0$, y behaves like $\dfrac{1}{x}$, and as $x \to \infty$, y behaves like $\sqrt{x}$.

c. $\dfrac{1}{2\sqrt{x}} - \dfrac{1}{x^2}$

d. $x > 0,\ y \geq 2^{\frac{1}{3}} + 2^{\frac{-2}{3}}$

18. a.

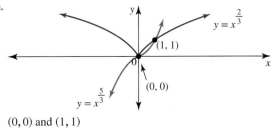

$(0, 0)$ and $(1, 1)$

b. At $(1, 1)$, the gradient of $y = x^{\frac{2}{3}}$ is $\dfrac{2}{3}$ and the gradient of $y = x^{\frac{5}{3}}$ is $\dfrac{5}{3}$, so $y = x^{\frac{5}{3}}$ is steeper; at $(0, 0)$, the gradient of $y = x^{\frac{2}{3}}$ is undefined and the gradient of $y = x^{\frac{2}{3}}$ is zero.

19. 1.323 seconds and 8.477 seconds

20. a. $\dfrac{dr}{dV} = \sqrt[3]{\dfrac{1}{36\pi V^2}}$ **b.** 1.784 m

11.7 Review

11.7 Exercise

1. C

2. 50 m/h

3. a. 1 kg/h b. 5 kg/h

4. A

5. B

6. D

7. D

8. A

9. a. 22 b. $-\dfrac{1}{9}, 1$

10. $y = 13x - 18$

11. $\dfrac{1}{2\sqrt{6}}x - \dfrac{1}{2}$ and $-\dfrac{1}{2\sqrt{6}}x - \dfrac{1}{2}$

12. a. $-\dfrac{1}{\sqrt{x}}$

    b. The gradient at $y = 0$ is $-\dfrac{2}{3}$.

13. a. $-1, \ -5, \ -3, 1, 0.5, 0$

    b. Decreasing, decreasing, decreasing, increasing, increasing, neither

14. a. $0.875 - 0.5x$ b. 1.75 m

15. a. $\dfrac{dy}{dx} = -0.000\,06x^2 + 0.012x$

    b. i. 0.384 ii. 0.6
       iii. 0.384 iv. 0.216

    c. $x = 50$ and $x = 150$

    d. $12.5 < y < 67.5$

16. a. The gradient is 0.16 at the beginning and -0.38 at the end.

    b. $(2, \ 2.04)$

    c. The maximum height is 2.04 km.

17. a.

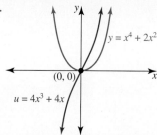

    b. The gradient of the derivative graph at the origin is not 0, so it is not a stationary point of inflection.

    c. $x = 3.7511$

    d. $4x^3 - 12x^2 + 4x - 4 = 0, \ x = 2.77$

18. 2014

19. $y = \dfrac{1}{2}x - \dfrac{1}{2}$

20. \$2549.08 with 6 employees

12 Graphical applications of derivatives

LESSON SEQUENCE

Fully worked solutions for this chapter are available online.

EXAM PREPARATION

Access exam-style questions in every lesson, available online.

on Resources

Solutions	Solutions – Chapter 12 (sol-1279)
Exam questions	Exam question booklet – Chapter 12 (eqb-0300)
Digital documents	Learning matrix – Chapter 12 (doc-41794)
	Chapter summary – Chapter 12 (doc-41807)

LESSON
12.1 Overview

12.1.1 Introduction

In the previous chapter, we saw that differential calculus can be applied to any situation where we are dealing with change. This chapter introduces the application of derivatives to four common fields: tangents, normal, kinematics and curve sketching.

These fields of study were the driving forces behind the development of calculus by its originators, Sir Isaac Newton and Gottfried Leibniz. Newton was interested in problems relating to motion, or kinematics, while Leibniz was interested in solving the tangent problem. This led both Newton and Leibniz to confront the problem of infinitesimals (infinitely small quantities) and their application to the study of change.

Scientists, mathematicians and engineers continue to use infinitesimals in their fields of study. They are used to help formulate equations for the relationships between all forms of energy, also known as the study of thermodynamics. For example, the internal combustion engine in a car can be explained using such equations.

12.1.2 Syllabus links

Lesson	Lesson title	Syllabus links
12.2	**Instantaneous rates of change**	○ Determine instantaneous rates of change.
12.3	**Tangents and normals of a graph**	○ Determine the equation of a tangent and a normal of the graph of $y = f(x)$.
12.4	**Kinematics**	○ Construct and interpret displacement–time graphs, with velocity as the slope of the tangent. ○ Recognise that velocity is the instantaneous rate of change of displacement with respect to time.
12.5	**Curve sketching**	○ Use the first derivative of a function to determine and identify the nature of stationary points. ○ Sketch curves associated with power functions and polynomials up to degree 4; find stationary points and local and global maxima and minima with and without technology; and examine the behaviour as $x \to \infty$ and $x \to -\infty$.

Source: Mathematical Methods Senior Syllabus 2024 © State of Queensland (QCAA) 2024; licensed under CC BY 4.0.

LESSON
12.2 Instantaneous rates of change

SYLLABUS LINKS

- Determine instantaneous rates of change.

Source: Mathematical Methods Senior Syllabus 2024 © State of Queensland (QCAA) 2024; licensed under CC BY 4.0.

12.2.1 Rates of change

Calculus enables us to analyse the behaviour of a changing quantity. Many of the fields of interest in the biological and physical sciences involve the study of **rates of change**. In Chapter 11, the concept of the derivative function and the instantaneous rate of change for a particular value were considered. In this lesson, we shall consider the application of calculus to rates of change in general contexts.

The derivative of a function measures the instantaneous rate of change.

For the function $y = f(x)$, the instantaneous rate of change of y with respect to x is given by $\dfrac{dy}{dx}$.

Instantaneous rates of change

For the function $y = f(x)$,

the instantaneous rate of change of y with respect to x when $x = a$ is $f'(a)$.

WORKED EXAMPLE 1 Determining the instantaneous rate of change of a function

Determine the rate at which the function $f(x) = x\sqrt{x}$ is changing when $x = 4$.

THINK	WRITE
1. Express the function in simplified form.	$f(x) = x\sqrt{x}$ $= x \times \left(x^{\frac{1}{2}} \right)$ $\therefore f(x) = x^{\frac{3}{2}}$
2. Determine the derivative of the function.	$f'(x) = \dfrac{3}{2} x^{\frac{1}{2}}$ $= \dfrac{3}{2}\sqrt{x}$
3. Evaluate the derivative at the given value.	When $x = 4$: $f'(4) = \dfrac{3}{2} \times \sqrt{4}$ $\therefore f'(4) = 3$

12.2.2 Applications of rates of change

As stated, the derivative of a function measures the instantaneous rate of change.

Some examples are:
- the derivative $\dfrac{dV}{dt}$ could be the rate of change of volume, V, with respect to time, t, with possible units being litres per minute
- the derivative $\dfrac{dV}{dh}$ could be the rate of change of volume, V, with respect to height, h, with possible units being litres per centimetre.

To determine these rates, V would need to be expressed as a function of one independent variable: either time or height. Some problems may require the variables to be connected to obtain a function of one variable.

When solving rates of change problems:
- draw a diagram of the situation, where appropriate
- identify the rate of change required and define the variables involved
- express the quantity that is changing as a function of one independent variable, the variable the rate is measured with respect to
- determine the derivative that measures the rate of change
- determine the required instantaneous rate of change by substituting into the derivative
- a negative value for the rate of change means the quantity is decreasing; a positive value for the rate of change means the quantity is increasing.

WORKED EXAMPLE 2 Determining rates of change in a problem

A container in the shape of an inverted right cone of radius 2 cm and depth 5 cm is being filled with water. When the depth of water is h cm, the radius of the water level is r cm.
a. Use similar triangles to express r in terms of h.
b. Express the volume of the water as a function of h.
c. Determine the rate, with respect to the depth of water, at which the volume of water is changing when its depth is 1 cm.

THINK

a. 1. Draw a diagram of the situation.

2. Obtain the required relationship between the variables and express r in terms of h.

WRITE

a.

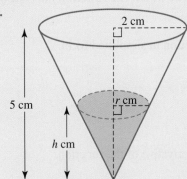

Using similar triangles,

$$\dfrac{r}{h} = \dfrac{2}{5}$$

$$\therefore r = \dfrac{2h}{5}$$

b. Express the function in the required form.

b. The volume of a cone is $V_{cone} = \frac{1}{3}\pi r^2 h$.

Therefore, the volume of water is $V = \frac{1}{3}\pi r^2 h$.

Substitute $r = \frac{2h}{5}$:

$$V = \frac{1}{3}\pi \left(\frac{2h}{5}\right)^2 h$$
$$= \frac{4\pi h^3}{75}$$

c. 1. Calculate the derivative of the function.

c. The derivative gives the rate of change at any depth.

$$\frac{dV}{dh} = \frac{4\pi}{75} \times 3h^2$$
$$= \frac{4\pi}{25}h^2$$

2. Evaluate the derivative at the given value.

When $h = 1$,
$$\frac{dV}{dh} = \frac{4\pi}{25}$$
$$= 0.16\pi$$

3. Write the answer in context, with the appropriate units.

At the instant the depth is 1 cm, the volume of water is increasing at the rate of 0.16π cm^3 per cm.

Exercise 12.2 Instantaneous rates of change

learn on

| 12.2 Exercise | 12.2 Exam practice on |

Simple familiar	Complex familiar	Complex unfamiliar
1, 2, 3, 4, 5, 6, 7, 8, 9, 10, 11, 12	13, 14, 15, 16, 17	18, 19, 20

These questions are even better in jacPLUS!
- Receive immediate feedback
- Access sample responses
- Track results and progress

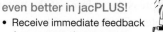
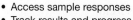

Find all this and MORE in jacPLUS ▶

Where appropriate in this exercise, express answers as multiples of π.

Simple familiar

1. **WE1** Determine the rate at which the function $f(x) = x + \frac{1}{x^2}$ is changing when $x = 3$.

2. Determine the rate at which the function $y = 4 - 3(5 - x)^2$ is changing when $x = 7$.

3. The formula for the area of a circle is given by $A = \pi r^2$.

   a. Determine the rate of change of A with respect to r.

   b. When the radius of a circle is 3 mm, determine the rate at which its area is changing.

4. The surface area, S, of a sphere of radius r is given by $S = 4\pi r^2$.
   a. Determine the rate of change of S with respect to r.
   b. Determine the rate at which the surface area is changing when the radius of the sphere is 5 cm.
   c. Determine the radius if the rate of change of the surface area is $16\pi\,\text{cm}^2/\text{cm}$.

5. The volume, V, of a sphere of radius r is given by $V = \dfrac{4}{3}\pi r^3$.
   a. Determine the rate of change of the volume of the sphere with respect to its radius.
   b. Determine the rate at which the volume is changing when the radius of the sphere is 0.5 m.

6. Water is poured into a container in such a way that the volume of water, $V\,\text{cm}^3$, is given by $V = \dfrac{1}{3}h^3 + 2h$, where $h\,\text{cm}$ is the depth of the water.

   a. Determine the rate of change of the volume of water with respect to the depth of water.
   b. Calculate the volume of water when its depth is 3 cm and the rate at which the volume is changing at that instant.

7. As the area covered by an ink spill from a printer cartridge grows, it maintains a circular shape. Calculate the rate at which the area of the circle is changing with respect to its radius at the instant its radius is 0.2 metres.

8. An ice cube melts in such a way that it maintains its shape as a cube. Calculate the rate at which its surface area is changing with respect to its side length at the instant the side length is 8 mm.

9. **WE2** A container in the shape of an inverted right cone of radius 5 cm and depth 10 cm is being filled with water. When the depth of water is $h\,\text{cm}$, the radius of the water level is $r\,\text{cm}$.
   a. Use similar triangles to express r in terms of h.
   b. Express the volume of the water as a function of h.
   c. Determine the rate, with respect to the depth of water, at which the volume of water is changing when its depth is 3 cm.

10. A container in the shape of an inverted right cone of radius 4 cm and depth 12 cm is being filled with water. When the depth of water is $h\,\text{cm}$, the radius of the water level is $r\,\text{cm}$.
    a. Use similar triangles to express r in terms of h.
    b. Express the volume of the water as a function of h.
    c. Determine the rate with respect to the depth of water at which the volume of water is changing when its depth is 5 cm.

11. An equilateral triangle has side length $x\,\text{cm}$.
    a. Express its area in terms of x.
    b. Determine the rate, with respect to x, at which the area is changing when $x = 2$.
    c. Determine the rate, with respect to x, at which the area is changing when the area is $64\sqrt{3}\,\text{cm}^2$.

12. The population, $P(t)$, of a colony of bacteria after t hours is given by $P(t) = 3t(200 - 2t)$, where $t \ge 0$.
    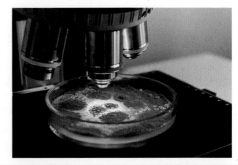
    a. Determine the rate of change of the population of bacteria with respect to time.
    b. Calculate the rate at which the population is changing after 20 hours. Identify whether the population is increasing or decreasing.

c. Calculate the rate at which the population is changing after 60 hours. Identify whether the population is increasing or decreasing.

d. Calculate the instant of time at which the population is:

    i. increasing at 240 bacteria per hour

    ii. decreasing at 240 bacteria per hour

    iii. neither increasing nor decreasing.

Complex familiar

13. A triangle has a base that is half the length of the other two equal sides (in cm) as shown.

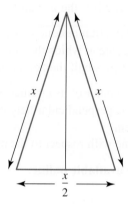

a. Express the area of the triangle in terms of x.

b. Determine the rate with respect to x at which the area is changing when $x = 8$ cm.

14. Water is being poured into a vase. The volume, V mL, of water in the vase after t seconds is given by

$$V = \frac{2}{3}t^2(15 - t), \ 0 \le t \le 10.$$

a. Calculate the volume after 10 seconds.

b. Determine the rate at which the water is flowing into the vase at t seconds.

c. Determine the rate of flow after 3 seconds.

d. Determine the time at which the rate of flow is the greatest, and the rate of flow at this time.

15. The number of rabbits on a farm is modelled by $N = \dfrac{110}{t}, t > 0$, where N is the number of rabbits present after t months.

a. Determine the rate at which the population of rabbits is changing after 5 months.

b. Calculate the average rate of change of the population over the interval $t \in [1, 5]$.

c. Describe what will happen to the population of rabbits in the long term.

16. The volume of water, V litres, in a bath t minutes after the plug is removed is given by $V = 0.4(8 - t)^3, \ 0 \le t \le 8$.

a. Determine the rate at which the water is leaving the bath after 3 minutes.

b. Determine the average rate of change of the volume for the first 3 minutes.

c. Determine when the rate of water leaving the bath is the greatest.

17. A tent in the shape of a square-based right pyramid has perpendicular height h metres, side length of its square base x metres and volume $\frac{1}{3}Ah$, where A is the area of its base. The slant height of the pyramid is 10 metres.

    a. Express the length of the diagonal of the square base in terms of x.

    b. Determine the value of h for which the rate of change of the volume equals zero. Explain what is significant about this value for h.

Complex unfamiliar

18. The surface area of a rectangular prism with a square base of length x cm is $300\,\text{cm}^2$. Determine the rate of change of the volume with respect to the length x, the instant the rectangular prism becomes a cube.

19. A container in the shape of an inverted right circular cone is being filled with water. The cone has a height of 12 cm and a radius of 6 cm. Determine the rate at which the volume of water is changing with respect to the depth of water when the depth of water reaches half the height of the cone.

20. A cone has a slant height of 10 cm. The diameter of its circular base is increased in such a way that the cone maintains its slant height of 10 cm while its perpendicular height decreases. When the base radius is r cm, the perpendicular height of the cone is h cm.
Determine the rate of change of the volume with respect to the perpendicular height when the height is 6 cm.

Fully worked solutions for this chapter are available online.

LESSON
12.3 Tangents and normals of a graph

SYLLABUS LINKS

- Determine the equation of a tangent and a normal of the graph of $y = f(x)$.

Source: Mathematical Methods Senior Syllabus 2024 © State of Queensland (QCAA) 2024; licensed under CC BY 4.0.

12.3.1 Tangents

Recall from Chapter 11 that the gradient of a curve at a point is equal to the gradient of the tangent line to the curve at that point. In section 11.5.2, 'Interpreting the derivative as the gradient of a tangent line', the functions were restricted to polynomial functions. In this section, both polynomial and power functions will be considered.

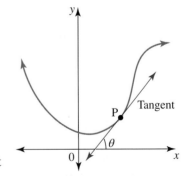

The gradient of a curve at a point, the derivative, is equal to the gradient of the tangent line to the curve at that point. The equation of the tangent line can be determined using the equation for a straight line.

The point, P, is called the point of tangency or the point of contact of the tangent line and the curve.

Recall that the equation of a straight line given a point (x_1, y_1) and gradient m is $y - y_1 = m(x - x_1)$.

Also recall that the angle a straight line makes with the positive x-axis is given by $\theta = \tan^{-1}(m)$.

WORKED EXAMPLE 3 Determining the equation of a tangent

Determine the equation of the tangent to the curve $y = 6\sqrt{x}$ at the point on the curve where $x = 4$.

THINK	WRITE
1. Obtain the coordinates of the point on the curve by substituting $x = 4$ into the equation of the curve.	$y = 6\sqrt{x}$ When $x = 4$: $y = 6\sqrt{4}$ $\quad = 6 \times 2$ $\quad = 12$ The point of contact on the curve is $(4, 12)$.
2. Determine the gradient function of the curve by differentiation.	$y = 6\sqrt{x}$ $\quad = 6x^{\frac{1}{2}}$ $\dfrac{dy}{dx} = 6 \times \dfrac{1}{2}x^{-\frac{1}{2}}$ $\quad = \dfrac{3}{\sqrt{x}}$
3. Determine the gradient of the tangent by substituting $x = 4$ into the derivative function.	When $x = 4$: $\dfrac{dy}{dx} = \dfrac{3}{\sqrt{4}}$ $\quad = \dfrac{3}{2}$
4. Form the equation of the tangent line.	Equation of tangent at $(4, 12)$, $m = \dfrac{3}{2}$: $y - y_1 = m(x - x_1)$ $y - 12 = \dfrac{3}{2}(x - 4)$ $y - 12 = \dfrac{3}{2}x - 6$ $y = \dfrac{3}{2}x + 6$ The equation of the tangent is $y = \dfrac{3}{2}x + 6$ or $3x - 2y + 12 = 0$.

Since the tangent line is a straight line, the properties of a straight line apply to the tangent line.

These include the following:
- The angle of inclination of the tangent to the horizontal can be calculated using $m = \tan(\theta)$.
- Tangents that are parallel have the same gradient.
- The gradient of perpendicular lines is found using $m_1 \times m_2 = -1$.
- The gradient of a horizontal tangent is zero.
- The gradient of a vertical tangent is undefined.
- Coordinates of points of intersections of tangents with other lines and curves can be found using simultaneous equations.

WORKED EXAMPLE 4 Applying the tangent line to a problem

The path of a car can be modelled by the function $f(x) = -0.002x^3 + 3x, 0 \leq x \leq 40$. The car is rounding a corner when it slides off the road at the point where $x = 25$. Assume that the car slides off the road in a straight line.

a. Calculate the gradient of the tangent to the curve at the point where $x = 25$.
b. Calculate the angle at which the vehicle slides, to the nearest degree, with respect to the positive x-axis.
c. Determine the equation of the tangent to the curve at the point where $x = 25$.
d. Determine if the car will hit a large tree located at the point $(40, 32)$ as it slides off the road.

THINK	WRITE
a. 1. Determine the instantaneous rate of change of $f(x)$ at the point where $x = 25$.	$f'(x) = -0.006x^2 + 3$ $f'(25) = -0.006(25)^2 + 3$ $\quad\quad = -0.75$
2. The gradient of the tangent to $f(x)$ at the point where $x = 25$ is equal to the instantaneous rate of change at $x = 25$.	The gradient of the tangent, m_T, at $x = 25$ is -0.75.
b. 1. The angle is related to the gradient by $m = \tan(\theta)$.	$\theta = \tan^{-1}(m_T)$ $\quad = \tan^{-1}(-0.75)$ $\quad = -37°$ or $143°$
c. 1. Determine the coordinates of the point where $x = 25$.	$f(x) = -0.002(25)^3 + 3(25)$ $\quad = 43.75$ $\quad \rightarrow (25, 43.75)$
2. Determine the equation of the tangent at $x = 25$ by substituting the values of the gradient and the known point at $(25, 43.75)$ into $y - y_1 = m(x - x_1)$.	$y - 43.75 = -0.75(x - 25)$ $\quad\quad y = -0.75x + 18.75 + 43.75$ $\quad\quad y = 62.5 - 0.75x$
d. 1. The car will only hit the tree if the tree is on its path. Since the car is sliding along the path given by the tangent to the curve at $x = 25$, substitute $x = 40$ into the equation of the tangent.	$y = 62.5 - 0.75(40)$ $\quad = 32.5$
2. Compare the coordinates of the car $(40, 32.5)$ and the coordinates of the tree $(40, 32)$.	The point $(40, 32)$ does not lie on a tangent (see **d1**), so the car will miss the tree.

TI | THINK	WRITE	CASIO | THINK	WRITE
a. 1. On a Graphs page, complete the entry line for function 1 as: $f1(x) = -0.002x^3 + 3x \| 0 \leq x \leq 40$ then press ENTER.	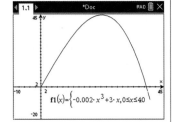	a. 1. On a Graph screen, complete the entry line for Y1 as: $y1 = -0.002x^3 + 3x, [0, 40]$ then press EXE. Select DRAW by pressing F6.	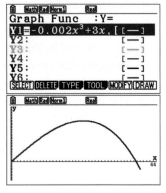

2. Press MENU, then
select:
6: Analyze Graph
5: dy/dx
Type '25', then press
ENTER.

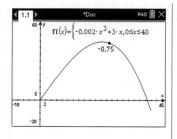

3. The answer appears on
the screen.

The gradient of the tangent of the
curve at $x = 25$ is -0.75.

b. 1. Click and hold on the
end of the tangent line
until the arrow head
turns into a hand, then
drag the tangent line
until it reaches the
x-axis.

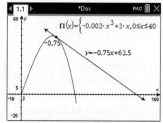

2. To determine the angle
between the tangent
and the positive x-axis,
press MENU, then
select:
8: Geometry
3: Measurement
4: Angle
Click on the point on
the tangent at $x = 25$,
then click on the point
where the tangent meets
the x-axis, then click on
a point on the x-axis to
the right of the tangent.
Note: Press MENU,
then select 9: Settings
to ensure that the
Graphing Angle is set
to Degree.

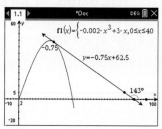

3. The answer appears on
the screen.

The vehicle slides at an angle of
$143°(-37°)$ with respect to the
positive x-axis.

c. 1. To draw the tangent
to the curve at $x = 25$,
press MENU, then
select:
8: Geometry
1: Points & Lines
7: Tangent
Click on the curve, then
click on the point on the
curve at $x = -25$.

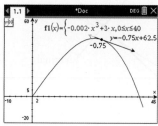

2. The answer appears on
the screen.

The equation of the tangent
to the curve at $x = 25$ is
$y = -0.75x + 62.5$.

2. Press SHIFT then
MENU to access the
set-up menu. Change the
Derivative Mode to On,
then press EXIT.
Select SKETCH by
pressing SHIFT then F4,
then select Tangent by
pressing F2.
Type '-2', then press
EXE.

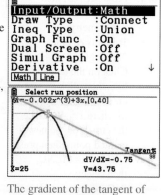

3. The answer appears on
the screen.

The gradient of the tangent of
the curve at $x = 25$ is -0.75.

b. 1. On a Run-Matrix screen,
complete the entry line
as:
$\tan^{-1}(-0.75)$
then press EXE.

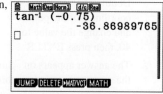

2. The answer appears on
the screen.

The vehicle slides at an angle of
$-37°$ with respect to the positive
x-axis.

c. 1. To draw the tangent
to the curve at $x = 25$,
select SKETCH by
pressing SHIFT then F4,
then select Tangent by
pressing F2.
Type '-2', then press
EXE twice.

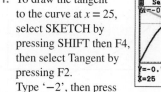
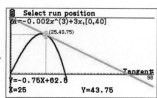

2. The answer appears on
the screen.

The equation of the tangent
to the curve at $x = 25$ is
$y = -0.75x + 62.5$.

d. 1. Press MENU, then select:
8: Geometry
1: Points & Lines
2: Point On
Click on the tangent line, then click a point on the tangent line close to where $x = 40$. Place the cursor on the newly created point, then press CRTL then MENU, then select:
7: Coordinates and Equations.
Double-click on the x-coordinate of the point and change the value to 40, then press ENTER.

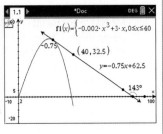

2. The answer appears on the screen.

Since the vehicle will pass through the point $(40, 32.5)$ and not $(40, 32)$, the vehicle will just miss the tree.

d. 1. On the Graph screen, complete the entry line for Y2 as:
$y2 = -0.75x + 62.5$
then press EXE. Select DRAW by pressing F6.

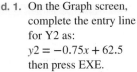

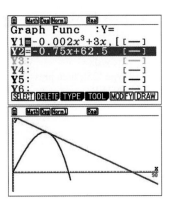

2. Select Trace by pressing SHIFT then F1. Use the up/down arrows to move the cursor to the graph of y2, then type '40' and press EXE twice.

3. The answer appears on the screen.

Since the vehicle will pass through the point $(40, 32.5)$ and not $(40, 32)$, the vehicle will just miss the tree.

12.3.2 Normals

Recall that the **normal** is a line that is perpendicular to the tangent at a point of tangency or point of contact. As such, the gradient of the normal is the negative reciprocal of the gradient of the tangent at the same point:

$$m_N = \frac{-1}{m_T}$$

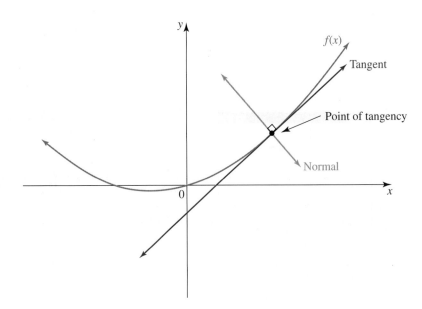

Determine the equation of the normal to the curve $y = \dfrac{6}{x}$ at the point on the curve where $x = -2$.

THINK	WRITE
1. Obtain the coordinates of the point on the curve by substituting $x = -2$ into the equation of the curve.	$y = \dfrac{6}{x}$ When $x = -2$: $y = \dfrac{6}{-2}$ $= -3$ The point of contact on the curve is $(-2, -3)$.
2. Determine the gradient function of the curve by differentiation.	$y = \dfrac{6}{x}$ $= 6x^{-1}$ $\dfrac{dy}{dx} = 6 \times (-1)x^{-2}$ $= \dfrac{-6}{x^2}$
3. Determine the gradient of the tangent by substituting $x = -2$ into the derivative function.	When $x = -2$: $\dfrac{dy}{dx} = \dfrac{-6}{(-2)^2}$ $= \dfrac{-6}{4}$ $m_T = \dfrac{-3}{2}$
4. Determine the gradient of the normal: $m_N = -\dfrac{1}{m_T}$.	$m_N = \dfrac{2}{3}$
5. Form the equation of the normal.	Equation of normal at $(-2, -3)$, $m = \dfrac{2}{3}$: $y - y_1 = m(x - x_1)$ $y - (-3) = \dfrac{2}{3}(x - (-2))$ $y + 3 = \dfrac{2}{3}x + \dfrac{4}{3}$ $y = \dfrac{2}{3}x - \dfrac{5}{3}$ The equation of the normal is $y = \dfrac{2}{3}x - \dfrac{5}{3}$ or $2x - 3y - 5 = 0$.

At the point $(2, 1)$ on the curve $y = \dfrac{1}{4}x^2$, a line is drawn perpendicular to the tangent to the curve. This line meets the curve $y = \dfrac{1}{4}x^2$ again at the point Q.

a. Calculate the coordinates of the point Q.
b. Calculate the magnitude of the angle that the line passing through Q and the point $(2, 1)$ makes with the positive direction of the x-axis.

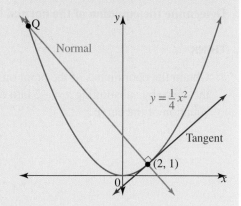

THINK

a. 1. Calculate the gradient of the tangent at the given point.

2. Calculate the gradient of the line perpendicular to the tangent.

3. Form the equation of the perpendicular line.

4. Use simultaneous equations to calculate the coordinates of Q.

WRITE

a. $y = \dfrac{1}{4}x^2$

$\dfrac{dy}{dx} = \dfrac{1}{2}x$

At the point $(2, 1)$,

$\dfrac{dy}{dx} = \dfrac{1}{2} \times 2$

$= 1$

The gradient of the tangent at the point $(2, 1)$ is 1.

For perpendicular lines, $m_1 m_2 = -1$.
Since the gradient of the tangent is 1, the gradient of the line perpendicular to the tangent is -1.

Equation of the line perpendicular to the tangent:
$y - y_1 = m(x - x_1), m = -1, (x_1, y_1) = (2, 1)$
$y - 1 = -(x - 2)$
$\therefore y = -x + 3$

Point Q lies on the line $y = -x + 3$ and the curve $y = \dfrac{1}{4}x^2$.

At Q:

$\dfrac{1}{4}x^2 = -x + 3$

$x^2 = -4x + 12$

$x^2 + 4x - 12 = 0$
$(x + 6)(x - 2) = 0$
$\therefore x = -6, x = 2$

$x = 2$ is the x-coordinate of the given point. Therefore, the x-coordinate of Q is $x = -6$.
Substitute $x = -6$ into $y = -x + 3$
$y = -(-6) + 3$
$= 9$
Point Q has coordinates $(-6, 9)$.

b. Calculate the angle of inclination required.

b. For the angle of inclination, $m = \tan\theta$.
As the gradient of the line passing through Q and the point $(2, 1)$ is -1, $\tan\theta = -1$.
Since the gradient is negative, the required angle is obtuse.
The second quadrant solution is:
$$\theta = 180° - \tan^{-1}(1)$$
$$= 180° - 45°$$
$$= 135°$$
The angle made with the positive direction of the x-axis is $135°$.

Exercise 12.3 Tangents and normals of a graph

learn

12.3 Exercise	12.3 Exam practice on

These questions are even better in jacPLUS!
- Receive immediate feedback
- Access sample responses
- Track results and progress

Find all this and MORE in jacPLUS ▶

Simple familiar	Complex familiar	Complex unfamiliar
1, 2, 3, 4, 5, 6, 7, 8, 9, 10, 11, 12, 13, 14, 15	16, 17, 18	19, 20

Simple familiar

1. Calculate the gradient of the tangent to the curve at $x = 3$ for the following functions.

   a. $f(x) = x^2 - 4x + 1$

   b. $f(x) = 2x^3 - 8x^2 + x$

   c. $f(x) = x^3 - 2x - \dfrac{4}{x}$

   d. $f(x) = 3\sqrt{x}$

2. Calculate the gradient of the normal to the curve at $x = 3$ for the functions given in question 1.

3. **WE3** Determine the equation of the tangent to the curve $y = 5x - \dfrac{1}{3}x^3$ at the point on the curve where $x = 3$.

4. For each of the following, determine the equation of the tangent to the curve at the given point.

   a. $y = 2x^2 - 7x + 3; (0, 3)$

   b. $y = 5 - 8x - 3x^2; (-1, 10)$

   c. $y = \dfrac{1}{2}x^3; (2, 4)$

   d. $y = \dfrac{1}{3}x^3 - 2x^2 + 3x + 5; (3, 5)$

   e. $y = \dfrac{6}{x} + 9; \left(-\dfrac{1}{2}, -3\right)$

   f. $y = 38 - 2x^{\frac{3}{4}}; (81, -16)$

5. **WE5** Determine the equation of the normal to the graph of $y = x^3 + 2x^2 - 3x + 1$ at $x = -2$.

6. Determine the equation of the tangent to $y = x^2 - 6x + 3$ that is:

   a. parallel to the line $y = 4x - 2$

   b. parallel to the x-axis

   c. perpendicular to the line $6y + 3x - 1 = 0$.

7. Consider the curve $y = x^2 - 2x + 5$

 a. Determine the equation of the tangent at the point where $x = 2$.
 b. Determine the equation of the normal at the point where $x = 2$.
 c. A tangent to the curve is inclined at $135°$ to the horizontal. Determine its equation.

8. For the function $f(x) = 4x^2 - 3x + \dfrac{2}{x}$:

 a. calculate the angle to the positive x-axis made by the tangent to the curve at $x = 7$
 b. calculate the angle to the positive x-axis made by the normal line at $x = -5$.

9. For the function $f(x) = 0.05x^3 - 0.4x^2 + x$:

 a. i. calculate the angle between the tangent to the curve at $x = -3$ and the positive x-axis
   ii. calculate the angle between the normal to the curve at $x = -3$ and the positive x-axis
 b. Describe what you notice about your answers from part a.

10. The function $y = -3x^2 + 4x + 5\sqrt{x}$ gives the path of a missile travelling through the air, where y is the vertical height above the ground in metres and x is the horizontal distance travelled in metres.

 a. Calculate the gradient of the curve at $x = 0$.
 b. Calculate the gradient of the tangent line as $x \to 0$.
 c. Calculate the angle of the tangent to the positive x-axis as $x \to 0$.
 d. Explain why there is no tangent at $x = 0$.

11. For the functions below, answer the following.

  i. $f(x) = x^2 + 4x - 3$ ii. $f(x) = 3.2x - 1.8x^2$
 iii. $f(x) = 190x^3 + 460x - 345$ iv. $f(x) = \dfrac{0.21}{x} + 4\sqrt{x} + 0.04x.$

 a. Calculate the average rate of change between $x = 2$ and $x = 4$.
 b. Calculate the value of x at which the gradient of the tangent is equal to the value found in part a.
 c. Draw a conclusion about the relationship between the average rate of change between two points and the tangent to the curve with gradient equal to the average rate of change.

12. **WE4** The function $y = -3x^2 + 4x + 5$ gives the height, y metres, and horizontal distance, x metres, of a particle travelling through the air relative to the position from which it is launched.

 a. Calculate the gradient of the tangent to the curve at the point where $x = 0$.
 b. Calculate the angle to the horizontal at which the particle is initially launched. Give your answer to the nearest degree.
 c. Determine the equation of the tangent to the curve at the point where $x = 0$.
 d. The particle is now launched in a vacuum rather than in the air, and travels in a straight line from its launching position. It is launched from the same initial position and at the same angle to the horizontal. Determine whether it passes through the point $(10, 50)$.

13. A pirate ship is making a sharp turn as it attempts to get into position to fire upon an English frigate. The path of the pirate ship is given by the function

$$y = 0.01x^3 - 0.3x^2, 0 \le x \le 30,$$

where x is the distance in metres north of the ship's initial position and y is the distance in metres east of the ship's initial position. The first gunshot is fired at the English frigate at an angle of 90 degrees to the side of the ship, when the ship is 15 m east from its initial position.

 a. Determine the equation of the path of the gunshot.
 b. If the frigate is at the position $(20, -31.530)$ at the time of the gunshot, determine whether it will be hit.

14. A small landing vehicle is being dropped from a height of 100 m to test its survivability. Its height is modelled by the function

$$h(t) = 100 - 4.9t^2.$$

After 2 seconds, a series of booster rockets are activated that cause the vehicle to continue to fall at a constant rate.

   a. Calculate the rate at which the landing vehicle is falling 2 seconds after being dropped.
   b. Determine the equation of the tangent to the curve at $t = 2$.
   c. Determine when the landing vehicle will reach the ground.
   d. Determine how much longer it took the landing vehicle to reach the ground compared to how long it would take if it didn't have the boosters.

15. **WE6** At the point $(-1, 1)$ on the curve $y = x^2$, a line is drawn perpendicular to the tangent to the curve at that point. This line meets the curve $y = x^2$ again at the point Q.

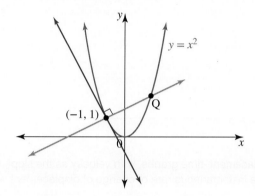

   a. Calculate the coordinates of the point Q.
   b. Calculate the magnitude of the angle that the line through Q and the point $(-1, 1)$ makes with the positive direction of the x-axis.

Complex familiar

16. The tangent to the curve $y = ax^2 + b$ at the point where $x = 1$ has the equation $y = 2x + 3$. Determine the values of a and b.

17. Consider the curve with equation $y = \dfrac{1}{3}x(x + 4)(x - 4)$.

   a. Sketch the curve and draw a tangent to the curve at the point where $x = 3$.
   b. Determine the equation of this tangent.
   c. i. The tangent meets the curve again at a point P. Show that the x-coordinate of the point P satisfies the equation $x^3 - 27x + 54 = 0$.
   ii. Explain why $(x - 3)^2$ must be a factor of this equation and hence calculate the coordinates of P.
   d. Show that the tangents to the curve at the points where $x = \pm 4$ are parallel.
   e. i. For $a \in R \backslash \{0\}$, show that the tangents to the curve $y = x(x + a)(x - a)$ at the points where $x = \pm a$ are parallel.
   ii. Calculate the coordinates, in terms of a, of the points of intersection of the tangent at $x = 0$ with each of the tangents at $x = -a$ and $x = a$.

18. The tangent to the curve $y = x^3 + 3x^2$ at point A where $x = 1$ meets the curve again at point B. Determine the coordinates of the midpoint of the line segment AB.

Complex unfamiliar

---

19. Show that the tangents drawn to the x-intercepts of the parabola $y = (x - a)(x - b)$ intersect on the parabola's axis of symmetry.

20. Determine where the parabola $y = x^2 + 2x - 8$ touches the curve $y = -\dfrac{4}{x} - 1$. Justify the procedures used to obtain your decision.

Fully worked solutions for this chapter are available online.

LESSON
12.4 Kinematics

SYLLABUS LINKS

- Construct and interpret displacement–time graphs, with velocity as the slope of the tangent.
- Recognise that velocity is the instantaneous rate of change of displacement with respect to time.

Source: Mathematical Methods Senior Syllabus 2024 © State of Queensland (QCAA) 2024; licensed under CC BY 4.0.

12.4.1 Position

Many quantities change over time, so many rates measure that change with respect to time. Motion is one such quantity. The study of the motion of a particle without considering the causes of the motion is called **kinematics**. Analysing motion requires interpretation of **position** and **velocity**, and this analysis depends on calculus. In this section, we will consider only objects moving in straight lines, either right and left or up and down. Motion in a straight line is known as **rectilinear motion**.

The position, x, gives the position of a particle by specifying both its distance and direction from a fixed origin, O.

Common units for position and distance are cm, m and km.

The commonly used conventions for motion along a horizontal straight line are:
- if $x > 0$, the particle is to the right of the origin
- if $x < 0$, the particle is to the left of the origin
- if $x = 0$, the particle is at the origin.

For example, if $x = -10$, this means the particle is 10 units to the left of origin O.

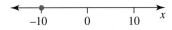

Similar conventions are used for motion along a vertical straight line, with the positive direction normally given as vertically upwards.

Distance and displacement

Distance is not concerned with the direction of motion and is always positive. This can have implications if there is a change of direction in a particle's motion.

Displacement describes the change in a particle's position; that is:

Displacement = change in position = final position − initial position

For example, suppose a particle initially 3 cm to the right of the origin travels 2 cm further to the right, then 2 cm to the left. The particle is then where it started.

The distance travelled is 4 cm, but its change in position is zero; hence, its displacement is zero.

12.4.2 Velocity

Velocity, v, is the rate of change of position with respect to time, so $v = \dfrac{dx}{dt}$, or the instantaneous rate of change of position with respect to time.

> **Velocity**
>
> $$v = \frac{dx}{dt}$$
>
> **Velocity is the instantaneous rate of change of position.**

Common units for velocity and speed include m/s and km/h.

For a particle moving in a horizontal straight line, the sign of the velocity indicates the direction of motion:
- if $v > 0$, the particle is moving to the right
- if $v < 0$, the particle is moving to the left
- if $v = 0$, the particle is stationary (instantaneously at rest).

Average velocity and speed

Average velocity is the average rate of change of the position. It is measured by the gradient of the line joining two points on the position-time graph and requires using the gradient formula, not calculus.

> **Average velocity**
>
> $$\text{Average velocity} = \frac{\text{change in position}}{\text{change in time}} = \frac{x_2 - x_1}{t_2 - t_1}$$

Speed, as for distance, is not concerned with the direction in which the particle is moving. It is always positive. For example, if $v = -5$ m/s, then the speed is 5 m/s.

> **Average speed**
>
> $$\text{Average speed} = \frac{\text{distance travelled}}{\text{time taken}}$$

WORKED EXAMPLE 7 Determining the average velocity and average speed of a particle

A particle moves in a straight line such that its position, x metres, from a fixed origin at time t seconds is modelled by $x = t^2 - 4t - 12$, $t \geq 0$.

a. Identify its initial position.
b. Determine its velocity function and hence state its initial velocity and describe its initial motion.
c. Determine the time and position at which the particle is momentarily at rest.
d. Show the particle is at the origin when $t = 6$ and calculate the distance it has travelled to reach the origin.
e. Calculate the average speed over the first 6 seconds.
f. Calculate the average velocity over the first 6 seconds.

THINK	WRITE
a. Calculate the value of x when $t = 0$.	a. $x = t^2 - 4t - 12, t \geq 0$ When $t = 0, x = -12$. Initially the particle is 12 metres to the left of the origin.
b. 1. Calculate the rate of change required.	b. $v = \dfrac{dx}{dt}$ $\therefore v = 2t - 4$
2. Calculate the value of v at the given instant.	When $t = 0, v = -4$. The initial velocity is -4 m/s.
3. Describe the initial motion.	Since the initial velocity is negative, the particle starts to move to the left with an initial speed of 4 m/s.
c. 1. Calculate when the particle is momentarily at rest. *Note:* This usually represents a change of direction of motion.	c. The particle is momentarily at rest when its velocity is zero. When $v = 0$, $2t - 4 = 0$ $\therefore t = 2$ The particle is at rest after 2 seconds.
2. Calculate where the particle is momentarily at rest.	The position of the particle when $t = 2$ is: $x = (2)^2 - 4(2) - 12$ $\quad = -16$ Therefore, the particle is momentarily at rest after 2 seconds at the position 16 metres to the left of the origin.
d. 1. Calculate the position to show the particle is at the origin at the given time.	d. When $t = 6$, $x = 6^2 - 4 \times 6 - 12$ $\quad = 0$ The particle is at the origin when $t = 6$.

2. Track the motion on a horizontal line and calculate the required distance.

The motion of the particle for the first 6 seconds is shown.

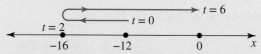

The distances travelled are 4 metres to the left then 16 metres to the right.
The total distance travelled is the sum of the distances in each direction.
The particle has travelled a total distance of 20 metres.

e. Calculate the value required.

e. Average speed = $\dfrac{\text{distance travelled}}{\text{time taken}}$

$= \dfrac{20}{6}$

$= 3\dfrac{1}{3}$

The average speed is $3\dfrac{1}{3}$ m/s.

f. Calculate the average rate of change required.
Note: As there is a change of direction, the average velocity will not be the same as the average speed.

f. Average velocity is the average rate of change of position.

For the first 6 seconds,
$(t_1, x_1) = (0, -12), (t_2, x_2) = (6, 0)$

Average velocity $= \dfrac{x_2 - x_1}{t_2 - t_1}$

$= \dfrac{0 - (-12)}{6 - 0}$

$= 2$

The average velocity is 2 m/s.

12.4.3 Position–time graphs

The position, x, of a particle can be plotted against time, t, to create a **position–time graph**, $x = f(t)$. Because $v = \dfrac{dx}{dt}$, the gradient of the tangent to the curve $x = f(t)$ at any point represents the velocity of the particle at that point: $v = \dfrac{dx}{dt} = f'(t)$.

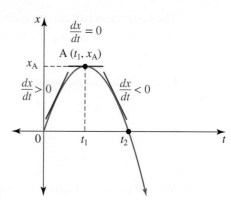

The position–time graph shown represents the position of a particle that starts at the origin and initially moves to the right, as the gradient of the graph, the velocity, is positive.

At point A, the tangent is horizontal and the velocity is zero, indicating the particle changes its direction of motion at this point.

The particle then starts to move to the left, which is indicated by the gradient of the graph, the velocity, having a negative sign. It returns to the origin and continues to move to the left, so its position becomes negative.

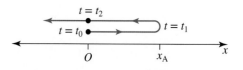

The same motion is also shown along the horizontal position line.

The graph shows the position, x, versus time, t, of a particle travelling horizontally according to the function

$$x(t) = \frac{t^3}{2} - 3t^2 + 4t + 5,$$

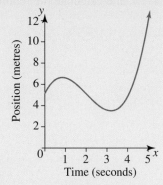

where x is the horizontal distance travelled in metres to the right of the particle's starting point and t is the time in seconds since the particle started moving.

a. Identify the direction in which the particle initially moved.
b. Determine the average velocity of the particle over the first 2 seconds.
c. Determine the speed at which the particle was travelling after 3 seconds.

THINK	WRITE
a. 1. Looking at the graph, it can be seen that initially the particle travels from 5 m towards 6 m.	**a.** The particle initially travels to the right of the observation point.
b. 1. Calculate the position of the particle at $t = 0$ and $t = 2$.	**b.** $x(0) = \dfrac{0^3}{2} - 3(0)^2 + 4(0) + 5$ $= 5$ $(0, 5)$ $x(2) = \dfrac{2^3}{2} - 3(2)^2 + 4(2) + 5$ $= 5$ $(2, 5)$
2. Calculate the gradient between the two points.	$m = \dfrac{x(2) - x(0)}{2 - 0}$ $= \dfrac{5 - 5}{2 - 0}$ $= 0$
3. Write the answer. Include appropriate units.	The average velocity over the first 2 seconds is 0 m/s.
c. 1. The instantaneous velocity function is the derivative of the displacement function, so determine the derivative of $x(t)$.	**c.** $v(t) = x'(t) = \dfrac{3t^2}{2} - 6t + 4$
2. Calculate the instantaneous velocity at $t = 3$ s.	$v(3) = \dfrac{3(3)^2}{2} - 6(3) + 4$ $= \dfrac{27}{2} - 18 + 4$ $= -\dfrac{1}{2}$
3. Write the answer. Include appropriate units.	The particle is travelling at 0.5 m/s at $t = 3$ s. (*Hint:* We do not need to include the direction as speed is a scalar quantity.)

The gradient of a position–time graph gives the velocity, because velocity is the rate of change of position with respect to time.

Therefore, by measuring the gradient of a position–time graph at various points, the velocity–time graph can be derived. Alternately, the derivative function represented by velocity can be graphed.

WORKED EXAMPLE 9 Sketching a velocity–time graph for a particle

The position–time graph for a particle moving in a straight line is shown.
The gradient of the curve at various times is indicated on the graph.
Use this information to draw a velocity–time graph for the particle.

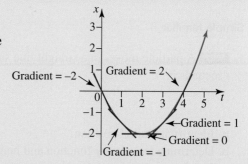

THINK

1. Set up a table of corresponding velocity and time values from the graph.

2. Use the table of values to plot the velocity–time graph for $t \geq 0$.

WRITE

t	0	1	2	3	4
v	-2	-1	0	1	2

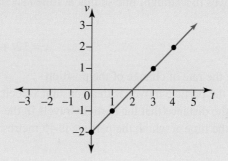

Exercise 12.4 Kinematics

12.4 Exercise	12.4 Exam practice on

Simple familiar	Complex familiar	Complex unfamiliar
1, 2, 3, 4, 5, 6, 7, 8, 9, 10	11, 12, 13, 14, 15, 16, 17, 18, 19, 20	N/A

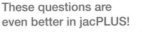

These questions are even better in jacPLUS!
- Receive immediate feedback
- Access sample responses
- Track results and progress

Find all this and MORE in jacPLUS ▶

Simple familiar

1. **WE7** A particle moves in a straight line such that its position, x metres, from a fixed origin at time t seconds is

$$x = 3t^2 - 6t, t \geq 0.$$

   a. Identify its initial position.
   b. Determine its velocity function and hence state its initial velocity and describe its initial motion.
   c. Determine the time and position at which the particle is momentarily at rest.
   d. Show that the particle is at the origin when $t = 2$ and calculate the distance it has travelled to reach the origin.
   e. Calculate the average speed over the first 2 seconds.
   f. Calculate the average velocity over the first 2 seconds.

2. A particle moves in a straight line so that at time t seconds, its position x metres from a fixed origin is given by

$$x = 12t + 9, t \geq 0.$$

   a. Determine the rate of change of the position.
   b. Show that the particle moves with a constant velocity.
   c. Calculate the distance that the particle travels in the first second.
   d. Calculate the time at which the particle is 45 metres to the right of the origin.

The following information and diagram relate to questions 3 and 4.

Consider the position and direction, at various times, of a particle travelling in a straight line as indicated at right.

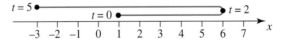

3. Determine:
   a. where the particle starts.
   b. where the particle finishes
   c. the direction in which the particle initially moves
   d. when the particle changes direction.

4. a. **MC** The total distance travelled in the first 5 seconds is:

    A. 4 m   **B.** 13 m   **C.** 9 m   **D.** 14 m

  b. **MC** The displacement of the particle after 5 seconds is:

    A. −3 m   **B.** 14 m   **C.** 4 m   **D.** −4 m

  c. **MC** The average speed in the first 2 seconds is:

    A. 3 m/s   **B.** −2.5 m/s   **C.** 6 m/s   **D.** 2.5 m/s

  d. **MC** The average velocity between $t = 2$ and $t = 5$ is:

    A. 3 m/s   **B.** −2 m/s   **C.** −3 m/s   **D.** 2 m/s

  e. **MC** The instantaneous speed when $t = 2$ is:

    A. 2.5 m/s   **B.** 0 m/s   **C.** 3 m/s   **D.** 2.8 m/s

5. A parachute ride takes people in a basket vertically up in the air from a platform 2 metres above the ground, then 'drops' them back to the ground. Use the illustration showing the position of the parachute basket at various times to determine:

  a. the total distance travelled by the parachute basket during a ride
  b. the displacement of the parachute basket after 80 s
  c. the average speed of the parachute basket during the ride
  d. the average velocity of the parachute basket during the ride.

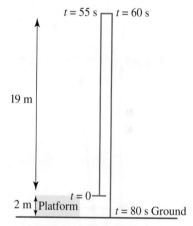

6. **WE8** The graph shows the position, x, versus time, t, of a particle travelling vertically according to the function

$$x(t) = 0.6t^3 - 1.2t^2 - 2.4t,$$

where x is the distance in metres above ground level and t is the time in seconds since the particle started moving.

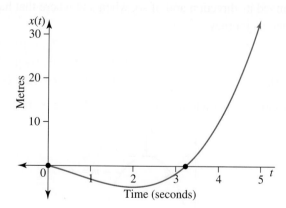

  a. Determine the direction in which the particle initially moves.
  b. Determine the average velocity of the particle over the first 4 seconds.
  c. Determine how fast the particle is travelling after 1 second.

7. The following position–time graphs show the journey of a particle travelling in a straight line. For each graph, determine:

   i. where the journey started
   ii. in which direction the particle moved initially
   iii. when and where the particle changed direction
   iv. when and where the particle finished its journey.

a.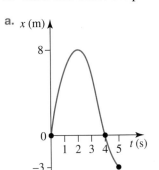

b.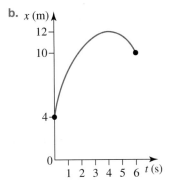

c.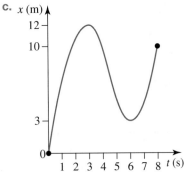

d.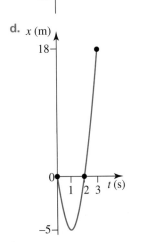

e.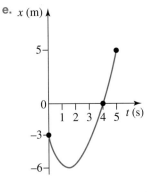

f.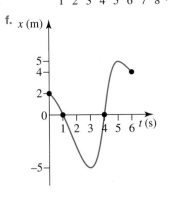

8. For each position function of a particle given below, sketch the position–time graph. In each case, explain:

   i. where the particle started its journey
   ii. in which direction it moved initially
   iii. whether the particle changed its direction and, if so, when and where that happened
   iv. where the particle finished its journey.

   a. $x(t) = 2t,\ t \in [0, 5]$
   b. $x(t) = 3t - 2,\ t \in [0, 6]$
   c. $x(t) = t^2 - 2t,\ t \in [0, 5]$
   d. $x(t) = 2t - t^2,\ t \in [0, 4]$
   e. $x(t) = t^2 - 4t + 4,\ t \in [0, 5]$
   f. $x(t) = t^2 + t - 12,\ t \in [0, 5]$

9. **WE9** The position–time graph for a particle moving in a straight line is shown.

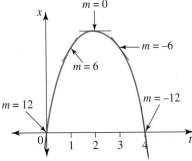

The gradient of the curve at various times is indicated on the graph. Use this information to draw a velocity–time graph for the particle.

10. A ball is projected vertically upwards from the top of a building 25 m high. Its position relative to the ground is given by the equation $x = 25 + 20t - 5t^2$, where t is the time in seconds.
   Sketch a position–time graph for the ball and hence determine:

   a. the greatest height reached
   b. when the ball reaches the ground
   c. when the velocity of the ball is zero
   d. an estimate for the velocity at which the ball is initially projected.

Complex familiar

11. The position from a fixed origin, x metres, at time t seconds, of a particle moving in a straight line is given by $x = t^2 - 6t - 7, t \geq 0$.

   a. Describe the initial position of the particle.
   b. Calculate the velocity at any time t and hence obtain the particle's initial velocity.
   c. Determine the time at which the particle is momentarily at rest.
   d. Determine the time at which the particle returns to its initial position and calculate its velocity at that time.

12. A particle travels in a straight line with its position from a fixed origin, x metres, at time t seconds, given by $x = t^3 - 4t^2 - 3t + 12, t \geq 0$.

   a. Determine the expression for the velocity at any time t.
   b. Determine the time at which the velocity is zero, and calculate its position at that instant.
   c. Calculate the average velocity over the first 3 seconds.

13. The position, x cm, relative to a fixed origin of a particle moving in a straight line at time t seconds is $x = 5t - 10, t \geq 0$.

   a. Give its initial position and its position after 3 seconds.
   b. Calculate the distance travelled in the first 3 seconds.
   c. Show that the particle is moving with a constant velocity.
   d. Sketch the x–t and v–t graphs, and explain their relationship.

14. The position x cm relative to a fixed origin of a particle moving in a straight line at time t seconds is $x = \frac{1}{3}t^3 - t^2, t \geq 0$.

   a. Show that the particle starts at the origin from rest.
   b. Determine the time and position at which the particle is next at rest.
   c. Determine when the particle returns to the origin.
   d. Calculate the particle's speed when it returns to the origin.
   e. Sketch the two motion graphs, x–t and v–t, and comment on their behaviour at $t = 2$.
   f. Describe the motion at $t = 1$.

15. A ball is thrown vertically upwards into the air so that after t seconds, its height h metres above the ground is $h = 40t - 5t^2$.

   a. Calculate the rate at which its height is changing after 2 seconds.
   b. Calculate its velocity when $t = 3$.
   c. Determine the time at which its velocity is -10 m/s and the direction in which the ball is then travelling.
   d. Determine when its velocity is zero.
   e. Determine the greatest height the ball reaches.
   f. Determine the time at which the ball strikes the ground and its speed at that time.

16. A particle moves in a straight line so that at time t seconds its position, x metres, from a fixed origin O is given by $x(t) = 3t^2 - 24t - 27, t \geq 0$.

  a. Calculate the distance the particle is from O after 2 seconds.
  b. Calculate the speed at which the particle is travelling after 2 seconds.
  c. Calculate the average velocity of the particle over the first 2 seconds of motion.
  d. Determine the time at which the particle reaches O and its velocity at that time.
  e. Calculate the distance the particle travels in the first 6 seconds of motion.
  f. Calculate the average speed of the particle over the first 6 seconds of motion.

17. Under emergency breaking, a truck's distance from when it started applying its breaks is given by the formula $d_t(t) = ut - 2.5t^2$, while for a car it is given by $d_c(t) = ut - 6t^2$, where u is the speed of the vehicle, in m/s, when it starts breaking. The vehicles stop when their speeds (the derivative of distance) reach 0.
Determine how much longer it takes for a truck to stop than a car if both are travelling at 60 km/h. Give your answer to the nearest second.

18. The position, x m, relative to a fixed origin of a particle moving in a straight line at time t seconds is

$$x = \frac{2}{3}t^3 - 4t^2, t \geq 0.$$

  a. Show that the particle starts at the origin from rest.
  b. Determine the time and position at which the particle is next at rest.
  c. Determine when the particle returns to the origin.
  d. Determine the particle's speed when it returns to the origin.

19. A particle. A. moves in a straight line so that at time t seconds, its position, x metres, from a fixed origin O is given by $x = 5t^2 - 40t - 12, t \geq 0$.

  a. Describe the initial position of the particle and the direction in which it starts to move.
  b. Calculate how far the particle has travelled by the time its velocity becomes zero.
  c. Calculate the distance the particle travels in the first 10 seconds of motion and its average speed over this time.

A second particle. B. moves on the same straight line so that at time t seconds, it has position s metres from the origin O, where $s = t^2 + 8t - 12, t \geq 0$.

  d. Show that the two particles start from the same initial position but move in opposite directions.
  e. Determine the time and position at which particle A overtakes particle B.
  f. Determine the times and positions at which the two particles are travelling with the same speed.

20. A particle. P. moving in a straight line has a position x metres from a fixed origin O of $x_P(t) = t^3 - 12t^2 + 45t - 34$ for time t seconds.

  a. Determine the time(s) at which the particle is stationary.
  b. Determine the time interval over which the velocity is negative.

A second particle. Q, also travels in a straight line with its position from O at time t seconds given by $x_Q(t) = -12t^2 + 54t - 44$.

  c. Determine the time at which P and Q are travelling with the same velocity.
  d. Determine the times at which P and Q have the same position from O.

Fully worked solutions for this chapter are available online.

LESSON
12.5 Curve sketching

SYLLABUS LINKS

- Use the first derivative of a function to determine and identify the nature of stationary points.
- Sketch curves associated with power functions and polynomials up to degree 4; find stationary points and local and global maxima and minima with and without technology; and examine the behaviour as $x \to \infty$ and $x \to -\infty$.

Source: Mathematical Methods Senior Syllabus 2024 © State of Queensland (QCAA) 2024; licensed under CC BY 4.0.

12.5.1 Sketching power functions and polynomials

When the graphs of power functions and polynomials are being sketched, features to be considered include:
- the basic shape, if possible
- stationary points
- axis intercepts
- increasing and decreasing functions
- global maxima and minima
- the behaviour of the function as $x \to \pm\infty$.

Stationary points

At a **stationary point** on a curve $y = f(x)$, $f'(x) = 0$. It is the point where the function momentarily stops rising or falling.

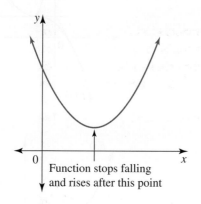

Function stops falling and rises after this point

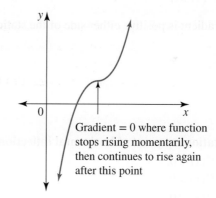

Gradient = 0 where function stops rising momentarily, then continues to rise again after this point

Types of stationary points

There are three types of stationary points:
- local minimum turning point
- local maximum turning point
- stationary point of horizontal inflection, classified by the gradient:
  - positive stationary point of inflection
  - negative stationary point of inflection.

A **local minimum** turning point at $x = a$:

If $x < a$, then $f'(x) < 0$ (immediately to the left of $x = a$, the gradient is negative).

If $x = a$, then $f'(x) = 0$ (at $x = a$ the gradient is zero).

If $x > a$, then $f'(x) > 0$ (immediately to the right of $x = a$, the gradient is positive).

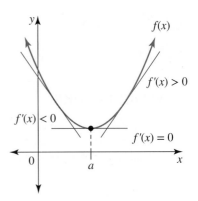

A **local maximum** turning point at $x = a$:

If $x < a$, then $f'(x) > 0$.

If $x = a$, then $f'(x) = 0$.

If $x > a$, then $f'(x) < 0$.

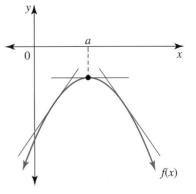

The two cases above can be called 'turning points' because the gradients each side of the stationary point are opposite in sign (that is, the graph turns).

The term 'local turning point at $x = a$' implies 'in the vicinity of $x = a$', as polynomials can have more than one stationary point.

A **positive stationary point of horizontal inflection** at $x = a$:

If $x < a$, then $f'(x) > 0$.

If $x = a$, then $f'(x) = 0$.

If $x > a$, then $f'(x) > 0$.

That is, the gradient is positive either side of the stationary point.

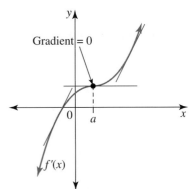

A **negative stationary point of horizontal inflection** at $x = a$:

If $x < a$, then $f'(x) < 0$.

If $x = a$, then $f'(x) = 0$.

If $x > a$, then $f'(x) < 0$.

That is, the gradient is negative either side of the stationary point.

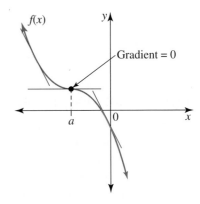

Determining the nature of a stationary point

At each of the different types of stationary points, $f'(x) = 0$. This means that the tangents to the curve at these points are horizontal. By examining the tangent to the curve immediately before and immediately after the stationary point, the nature or type of stationary point can be determined.

Completing a 'gradient table', or alternately a sign diagram of the gradient or derivative function, allows the nature to be determined and is illustrated in the following table.

	Minimum turning point	Maximum turning point	Stationary point of inflection
Stationary point			or
Slope of tangent			or
Sign diagram of gradient function $f'(x)$			or

To identify stationary points and their nature:
- establish where $f'(x) = 0$
- determine the nature by drawing the sign diagram of $f'(x)$ around the stationary point or, alternatively, by testing the slope of the tangent at selected points either side of, and in the neighbourhood of, the stationary point.

WORKED EXAMPLE 10 Determining stationary points and their nature

a. **Determine the stationary points of $f(x) = 2 + 4x - 2x^2 - x^3$ and justify their nature.**
b. **The curve $y = ax^2 + bx - 24$ has a stationary point at $(-1, -25)$.**
   Calculate the values of a and b.

THINK

a. 1. Calculate the x-coordinates of the stationary points.
   Note: Always include the reason why $f'(x) = 0$.

2. Calculate the corresponding y-coordinates.

WRITE

a. $f(x) = 2 + 4x - 2x^2 - x^3$

$f'(x) = 4 - 4x - 3x^2$

At stationary points, $f'(x) = 0$, so:
$$4 - 4x - 3x^2 = 0$$
$$(2 - 3x)(2 + x) = 0$$
$$x = \frac{2}{3} \text{ or } x = -2$$

When $x = \frac{2}{3}$,

$$f(x) = 2 + 4\left(\frac{2}{3}\right) - 2\left(\frac{2}{3}\right)^2 - \left(\frac{2}{3}\right)^3$$

$$= \frac{94}{27}$$

When $x = -2$,
$$f(x) = 2 + 4(-2) - 2(-2)^2 - (-2)^3$$
$$= -6$$

The stationary points are $\left(\frac{2}{3}, \frac{94}{27}\right)$ and $(-2, -6)$.

3. To justify the nature of the stationary points, draw the sign diagram of $f'(x)$.
Note: The shape of the cubic graph would suggest the nature of the stationary points.

Since $f'(x) = 4 - 4x - 3x^2$
$$= (2 - 3x)(2 + x)$$
the sign diagram of $f'(x)$ is that of a concave down parabola with zeros at $x = -2$ and $x = \dfrac{2}{3}$.

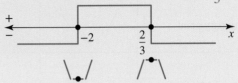

4. Identify the nature of each stationary point by examining the sign of the gradient before and after each point.

At $x = -2$, the gradient changes from negative to positive, so $(-2, -6)$ is a minimum turning point.
At $x = \dfrac{2}{3}$, the gradient changes from positive to negative, so $\left(\dfrac{2}{3}, \dfrac{94}{27}\right)$ is a maximum turning point.

b. 1. Use the coordinates of the given point to form an equation.
Note: As there are two unknowns to determine, two pieces of information are needed to form two equations in the two unknowns.

b. $y = ax^2 + bx - 24$
The point $(-1, -25)$ lies on the curve.
$$-25 = a(-1)^2 + b(-1) - 24$$
$$\therefore a - b = -1 \quad [1]$$

2. Use the other information given about the point to form a second equation.

The point $(-1, -25)$ is a stationary point, so $\dfrac{dy}{dx} = 0$ at this point.
$$y = ax^2 + bx - 24$$
$$\dfrac{dy}{dx} = 2ax + b$$
At $(-1, -25)$, $\dfrac{dy}{dx} = 2a(-1) + b$
$$= -2a + b$$
$$\therefore -2a + b = 0 \quad [2]$$

3. Solve the simultaneous equations and write the answer.

$$a - b = -1 \quad [1]$$
$$-2a + b = 0 \quad [2]$$

Adding the equations,
$$-a = -1$$
$$\therefore a = 1$$

Substitute $a = 1$ into equation [2]:
$$-2 + b = 0$$
$$\therefore b = 2$$
The values are $a = 1$ and $b = 2$.

12.5.2 Increasing and decreasing functions

If the tangents to a function's curve have a positive gradient over some domain interval, then the function is increasing over that domain; if the gradients of the tangents are negative, then the function is decreasing. From this it follows that for a function $y = f(x)$:

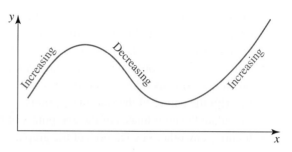

- if $f'(x) > 0$, then the function increases as x increases over the interval
- if $f'(x) < 0$, then the function decreases as x increases over the interval.

WORKED EXAMPLE 11 Increasing and decreasing functions

If $f(x) = x^3 + 4x^2 - 3x - 7$:
a. sketch the graph of $f'(x)$
b. identify the values of x where $f(x)$ is
   i. increasing   ii. decreasing.

THINK

a. 1. Write the rule for $f(x)$.

   2. Differentiate $f(x)$ to determine $f'(x)$.

   3. Solve $f'(x) = 0$ to determine the x-intercepts of $f'(x)$.

   4. Evaluate $f'(0)$ to determine the y-intercept of $f'(x)$.

   5. Sketch the graph of $f'(x)$ (an upright or concave up parabola).

b. i. By inspecting the graph of $f'(x)$, deduce where $f'(x)$ is positive (that is above the x-axis).

  ii. By inspecting the graph of $f'(x)$, deduce where $f'(x)$ is negative (that is below the x-axis).

WRITE

a. $f(x) = x^3 + 4x^2 - 3x - 7$

   $f'(x) = 3x^2 + 8x - 3$

   x-intercepts: when $f'(x) = 0$,
     $3x^2 + 8x - 3 = 0$
    $(3x - 1)(x + 3) = 0$
            $x = \dfrac{1}{3}$ or -3

   The x-intercepts of $f'(x)$ are $\left(\dfrac{1}{3}, 0\right)$ and $(-3, 0)$.

   y-intercept: when $x = 0$,
   $f'(0) = -3$
   so the y-intercept of $f'(x)$ is $(0, -3)$.

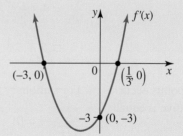

b. i. $f'(x) > 0$ where $x < -3$ and $x > \dfrac{1}{3}$

     so $f(x)$ is increasing where $x < -3$ and $x > \dfrac{1}{3}$.

  ii. $f'(x) < 0$ where $-3 < x < \dfrac{1}{3}$

      so $f(x)$ is decreasing where $-3 < x < \dfrac{1}{3}$.

12.5.3 Curve sketching

To sketch the graph of any function $y = f(x)$, perform the following steps:

- Obtain the y-intercept by evaluating $f(0)$.
- Obtain any x-intercepts by solving, if possible, $f(x) = 0$.
- Calculate the x-coordinates of the stationary points by solving $f'(x) = 0$ and use the equation of the curve to obtain the corresponding y-coordinates.
- Identify the nature of the stationary points.
- Calculate the coordinates of the end points of the domain, where appropriate.
- Identify any other key features of the graph, where appropriate.

WORKED EXAMPLE 12 Sketching a curve

Sketch the curve $y = 4\sqrt{x} - x$, $x \geq 0$, locating any intercepts with the coordinate axes and any stationary points, and justify their nature.

THINK	WRITE
1. State the y-intercept.	$y = 4\sqrt{x} - x$ When $x = 0$: $y = 0$ y-intercept: $(0, 0)$
2. Calculate any x-intercepts.	When $y = 0$: $4\sqrt{x} - x = 0$ $\quad 4\sqrt{x} = x$ $\quad\quad 16x = x^2$ $x^2 - 16x = 0$ $x(x - 16) = 0$ $x = 0$ or $x = 16$ x-intercepts: $(0, 0)$ and $(16, 0)$
3. Obtain the derivative to locate any stationary points.	$y = 4\sqrt{x} - x$ $\quad = 4x^{\frac{1}{2}} - x$ $\dfrac{dy}{dx} = 4 \times \dfrac{1}{2}x^{-\frac{1}{2}} - 1$ $\quad\quad = \dfrac{2}{\sqrt{x}} - 1$
4. The stationary points occur when the derivative is equal to zero.	For stationary points: $\dfrac{dy}{dx} = 0$ $\dfrac{2}{\sqrt{x}} - 1 = 0$ $\dfrac{2}{\sqrt{x}} = 1$ $\quad 2 = \sqrt{x}$ $\quad x = 4$ Substitute into the equation of the function: $y = 4\sqrt{4} - 4$ $\quad = 4$ The stationary point is $(4, 4)$.

5. Identify the type of stationary point using a gradient table or a sign diagram of the gradient function.

The point (4, 4) is a maximum turning point.

6. Sketch the curve, showing the intercepts with the axes and the stationary point.

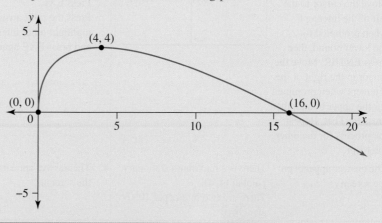

TI	THINK	WRITE

1. On a Graphs page, complete the entry line as:
$f1(x) = 4\sqrt{x} - x$
then press ENTER.

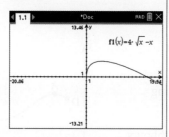

2. To determine the maximum, press MENU, then select:
6: Analyze Graph
3: Maximum
Move the cursor to the left of the maximum when prompted for the lower bound, then press ENTER. Move the cursor to the right of the maximum when prompted for the upper bound, then press ENTER.

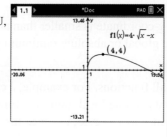

CASIO	THINK	WRITE

1. On a Graph screen, complete the entry line for Y1 as:
$y1 = 4\sqrt{x} - x$
then press EXE.
Select DRAW by pressing F6.
Note: You may need to zoom out using the V-Window function by pressing F3.

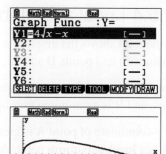

2. To determine the maximum, select G-Solv by pressing SHIFT F5, then select MAX by pressing F2. Press EXE.

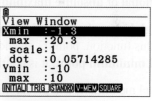

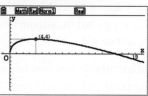

| 3. To determine the x-intercepts, press MENU, then select:
6: Analyze Graph
1: Zero
Move the cursor to the left of the intercept when prompted for the lower bound, then press ENTER. Move the cursor to the right of the intercept when prompted for the upper bound, then press ENTER. Repeat the process for the other intercept. | 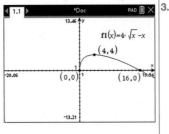 | 3. To determine the intercepts, select G-Solv by pressing SHIFT F5, then select ROOT by pressing F1. Press EXE.
Press the right arrow to highlight other intercept and press EXE again. | 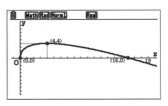 |
| 4. The answer appears on the screen. | There is a maximum stationary point at $(4, 4)$.
There are two intercepts, $(0, 0)$ and $(16, 0)$. | 4. The answer appears on the screen. | There is a maximum stationary point at $(4, 4)$.
There are two intercepts, $(0, 0)$ and $(16, 0)$. |

12.5.4 Local and global maxima and minima

The diagram shows the graph of a function sketched over a domain with end points D and E.

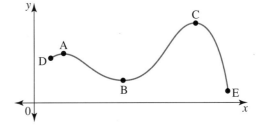

There are three turning points: A and C are maximum turning points, and B is a minimum turning point.

The y-coordinate of point A is greater than those of its neighbours, so A is a **local maximum** point. At point C, not only is its y-coordinate greater than those of its neighbours, it is greater than that of any other point on the graph. For this reason, C is called the **global** or **absolute maximum** point.

The **global** or **absolute minimum** point is the point whose y-coordinate is smaller than any others on the graph. For this function, point E, an end point of the domain, is the global or absolute minimum point. Point B is a **local minimum point**; it is not the global minimum point.

Global maximums and global minimums may not exist for all functions. For example, a cubic function on its maximal domain may have one local maximum turning point and one local minimum turning point, but there is neither a global maximum nor a global minimum point, since as $x \to \pm\infty, y \to \pm\infty$ (assuming a positive coefficient of x^3).

If a differentiable function has a global maximum or a global minimum value, then this will either occur at a turning point or at an end point of the domain. The y-coordinate of such a point gives the value of the global maximum or the global minimum.

Maxima and minima

- A function $y = f(x)$ has a global maximum $f(a)$ if $f(a) \geq f(x)$ for all x-values in its domain.
- A function $y = f(x)$ has a global minimum $f(a)$ if $f(a) \leq f(x)$ for all x-values in its domain.
- A function $y = f(x)$ has a local maximum $f(x_0)$ if $f(x_0) \geq f(x)$ for all x-values in the neighbourhood of x_0.
- A function $y = f(x)$ has a local minimum $f(x_0)$ if $f(x_0) \leq f(x)$ for all x-values in the neighbourhood of x_0.

A function defined on a restricted domain has the rule $y = \dfrac{3}{8}x(x-3)^2$, $x \in [-1, 4]$.

a. Determine the coordinates of the end points of the domain.
b. Determine the coordinates of any stationary points and determine their nature.
c. Sketch the graph of the function.
d. Identify the global, or absolute, maximum and global minimum values of the function.

THINK	WRITE
a. 1. Use the given domain to calculate the coordinates of the end points.	$y = \dfrac{3}{8}x(x-3)^2$
2. Substitute each of the end values of the domain in the function's rule.	For the domain $-1 \le x \le 4$: Left end point: when $x = -1$, $y = \dfrac{3}{8}(-1)(-1-3)^2$ $= -\dfrac{3}{8} \times 16$ $= -6$ Right end point: when $x = 4$, $y = \dfrac{3}{8}(4)(4-3)^2$ $= \dfrac{3}{2} \times 1$ $= \dfrac{3}{2}$
3. Write the answer.	The end points are $(-1, -6)$, $\left(4, \dfrac{3}{2}\right)$.
b. 1. Expand the function to calculate the derivative.	$y = \dfrac{3}{8}x(x-3)^2$ $= \dfrac{3}{8}x\left(x^2 - 6x + 9\right)$ $= \dfrac{3}{8}\left(x^3 - 6x^2 + 9x\right)$
2. Calculate the derivative of the function.	$\dfrac{dy}{dx} = \dfrac{3}{8}\left(3x^2 - 12x + 9\right)$
3. Calculate the coordinates of any stationary points.	At a stationary point, $\dfrac{dy}{dx} = 0$, so: $\dfrac{3}{8}\left(3x^2 - 12x + 9\right) = 0$ $3x^2 - 12x + 9 = 0$ $3\left(x^2 - 4x + 3\right) = 0$ $3(x-1)(x-3) = 0$ $x = 1, 3$

When $x = 1$,

$$y = \frac{3}{8}(1)(1-3)^2$$

$$= \frac{3}{8} \times 4$$

$$= \frac{3}{2}$$

When $x = 3$,

$$y = \frac{3}{8}(3)(3-3)^2$$

$$= 0$$

4. Test the gradient either side of the stationary points to determine their nature.

Stationary points: $\left(1, \dfrac{3}{2}\right)$, $(3, 0)$

x	0	1	2	3	4
$\dfrac{dy}{dx}$	$\frac{3}{8} \times (9) > 0$	0	$\frac{3}{8} \times (12-24+9) < 0$	0	$\frac{3}{8} \times (48-48+9) > 0$
Slope of the tangent	/	—	\	—	/

5. State the nature of the stationary points

The point $\left(1, \dfrac{3}{2}\right)$ is a maximum turning point and the point $(3, 0)$ is a minimum turning point.

c. 1. Calculate any intercepts with the coordinate axes.

For the y-intercept, $x = 0$: point $(0, 0)$

For the x-intercepts, $y = 0$:

$$\frac{3}{8}x(x-3)^2 = 0$$

$$\therefore x = 0 \ \text{ or } \ x = 3$$

Axis intercepts: $(0, 0)$ and $(3, 0)$

2. Sketch the graph using the end points, axis intercepts and stationary points.

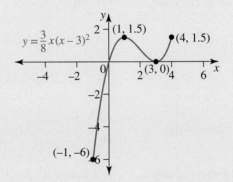

d. Examine the graph and the y-coordinates to identify the global, or absolute extreme values.

The function has a global maximum of $\dfrac{3}{2}$ at the maximum turning point and the right end point, and a global minimum of -6 at the left end point.

12.5.5 End behaviour of a function

The end behaviour of a function indicates what the function does as x approaches positive or negative infinity. We examine end behaviour when sketching functions because it gives us further information about the shape of the graph.

For a polynomial function, the term with the degree of the polynomial has the biggest impact. We use the degree of the polynomial and the sign of the leading coefficient to determine the end behaviour of a function.

Degree	Leading coefficient	End behaviour of the function	Example graphs of the functions
2 or 4	Positive	$f(x) \to \infty$ as $x \to -\infty$ $f(x) \to \infty$ as $x \to \infty$	
	Negative	$f(x) \to -\infty$ as $x \to -\infty$ $f(x) \to -\infty$ as $x \to \infty$	
3	Positive	$f(x) \to -\infty$ as $x \to -\infty$ $f(x) \to \infty$ as $x \to \infty$	
	Negative	$f(x) \to \infty$ as $x \to -\infty$ $f(x) \to -\infty$ as $x \to \infty$	

Exercise 12.5 Curve sketching

12.5 Exercise	12.5 Exam practice on

Simple familiar	Complex familiar	Complex unfamiliar
1, 2, 3, 4, 5, 6, 7, 8, 9, 10, 11, 12, 13, 14	15, 16, 17	18, 19, 20

These questions are even better in jacPLUS!
- Receive immediate feedback
- Access sample responses
- Track results and progress

Find all this and MORE in jacPLUS ▶

Simple familiar

1. Determine the stationary points and their nature for each of the following functions.

   a. $y = 8 - x^2$
   b. $f(x) = x^3 - 3x$
   c. $g(x) = 2x^2 - 8x$
   d. $f(x) = 4x - 2x^2 - x^3$
   e. $g(x) = 4x^3 - 3x^4$
   f. $y = x^2(x + 3)$

2. **MC** A continuous polynomial function, $y = f(x)$, has a local maximum turning point at the point where $x = 7$. Select the statement that describes how the gradient $f'(x)$ changes in the neighbourhood of $x = 7$.

   A. $f'(x) > 0$ just before the turning point, $f'(7) = 0$ and $f'(x) > 0$ just after the turning point.
   B. $f'(x) > 0$ just before the turning point, $f'(7) = 0$ and $f'(x) < 0$ just after the turning point.
   C. $f'(x) < 0$ just before the turning point, $f'(7) = 0$ and $f'(x) > 0$ just after the turning point.
   D. $f'(x) < 0$ just before the turning point, $f'(7) = 0$ and $f'(x) < 0$ just after the turning point.

3. **MC** If $f'(x) < 0$ where $x > 2$ and $f'(x) > 0$ where $x < 2$, then $x = 2$, $f(x)$ has a:

   A. local minimum
   B. local maximum
   C. point of inflection
   D. discontinuous point.

4. **MC** The graph of $y = x^4 + x^3$ has:

   A. a local maximum where $x = 0$
   B. a local minimum where $x = 0$
   C. a local minimum where $x = -\dfrac{3}{4}$
   D. a local maximum where $x = -\dfrac{3}{4}$.

5. **WE10** **a.** Determine the stationary points of $f(x) = x^3 + x^2 - x + 4$ and justify their nature.

   b. The curve $y = ax^2 + bx + c$ contains the point $(0, 5)$ and has a stationary point at $(2, -14)$. Calculate the values of a, b and c.

6. Determine any stationary points of the following curves and justify their nature using the sign of the derivative.

   a. $y = \dfrac{1}{3}x^3 + x^2 - 3x - 1$
   b. $y = -x^3 + 6x^2 - 12x + 18$
   c. $y = \dfrac{23}{6}x(x - 3)(x + 3)$
   d. $y = 4x^3 + 5x^2 + 7x - 10$

7. **a.** There is a stationary point on the curve $y = x^3 + x^2 + ax - 5$ at the point where $x = 1$. Determine the value of a.

   b. There is a stationary point on the curve $y = -x^4 + bx^2 + 2$ at the point where $x = -2$. Determine the value of b.

   c. The curve $y = ax^2 - 4x + c$ has a stationary point at $(1, -1)$. Calculate the values of a and c.

8. **WE11** If $h(x) = x^4 + 4x^3 + 4x^2$:

   a. sketch the graph of $h'(x)$

   b. identify the values of x where $h(x)$ is:

   i. increasing
   ii. decreasing.

9. Describe and sketch the graph of $g(x) = x^4 - 4x^2$ by examining its stationary points and end behaviour.

10. **WE12** Sketch the function $y = 2x^2 - x^3$. Identify any intercepts with the coordinate axes and any stationary points, and justify their nature.

11. Sketch the graphs of each of the following functions and label any axis intercepts and any stationary points with their coordinates. Justify the nature of the stationary points.

   a. $f: R \rightarrow R, f(x) = 2x^3 + 6x^2$
   c. $p: [-1, 1] \rightarrow R, p(x) = x^3 + 2x$

   b. $g: R \rightarrow R, g(x) = -x^3 + 4x^2 + 3x - 12$
   d. $\{(x, y): y = x^4 - 6x^2 + 8\}$

12. The graphs of $f'(x)$ are shown below. Identify all values of x for which $f(x)$ has stationary points and state their nature.

   a.
   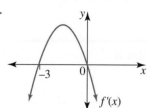

   b.
   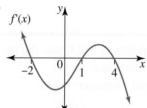

   c.
   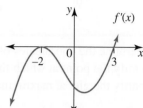

   d.
   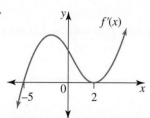

   e.
   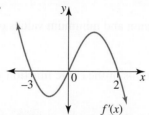

   f.
   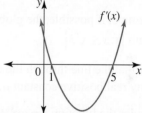

13. **WE13** A function defined on a restricted domain has the rule $y = 16x(x + 1)^3$, $x \in [-2, 1]$.

   a. Determine the coordinates of the end points of the domain.
   b. Determine the coordinates of any stationary point and determine its nature.
   c. Sketch the graph of the function.
   d. Identify the global, or absolute, maximum and global minimum values of the function.

14. Sketch the graphs of each of the following functions and identify any local and global maximum or minimum values over the given domain.

   a. $y = 4x^2 - 2x + 3, -1 \leq x \leq 1$
   c. $y = 3 - 2x^3, x \leq 1$

   b. $y = x^3 + 2x^2, -3 \leq x \leq 3$
   d. $f: [0, \infty) \rightarrow R, f(x) = x^3 + 6x^2 + 3x - 10$

Complex familiar

15. Sketch a possible graph of the function $y = f(x)$ for which:

   a. $f'(-4) = 0, f'(6) = 0, f'(x) < 0$ for $x < -4, f'(x) > 0$ for $-4 < x < 6$ and $f'(x) < 0$ for $x > 6$
   b. $f'(1) = 0, f'(x) > 0$ for $x \in R \setminus \{1\}$ and $f(1) = -3$.

16. A function defined on a restricted domain has the rule:

$$y = \frac{1}{16}x^2 + \frac{1}{x}, x \in \left[\frac{1}{4}, 4\right]$$

   a. Determine the coordinates of the end points of the domain.
   b. Determine the coordinates of any stationary point and determine its nature.
   c. Sketch the graph of the function.
   d. Identify the global maximum and global minimum values of the function, if they exist.

17. The graph of $f(x) = 2\sqrt{x} + \dfrac{1}{x}$, $0.25 \le x \le 5$ is shown.

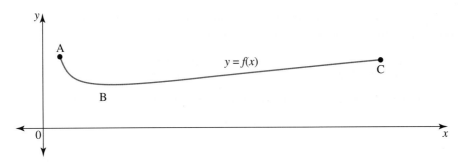

    a. Determine the coordinates of the end points A and C and the stationary point B.
    b. Identify the point at which the global maximum occurs.
    c. Identify the global maximum and global minimum values.

Complex unfamiliar

18. Determine, if possible, the global maximum and minimum values of the function $f(x) = 4x^3 - 12x$ over the domain $\{x : x \le \sqrt{3}\}$.

19. Show that the line through the turning points of the cubic function $y = xa^2 - x^3$ must pass through the origin for any real positive constant a.

20. Consider the function $y = ax^3 + bx^2 + cx + d$. The graph of this function passes through the origin at an angle of $135°$ with the positive direction of the x-axis.
Sketch the curve if the point $(2, -2)$ is a stationary point.

Fully worked solutions for this chapter are available online.

LESSON
12.6 Review

doc-41807

12.6.1 Summary

Hey students! Now that it's time to revise this chapter, go online to:

 Access the chapter summary

 Review your results

 Practise exam questions

Find all this and MORE in jacPLUS

12.6 Exercise

learn on

12.6 Exercise	12.6 Exam practice on

Simple familiar	Complex familiar	Complex unfamiliar
1, 2, 3, 4, 5, 6, 7, 8, 9, 10, 11, 12	13, 14, 15, 16, 17	18, 19, 20

These questions are even better in jacPLUS!
- Receive immediate feedback
- Access sample responses
- Track results and progress

Find all this and MORE in jacPLUS

Simple familiar

1. Determine the equation of the tangent line to the function $y = 2x^2 - 3x + 1$ at $x = 3$.

2. For the function $y = 0.5x^4 - x^2 - 4$:

    a. calculate the angle between the tangent to the curve at $x = 0.8$ and the positive x-axis
    b. calculate the angle between the normal to the curve at $x = -1.2$ and the positive x-axis.

3. Determine the equation of the normal line to the function $y = x^3 + 7x^2 - 2x + 3$ at $x = -2$.

4. For the position–time function $x(t) = 2t^2 - 4t + 1$:

    a. calculate the gradient at:
        i. $t = 0$
        ii. $t = 2$
        iii. $t = 4$

    b. sketch the velocity–time graph from $t = 0$ to $t = 4$.

5. Consider the function $f(x) = 2x + \dfrac{1}{3x}$.

    Determine the equations of the tangents to the curve with a gradient of -1.

6. The position, x metres, of a particle after t seconds is given by $x(t) = -\dfrac{1}{3}t^3 + t^2 + 8t + 1$, $t \geq 0$.

    a. Determine its initial position and initial velocity.
    b. Calculate the distance travelled before it changes its direction of motion.
    c. Determine the position of the particle at the instant it changes direction.

7. Use the first derivative test to determine the nature of the stationary points for each of the following functions.

   a. $f(x) = 2x^3 + 6x^2$

   b. $g(x) = -x^3 + 4x^2 + 3x - 12$

   c. $h(x) = 9x^3 - 117x + 108$

   d. $p(x) = x^3 + 2x$

   e. $y = x^4 - 6x^2 + 8$

   f. $y = 2x(x+1)^3$

8. Determine the stationary points of $f(x) = x^3 + x^2 - x + 4$ and justify their nature.

9. Consider the function defined by $f(x) = x^3 + 3x^2 + 8$.

   a. Show that $(-2, 12)$ is a stationary point of the function.
   b. Determine the nature of this stationary point.
   c. Identify the coordinates of the other stationary point.
   d. Justify the nature of the second stationary point.

10. The cost in dollars of employing n people per hour in a small distribution centre is modelled by
    $C = n^3 - 10n^2 - 32n + 400$, $5 \le n \le 10$.
    Calculate the number of people who should be employed in order to minimise the cost and justify your answer.

11. A batsman opening the innings in a cricket match strikes the ball so that its height y metres above the ground after it has travelled a horizontal distance x metres is given by $y = 0.0001x^2(625 - x^2)$.

    a. Calculate, to 2 decimal places, the greatest height the ball reaches and justify the maximum nature.
    b. Determine how far the ball travels horizontally before it strikes the ground.

12. The curve $y = x^4 - 4x^3 + 16x + 16$ has two stationary points, one of which occurs at $x = -1$.

    a. Show that there is a stationary point at $x = 2$ and determine its nature using the slope of the tangent.
    b. Determine the nature of the stationary point at $x = -1$.
    c. Determine the coordinates of the two stationary points and the y-intercept.
    d. Hence, draw a sketch of the curve $y = x^4 - 4x^3 + 16x + 16$.

Complex familiar

13. For the function $f(x) = 0.8x^2 + 0.4x - 3$, calculate the value of x at which the gradient of the tangent is equal to the average rate of change between $x = -1$ and $x = 2$.

14. A ball is thrown vertically upwards into the air so that after t seconds, its height h metres above the ground is $h = 40t - 5t^2$.

    a. Determine the rate at which its height is changing after 2 seconds.
    b. Calculate its velocity when $t = 3$.
    c. Determine the time at which its velocity is -10 m/s and the direction in which the ball is then travelling.
    d. Determine when its velocity is zero.
    e. Determine the greatest height the ball reaches.
    f. Determine the time and speed at which the ball strikes the ground.

15. Sketch the function $y = x^4 + 2x^3 - 2x - 1$. Determine any intercepts with the coordinate axes and any stationary points, and justify their nature.

16. Determine the values of k so the graph of $y = 3x^3 + 6x^2 + kx + 6$ will have no stationary points.

17. A particle, P, moving in a straight line has displacement, x metres, from a fixed origin O of $x_P(t) = t^3 - 12t^2 + 45t - 34$ for time t seconds.

   a. Determine the time(s) at which the particle is stationary.
   b. Determine the time interval over which the velocity is negative.

   A second particle, Q, also travels in a straight line with its position from O at time t seconds given by $x_Q(t) = -12t^2 + 54t - 44$.

   c. Determine the time at which P and Q are travelling with the same velocity.
   d. Determine the times at which P and Q have the same displacement from O.

Complex unfamiliar

18. A skier is following a trail that curves according to the function

$$y = -0.09x^2 + 2x, \, 0 \le x \le 30,$$

when she loses control and slides off the trail at the point where $x = 15$. Assuming that the skier slides off the path in a straight line, determine if the skier is likely to hit a large tree located at the point $(20, 6.25)$.

19. The point $(2, -54)$ is a stationary point of the curve $y = x^3 + bx^2 + cx - 26$. Sketch the curve and label all key points with their coordinates.

20. Sketch the function $y = 10\sqrt{x} - x^2\sqrt{x}$, giving the coordinates of any stationary points in exact form. Discuss the end behaviour of the function.

Fully worked solutions for this chapter are available online.

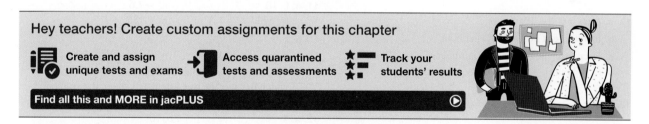

Answers

Chapter 12 Graphical applications of derivatives

12.2 Instantaneous rates of change

12.2 Exercise

1. $f'(3) = \dfrac{25}{27}$

2. $\dfrac{dy}{dx} = -12$

3. a. $\dfrac{dA}{dr} = 2\pi r$ b. 6π mm^2/mm

4. a. $\dfrac{dS}{dr} = 8\pi r$ b. 40π cm^2/cm   c. 2 cm

5. a. $\dfrac{dV}{dr} = 4\pi r^2$ b. π m^3/m

6. a. $\dfrac{dV}{dh} = h^2 + 2$ b. 15 cm^3, 11 cm^3/cm

7. 0.4π m^2/m

8. 96 mm^2/mm

9. a. $r = \dfrac{1}{2}h$ b. $V = \dfrac{1}{12}\pi h^3$ c. $\dfrac{9\pi}{4}$ cm^3/cm

10. a. $r = \dfrac{h}{3}$

    b. $V = \dfrac{\pi h^3}{27}$

    c. $\dfrac{dV}{dh} = \dfrac{25\pi}{9}$ cm^3/cm

11. a. $A = \dfrac{\sqrt{3}}{4}x^2$ b. $\sqrt{3}$ cm^2/cm   c. $8\sqrt{3}$ cm^2/cm

12. a. $P'(t) = 600 - 12t$

    b. Increasing at 360 bacteria per hour

    c. Decreasing at 120 bacteria per hour

    d. i. 30 hours   ii. 70 hours   iii. 50 hours

13. a. $A = \dfrac{\sqrt{15}}{16}x^2$ b. $\sqrt{15}$ cm^2/cm

14. a. $V = 333\dfrac{1}{3}$ mL   b. $\dfrac{dV}{dt} = 20t - 2t^2$

    c. $\dfrac{dV}{dt} = 42$ mL/s   d. $t = 5$ s, $\dfrac{dV}{dt} = 50$ mL/s

15. a. -4.4 rabbits/month

    b. -22 rabbits/month

    c. $(t \to \infty, N \to 0)$. Effectively, the population of rabbits will be zero in the long term.

16. a. $\dfrac{dV}{dt} = -30$ L/min   b. -51.6 L/min

    c. $t = 0$

17. a. $\sqrt{2}x$ m

    b. $h = \dfrac{10\sqrt{3}}{3}$; gives the height for maximum volume

18. 0 cm^3/cm

19. 9π cm^3/cm

20. The volume is decreasing at the rate of $\dfrac{8\pi}{3}$ cm^3/cm.

12.3 Tangents and normals of a graph

12.3 Exercise

1. a. 2   b. 7   c. $\dfrac{229}{9}$ d. $\dfrac{\sqrt{3}}{2}$

2. a. $-\dfrac{1}{2}$ b. $-\dfrac{1}{7}$

    c. $-\dfrac{9}{229}$ d. $-\dfrac{2}{\sqrt{3}}$ or $\dfrac{-2\sqrt{3}}{3}$

3. $y = -4x + 18$

4. a. $y = -7x + 3$ b. $y = -2x + 8$
    c. $y = 6x - 8$ d. $y = 5$
    e. $y = -24x - 15$ f. $2y + x = 49$

5. $y = 5 - x$

6. a. $y = 4x - 22$ b. $y = -6$
    c. $y = 2x - 13$

7. a. $y = 2x + 1$ b. $y = -\dfrac{1}{2}x + 6$

    c. $y = -x + \dfrac{19}{4}$

8. a. $88.9°$ b. $1.33°$

9. a. i. $78.11°$ ii. $-11.89°$

    b. The difference is $90°$.

10. a. Undefined

    b. ∞

    c. $90°$

    d. The derivative of the function is undefined at $x = 0$.

11. a. i. 10   ii. -7.6
     iii. 5780   iv. 1.185

    b. i. 3   ii. 3
     iii. $-3.055, 3.055$ iv. $0.284, 2.923$

    c. There is always a point between the two points over which the average was calculated where the gradient of the tangent is equal to the average rate of change.

12. a. 4   b. $76°$
    c. $y = 4x + 5$ d. No

13. a. $y = 0.444x - 40.417$

    b. Yes

14. a. -19.6 m/s   b. $y = -19.6t + 119.6$
    c. 6.102s   d. 1.585s

15. a. $\left(\dfrac{3}{2}, \dfrac{9}{4}\right)$ b. $26.6°$

16. $a = 1$ and $b = 4$

17. a.

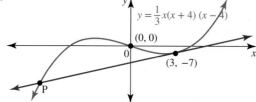

$y = \frac{1}{3}x(x+4)(x-4)$

(0, 0)

(3, −7)

P

b. $y = \frac{11}{3}x - 18$

c. i. Sample responses can be found in the worked solutions in the online resources.

ii. $(-6, -40)$

d. Same gradient

e. i. Same gradient

ii. $\left(-\frac{2a}{3}, \frac{2a^3}{3}\right), \left(\frac{2a}{3}, -\frac{2a^3}{3}\right)$

18. $(-2, -23)$

19. Sample responses can be found in the worked solutions in the online resources.

20. $(1, -5)$

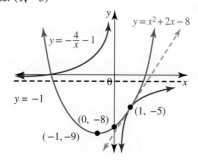

$y = -\frac{4}{x} - 1$

$y = x^2 + 2x - 8$

$y = -1$

(0, −8)

(1, −5)

(−1, −9)

12.4 Kinematics

12.4 Exercise

1. a. At the origin

b. $v = 6t - 6$; initial velocity -6 m/s; move to the left

c. After 1 second and 3 metres to the left of the origin

d. 6 metres

e. 3 m/s

f. 0 m/s

2. a. $\frac{dx}{dt} = 12$

b. The velocity is constant at 12 m/s.

c. 12 m

d. 3 seconds

3. a. $x = 1$ **b.** $x = -3$

c. Right   **d.** $t = 2$

4. a. D   **b.** D   **c.** D

d. C   **e.** B

5. a. 40 m

b. -2 m (or 2 m below the platform)

c. 0.5 m/s

d. 0.025 m/s (or 0.025 m/s downwards)

6. a. Downward   **b.** 2.4 m/s   **c.** -3 m/s

7. a. i. $x = 0$ **ii.** Right

iii. $t = 2, x = 8$ **iv.** $t = 5, x = -3$

b. i. $x = 4$ **ii.** Right

iii. $t = 4, x = 12$ **iv.** $t = 6, x = 10$

c. i. $x = 0$

ii. Right

iii. $t = 3, x = 12$ and $t = 6, x = 3$

iv. $t = 8, x = 10$

d. i. $x = 0$ **ii.** Left

iii. $t = 1, x = -5$ **iv.** $t = 3, x = 18$

e. i. $x = -3$ **ii.** Left

iii. $t = 1\frac{1}{2}, x = -6$ **iv.** $t = 5, x = 5$

f. i. $x = 2$

ii. Left

iii. $t = 3, x = -5$ and $t = 5, x = 5$

iv. $t = 6, x = 4$

8. a.

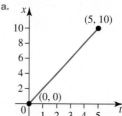

(5, 10)

(0, 0)

i. $x = 0$ **ii.** Right

iii. No   **iv.** $x = 10$

b.

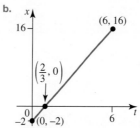

(6, 16)

$\left(\frac{2}{3}, 0\right)$

(0, −2)

i. $x = -2$ **ii.** Right

iii. No   **iv.** $x = 16$

c.

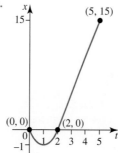

(5, 15)

(0, 0)   (2, 0)

i. $x = 0$

ii. Left

iii. Yes, $t = 1, x = -1$

iv. $x = 15$

d.

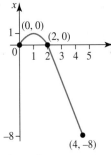

i. $x = 0$

ii. Right

iii. Yes, $t = 1, x = 1$

iv. $x = -8$

e.

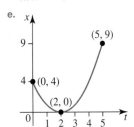

i. $x = 4$

ii. Left

iii. Yes, $t = 2, x = 0$

iv. $x = 9$

f.

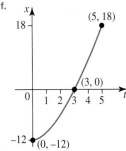

i. $x = -12$ ii. Right

iii. No   iv. $x = 18$

9.

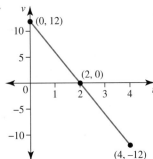

10. x (m)

(2, 45)
45
25 (0, 25)
(5, 0)
0   1   2   3   4   5 t (s)

a. 45 m   b. $t = 5$ c. $t = 2$ d. 20 m/s

11. a. 7 m to the left of the origin

b. $v = 2t - 6$, -6 m/s

c. 3 seconds

d. 6 s, 6 m/s

12. a. $v = \dfrac{dx}{dt} = 3t^2 - 8t - 3$

b. $t = 3$ s, 6 m to the left of origin.

c. -6 m/s

13. a. 10 cm to left of origin, 5 cm to right of origin

b. 15 cm

c. $v = 5$ cm/s

d.

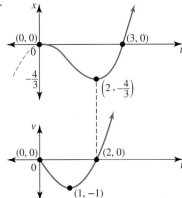

The velocity graph is the gradient graph of the position.

14. a. $t = 0, x = 0, v = 0$

b. 2 seconds; $1\dfrac{1}{3}$ cm left of origin

c. 3 seconds

d. 3 cm/s, 4 cm/s^2

e.

At $t = 2$, position is most negative, Velocity is zero

f. At $t = 1$, velocity is most negative and acceleration is zero.

15. a. 20 m/s

b. 10 m/s

c. 5 seconds, travelling down towards the ground

d. 4 seconds

e. 80 metres

f. 8 seconds, 40 m/s

16. a. 63 m   **b.** 12 m/s

c. -18 m/s   **d.** 9 seconds; 30 m/s

e. 60 m   **f.** 10 m/s

17. 2 seconds

18. a. $t = 0,\ x = \frac{2}{3}(0)^3 - 4(0)^2$

   $= 0\,\text{m}$

   $v = \frac{dx}{dt} = 2t^2 - 8t$

   $t = 0,\ v = 2(0)^2 - 8(0)$
   $= 0\,\text{m/s}$

b. 4 seconds, $21\frac{1}{3}$ metres left of origin

c. 6 seconds

d. 24 m/s

19. a. 12 metres to left of O; moves to the left

b. 80 metres

c. 260 metres; 26 m/s

d. For particle B, $s = t^2 + 8t - 12,\ t \geq 0$.
   Initial position: when $t = 0,\ s = -12$.
   Therefore, both particles A and B start from the same initial position 12 metres to the left of O.

   B's velocity: $v = \frac{ds}{dt}$

   $= 2t + 8$

   When $t = 0,\ v = 8$

   Since particle B's initial velocity is positive, it starts to move to the right toward O. Hence, particles A and B start from the same initial position, but they start to move in opposite directions.

e. 12 seconds; 228 metres to right of O

f. Time 6 seconds, positions $x = -72,\ s = 72$; time $2\frac{2}{3}$

   seconds, positions $x = -83\frac{1}{9},\ s = 16\frac{4}{9}$

20. a. 3 seconds and 5 seconds

b. $t \in (3, 5)$

c. $\sqrt{3}$ seconds

d. $\left(\sqrt{6} - 1\right)$ seconds and 2 seconds

12.5 Curve sketching

12.5 Exercise

1. a. $(0, 8)$ maximum

b. $(-1, 2)$ maximum and $(1, -2)$ minimum

c. $(2, -8)$ minimum

d. $(-2, -8)$ minimum and $\left(\frac{2}{3}, \frac{40}{27}\right)$ maximum

e. $(0, 0)$ point of horizontal inflection and $(1, 1)$ maximum

f. $(-2, 4)$ maximum and $(0, 0)$ minimum

2. B

3. B

4. C

5. a. $(-1, 5)$ is a maximum turning point; $\left(\frac{1}{3}, \frac{103}{27}\right)$ is a

   minimum turning point.

b. $a = \frac{19}{4},\ b = -19,\ c = 5$

6. a. $(-3, 8)$ is a maximum turning point; $\left(1, -\frac{8}{3}\right)$ is a

   minimum turning point.

b. $(2, 10)$ is a stationary point of inflection.

c. $\left(-\sqrt{3}, 23\sqrt{3}\right)$ is a maximum turning point;

   $\left(\sqrt{3}, -23\sqrt{3}\right)$ is a minimum turning point.

d. There are no stationary points.

7. a. $a = -5$ **b.** $b = 8$
   c. $a = 2,\ c = 1$

8. a.

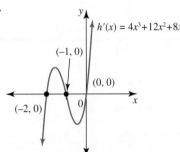

b. i. $-2 < x < -1$ and $x > 0$
   ii. $x < -2$ and $-1 < x < 0$

9. Local minimums at $(-\sqrt{2}, -4)$ and $(\sqrt{2}, -4)$, and a maximum at $(0, 0)$; $f(x) \to \infty$ as $x \to -\infty$, $f(x) \to \infty$ as $x \to \infty$.

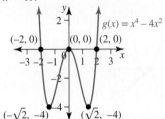

10.

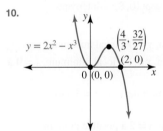

   x-intercepts when $x = 2,\ x = 0$; $(0, 0)$ is a minimum turning

   point; $\left(\frac{4}{3}, \frac{32}{27}\right)$ is a maximum turning point.

11. a.

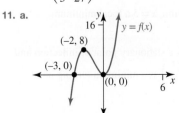

   Minimum turning point $(0, 0)$; maximum turning point $(-2, 8)$; intercept $(-3, 0)$

b.

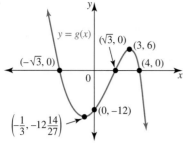

Minimum turning point $\left(-\dfrac{1}{3}, -12\dfrac{14}{27}\right)$; maximum turning point $(3, 6)$; intercepts $\left(\pm\sqrt{3}, 0\right), (4, 0), (0, -12)$

c.

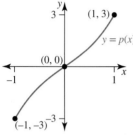

End points $(-1, -3), (1, 3)$; intercept $(0, 0)$; no stationary points

d.

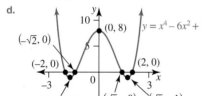

Minimum turning points $\left(\pm\sqrt{3}, -1\right)$; maximum turning point and y-intercept $(0, 8)$; x-intercepts $\left(\pm\sqrt{2}, 0\right), (\pm 2, 0)$

12. **a.** $x = -3$ a local minimum, $x = 0$ a local maximum

  b. $x = -2$ a local maximum, $x = 1$ a local minimum, $x = 4$ a local maximum

  c. $x = -2$ a negative point of inflection, $x = 3$ a local minimum

  d. $x = -5$ a local minimum, $x = 2$ a positive point of inflection

  e. $x = -3$ a local maximum, $x = 0$ a local minimum, $x = 2$ a local maximum

  f. $x = 1$ a local maximum, $x = 5$ a local minimum

13. **a.** $(-2, 32), (1, 128)$

  b. The point $(-1, 0)$ is a stationary point of inflection and the point $\left(-\dfrac{1}{4}, -\dfrac{27}{16}\right)$ is a minimum turning point.

c.

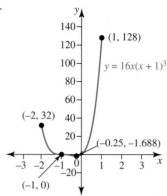

d. Absolute maximum value: 128; absolute minimum value: $-\dfrac{27}{16}$

14. **a.**

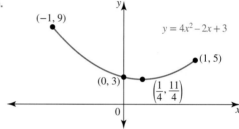

Local and global minimum $\dfrac{11}{4}$; no local maximum; global maximum 9

b.

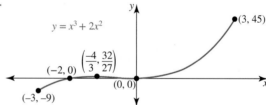

Local maximum $\dfrac{32}{27}$; local minimum 0; global maximum 45; global minimum -9

c.

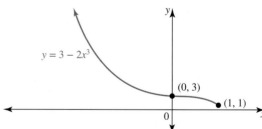

No local or global maximum; no local minimum; global minimum 1

d.

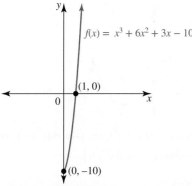

No local or global maximum; no local minimum; global minimum -10

15. a.

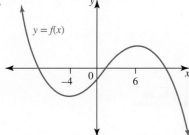

Negative cubic with minimum turning point where $x = -4$ and maximum turning point where $x = 6$

b.

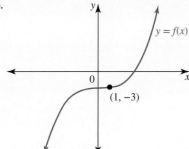

Positive cubic with stationary point of inflection at $(1, -3)$

16. a. $\left(\dfrac{1}{4}, \dfrac{1025}{256}\right),\ \left(4, \dfrac{5}{4}\right)$

b. $\left(2, \dfrac{3}{4}\right)$ is a minimum turning point.

c.

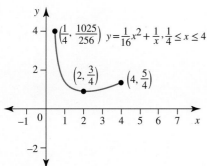

d. Global maximum $\dfrac{1025}{256}$; global minimum $\dfrac{3}{4}$

17. a. A $(0.25, 5)$, B$(1, 3)$, C$(5, 2\sqrt{5} + 0.2)$

  b. A

  c. $5, 3$

18. No global minimum; global maximum 8

19. The origin lies on the line $y = \dfrac{2a^2 x}{3}$.

20. The curve is a positive cubic passing through the origin with a maximum turning point at $\left(\dfrac{-2}{3}, \dfrac{10}{27}\right)$ and a minimum turning point at $(2, -2)$.

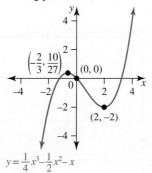

12.6 Review

12.6 Exercise

1. $y = 9x - 17$

2. a. $-29.9°$ or $150.1°$ **b.** $43.4°$

3. $y = \dfrac{1}{18}x + \dfrac{244}{9}$

4. a. i. -4 **ii.** 4 **iii.** 12

  b.

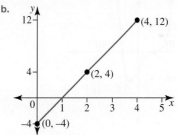

5. $y = -x + 2$ and $y = -x - 2$

6. a. 1 metre to right of origin; 8 m/s

  b. $26\dfrac{2}{3}$ metres

  c. $27\dfrac{2}{3}$ metres to the right of O

7. a. Maximum at $x = -2$ and minimum at $x = 0$

  b. Minimum at $x = \dfrac{-1}{3}$ and maximum at $x = 3$

  c. Maximum at $x = -\sqrt{\dfrac{13}{3}}$ and minimum at $x = \sqrt{\dfrac{13}{3}}$

  d. None

  e. Minimum at $x = -\sqrt{3}$, maximum at $x = 0$ and minimum at $x = \sqrt{3}$

  f. Point of horizontal inflection at $x = -1$ and minimum at $x = -\dfrac{1}{4}$

8. Maximum at $(-1, 5)$ and minimum at $\left(\dfrac{1}{3}, \dfrac{103}{27}\right)$

9. a. Sample responses can be found in the worked solutions in the online resources.

   b. Maximum

   c. $(0, 8)$

   d. Minimum

10. 8 people

11. a. 9.77 m   **b.** 25 m

12. a. Please see the worked solution in the online resources.

   b. Minimum turning point

   c. Point of inflection $(2, 32)$, minimum turning point $(-1, 5)$, y-intercept $(0, 16)$.

   d.

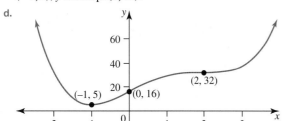

13. $x = 0.5$

14. a. 20 m/s, upwards

   b. 10 m/s upwards

   c. 5 seconds, travelling downward

   d. 4 seconds

   e. 80 m

   f. 8 seconds, 40 m/s

15.

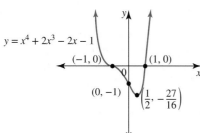

$(0, -1)$, $(\pm 1, 0)$ intercepts with axes; $(-1, 0)$ is a stationary point of inflection; $\left(\dfrac{1}{2}, -\dfrac{27}{16}\right)$ is a minimum turning point.

16. $k > 4$

17. a. 3 seconds and 5 seconds

   b. $t \in (3, 5)$

   c. $\sqrt{3}$ seconds

   d. 2 seconds and $(\sqrt{6} - 1)$ seconds

18. If the skier continues in that direction they will reach the point $(20, 6.25)$, so they are likely to hit the tree at $(20, 6.25)$.

19.

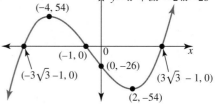

The key points are $(-1 - 3\sqrt{3}, 0)$, $(-4, 54)$, $(-1, 0)$, $(0, -26)$, $(2, -54)$, $\left(-1 + 3\sqrt{3}, 0\right)$.

20.

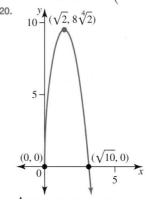

As $x \to \infty$, $y \to -\infty$.

13 Differentiation rules

LESSON SEQUENCE

Fully worked solutions for this chapter are available online.

EXAM PREPARATION

Access exam-style questions in every lesson, available online.

 Resources

Solutions	Solutions — Chapter 13 (sol-1280)
Exam questions	Exam question booklet — Chapter 13 (eqb-0301)
Digital documents	Learning matrix — Chapter 13 (doc-41795)
	Chapter summary — Chapter 13 (doc-41808)

LESSON
13.1 Overview

13.1.1 Introduction

Although there is some debate, the discovery of the power rule is usually credited to Leibniz. The first known use of the chain rule was also by Leibniz. He used it to solve functions of the form $\sqrt{a + bz + cz^2}$.

As we saw in the previous chapter, there are many functions that can be differentiated using the power rule after they have been manipulated in some manner (e.g. expanded or simplified). For many functions, the process of manipulation is quite time consuming and can increase the likelihood of errors being made during calculations.

There are also many functions that cannot be differentiated using the power rule. As such, we need other methods to deal with a wider variety of differentiable functions. In this chapter, we will examine three more differentiation rules.

13.1.2 Syllabus links

Lesson	Lesson title	Syllabus links
13.2	**The chain rule**	○ Use the chain rule, if $y = f(u)$ and $u = g(x)$ then $\dfrac{dy}{dx} = \dfrac{dy}{du} \times \dfrac{du}{dx}$, to determine the derivative of composite functions involving power and polynomial functions.
13.3	**The product rule**	○ Use the product rule, $\dfrac{d(uv)}{dx} = u\dfrac{dv}{dx} + v\dfrac{du}{dx}$, to determine the derivative of products of functions involving power and polynomial functions.
13.4	**The quotient rule**	○ Use the quotient rule, $\dfrac{d\left(\frac{u}{v}\right)}{dx} = \dfrac{v\frac{du}{dx} - u\frac{dv}{dx}}{v^2}$, to determine the derivative of quotients of functions involving power and polynomial functions.
13.5	**Combinations of the chain rule, product rule and quotient rule**	○ Solve problems that involve combinations of the chain rule, product rule and quotient rule to differentiate functions involving power and polynomial functions, expressing derivatives in simplest and factorised form.

Source: Mathematical Methods Senior Syllabus 2024 © State of Queensland (QCAA) 2024; licensed under CC BY 4.0.

LESSON
13.2 The chain rule

SYLLABUS LINKS

- Use the chain rule, if $y = f(u)$ and $u = g(x)$ then $\dfrac{dy}{dx} = \dfrac{dy}{du} \times \dfrac{du}{dx}$, to determine the derivative of composite functions involving power and polynomial functions.

Source: Mathematical Methods Senior Syllabus 2024 © State of Queensland (QCAA) 2024; licensed under CC BY 4.0.

13.2.1 Composite functions

A **composite function**, also known as a function of a function, consists of two or more functions nested within each other or chained together.

For example, if $f(x) = 3x + 4$ and $g(x) = x^2$, then $g(f(x)) = (3x + 4)^2$ is a composite function.

Other examples of composite functions are $\sqrt{5x - 2}$, $\sin(3x + 2)$ and $\dfrac{1}{(5x + 2)^3}$.

Consider the function $y = (3x + 4)^2$.

Currently, if we want to derive this, we first have to expand the function. Once it is expanded, it is possible to find the derivative.

$$y = 9x^2 + 24x + 16$$

$$\frac{dy}{dx} = 18x + 24$$

$$= 6(3x + 4)$$

The chain rule allows us to reach this same outcome without having to expand the function first.

13.2.2 The chain rule

The chain rule states the following.

> **The chain rule**
>
> If $y = f(u)$ and $u = g(x)$, then:
>
> $$\frac{dy}{dx} = \frac{dy}{du} \times \frac{du}{dx}$$

It is known as the chain rule because $u = f(g(x))$, meaning there are two functions chained together; u provides the 'link' between y and x.

It is known as the chain rule because u provides the 'link' between y and x.

Consider again the function $y = (3x + 4)^2$.

Let $u = 3x + 4$; therefore, $\dfrac{du}{dx} = 3$.

Let $y = u^2$; therefore, $\dfrac{dy}{du} = 2u$.

By the chain rule, $\dfrac{dy}{dx} = \dfrac{dy}{du} \times \dfrac{du}{dx} = 2u \times 3 = 6u$.

Since $u = 3x + 4$, $\dfrac{dy}{dx} = 6(3x + 4)$.

If functions are written in function notation, the chain rule can still be applied.

If $y = f(g(x))$, then $y' = f'(g(x)) \times g'(x)$.

WORKED EXAMPLE 1 Applying the chain rule

Determine $\dfrac{dy}{dx}$ if $y = (3x - 2)^3$.

THINK	WRITE
1. Write the equation.	$y = (3x - 2)^3$
2. Express y as a function of u.	Let $y = u^3$ where $u = 3x - 2$.
3. Differentiate y with respect to u.	$\dfrac{dy}{du} = 3u^2$
4. Express u as a function of x.	$u = 3x - 2$
5. Differentiate u with respect to x.	$\dfrac{du}{dx} = 3$
6. Determine $\dfrac{dy}{dx}$ using the chain rule.	$\dfrac{dy}{dx} = \dfrac{dy}{du} \times \dfrac{du}{dx}$ $= 3u^2 \times 3$ $= 9u^2$
7. Replace u as a function of x.	$= 9(3x - 2)^2$

WORKED EXAMPLE 2 Applying the chain rule and function notation

If $f(x) = \dfrac{1}{\sqrt{2x^2 - 3x}}$, find $f'(x)$.

THINK	WRITE
1. Write the equation.	$f(x) = \dfrac{1}{\sqrt{2x^2 - 3x}}$
2. Express $f(x)$ in index form, that is, as $y = (g(x))^n$.	$y = (2x^2 - 3x)^{-\frac{1}{2}}$
3. Express y as a function of u.	Let $y = u^{-\frac{1}{2}}$ where $u = 2x^2 - 3x$.
4. Differentiate y with respect to u.	$\dfrac{dy}{du} = -\dfrac{1}{2}u^{-\frac{3}{2}}$
5. Express u as a function of x.	$u = 2x^2 - 3x$
6. Differentiate u with respect to x.	$\dfrac{du}{dx} = 4x - 3$
7. Find $f'(x)$ using the chain rule.	$\dfrac{dy}{dx} = f'(x) = -\dfrac{1}{2}u^{-\frac{3}{2}} \times (4x - 3)$ $= -\dfrac{1}{2}(2x^2 - 3x)^{-\frac{3}{2}}(4x - 3)$

8. Replace u as a function of x and simplify.

$$= \frac{-(4x-3)}{2(2x^2-3x)^{\frac{3}{2}}}$$

$$= \frac{3-4x}{2\sqrt{(2x^2-3x)^3}}$$

WORKED EXAMPLE 3 Applying the chain rule

Determine the minimum distance from the curve $y = 2x^2$ to the point (4, 0), correct to 2 decimal places. You do not need to justify your answer.

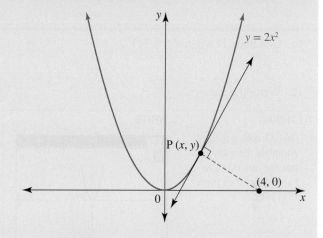

THINK

1. Let P be the point on the curve such that the distance from P to the point (4, 0) is a minimum.

2. Write the formula for the distance, D, between the two points.

3. Express the distance between the two points as a function of x only.

4. Express D as a function of u.

5. Differentiate D with respect to u.

6. Express u as a function of x.

7. Differentiate u with respect to x.

8. Determine $\dfrac{dD}{dx}$ using the chain rule

WRITE

$P = (x, y)$

$D = \sqrt{(x_2 - x_1)^2 + (y_2 - y_1)^2}$

$\quad = \sqrt{(x-4)^2 + (y-0)^2}$

$\quad = \sqrt{(x-4)^2 + y^2}$

$y = 2x^2$

$\therefore D = \sqrt{(x-4)^2 + (2x^2)^2}$

$\quad = (x^2 - 8x + 16 + 4x^4)^{\frac{1}{2}}$

Let $D = (u)^{\frac{1}{2}}$ where $u = 4x^4 + x^2 - 8x + 16$.

$\dfrac{dD}{du} = \dfrac{1}{2}u^{-\frac{1}{2}}$

$\quad = \dfrac{1}{2\sqrt{u}}$

$u = 4x^4 + x^2 - 8x + 16$

$\dfrac{du}{dx} = 16x^3 + 2x - 8$

$\dfrac{dD}{dx} = \dfrac{dD}{du} \times \dfrac{du}{dx}$

$\quad = \dfrac{1}{2\sqrt{u}} \times \left(16x^3 + 2x - 8\right)$

9. Replace u as a function of x.

$$= \frac{2\left(8x^3 + x - 4\right)}{2\sqrt{u}}$$

$$= \frac{\left(8x^3 + x - 4\right)}{\sqrt{4x^4 + x^2 - 8x + 16}}$$

10. To determine the minimum distance, solve $\dfrac{dD}{dx} = 0$ using technology.

$$\frac{\left(8x^3 + x - 4\right)}{\sqrt{4x^4 + x^2 - 8x + 16}} = 0$$

$$8x^3 + x - 4 = 0$$

$$x = 0.741$$

11. Calculate D when $x = 0.741$.

$$D = \sqrt{4\left(0.741\right)^4 + \left(0.741\right)^2 - 8\left(0.741\right) + 16}$$

$$= 3.439$$

12. Write the answer.

The minimum distance is 3.44 units.

TI	THINK	WRITE	CASIO	THINK	WRITE
1. On a Graphs page, complete the entry line for function 1 as $f1(x) = 2x^2$ then press ENTER.	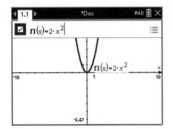	1. Let P be the point on the curve such that the distance from P to the point $(4, 0)$ is a minimum.	$P = (x, y)$
2. Press MENU, then select 8: Geometry 1: Points & Lines 2: Point On Click on the graph of function 1 then click again on the graph to create a point on the curve. Click on the x-axis and then click on the point at $(4, 0)$ to create another point.	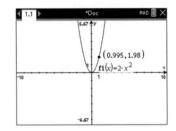	2. Write the formula for the distance between the two points.	$D = \sqrt{(x_2 - x_1)^2 + (y_2 - y_1)^2}$ $= \sqrt{(x - 4)^2 + (y - 0)^2}$ $= \sqrt{(x - 4)^2 + y^2}$
3. Press MENU, then select 8: Geometry 1: Points & Lines 6: Segment Click on the point created on the graph of function 1 and then click on the point at $(4, 0)$ to create a line segment between the two points.	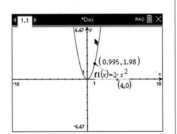	3. Express the distance between the two points as a function of x only.	$y = 2x^2$ $\therefore D = \sqrt{(x - 4)^2 + (2x^2)^2}$

4. Press MENU, then select 8: Geometry 3: Measurement 1: Length Click on the line segment joining the two points, then click next to the line segment to display the length of the line segment.

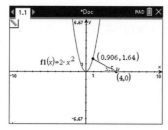

5. Click and drag the point on the graph of function 1, finding the position that results in the length of the line segment being as small as possible.

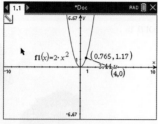

6. The answer appears on the screen.

The minimum distance is 3.44 units.

4. On a Run-Matrix screen, press OPTN then select CALC by pressing F4, press F6 to scroll across to more menu options, then select FMin by pressing F1. Complete the entry line as

$$\text{FMin}\left(\sqrt{(x-4)^2 + (2x^2)^2}, 0, 10, 5\right)$$

then press EXE.

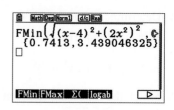

5. The answer appears on the screen.

The minimum distance is 3.44 units.

Exercise 13.2 The chain rule

learn on

13.2 Exercise	13.2 Exam practice on

Simple familiar	Complex familiar	Complex unfamiliar
1, 2, 3, 4, 5, 6, 7, 8, 9, 10, 11, 12	13, 14, 15, 16	17, 18, 19, 20

Simple familiar

1. **WE1** For each of the following composite functions, determine $\dfrac{dy}{dx}$ if:

   a. $y = (3x + 2)^2$

   b. $y = (7 - x)^3$

   c. $y = \dfrac{1}{2x - 5}$

   d. $y = \dfrac{1}{(4 - 2x)^4}$

   e. $y = \sqrt{5x + 2}$

   f. $y = \dfrac{3}{\sqrt{3x - 2}}$

2. Use the chain rule to determine the derivatives of the following functions.

   a. $y = (8x + 3)^4$

   b. $y = (2x - 5)^3$

   c. $f(x) = (4 - 3x)^5$

   d. $y = \sqrt{3x^2 - 4}$

   e. $f(x) = (x^2 - 4x)^{\frac{1}{3}}$

   f. $g(x) = (2x^3 + x)^{-2}$

3. **WE2** If $f(x) = \dfrac{1}{\sqrt{4x + 7}}$, determine $f'(x)$.

4. Determine the derivative of:

a. $f(x) = (x^2 + 5x)^8$

b. $y = (x^3 - 2x)^2$

c. $f(x) = (x^3 + 2x^2 - 7)^{\frac{1}{5}}$

d. $y = (2x^4 - 3x^2 + 1)^{\frac{3}{2}}$.

5. Match the following functions to their derivatives.

a. $(3x - 2)^3$ A $\quad 3(x - 2)^2$

b. $3(3x - 2)^2$ B $\quad 9(x - 2)^2$

c. $3(x - 2)^3$ C $\quad 9(3x - 2)^2$

d. $(x - 2)^3$ D $\quad 18(3x - 2)$

6. The length of a snake, L cm, at time t weeks after it is born is modelled as:

$$L = 12 + 6t + 0.01(20 - t)^2, \; 0 \le t \le 20.$$

Determine:

a. the length at:

   i. birth

   ii. 20 weeks

b. R, the rate of growth, at any time, t

c. the maximum and minimum growth rates.

7. Use the chain rule to determine the derivatives of the following functions.
(*Hint:* Simplify first using index notation and the laws of indices.)

a. $y = \dfrac{\sqrt{6x - 5}}{6x - 5}$

b. $f(x) = \dfrac{(x^2 + 2)^2}{\sqrt{x^2 + 2}}$

8. If $f(x) = \sqrt{x^2 - 2x + 1}$, determine:

a. $f(3)$

b. $f'(x)$

c. $f'(3)$

d. $f'(x)$ when $x = 2$.

9. Differentiate the function $f(x) = 2(5x + 1)^3$ using the chain rule and by expanding. Confirm they produce the same result.

10. a. If $f(x) = (2x - 1)^6$, calculate $f'(3)$.

b. If $g(x) = (x^2 - 3x)^{-2}$, calculate $g'(-2)$.

11. The profit, $\$P$, per item that a store makes by selling n items of a certain type each day is

$$P = 40\sqrt{n + 25} - 200 - 2n.$$

a. Calculate the number of items that need to be sold to maximise the profit on each item.

b. Determine the maximum profit per item.

c. Hence, determine the total profit per day that would be made by selling this number of items.

12. A particle moves in a straight line so that its displacement from a point, O, at any time, t, is $x = \sqrt{3t^2 + 4}$. Calculate:

a. the velocity as a function of time

b. the velocity when $t = 2$.

13. **WE3** Calculate the minimum distance from the line $y = 2x - 5$ to the origin. You do not need to justify your answer.

14. Calculate the minimum distance from the line $y = 2\sqrt{x}$ to the point (5, 0).

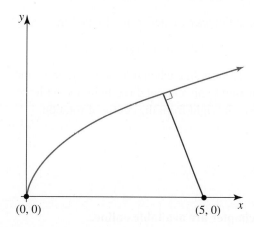

15. Consider the function $f(x) = \dfrac{1}{x-1} + \dfrac{1}{2-x}$ where $x \in R$, $x \neq 1$ or 2.

 a. Determine $f'(x)$.
 b. Determine the coordinates of the stationary point of $f(x)$.

16. For the function $y = \dfrac{1}{(2x-1)^2}$, determine:

 a. $\dfrac{dy}{dx}$

 b. the equation of the tangent to the curve at the point where $x = 1$.

17. A population of butterflies in an enclosure at a zoo is modelled by

$$N = 220 - \frac{150}{t+1},\ t \geq 0$$

where N is the number of butterflies t years after observations of the butterflies commenced. Determine how long it took for the butterfly population to reach 190 butterflies and the rate at which the population was growing at that time.

18. A rower is in a boat 4 km from the nearest point, O, on a straight beach. His destination is 8 km along the beach from O.
If he is able to row at 5 km/h and walk at 8 km/h, determine the point on the beach that he should row to in order to reach his destination in the least possible time. Give your answer correct to 1 decimal place.

19. An internet cable is being constructed to join two remote properties, A and B, to an existing cable at a connection point C. The existing cable runs along a straight road.

The direct distances from the properties to the road are 2 km and 1 km, respectively. The distance between these two points along the road is 5 km.

Determine where the connecting point should be positioned so that the cable is as short as possible. Calculate the length of the cable correct to 2 decimal places.

20. A slingshot is spinning a stone anticlockwise according to the functions

$$y = \pm \sqrt{0.09 - x^2}, \quad -0.3 \le x \le 0.3 \text{ (distances in metres)}.$$

Assuming the stone travels in a straight line when released, determine the point at which it should be released to hit a target directly in line with the x-axis at a distance of 1 metre to the right from the centre of rotation.

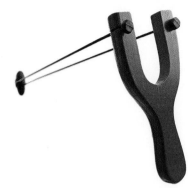

Fully worked solutions for this chapter are available online.

LESSON
13.3 The product rule

SYLLABUS LINKS

• Use the product rule, $\dfrac{d(uv)}{dx} = u\dfrac{dv}{dx} + v\dfrac{du}{dx}$, to determine the derivative of products of functions involving power and polynomial functions.

Source: Mathematical Methods Senior Syllabus 2024 © State of Queensland (QCAA) 2024; licensed under CC BY 4.0.

13.3.1 Products of functions

There are many functions that have rules that are the product of two simpler functions.

Consider the function $y = (2x + 3)(4x - 5)$.

If this is expanded, it is possible to find the derivative.

$$y = 8x^2 + 2x - 15$$

$$\frac{dy}{dx} = 16x + 2$$

Given the simplicity of this function, it makes sense to expand and then derive. However, expanding before deriving is not always the most efficient method. The product rule allows us to reach this same outcome without having to expand the function first.

13.3.2 The product rule

The product rule states the following:

> ## The product rule
>
> **If** $y = uv$
>
> **then**
>
> $$\frac{d(uv)}{dx} = u\frac{dv}{dx} + v\frac{du}{dx}$$

Consider again the function $y = (2x + 3)(4x - 5)$.

Let $u = 2x + 3$; therefore, $\dfrac{du}{dx} = 2$.

Let $v = 4x - 5$; therefore, $\dfrac{dv}{dx} = 4$.

By the product rule: $\dfrac{dy}{dx} = u\dfrac{dv}{dx} + v\dfrac{du}{dx}$

$$= (2x + 3) \times 4 + (4x - 5) \times 2$$
$$= 8x + 12 + 8x - 10$$
$$= 16x + 2$$

The product rule gives the same result as earlier.

If functions are written in function notation, the product rule can still be applied.

If $f(x) = g(x) \times h(x)$, then $f'(x) = g(x) \times h'(x) + h(x) \times g'(x)$.

Alternatively: if $y = uv$, then $y' = uv' + vu'$.

WORKED EXAMPLE 4 Applying the product rule

Determine the derivative of $y = (3x - 1)(x^2 + 4x + 3)$.

THINK	WRITE
1. Write the equation.	$y = (3x - 1)(x^2 + 4x + 3)$
2. Identify u and v, two functions of x that are multiplied together.	Let $u = 3x - 1$ and $v = x^2 + 4x + 3$.
3. Differentiate u with respect to x.	$\dfrac{du}{dx} = 3$
4. Differentiate v with respect to x.	$\dfrac{dv}{dx} = 2x + 4$
5. Apply the product rule to find $\dfrac{dy}{dx}$.	$\dfrac{d}{dx}(uv) = u\dfrac{dv}{dx} + v\dfrac{du}{dx}$
6. Expand and simplify where possible.	$= (3x - 1)(2x + 4) + (x^2 + 4x + 3) \times 3$ $= 6x^2 + 10x - 4 + 3x^2 + 12x + 9$ $\dfrac{dy}{dx} = 9x^2 + 22x + 5$

The product rule may need to be used first before an application problem can be solved.

WORKED EXAMPLE 5 Applying the product rule

The sides of a rectangle, in millimetres, are varying with time, in seconds, as shown.
a. State the function that expresses the area, A, of the rectangle at any time, t.
b. Calculate the rate of change of area against time at $t = 2$ seconds.

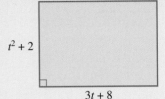

$t^2 + 2$

$3t + 8$

THINK	WRITE
a. Area of a rectangle $=$ length $\times$ width.	a. $A = l \times w$ $= (3t + 8)(t^2 + 2)$
b. 1. Apply the product rule to find the derivative.	b. Let $u = 3t + 8$ and $v = t^2 + 2$. $u' = 3$ and $v' = 2t$ $A' = uv' + vu'$ $= (3t + 8) \times 2t + (t^2 + 2) \times 3$ $= 6t^2 + 16t + 3t^2 + 6$ $= 9t^2 + 16t + 6$
2. Solve the derivative at $t = 2$.	$A'(2) = 9(2)^2 + 16(2) + 6$ $= 74$
3. Write the answer.	The area is changing at a rate of $74 \text{ mm}^2/\text{s}$ at $t = 2$.

Exercise 13.3 The product rule

learn**on**

13.3 Exercise	13.3 Exam practice **on**	These questions are even better in jacPLUS! • Receive immediate feedback • Access sample responses • Track results and progress **Find all this and MORE in jacPLUS**

Simple familiar	Complex familiar	Complex unfamiliar
1, 2, 3, 4, 5, 6, 7, 8, 9, 10	11, 12, 13, 14, 15, 16	17, 18, 19, 20

Simple familiar

1. **WE4** Determine the derivative of $y = (x + 3)\left(2x^2 - 5x\right)$.

2. Determine the derivatives of the following functions.
   a. $h(x) = (x + 2)(x - 3)$
   b. $h(x) = 3x^2(x^2 - 4x + 1)$

3. Differentiate $2\sqrt{x}(4 - x)$, using the product rule.

4. Determine $f'(x)$ given $f(x) = 4x^{-1}(3 + x^2)$, using the product rule.

5. Determine the derivatives of the following functions using the product rule, expressing your answers in simplified form.

   a. $k(x) = x^{-1}(x+2)$

   b. $p(x) = (\sqrt{x} + 3x)(x^2 - 4)$

6. **WE5** The sides of a rectangle, in millimetres, are varying with time, in seconds, as shown.

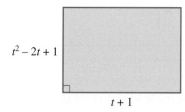

$t^2 - 2t + 1$

$t + 1$

   a. State the function that expresses the area of the rectangle at any time (t).
   b. Determine the rate of change of area at $t = 5$ s.

7. Given $f(x) = 2x^2(x - x^2)$, use the product rule to determine the coordinates where $f'(x) = 0$.

8. Determine the derivative of the function $f(x) = \left(2x^2 - 5\right)\left(x^3 - 6x + 3\right)$ at $x = 2$.

9. On a warm day in a garden, water in a bird bath evaporates in such a way that the volume, V mL, at time t hours is given by $V = t^{-1}(30t + 7)$, $t > 0$.
   Calculate the rate at which the volume is changing after 2 hours.

10. Determine the derivative of $y = \left(x^2 + 4x - 5\right)\left(x^2 + 3\right)$ by expanding and by applying the product rule. Discuss the two methods for determining the derivative in this case.

Complex familiar

11. Determine the equation of the tangent and the normal to the line $y = (x^2 - 2)(4 - 3x)$ at $x = 2$.

12. The position of a particle from its starting point when moving in a straight line is given by the function $x(t) = (t + 2)(2t - t^2)$, where x is in metres and t is in seconds.

   a. Sketch the position–time function over $0 \le t \le 2$.
   b. Sketch the velocity–time function over $0 \le t \le 2$.
   c. Describe the behaviour of the particle during the first 2 seconds.

13. Expand two terms and then apply the product rule to find the derivatives of:

   a. $y = x(2x - 5)(3x - 1)$

   b. $y = (x - 2)(2x + 1)(3x + 3)$.

14. Sketch the function $y = (x^2 + 1)(x + 2)$, examining end behaviours and identifying all intercepts and stationary points.

15. Find the equation of the tangent to the curve $y = \left(x^2 - 1\right)(x - 2)$ at the point where the curve crosses the negative x-axis.

16. The revenue generated (in $) by selling a product is calculated using the formula $R(x) = x\rho(x)$, where x is the number sold and ρ is the demand function, which indicates the price per item. The number of items sold is often closely linked to the price. If the price of an item is lower, then more items are likely to be sold. A manufacturer knows that its demand function varies linearly (i.e. $\rho(x) = ax + b$) with 1250 units being sold if the price is $500 per unit and 1500 units if the price is $400.

   a. Determine the revenue function, R.
   b. The marginal revenue of a product indicates the rate of change in total revenue with respect to the quantity demanded. It is calculated by finding the derivative of the revenue function.
      Using the product rule, determine the marginal revenue for this product at $x = 50$.

17. Sketch the function $y = (2x + 1)(x^2 - 3x)$ and its derivative on the same axes, and compare the two graphs.

18. A rectangular gutter is formed by bending a 30-cm-wide sheet of metal as shown in the diagram.

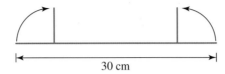

30 cm

Determine where the bends should be made to maximise the amount of water the gutter can hold and, if the gutter is 12 metres in length, determine the greatest volume of water it is able to hold. Express your answer for the volume in litres.
Justify your procedures and decisions.

19. Determine the area of the largest rectangle that can be inscribed in the region bounded by the parabola $y = 4 - x^2$ and the x-axis.

20. The opening batter in a cricket match strikes the ball so that its height y metres above the ground after it has travelled a horizontal distance x metres is given by $y = 0.0001x^2 \left(625 - x^2\right)$.
Calculate, to 2 decimal places, the greatest height the ball reaches and justify the maximum nature.

Fully worked solutions for this chapter are available online.

LESSON
13.4 The quotient rule

SYLLABUS LINKS

- Use the quotient rule, $\dfrac{d\left(\frac{u}{v}\right)}{dx} = \dfrac{v\frac{du}{dx} - u\frac{dv}{dx}}{v^2}$, to determine the derivative of quotients of functions involving power and polynomial functions.

Source: Mathematical Methods Senior Syllabus 2024 © State of Queensland (QCAA) 2024; licensed under CC BY 4.0.

13.4.1 Quotients of functions

Expressions such as $\dfrac{x^2}{4 - 3x}$, $\dfrac{2\sqrt{x}}{1 - 3x}$, $\dfrac{1}{x^2 + 5}$ and $\dfrac{2x + 3}{x^2}$ are called quotients.

Consider the last function: $y = \dfrac{2x + 3}{x^2}$

It can be expressed as:
$$y = \frac{2x}{x^2} + \frac{3}{x^2}$$
$$= 2x^{-1} + 3x^{-2}$$

The derivative is $\dfrac{dy}{dx} = -2x^{-2} - 6x^{-3}$

$$= \dfrac{-2}{x^2} - \dfrac{6}{x^3}$$

$$= \dfrac{-2x - 6}{x^3}$$

$$= \dfrac{-2(x+3)}{x^3}$$

The quotient rule allows us to reach this same outcome without separating the expression into fractions. The other expressions above could not be separated, necessitating the use of the quotient rule.

13.4.2 The quotient rule

The quotient rule states the following.

The quotient rule

If $\qquad\qquad y = \dfrac{u}{v}$

then

$$\dfrac{d\left(\frac{u}{v}\right)}{dx} = \dfrac{v\frac{du}{dx} - u\frac{dv}{dx}}{v^2}$$

Consider again the function $y = \dfrac{2x+3}{x^2}$.

Let $u = 2x + 3$; therefore, $\dfrac{du}{dx} = 2$.

Let $v = x^2$; therefore, $\dfrac{dv}{dx} = 2x$.

By the quotient rule: $\dfrac{d(\frac{u}{v})}{dx} = \dfrac{v\frac{du}{dx} - u\frac{dv}{dx}}{v^2}$

$$= \dfrac{x^2 \times 2 - (2x+3) \times 2x}{(x^2)^2}$$

$$= \dfrac{2x^2 - 4x^2 - 6x}{x^4}$$

$$= \dfrac{-2x^2 - 6x}{x^4}$$

$$= \dfrac{-2x(x+3)}{x^4}$$

$$= \dfrac{-2(x+3)}{x^3}$$

The quotient rule gives the same result as earlier.

The quotient rule is used to differentiate functions that are rational expressions — one function divided by another. When applying the quotient rule, the derivative should be expressed in simplest factorised form.

If functions are written in function notation, the quotient rule can still be applied.

If $f(x) = \dfrac{g(x)}{h(x)}$, then $f'(x) = \dfrac{h(x) \times g'(x) - g(x) \times h'(x)}{(h(x))^2}$.

Alternatively: if $y = \dfrac{u}{v}$, then $y' = \dfrac{vu' - uv'}{v^2}$.

WORKED EXAMPLE 6 Applying the quotient rule

Determine the derivative of $y = \dfrac{3-x}{x^2 + 4x}$.

THINK	WRITE
1. Write the equation.	$y = \dfrac{3-x}{x^2+4x}$
2. Identify u and v.	Let $u = 3 - x$ and $v = x^2 + 4x$.
3. Differentiate u with respect to x.	$\dfrac{du}{dx} = -1$
4. Differentiate v with respect to x.	$\dfrac{dv}{dx} = 2x + 4$
5. Apply the quotient rule to obtain $\dfrac{dy}{dx}$.	$\dfrac{d\left(\frac{u}{v}\right)}{dx} = \dfrac{v\frac{du}{dx} - u\frac{dv}{dx}}{v^2}$
	$= \dfrac{\left(x^2+4x\right) \times -1 - (3-x)(2x+4)}{\left(x^2+4x\right)^2}$
6. Simplify $\dfrac{dy}{dx}$ where possible, factorising the final answer where appropriate.	$= \dfrac{-x^2 - 4x - 12 - 2x + 2x^2}{\left(x^2+4x\right)^2}$
	$= \dfrac{x^2 - 6x - 12}{\left(x^2+4x\right)^2}$
	$= \dfrac{x^2 - 6x - 12}{x^2(x+4)^2}$

WORKED EXAMPLE 7 Applying the quotient rule with tangents

Determine the equation of the tangent to the curve $y = \dfrac{x^2}{3-2x}$ at the point where $x = 2$.

THINK	WRITE
1. Write the equation.	$y = \dfrac{x^2}{3-2x}$
2. Identify u and v.	Let $u = x^2$ and $v = 3 - 2x$.
3. Differentiate u and v with respect to x.	$\dfrac{du}{dx} = 2x \qquad \dfrac{dv}{dx} = -2$

4. Apply the quotient rule to obtain $\dfrac{dy}{dx}$.

$$\dfrac{d\left(\frac{u}{v}\right)}{dx} = \dfrac{v\frac{du}{dx} - u\frac{dv}{dx}}{v^2}$$

$$\dfrac{dy}{dx} = \dfrac{(3-2x) \times 2x - x^2 \times (-2)}{(3-2x)^2}$$

$$= \dfrac{6x - 4x^2 + 2x^2}{(3-2x)^2}$$

$$= \dfrac{6x - 2x^2}{(3-2x)^2}$$

Note that you can simplify this expression.

$$= \dfrac{2x(3-x)}{(3-2x)^2}$$

5. Determine the gradient of the tangent, m, at the point where $x = 2$.

$$m = \dfrac{2 \times 2 \times (3-2)}{(3 - 2 \times 2)^2}$$

$$= 4$$

6. Determine the y-coordinate of the point where $x = 2$.

$$y = \dfrac{x^2}{(3-2x)}$$

$$= \dfrac{2^2}{3 - 2 \times 2}$$

$$= -4$$

7. State the coordinates of the point and the gradient of the tangent at that point.

Point $(2, -4)$, gradient $= 4$

8. Determine the equation of the tangent to the curve.

$$y - (-4) = 4(x - 2)$$
$$y + 4 = 4x - 8$$
$$y = 4x - 12$$

Exercise 13.4 The quotient rule

learn on

13.4 Exercise	13.4 Exam practice on

Simple familiar	Complex familiar	Complex unfamiliar
1, 2, 3, 4, 5, 6, 7, 8, 9, 10, 11, 12	13, 14, 15, 16	17, 18, 19, 20

These questions are even better in jacPLUS!
• Receive immediate feedback
• Access sample responses
• Track results and progress

Find all this and MORE in jacPLUS ▶

Simple familiar

1. **WE6** Determine the derivative of $y = \dfrac{x+3}{x+7}$ using the quotient rule.

2. Determine the derivative of $h(x) = \dfrac{x^2 + 2x}{5 - x}$ using the quotient rule.

3. Determine the derivative of $\dfrac{x+1}{x^2-1}$.

4. Calculate $f'(x)$, given $f(x) = \dfrac{2x^2+3x-1}{5-2x}$.

5. Determine the derivative of each of the following.

   a. $\dfrac{2x}{x^2-4x}$

   b. $\dfrac{x^2+7x+6}{3x+2}$

   c. $\dfrac{4x-7}{10-x}$

   d. $\dfrac{5-x^2}{x^{\frac{3}{2}}}$

6. Explain how you could derive the following rational functions without using the quotient rule (or first principles).

   a. $y = \dfrac{7+x}{x}$

   b. $y = \dfrac{x^3-3x^2+4}{x^2}$

   c. $y = \dfrac{x^2+5x+6}{x+2}$

   d. $y = \dfrac{x^2-16}{x+4}$

7. Determine, using the quotient rule, the derivative of each of the following rational functions.

   a. $y = \dfrac{7+x}{x}$

   b. $y = \dfrac{x^3-3x^2+4x}{x^2}$

   c. $y = \dfrac{x^2+5x+6}{x+2}$

   d. $y = \dfrac{x^2-16}{x+4}$

8. Compare the derivative of $\dfrac{8-9x^2}{x^2}$ using the quotient rule and the derivative of $\dfrac{8}{x^2}-9$ using the power rule. Comment on the result.

9. If $y = \dfrac{4\sqrt{x}}{1-x}$, show that $\dfrac{dy}{dx} = \dfrac{2(x+1)}{\sqrt{x}(x-1)^2}$.

10. Determine the derivative of $\dfrac{\sqrt{x}+1}{\sqrt{x}-1}$.

11. The average profit of a product, $AP(x)$, is the profit generated per unit sold. It is calculated using the formula $AP(x) = \dfrac{P(x)}{x}$, where $P(x)$ is the profit and x is the number of units sold.

    The profit generated by a certain product is given by the function $P(x) = 2x^2 + 12x + 4$.

    a. Determine the average profit function for the product.
    b. Determine the rate of change of average profit at $x = 100$.

12. Consider the curve defined by the rule $y = \dfrac{2x-1}{3x^2+1}$.

    a. Determine the rule for the gradient.
    b. Determine the value(s) of x for which the gradient is equal to 0.875. Give your answers correct to 4 decimal places.

Complex familiar

13. **WE7** Determine the equation of the tangent and the normal to the line $y = \dfrac{1}{x^2-9}$ at $x = -1$.

14. The position of a particle from its starting point, when moving in a straight line, is given by the function $x(t) = \dfrac{t+2}{t+1}$, where x is in metres and t is in seconds.

    a. Sketch the position–time function over $0 \le t \le 2$.
    b. Sketch the velocity–time function over $0 \le t \le 2$.
    c. Describe the behaviour of the particle during the first 2 seconds.

15. Differentiate the function $\dfrac{(2x-3)(3x+4)}{x-2}$ using the product and quotient rules.

16. The yearly average cost function for a large Australian gold mine is given by the function

$$AvgC(m) = \frac{700\,m^3 - 1.8 \times 10^6\,m}{m + 10^5} + 6 \times 10^6$$

where m is the mass of gold in tonnes, t, and $AvgC$ is the cost in dollars per tonne. Determine the optimal amount of gold required to mine in a year to have the lowest average cost. Justify your procedures and decisions.

Complex unfamiliar

17. Calculate the minimum volume possible for a cylinder with a height given by $h = \dfrac{2}{r-3}, r > 3$.

    Justify your procedures and decisions.

18. A veterinarian has administered a painkiller by injection to a sick horse. The concentration of painkiller in the blood, y mg/L, can be defined by the rule

$$y = \frac{3t}{\left(4 + t^2\right)}$$

where t is the number of hours since the medication was administered. Determine the maximum concentration of painkiller in the blood and the time at which this is achieved.

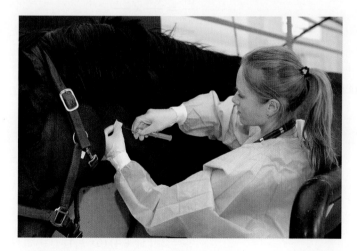

19. Sketch the function $f(x) = \dfrac{x^2 - 1}{x^2 + 1}$, showing all important features.

20. Part of a curve has been sketched but the equation has been lost. The curve is of the form $f(x) = \dfrac{x^2 + a}{x - b}, x > b$ where a and b are constants. If $f(6) = 10$ and $f(10) = 13$, determine the minimum value of the curve.

Fully worked solutions for this chapter are available online.

LESSON
13.5 Combinations of the chain rule, product rule and quotient rule

SYLLABUS LINKS

- Solve problems that involve combinations of the chain rule, product rule and quotient rule to differentiate functions involving power and polynomial functions, expressing derivatives in simplest and factorised form.

Source: Mathematical Methods Senior Syllabus 2024 © State of Queensland (QCAA) 2024; licensed under CC BY 4.0.

We can combine the product, quotient and chain rules to derive more complicated functions.

WORKED EXAMPLE 8 Combining the rules of differentiation

Determine the derivative of $y = x^2(3x + 4)^2$.

THINK	WRITE
1. Recognise that the product rule can be applied as $y = uv$.	$u = x^2, v = (3x + 4)^2$ $u' = 2x, v' = \ldots$
2. Identify that to derive v the chain rule needs to be applied.	Let $w = 3x + 4$, $\therefore v = w^2$ $\dfrac{dv}{dx} = \dfrac{dv}{dw} \times \dfrac{dw}{dx}$ $\quad = 2w \times 3$ $\quad = 6(3x + 4)$
3. Complete the product rule.	$u = x^2, \ v = (3x + 4)^2$ $u' = 2x, \ v' = 6(3x + 4)$ $y' = uv' + vu'$ $\quad = x^2 \times 6(3x + 4) + 2x(3x + 4)^2 \times 2x$ $\quad = 6x^2(3x + 4) + 2x(3x + 4)^2$ $\quad = 2x(3x + 4)[3x + (3x + 4)]$ $\quad = 2x(3x + 4)(6x + 4)$ $\quad = 2x(3x + 4) \cdot 2(3x + 2)$ $\quad = 4x(3x + 4)(3x + 2)$

Exercise 13.5 Combinations of the chain rule, product rule and quotient rule

learn

13.5 Exercise	**13.5 Exam practice** on

These questions are even better in jacPLUS!
- Receive immediate feedback
- Access sample responses
- Track results and progress

Find all this and MORE in jacPLUS ▶

Simple familiar	Complex familiar	Complex unfamiliar
1, 2, 3, 4, 5, 6, 7, 8, 9, 10, 11, 12	13, 14, 15, 16	17, 18, 19, 20

Simple familiar

1. **WE8** Determine the derivative of each of the following.

   a. $x^2(x+1)^3$
   b. $x^3(x+1)^2$
   c. $\sqrt{x}(x+1)^5$
   d. $x^{\frac{3}{2}}(x-2)^3$

   e. $x(x-1)^{-2}$
   f. $x\sqrt{x+1}$

2. Differentiate the following.

   a. $x^{-2}(2x+1)^3$
   b. $2\sqrt{x}(4-x)$
   c. $(x-1)^4(3-x)^{-2}$
   d. $(3x-2)^2(4x+5)^3$

3. Differentiate the following.

   a. $h(x) = \dfrac{(5-x)^2}{\sqrt{5-x}}$
   b. $y = \dfrac{3x-1}{2x^2-3}$
   c. $h(x) = \dfrac{x-4x^2}{2\sqrt{x}}$
   d. $y = \dfrac{3\sqrt{x}}{x+2}$

4. Determine the derivative of each of the following.

   a. $x\left(x^2+1\right)^3$
   b. $\dfrac{\left(x^2+1\right)^3}{x}$
   c. $\dfrac{1}{(x^2-3)^5}$
   d. $\dfrac{\sqrt{x}(x+1)^3}{x-1}$

5. Determine the gradient at the stated point for each of the following functions.

   a. $y = \dfrac{2x}{x^2+1}, x=1$
   b. $y = \dfrac{x+1}{\sqrt{3x+1}}, x=5$

6. Consider the function $f(x) = \dfrac{2x}{(x^2-3x)}$.

   Differentiate the function using:

   a. the quotient rule
   b. the product and chain rule.

7. Develop a general rule for functions of the form $y = (ax+b)^n$.

8. Determine $f'(1)$ if $f(x) = x\left(x^2+x\right)^4$.

9. A colony of viruses can be modelled by the rule

   $$N(t) = \dfrac{2t}{(t+0.5)^2} + 0.5$$

   where N hundred thousand is the number of viruses on the nutrient plate t hours after they started multiplying.

   a. State how many viruses were present initially.
   b. Determine $N'(t)$.
   c. Determine the maximum number of viruses and when this maximum will occur.
   d. Calculate the rate at which the virus numbers were changing after 10 hours.

CHAPTER 13 Differentiation rules **621**

10. Calculate the gradient of the tangent to the curve with equation $y = \dfrac{2x}{(3x+1)^{\frac{3}{2}}}$ at the point where $x = 1$.

11. For the curve with the rule $y = \dfrac{x-5}{x^2+5x-14}$:

   a. state when the function is undefined
   b. determine the coordinates when the gradient is zero
   c. determine the equation of the tangent to the curve at the point where $x = 1$.

12. Given that $f(x) = \dfrac{\sqrt{2x-1}}{\sqrt{2x+1}}$, determine the value of m such that $f'(m) = \dfrac{2}{5\sqrt{15}}$.

Complex familiar

13. For each of the following functions, determine the equation of:

   i. the tangent at the given value of x
   ii. the normal at the given value of x.

   a. $y = x^2 + 1$, $x = 1$
   b. $y = (x-1)(x^2+2)$, $x = -1$
   c. $y = \sqrt{2x+3}$, $x = 3$
   d. $y = x(x+2)(x-1)$, $x = -1$

14. Sketch the function $3x(x-6)^3$ and its derivative. Comment on any points of significance.

15. Determine any stationary points of the following curves and justify their nature.

   a. $y = x(x+2)^2$
   b. $y = \dfrac{x^2}{x+1}$

16. Determine the derivative of the function $f(x) = \dfrac{10x}{x^2+1}$ and calculate when the gradient is negative.

Complex unfamiliar

17. Calculate the area of the largest rectangle with its base on the x-axis that can be inscribed in the semicircle $y = \sqrt{4-x^2}$.

18. A bushwalker can walk at 5 km/h through clear land and 3 km/h through bushland. If she has to get from point A to point B following the route indicated in the diagram, determine the value of x so that the route is covered in a minimum time.

$$\left(Note: \text{time} = \dfrac{\text{distance}}{\text{speed}} \right)$$

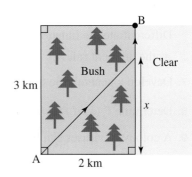

19. A cylinder has a surface area of 220π cm^2. Calculate the height and radius of each end of the cylinder so that the volume of the cylinder is maximised, and determine the maximum volume for the cylinder. Give answers correct to 2 decimal places.

20. Determine the dimensions of the cylinder of largest volume that can be cut from a sphere with a diameter of 10 cm.

Fully worked solutions for this chapter are available online.

LESSON
13.6 Review

doc-
41808

13.6.1 Summary

13.6 Exercise

learn on

Simple familiar

1. Determine the derivatives of the following functions.

   a. $h(x) = x^3(x^2 + 2x)$

   b. $h(x) = \dfrac{2}{x}(x^3 + 7)$

2. Determine the gradient at $x = 4$ for the function $y = 4x^2(3 - 5x)$.

3. Determine the derivative of $y = \left(2x^2 - 3 + \dfrac{1}{x}\right)\left(1 + \dfrac{3}{x}\right)$.

4. Determine the derivative of $\dfrac{x+1}{x^2 - 1}$.

5. Consider the curve defined by the rule $y = \dfrac{2x - 1}{3x^2 + 1}$.

   a. Determine the rule for the gradient.
   b. Determine the value(s) of x for which the gradient is equal to 0.875. Give your
      answers correct to 4 decimal places. Use technology of your choice to answer
      the question.

6. The amount of chlorine in a jug of water t hours after it was filled from a tap
   is $C = \dfrac{20}{t + 1}$, where C is in millilitres. Evaluate the rate of decrease of chlorine
   9 hours after being poured.

7. Determine the derivatives of the following functions.

   a. $y = \sqrt{x^2 - 7x + 1}$

   b. $y = (3x^2 + 2x - 1)^3$

8. Determine the derivatives of the following functions.

   a. $g(x) = 3(x^2 + 1)^{-1}$

   b. $g(x) = \sqrt{(x+1)^2 + 2}$

   c. $f(x) = \sqrt{x^2 - 4x + 5}$

   d. $f(x) = \left(x^3 - \dfrac{2}{x^2}\right)^{-2}$

9. The function h has a rule $h(x) = \sqrt{x^2 - 16}$ and the function g has the rule $g(x) = x - 3$. Calculate the gradient of the function $h(g(x))$ at the point when $x = -2$.

10. Determine the derivatives of the following.

   a. $(4 - x^2)^3$

   b. $x^2(x + 3)^4$

   c. $\dfrac{x^3}{x^2 + 1}$

11. Given $f(x) = 2x^2(1 - x)^3$, use calculus to determine the coordinates where $f'(x) = 0$.

12. Determine the derivative of the following function, and hence calculate the gradient at the given x-value.

$$f(x) = \sqrt[3]{(3x^2 - 2)^4}; \; x = 1$$

Complex familiar

13. A pen for holding farm animals has dimensions $l \times w$ metres. This pen is to be partitioned so that there are four spaces of equal area as shown.

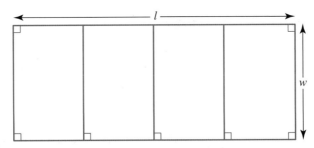

The farmer has 550 metres of fencing material to construct this pen.

   a. Calculate the required length and width in order to maximise the area of the pen.

   b. Hence, calculate the maximum area.

14. Water is being poured into a vase. The volume, V mL, of water in the vase after t seconds is given by:

$$V = \frac{2}{3}t^2(15 - t), \; 0 \le t \le 10.$$

   a. Calculate the volume after 10 seconds.

   b. Determine the rate at which the water is flowing into the vase at t seconds.

   c. Determine the rate of flow after 3 seconds.

   d. Determine when the rate of flow is the greatest and what the rate of flow is at this time.

15. The volume of water, V litres, in a bath t minutes after the plug is removed is given by

$$V = 0.4(8 - t)^3, \; 0 \le t \le 8$$

Use technology of your choice to answer the following.

   a. Determine the rate at which the water is leaving the bath after 3 minutes.

   b. Determine the average rate of change of the volume for the first 3 minutes.

   c. Determine when the rate of water leaving the bath is the greatest.

16. Determine the minimum distance from the line $y = 2x + 3$ to the point $(1, 0)$.

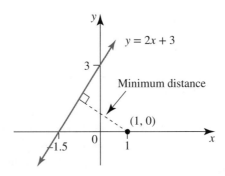

Complex unfamiliar

17. A rectangular box with an open top is to be constructed from a rectangular sheet of cardboard measuring 16 cm by 10 cm. The box will be made by cutting equal squares of side length x cm out of the four corners and folding the flaps up.
Determine the dimensions of the box with the greatest volume, and give this maximum volume.

18. A veterinarian has administered a painkiller by injection to a sick horse.
The concentration of painkiller in the blood, y mg/L, can be defined by the rule

$$y = \frac{3t}{(4 + t^2)}$$

where t is the number of hours since the medication was administered.
Determine when the rate of change of painkiller in the blood is equal to -0.06 mg/L/hour. Give your answer correct to 2 decimal places.

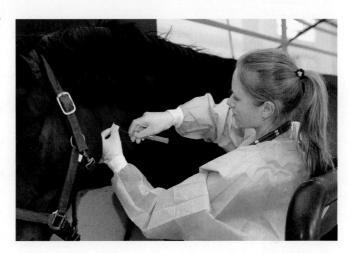

19. Consider the function $f(x) = (a - x)^2(x - 2)$ where $a > 2$.
Determine the value of a if the graph of $y = f(x)$ has a turning point at $(3, 4)$.

20. A cone is 10 cm high and has a base radius of 8 cm. Calculate the radius and height of a cylinder that is inscribed in the cone such that the volume of the cylinder is a maximum. Determine the maximum volume of the cylinder, correct to the nearest cubic centimetre.

Fully worked solutions for this chapter are available online.

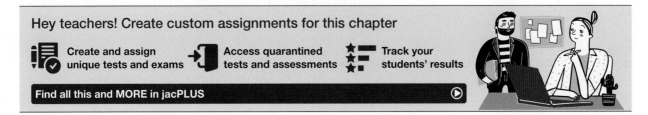

Answers

Chapter 13 Differentiation rules

13.2 The chain rule

13.2 Exercise

1. a. $6(3x + 2)$ b. $-3(7 - x)^2$
   c. $-\dfrac{2}{(2x - 5)^2}$ d. $\dfrac{8}{(4 - 2x)^5}$
   e. $\dfrac{5}{2\sqrt{5x + 2}}$ f. $\dfrac{-9}{2(3x - 2)^{\frac{3}{2}}}$

2. a. $32(8x + 3)^3$ b. $6(2x - 5)^2$
   c. $-15(4 - 3x)^4$ d. $\dfrac{3x}{\sqrt{3x^2 - 4}}$
   e. $\dfrac{2}{3}(x - 2)(x^2 - 4x)^{-\frac{2}{3}}$ f. $-2(6x^2 + 1)(2x^3 + x)^{-3}$

3. $\dfrac{-2}{\sqrt{(4x + 7)^3}}$

4. a. $8x^7(2x + 5)(x + 5)^7$
   b. $2x(3x^2 - 2)(x^2 - 2)$
   c. $\dfrac{3x^2 + 4x}{5(x^3 + 2x^2 - 7)^{\frac{4}{3}}}$
   d. $3x(4x^2 - 3)\sqrt{2x^4 - 3x^2 + 1}$

5. a. C   b. D   c. B   d. A

6. a. i. $16\,\text{cm}$
      ii. $132\,\text{cm}$
   b. $R = 0.02t + 5.6$
   c. Maximum $= 6$, minimum $= 5.6$

7. a. $\dfrac{-3}{(6x - 5)^{\frac{3}{2}}}$ b. $3x\sqrt{x^2 + 2}$

8. a. 2 b. $\dfrac{x - 1}{\sqrt{x^2 - 2x + 1}}$
   c. 1 d. 1

9. $f'(x) = 30(5x + 1)^2$

10. a. $37\,500$ b. $\dfrac{7}{500}$ or 0.014

11. a. 75 b. $\$50$ c. $\$3750$

12. a. $v = \dfrac{3t}{(3t^2 + 4)^{\frac{1}{2}}} = \dfrac{3t}{\sqrt{3t^2 + 4}}$
    b. $v = 1.5$

13. The point on the curve is $(2, -1)$. The minimum distance is $\sqrt{5}$ units.

14. 4

15. a. $f'(x) = \dfrac{-1}{(x - 1)^2} + \dfrac{1}{(2 - x)^2}$
    b. $\left(\dfrac{3}{2}, 4\right)$

16. a. $\dfrac{dy}{dx} = -\dfrac{4}{(2x - 1)^3}$
    b. $y = -4x + 5$

17. 4 years, 6 butterflies per year

18. 3.2 km to the right

19. The connection point, C, should be positioned 3.33 km from point P, giving the shortest length of the cable of 5.83 km.

20. At $x = 0.09$

13.3 The product rule

13.3 Exercise

1. $\dfrac{dy}{dx} = 6x^2 + 2x - 15$

2. a. $h'(x) = 2x - 1$
   b. $h'(x) = 6x(2x^2 - 6x + 1)$

3. $\dfrac{4}{\sqrt{x}} - 3\sqrt{x}$

4. $4 - \dfrac{12}{x^2}$

5. a. $\dfrac{-2}{x^2}$
   b. $9x^2 + \dfrac{5}{2}\sqrt{x^3} - 12 - \dfrac{2}{\sqrt{x}}$

6. a. $A(t) = (t^2 - 2t + 1)(t + 1)$
   b. $64\,\text{mm/s}$

7. $(0, 0)$ and $\left(\dfrac{3}{4}, \dfrac{27}{128}\right)$

8. 10

9. $-\dfrac{7}{4}\,\text{mL/hr}$

10. $\dfrac{dy}{dx} = 4x^3 + 12x^2 - 4x + 12$

    Expanding is usually easier if the two terms are simple polynomials. Both methods use expanding and simplifying to find the derivative.

11. $y_T = -14x + 24$, $y_N = \dfrac{1}{14}x - \dfrac{29}{7}$

12. a.
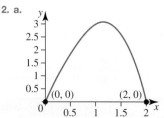

b.

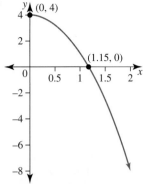

c. The particle travels away from the starting point for about 1.15 seconds (clearly seen on the velocity graph) before returning to the initial position. It starts with an initial velocity of 4 m/s and slows as it moves away from the starting point.

13. a. $18x^2 - 34x + 5$ **b.** $3(6x^2 - 2x - 5)$

14.

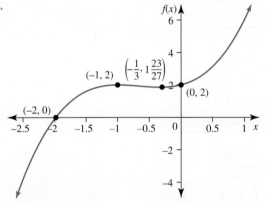

15. $y = 6x + 6$

16. a. $R(x) = x\left(-\dfrac{2}{5}x + 1000\right)$

   b. \$960/unit

17. $\dfrac{dy}{dx} = 6x^2 - 10x - 3$

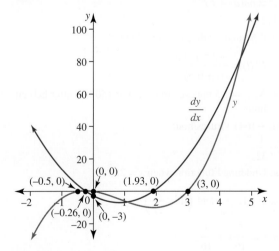

The function is a cubic and the derivative is a quadratic. The derivative crosses the x-axis when the function is at its maximum and minimum.

18. 7.5 cm from both ends; 135 litres

19. $\dfrac{32\sqrt{3}}{9}$ units squared

20. 9.77 metres

13.4 The quotient rule

13.4 Exercise

1. $\dfrac{d}{dx}\left(\dfrac{u}{v}\right) = \dfrac{4}{(x+7)^2}$

2. $h'(x) = \dfrac{-x^2 + 10x + 10}{(5-x)^2}$

3. $-\dfrac{x^2 + 2x + 1}{(x^2 - 1)^2}$

4. $\dfrac{-4x^2 + 20x + 13}{(5 - 2x)^2}$

5. a. $\dfrac{-2}{(x-4)^2}$ **b.** $\dfrac{3x^2 + 4x - 4}{(3x+2)^2}$

   c. $\dfrac{33}{(10-x)^2}$ **d.** $-\dfrac{(x^2 + 15)}{2x^{\frac{5}{2}}}$

6. a. Change form to $\dfrac{7}{x} + 1 = 7x^{-1} + 1$, then apply the power rule.

   b. Change form to $x - 3 + \dfrac{4}{x^2} = x - 3 + 4x^{-2}$, then apply the power rule.

   c. Factorise $\dfrac{(x+2)(x+3)}{x+2} = x + 3,\ x \neq -2$, then apply the power rule.

   d. Factorise using difference of two squares $\dfrac{(x-4)(x+4)}{x+4} = x - 4,\ x \neq -4$, then apply the power rule.

7. a. $-\dfrac{7}{x^2}$ **b.** $\dfrac{x^3 - 8}{x^3}$ **c.** 1   **d.** 1

8. $y' = \dfrac{-16}{x^3}$; both methods produce the same result as they are the same function represented differently.

9. $\dfrac{dy}{dx} = \dfrac{2(x+1)}{\sqrt{x}(x-1)^2}$

10. $y' = -\dfrac{1}{\sqrt{x}(\sqrt{x}-1)^2}$

11. a. $AP(x) = \dfrac{2x^2 + 12x + 4}{x}$

   b. $\dfrac{4999}{2500}$

12. a. $\dfrac{-6x^2 + 6x + 2}{(3x^2 + 1)^2}$

   b. $x = -0.1466, 0.5746$

13. $y_T = \dfrac{1}{32}x + \dfrac{5}{32}$

$y_N = -32x - \dfrac{255}{8}$

14. a.

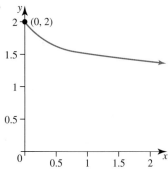

b. $v(t) = \dfrac{-1}{(t+1)^2}$

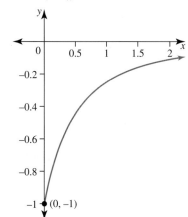

c. The particle starts 2 m away from the reference point and then moves towards the reference point.

15. $\dfrac{2(3x^2 - 12x + 7)}{(x-2)^2}$

16. 29.27 t $(2\ \text{d.p})$

17. At $r = 6$, $V = 24\pi$ unit^3

18. $y_{\max} = 0.75\,\text{mg/L}$ after 2 hours

19.

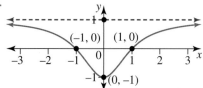

20. $4 + 4\sqrt{2}$

13.5 Combinations of the chain rule, product rule and quotient rule

13.5 Exercise

1. a. $x(5x + 2)(x + 1)^2$

  b. $(x + 1)(5x + 3)x^2$

  c. $\dfrac{1}{2\sqrt{x}}(x + 1)^4(11x + 1)$

d. $\dfrac{3}{2}\sqrt{x}(x - 2)^2(3x - 2)$

e. $-(x + 1)(x - 1)^{-3}$

f. $\dfrac{1}{2\sqrt{x + 1}}(3x + 2)$

2. a. $\dfrac{2(x - 1)(2x + 1)^2}{x^3}$

  b. $\dfrac{4 - 3x}{\sqrt{x}}$

  c. $\dfrac{2(x - 5)(x - 1)^3}{(x - 3)^3}$

  d. $6(3x - 2)(4x + 5)^2(10x + 1)$

3. a. $-\dfrac{3\sqrt{5 - x}}{2}$

  b. $\dfrac{-6x^2 + 4x - 9}{(2x^2 - 3)^2}$

  c. $\dfrac{1}{4\sqrt{x}} - 3\sqrt{x}$

  d. $\dfrac{6 - 3x}{2\sqrt{x}(x + 2)^2}$

4. a. $(x^2 + 1)^2(7x^2 + 1)$

  b. $\dfrac{(x^2 + 1)^2(5x^2 - 1)}{x^2}$

  c. $\dfrac{-10x}{(x^2 - 3)^6}$

  d. $\dfrac{(x + 1)^2(5x^2 - 8x - 1)}{2\sqrt{x}(x - 1)^2}$

5. a. 0   **b.** $\dfrac{7}{64}$

6. All equal $\dfrac{-2}{(x - 3)^2}$ if fully simplified.

7. $y' = na(ax + b)^{n-1}$

8. 112

9. a. 0.5 hundred thousand or 50 000

  b. $N'(t) = \dfrac{-2t^2 + 0.5}{(t + 0.5)^4}$

  c. $N_{\max} = 1.5$ hundred thousand or 150 000 after half an hour.

  d. -1641 viruses/hour

10. $-\dfrac{1}{32}$

11. a. Undefined function when $x = 2, -7$

  b. $\left(-1, \dfrac{1}{3}\right)$ and $\left(11, \dfrac{1}{27}\right)$

  c. $y = \dfrac{5}{16}x + \dfrac{3}{16}$

12. $m = 2$

13. a. i. $y = 2x$

    ii. $x + 2y = 5$

    b. i. $y = 7x + 1$

    ii. $x + 7y + 43 = 0$

    c. i. $3y = x + 6$

    ii. $y + 3x = 12$

    d. i. $x + y = 1$

    ii. $y = x + 3$

14.

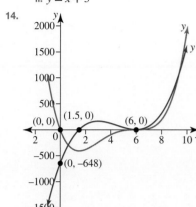

When the function (blue) has a stationary point, the derivative (orange) intercepts the x-axis. When the function's gradient is downward, the derivative is below the x-axis, and when the function's gradient is upward, the derivative is above the x-axis.

15. a. $(-2, 0)$ maximum turning point, $\left(-\dfrac{2}{3}; -\dfrac{32}{27}\right)$ minimum turning point

    b. $(-2, -4)$ maximum turning point, $(0, 0)$ minimum turning point

16. $\dfrac{dy}{dx} = \dfrac{10\left(1 - x^2\right)}{\left(x^2 + 1\right)^2}$; negative gradient when $x \in (-\infty, -1) \cup (1, \infty)$

17. 4 units^2

18. 1.5 km

19. $r = 6.06 \text{ cm}, h = 12.11 \text{ cm}$ and $V_{\max} = 1395.04 \text{ cm}^3$

20. The cylinder with maximum volume would have a radius of $\dfrac{5\sqrt{6}}{3}$ cm and a height of $\dfrac{10\sqrt{3}}{3}$ cm.

13.6 Review

13.6 Exercise

1. a. $x^3(5x + 8)$ b. $4x - \dfrac{14}{x^2}$

2. -864

3. $4x + 6 + \dfrac{8}{x^2} - \dfrac{6}{x^3}$

4. $-\dfrac{1}{(x - 1)^2}$

5. a. $\dfrac{-6x^2 + 6x + 2}{(3x^2 + 1)^2}$

   b. $x = -0.1466, 0.5746$

6. 0.2 mL/h

7. a. $\dfrac{2x - 7}{2\sqrt{x^2 - 7x + 1}}$

   b. $6(3x + 1)(3x^2 + 2x - 1)^2$

8. a. $-\dfrac{6x}{(x^2 + 1)^2}$ b. $\dfrac{x + 1}{\sqrt{x^2 + 2x + 3}}$

   c. $\dfrac{x - 2}{\sqrt{x^2 - 4x + 5}}$ d. $-\dfrac{6x^5 + 8}{x^3\left(x^3 - \dfrac{2}{x^2}\right)^3}$

9. $-\dfrac{5}{3}$

10. a. $-6x(4 - x^2)^2$

    b. $6x(x + 1)(x + 3)^3$

    c. $\dfrac{x^4 + 3x^2}{(x^2 + 1)^2}$

11. $(0, 0), (1, 0), \left(\dfrac{2}{5}, \dfrac{216}{3125}\right)$

12. $f'(x) = 8x \sqrt[3]{3x^2 - 2}, f'(1) = 8$

13. a. $l = 137.5 \text{ m}, w = 55 \text{ m}$

    b. $A_{\max} = 7562.5 \text{ m}^2$

14. a. $V = 333\dfrac{1}{3} \text{ mL}$

    b. $\dfrac{dV}{dt} = 20t - 2t^2$

    c. $\dfrac{dV}{dt} = 42 \text{ mL/s}$

    d. $t = 5s, \dfrac{dV}{dt} = 50 \text{ mL/s}$

15. a. $\dfrac{dV}{dt} = -30 \text{ L/min}$

    b. -51.6 L/min

    c. $t = 0$

16. $\sqrt{5} \text{ units}$

17. Height is 2 cm, length is 12 cm and width is 6 cm. Volume is 144 cm^3.

18. $t = 2.45 \text{ h}$ and $t = 6 \text{ h}$

19. $a = 5$

20. $h = \dfrac{10}{3}$ cm, $r = \dfrac{16}{3}$ cm and $V_{\max} = 298 \text{ cm}^3$.

GLOSSARY

absolute maximum *see* global maximum

absolute minimum *see* global minimum

amplitude the distance a sine or cosine graph oscillates up or down from its equilibrium, or mean, position

anticlockwise moving around a circle in the opposite direction to the hands of a clock

arc a section of the circumference of a circle. A *minor arc* is shorter than the circumference of the semicircle; a *major arc* is longer than the circumference of the semicircle

asymptote a line that a graph approaches but never reaches. A horizontal asymptote shows the long-term behaviour as, for example, $x \to \infty$; a vertical asymptote may occur where a function is undefined, such as at $x = 0$ for the hyperbola $y = \dfrac{1}{x}$.

average rate of change The average rate of change of a function f over an interval $[a, b]$ is measured by $\dfrac{f(b) - f(a)}{b - a}$; this is the gradient of the chord joining the end points of the interval on the curve $y = f(x)$.

average velocity average rate of change of position with respect to time or $\dfrac{\text{change in position}}{\text{change in time}}$

axis of symmetry a line about which a graph is symmetrical

binomial coefficients the coefficients of the terms in the binomial expansion of $(x + y)^n$, with $\dbinom{n}{r}$ or $^{n}C_r$ the coefficient of the $(r + 1)$th term. For r and n non-negative whole numbers, $\dbinom{n}{r} = \dfrac{n!}{r!\,(n - r)!}$, $0 \leq r \leq n$.

boundary points in trigonometry, the four points $(1, 0), (0, 1), (-1, 0), (0, -1)$ on the circumference of the unit circle that lie on the boundaries between the four quadrants

circle a relation with the rule $(x - a)^2 + (y - b)^2 = r^2$, where the centre is (a, b) and the radius is r

circular functions sine, cosine, tangent. On a unit circle, $\cos(\theta)$ is the x-coordinate of the trigonometric point $[\theta]$; $\sin(\theta)$ is the y-coordinate of the trigonometric point $[\theta]$; $\tan(\theta)$ is the length of the intercept the line through the origin and the trigonometric point $[\theta]$ cuts off on the tangent drawn to the unit circle at $(1, 0)$.

clockwise moving around a circle in the same direction as the hands of a clock

codomain the set of all y-values available for pairing with x-values to form a mapping according to a function rule $y = f(x)$

combinatoric coefficients *see* binomial coefficients

complement The complement of set A is the set A' containing all the elements of the sample space that are not in set A.

completing the square technique used to express a quadratic as a difference of two squares. To apply this technique to a monic quadratic expression in x, add the square of half the coefficient of the term in x and then subtract this. This enables $x^2 \pm mx$ to be expressed as $\left(x \pm \dfrac{m}{2}\right)^2 - \left(\dfrac{m}{2}\right)^2$.

composite function a function that can be expressed as a composition of two simpler functions

concave down describes parabolas that open downwards

concave up describes parabolas that open upwards

conditional probability the probability of an event given that another event has occurred

constant of proportionality the constant value of the ratio between two proportional quantities, for example the number k, the gradient of the straight-line graph $y = kx$

constant term (of a polynomial or other expression) the term that does not contain the variable

continuous all values within a specified interval are permitted

degree of a polynomial the highest power of a variable; for example, polynomials of degree 1 are linear, degree 2 are quadratic and degree 3 are cubic

denominator the bottom expression of a fraction

dependent variable the second value, or y-value, in a set of ordered pairs

derivative function the rate of change of a function with respect to a variable; the derivative at a certain point gives the slope of the tangent to the function at that point; expressed as $f'(x)$ or $\dfrac{dy}{dx}$

differentiation by first principles requires the derivative to be obtained from its limit definition, either $f'(x) = \lim\limits_{h \to 0} \dfrac{f(x+h) - f(x)}{h}$ or $\dfrac{dy}{dx} = \lim\limits_{\delta x \to 0} \dfrac{\delta y}{\delta x}$

dilation factor measures the amount of stretching or compression of a graph from an axis

discontinuous describes a point in a graph at which there is a break

discrete only fixed values are permitted

discriminant For the quadratic expression $ax^2 + bx + c$, the discriminant is $\Delta = b^2 - 4ac$.

displacement measures both distance and direction from a fixed origin; represents the position of an object relative to a fixed origin

distance how far an object is from a fixed origin or another object; the magnitude of displacement

domain the set of all x-values of the ordered pairs (x, y) that make up a relation

element a member of a set; $a \in A$ means a is an element of, or belongs to, the set A. If a is not an element of the set A, this is written as $a \notin A$.

equating coefficients If $ax^2 + bx + c \equiv 2x^2 + 5x + 7$, then $a = 2$, $b = 5$ and $c = 7$ for all values of x.

equilibrium or **mean position** the position about which a trigonometric graph oscillates

equiprobable having an equal likelihood of occurring

event a set of outcomes that is a subset of the sample space in an experiment

exponential form also known as index form; a way of expressing a standard number n using a base a and an exponent or index (power) x: $n = a^x$

exponential functions functions of the form $f: R \to R, f(x) = r^x, r \in R^+ \setminus \{1\}$

function a relation in which the set of ordered pairs (x, y) have each x-coordinate paired to a unique y-coordinate

global or **absolute maximum** a point on a curve where the y-coordinate is greater than that of any other point on the curve for the given domain

global or **absolute minimum** a point on a curve where the y-coordinate is smaller than that of any other point on the curve for the given domain

horizontal translation a transformation that moves a figure a given distance in a horizontal direction where $(x, y) \to (x + a, y)$

hyperbola a smooth curve with two branches, formed by the intersection of a plane with both halves of a double cone

identically equal describes polynomials with the same coefficients of corresponding like terms; the $\equiv$ symbol is used to emphasise the fact that the equality must hold for all values of the variable

image a figure after a transformation; for a function $x \to f(x)$, $f(x)$ is the image of x under the mapping f

implied domain the set of x-values for which a rule has meaning

independent events events that have no effect on each other

independent variable the first value, or x-value, in a set of ordered pairs

index *see* exponent

instantaneous rate of change the derivative of a function evaluated at a particular instant

interval a set of numbers described by two numbers where any number that lies between those two numbers is also included in the set

interval notation a convenient alternative way of describing sets of numbers: $[a, b] = \{x: a \leq x \leq b\}$; $(a, b) = \{x: a < x < b\}$; $[a, b) = \{x: a \leq x < b\}$; $(a, b] = \{x: a < x \leq b\}$

irrational describes a number that cannot be expressed as a ratio of two integers

irrational number a number that cannot be expressed as a ratio of two integers

Karnaugh map *see* probability table

kinematics the study of motion

leading term (of a polynomial or other expression) the term containing the highest power of a variable

local maximum a stationary point on a curve where the gradient is zero and the y-coordinate is greater than those of neighbouring points

local minimum a stationary point on a curve where the gradient is zero and the y-coordinate is lower than those of neighbouring points

logarithm an index or power. If $n = a^x$, then $x = \log_a(n)$ is an equivalent statement.

mapping a function that pairs each element of a given set (the domain) with one or more elements of a second set (the range)

mean position *see* equilibrium position

modelling using mathematical concepts to describe the behaviour of a system, usually in the form of equations

monic For quadratics, 'monic' indicates the coefficient of x^2 equals 1.

mutually exclusive describes events that cannot occur simultaneously

negative definite describes a quadratic expression $ax^2 + bx + c$ if $ax^2 + bx + c < 0$ for all $x \in R$

negative stationary point of horizontal inflection a point on a curve where the tangent to the curve is horizontal and the curve changes concavity so that the gradient is negative either side of the stationary point. The graph of $y = -x^3$ has a negative stationary point of inflection at (0, 0).

normal a line that is perpendicular to the tangent at a point of tangency

Null Factor Law mathematical law stating that if $ab = 0$, then either $a = 0$ or $b = 0$ or both a and b equal zero. This allows equations for which the product of two or more terms equals zero to be solved.

number of elements how many elements are contained in a set; the number of elements in set A is denoted by $n(A)$

out of phase *see* phase difference

outcome the result of an experiment

parabola (with a vertical axis of symmetry) the graph of a quadratic function; its equation may be expressed in turning point form $y = a(x - h)^2 + k$ with turning point (h, k), factored form $y = a(x - x_1)(x - x_2)$ with x-intercepts at $x = x_1, x = x_2$, or general form $y = ax^2 + bx + c$

Pascal's triangle a triangle formed by rows of the binomial coefficients of the terms in the expansion of $(a + b)^n$, with n as the row number

period on a trigonometric graph, the length of the domain interval required to complete one full cycle. For the sine and cosine functions, the period is 2π since $\sin(x + 2\pi) = \sin(x)$ and $\cos(x + 2\pi) = \cos(x)$. The tangent function has a period of π since $\tan(x + \pi) = \tan(x)$.

periodicity a function that repeats itself in a regular pattern

phase difference or **phase shift** the horizontal translation of a trigonometric graph. For example, $\sin(x - a)$ has a phase shift or a phase change of a or a horizontal translation of a from $\sin(x)$; and $\sin(x)$ and $\cos(x)$ are out of phase by $\dfrac{\pi}{2}$ or have a phase difference of $\dfrac{\pi}{2}$.

polynomial an algebraic expression in which the power of the variable is a positive whole number

position measures both distance and direction from a fixed origin

position–time graph a graph of the position of a particle plotted against time

positive definite describes a quadratic expression $ax^2 + bx + c$ if $ax^2 + bx + c > 0$ for all $x \in R$

positive stationary point of horizontal inflection a point on a curve where the tangent to the curve is horizontal and the curve changes concavity so that the gradient is positive either side of the stationary point. The graph of $y = x^3$ has a positive stationary point of inflection at (0, 0).

power function a function with a rule of the form $f(x) = x^n$, $n \in Q$

probability the long-term proportion, or relative frequency, of the occurrence of an event

probability table a format for presenting all known probabilities of events in rows and columns

Pythagorean identity states that, for any θ, $\sin^2(\theta) + \cos^2(\theta) = 1$

quadrants the four sections into which the Cartesian plane is divided by the coordinate axes

quadratic formula the formula $x = \dfrac{-b \pm \sqrt{b^2 - 4ac}}{2a}$

radian measure the measure of the angle subtended at the centre of a circle by an arc of length equal to the radius of the circle

random number generators methods of generating random numbers, for example rolling dice or using algorithms

range the set of all y-values of the ordered pairs (x, y) that make up a relation

rate of change (instantaneous) the derivative of a function evaluated at a given value or instant

rational describes a number that can be expressed as a ratio of two integers

rational function a function expressed as the quotient of two polynomials $\dfrac{P(x)}{Q(x)}, Q(x) \neq 0$

rational number any real number that can be expressed exactly as a ratio of two integers

real number any number from the set of all rational and irrational numbers

rectangular hyperbola the function $f: R \backslash \{0\} \rightarrow R, f(x) = x^{-1}$

rectilinear motion motion in a straight line

reflections transformations of a figure defined by the line of reflection where the image point is a mirror image of the pre-image point

relation any set of ordered pairs

restricted domain a subset of a function's maximal domain, often due to practical limitations on the independent variable in modelling situations

roots the solutions to an equation

sample space the set of all possible outcomes in an experiment, denoted as ξ

sampling with replacement a type of sampling whereby the object is replaced, resulting in an independent trial

sampling without replacement a type of sampling whereby the object is not replaced, resulting in a dependent trial

scientific notation or **standard form** expression of a number in the form $a \times 10^b$, the product of a number a, where $1 \leq a < 10$, and a power of 10

sector a fraction of the area of a circle, enclosed by two radii and an arc

semicircle the area with radius r and centre $(0, 0)$ described by the function $f: [-r, r] \rightarrow R, f(x) = \sqrt{r^2 - x^2}$

set a collection of elements

set of integers the positive and negative whole numbers and the number zero; $Z = \{\ldots -2, -1, 0, 1, 2, \ldots\}$

set of irrational numbers numbers that cannot be expressed in fraction form as the ratio of two integers, including numbers such as $\sqrt{2}$ and π; $Q' = R \backslash Q$

set of natural numbers the positive whole numbers (counting numbers); $N = \{1, 2, 3, \ldots\}$

set of rational numbers the numbers that can be expressed as quotients in the form $\dfrac{p}{q}$, including finite and recurring decimals as well as fractions and integers; Q is the set of quotients $\dfrac{p}{q}, q \neq 0$ with $p, q \in Z$

set of real numbers the union of the set of rational and irrational numbers; $R = Q \cup Q'$

sideways parabola a parabola with a horizontal axis of symmetry; a relation with the rule $y^2 = x$; the inverse of the parabola $y = x^2$

significant figures the number of digits that would occur in a if the number was expressed in scientific notation as $a \times 10^b$ or as $a \times 10^{-b}$

solving the triangle the process of calculating all side lengths and all angle magnitudes of a triangle

speed the rate of change of position of an object in any direction; how fast an object is moving without regard to direction; average speed $= \dfrac{\text{distance travelled}}{\text{time taken}}$

square root function a function of the form $y = a\sqrt{x - h} + k$

standard form *see* scientific notation

stationary point a point where a function has a gradient of zero

stationary point of inflection a point on a curve where the tangent to the curve is horizontal and the curve changes concavity; the graph of $y = x^3$ has a stationary point of inflection at $(0, 0)$

surd irrational number containing a radical sign such as a square or cube root (but not all numbers with radical signs are surds)

surds roots of numbers that do not have an exact answer, so they are irrational numbers. Surds themselves are exact numbers; for example, $6 - \sqrt{6}$ or $5 - \sqrt{353}$.

symmetry properties indicate the relationships between trigonometric values of symmetric points in the four quadrants. If $[\theta]$ lies in the first quadrant, the symmetric points to it could be expressed as $[\pi - \theta]$, $[\pi + \theta]$, $[2\pi - \theta]$. This gives, for example, the symmetric property $\cos(\pi - \theta) = -\cos(\theta)$.

tangent line a line that touches a curve at a single point; for a circle, the tangent is perpendicular to the radius drawn to the point of contact

transcendental describes a number that cannot be expressed as the root of a polynomial equation, such as π

trigonometric functions sine, cosine, tangent. On a unit circle, $\cos(\theta)$ is the x-coordinate of the trigonometric point $[\theta]$; $\sin(\theta)$ is the y-coordinate of the trigonometric point $[\theta]$; $\tan(\theta)$ is the length of the intercept the line through the origin and the trigonometric point $[\theta]$ cuts off on the tangent drawn to the unit circle at $(1, 0)$.

trigonometric point On a unit circle, the end point of an arc, drawn from the initial point $(1, 0)$ and of length θ, where θ is any real number, is the trigonometric point $[\theta]$. The point has Cartesian coordinates $(\cos\theta, \sin\theta)$.

unit circle in trigonometry, the circle $x^2 + y^2 = 1$ with the centre at the origin and a radius of 1 unit

velocity the rate of change of displacement with respect to time

Venn diagram a diagram that displays the union and intersection of sets

vertex (plural vertices) the turning point of a curve; the point where two lines meet to form an angle; the meeting point of two or more sides of a polygon or three or more edges of a solid

vertical translation a transformation that moves a figure a given distance in a vertical direction where $(x, y) \rightarrow (x, y + a)$

zero (of a function) a value of x for which $f(x) = 0$